S0-EKX-239

Developmental Psychology

William R. Looft

Developmental Psychology

A Book of Readings

The Dryden Press Inc.
Hinsdale, Illinois

Copyright © 1972 by the Dryden Press Inc.
All rights reserved
Library of Congress Catalog Card Number: 72-87737
ISBN: 0-03-089113-2
Printed in the United States of America

Preface

This book is of such a nature that it can beneficially supplement most textbooks of developmental psychology, or it could even serve as the sole text for courses in human development. Obviously not every topic discussed in a text can be accompanied by a separate article in a book of readings. I have attempted to select items which appear to be of major importance to the intended audience—undergraduate students in developmental and educational psychology courses—and which would seem to have high interest potential. I do not wish to imply that I have paternalistically decided what is important and of interest to undergraduate students; quite to the contrary, the articles herein were selected primarily on the basis of the interests and proclivities of students as they have been revealed to me after many quarters and semesters of teaching courses in human development.

A perusal of the Contents will quickly reveal the conceptual and topical orientation of this book. Although the book is intended for use in courses on developmental psychology, a few selected aspects of human development are emphasized. The emphases are these: a historical perspective, which points to the parallel between the development of the individual and the development of the society in which the individual lives; generational issues—the flow of generations or cohorts through historical time; a qualitative, structural view of intelligence and thinking, in contrast to the overemphasized quantitative, standardized-test approach; education and schooling as a developmental influence; a view that human development should be studied within a life-span framework, for important life changes do not end with the "termination" of adolescence.

Each article is preceded by an introduction which is intended both to explain why that article was selected and to outline a few of its central issues. It will be clear that some of the articles in this book were included not because I agree with the point of view taken by the authors, but precisely for the opposite reason. Much can be learned from a careful consideration of arguments not in accord with one's own biases. Each introduction also suggests a few additional articles or books that are related to the immediate topic of interest and are of exemplary caliber.

Another bias reflected in this set of readings has developed as a result of my experience in teaching undergraduates and frequently assigning books of readings for them. Undergraduate students, particularly freshmen and sophomores, seem to be not much interested in papers that are primarily reports of research; in fact, the evidence available to me (I ask the students) indicates that students not only do not like this kind of reading, they do not read it. In this light, the inclusion of elaborate graphs, charts, statistical tables, and the like is an expensive waste of effort and space. Thus, this book consists primarily of review papers, essays on developmental topics, and similar materials. Nevertheless, a few articles which are research reports, with the typical charts and tables, have been included, for I do feel that students should at least gain familiarity with this form of scientific presentation; however, these are kept to a minimum. No purpose is served in presenting to students at this point in their development a collection of reports on methodologically complex experiments if (a) the articles will thereby not be read, or (b) the students' interest and enthusiasm in the course of human growth and development might possibly be diminished.

Appreciation is extended to the authors whose erudition, insights, and creativity appear in this book. Special appreciation must be expressed to Barbara J. King, whose speed and accuracy in assisting me with my work is nothing less than astounding. And Barbara is a really fine person.

Any royalties that may accrue from the sale of this book will be deposited into the Human Development Fund, which supports both graduate and undergraduate students engaged in the study of human development.

August 1972   *W. R. L.*

Contents

	Part One	The Psychological Study of Human Development	1
1.	William R. Charlesworth	*Developmental Psychology: Does It Offer Anything Distinctive?*	3
2.	S. Anandalakshmy Robert E. Grinder	*Conceptual Emphasis in the History of Developmental Psychology: Evolutionary Theory, Teleology, and the Nature-Nurture Issue*	24
3.	L. Joseph Stone Joseph Church	*Some Representative Theoretical Orientations in Developmental Psychology*	35
4.	K. Warner Schaie	*Age Changes and Age Differences*	60
5.	Charlotte Buhler	*The Course of Human Life as a Psychological Problem*	68
6.	Hermine H. Marshall	*Behavior Problems of Normal Children: A Comparison Between the Lay Literature and Developmental Research*	85

Part Two	**Influences on Development**	97
7. Jerry Hirsch	*Behavior-Genetic Analysis and Its Biosocial Consequences*	99
8. L. Joseph Stone Joseph Church	*Intelligence, Intelligence Tests, Intellectual Differences, and Intellectual Development*	118
9. Alex Inkeles	*Social Structure and the Socialization of Competence*	129
10. Heinz F. Eichenwald Peggy Crooke Fry	*Nutrition and Learning*	148
11. Bettye M. Caldwell	*What Is the Optimal Learning Environment for the Young Child?*	160
12. Lawrence B. Schiamberg	*Some Socio-Cultural Factors in Adolescent-Parent Conflict: A Cross-Cultural Comparison of Selected Cultures*	176
13. Thomas J. Cottle	*Of Youth and the Time of Generations*	196
14. L. Woiwode	*The Contest*	210
15. Bernice L. Neugarten	*The Old and the Young in Modern Societies*	223
16. Herbert J. Walberg	*Physics, Femininity, and Creativity*	234
17. Alan Ziajka	*The Black Youth's Self-Concept*	249
Part Three	**The Course of Human Development**	269
18. Harriet L. Rheingold Carol O. Eckerman	*The Infant Separates Himself from His Mother*	271
19. J. Robert Staffieri	*A Study of Social Stereotype of Body Image in Children*	289
20. Guy R. Lefrancois	*Jean Piaget's Developmental Model: Equilibration-Through-Adaptation*	297
21. William R. Looft Warne H. Bartz	*Animism Revived*	309
22. Marie Dellas Eugene L. Gaier	*Identification of Creativity: The Individual*	337
23. Stephen J. Fitzsimmons Julia Cheever Emily Leonard Diane Macunovich	*School Failures: Now and Tomorrow*	367

24. Robert M. Krauss	*Language as a Symbolic Process in Communication: A Psychological Perspective*	387
25. Riley W. Gardner	*Individuality in Development*	402
26. Robert D. Hess	*Political Socialization in the Schools*	415
27. Bruno Bettelheim	*Psychoanalysis and Education*	423
28. Kenneth Keniston	*Student Activism, Moral Development, and Morality*	437
29. William Simon John H. Gagnon	*On Psychosexual Development*	456
30. Paul Cameron	*The Generation Gap: Beliefs about Sexuality and Self-Reported Sexuality*	479
31. Robert Kastenbaum	*The Foreshortened Life Perspective*	485

Developmental Psychology

Part One　The Psychological Study of Human Development

1 Developmental Psychology: Does It Offer Anything Distinctive?

William R. Charlesworth

Orientation

Why developmental psychology? Are the subdivisions of the field of psychology necessary divisions, or are they only the outgrowths of the specific and narrow interests of individual psychologists? "Developmental Psychology: Does It Offer Anything Distinctive?" addresses this question. Professor Charlesworth offers a convincing argument that, indeed, developmental psychology is a distinct field and that this distinction is an important one.

Living organisms, like social systems, have a history. And this is what developmental psychology is all about. This viewpoint insists that an understanding of an individual person is not possible without a knowledge of what has taken place previously in his life. Charlesworth has phrased this orientation most succinctly: "Human behavior takes place in a context of historical givens as well as in a context of what is currently being given."

Despite the distinctiveness of developmental psychology—and other disciplines as well—and the usefulness of these distinctions, it is clear that our understanding of the complexity of growth and development of individuals can only be advanced through integration of many fields of study. Each of these fields, with its special concepts and methodologies, will contribute to the others, resulting in a broader, more comprehensive picture of the course of human development. It will be clear that Charlesworth emphasizes the special importance of integrating biological sciences—genetics, endocrinology, neurophysiology, ethology—into our psychological and social sciences.

The reader will want to pay close attention to Charlesworth's comments about the implications of these ideas for the teaching of developmental psychology. Perhaps his most important point is that both the teacher and the student must adopt a historical perspective; our understanding of anything will always be limited if we reject the past.

Some of the terms, concepts, and principles discussed in this article may be foreign to the new

Invited address presented at the meeting of the Division of Teaching Psychology, American Psychological Association, Miami, Florida, September 1970. Reprinted by permission of the author. The author is grateful to Professor John Flavell, Institute of Child Development, University of Minnesota, and Professor Irving Gottesman, Department of Psychology, University of Minnesota, for their helpful comments and encouragement, to Rev. Ernan McMullin, Philosophy Department, University of Notre Dame, for many useful discussions, and to Mrs. Ruth Dickie and Miss Liz Burkhardt for their patience in typing and proofreading the manuscript.

student in this area. Perhaps the best procedure is to read through the article at the outset of the course of study to obtain an overview of developmental psychology; later in the course, reread this article, for by then most of the issues and problems discussed here will have attained greater meaning and significance for the reader.

William R. Charlesworth is associated with the Institute of Child Development at the University of Minnesota. His professional work has focused primarily upon cognitive development in infancy.

1 Introduction

From all indications it appears that developmental psychology has come into its own; and as Bijou (1968) pointed out in his presidential address to Division 7 the increase in its activity has been much desired. Society has become convinced, as it has on a number of earlier occasions, that research in developmental psychology will provide answers to many of the most urgent problems having to do with the education, health and welfare of children. So at least from a practical point of view the study of development has been selected out as important. But, as we all know, practical considerations are not enough in science, especially if they are unexamined. Bijou (1968) foresaw what would happen if developmentalists could not deliver what was expected of them. Therefore, in an attempt to forestall this possibility he set himself the task of analyzing what it was that made developmental different from the rest of psychology. As a result, he did a good job of explicating the internal conceptual and methodological problems developmentalists have to face if they wish to have a rigorous discipline which would ultimately lead to the solid achievements society desired.

I would like to continue this self-examination in the form of an answer to a question. The question is whether the current emphasis on developmental psychology as a distinct discipline is justified and, if so, whether it makes any difference for the teaching of psychology.

The answer will consist of three points: (1) metatheoretical and methodological issues that have historically characterized distinctions between developmental psychology and general psychology, including a brief historical treatment of two different phases of developmental psychology and the need to move on to a third phase, (2) suggestions as to how this third phase can be brought about through greater collaboration with disciplines generally associated with the biological sciences and (3) the implications that (1) and (2) have for teaching developmental psychology. For it is in teaching a course, especially an introductory course, where one has to have the greatest clarity and the greatest warrant for the existence of one's scientific discipline.

Conceptual Status

The conceptual status of developmental psychology, despite its relatively short history as an independent science, has already been an object of a number of discussions (Harris, 1957a; Stevenson, 1966). Numerous attempts have been made to spell out what is implied by the term development and what features distinguish developmental research from other research efforts. On the basis of definitions culled from a variety of sources, it seems safe to say that development generally refers to a phenomenon which involves a change in form or organization over time. This change is in the direction from relatively simple to more complex forms; it usually proceeds through stages and transitions between stages viewed as relatively irreversible. The force behind the change is believed to come from maturational factors within the organism as well as from an interaction between such factors and environmental stimulation. Because of this latter characteristic, developmental change is not to be identified with learning.

As for distinguishing features, developmental psychology, according to such writers as Harris (1957b), Russell (1957), Spiker (1966), and Elkind (1970), tends to be more nativistic, holistic, biologically oriented, and more teleological than general psychology. Developmental psychologists tend to see changes as epigenetic and therefore essentially irreversible; they tend to consider the child as possessing unlearned abilities or predispositions to act in certain ways and therefore as an active as well as passive element in his environment. They also are willing to admit to unseen maturational forces controlling behavior. And they tend to rely heavily upon induction based on naturalistic observations and descriptions and to look at both process and content as research objects. Last, but not least, developmentalists are reputedly not averse to lending the results of their efforts to practical application.

The general behavior psychologist (or the learning psychologist) is most often pitted against the developmentalist in situations like this. He, in contrast, tends to be empiricistic, elementaristic (or anti holistic), more interested in causes than in purposes. He sees changes as accumulations of some kind of units and therefore for the most part decomposable or reversible; he considers the child as basically naive, incompetent, and passive, and while he does not totally ignore maturation, he tends to play it down. He relies heavily, but not exclusively, upon deduction based on controlled experimentation, tends to be mostly preoccupied with process rather than content, and is willing to resist immediate and direct application of his research efforts.

If all this does not completely characterize the beliefs and activities of developmental psychologists at present, there is at least one distinctive feature about developmental psychology that cannot be overlooked. It requires that behavior be looked at in terms of greater time spans than minutes, hours, or days as is the case for much of general psychology. At first glance this feature may seem relatively insignificant, but it is not. It means that the developmentalist, like many biologists, must perforce work both in a context of history (ontogenetic as well as phylogenetic) and in a context of continual change. This change does not only involve the gradual accumulation of experiences and subsequent alterations of behavior, but also alterations in the total physical substrata which mediate such behaviors. It is these latter alterations that make the developmentalist's problem different and have a profound effect on how he should do research and what for him should constitute an adequate explanation of behavior. More will be said later about the significance of this historical and physical dimension. What is important at the moment is whether such features are functional in the sense that they actually distinguish the research problems and methodologies of developmental psychology from general psychology. To answer this requires a brief glance at some history.

History: The "First" and "Second Phases"

If one looks at the relatively short history of developmental psychology (it cannot be said to have become a scientific discipline before the beginning of this century), it becomes clear that its identity as a unique discipline of the behavioral sciences was never quite certain. (See Palermo and Lipsitt, 1963, for a good, concise account of this.) There was a fairly prolonged period before the 1950s (we can call it the first phase) which was characterized by much naturalistic observation and description and what appears today to some (quite unfairly, in a way) as an excess of mindless mental and motor testing. All this was accompanied by some fairly strong beliefs in the fixedness of intelligence and personality traits. And maturationally controlled processes were viewed as responsible for the inexorable unfolding of many important behaviors. In many ways this approach was in line with most of the connotations of the term developmental.

The second phase was different. Experimentally inclined psychologists with strong interests in both general learning theory as well as children reacted strongly to what in their minds was just barely passing as science. They injected a strong sense of conceptual and methodological rigor into the field in an attempt to make it more respectable. As a result, distinctions between developmental psychology and general psychology began to blur. For many of the new people in the field, child psychology and child development became obviously interchangeable terms. And in a sense, a slow process of reductionism began taking place. Spiker (1966), for example, felt that the connotative meanings of developmental having to do with pre-existing capacities influencing structures and the structures in turn undergoing sequential, more or less irreversible changes were not only unnecessary meanings but also undesirable. General behavior theory and its particular research paradigm was considered quite capable of assimilating developmental phenomena. Zigler (1963) felt that learning psychologists would probably make the most significant contribution to developmental psychology and in a way he was correct. Those trained in learning were also very well trained in research methodology. They brought critical skills the field needed badly, skills that not only aided in establishing a more rigorous approach to research methodology, but in pointing out the conceptual weaknesses in the existing theoretical frameworks erected by Gesell, Werner, and Piaget. This second phase, it should be noted, had considerable outside support. As one can well guess, it came from what was, and to some extent still is, the prevalent American Zeitgeist—environmentalism in all its various forms. For complex historical, political, and social reasons, Locke, Hume, Pavlov, Watson, Hull, Thorndike, and Skinner succeeded in having a greater overall effect on American psychology than Rousseau, Kant, Darwin, Gesell, Carmichael, Werner, and Piaget.

A major feature of environmentalism is the way its proponents view changes in behavior. In its most simple form such a view perceives most significant human behaviors as being constructed over time mainly as a result of numerous environmental contacts. This view has great methodological as well as explanatory advantages. The environment can be made to yield variables which can be identified and manipulated. This, in turn, makes it possible to specify independent variables and thereby conduct rigorous experiments to build and test theories and, according to some writers, to get child psychology out of a stage comparable to the one physics was in before Galileo's time (Lewin, 1931).

Nevertheless, despite the attractiveness of the environmentalist view, its hegemony was never complete. Munn's (1955) book emphasizing the biological basis of developmental phenomena, in contrast to a strictly environmentalistic one, appeared in the middle of the environmentalist phase and had a positive reception. In addition, there were pockets of resistance to what was conceived as a premature attempt to conduct experimental research with developmental phenomena before such phenomena were properly observed and described. Ausubel (1958), for example, argued that developmental problems were so complex that the first methodological step toward them must be naturalistic—observational, descriptive, and correlational. The experimental method he felt was inappropriate at this early stage and may actually be inapplicable because the significant causes of most developmental phenomena, namely the subsurface maturational changes, could not be manipulated.

The Problem for Developmental Psychology

Despite the reasonableness of the arguments on both sides, the real problem for the developmentalist remained pretty much untouched during the two phases and as a consequent is still with us as strong as ever. The problem is simple to state. What does one do about the neurophysiological, anatomical, and endocrinological processes that are correlated with behavioral changes for at least the first 16 years of life and most probably are causally related to them? The rigorous experimental learning studies can say little or nothing about the organismic or physical reasons underlying observed age differences in behavior. The naturalistic-descriptive and semi-structured test studies do no better. And this is true even for relatively new areas of research which deal with phenomena that seem to have a strong organismic basis.

An example of this is the well-known weakness of many conservation training studies inspired by Piaget's work. A source of this weakness is due to the fact that investigators have no idea what maturational processes mediated by neurophysiological mechanisms are working with environmental inputs to make nonconservation or conservation possible. Those who conduct training studies know the problems of deciding when training should start and how long it should last. If it is started too early, the child may be too "immature" to understand the instructions; if training, once started, draws out too long, some subjects may acquire conservation "spontaneously." This helplessness is hardly a sign that the investigator is in rigorous control of what accounts for a major part of the variance of his dependent variable.

What is more depressing for those who want to create in the laboratory the cognitive structures observed by Piaget is the question of what stimulus inputs are responsible for such structures. One is hard put to specify what it is in the preschooler's environment, for example, that could be identified as a necessary, if not sufficient, environmental input for the construction of those concrete operations Piaget has observed to appear around seven years of age. Such inputs may actually be nonspecific in their effects—the way most food is. Piaget said this years ago. If this is really the case, it would be much more interesting and useful to consider different stimuli simply as fuel and try instead to find the specific mechanisms that use such fuel to produce behavioral structures according to their own set of rules.

To sum up, then, if most behavioral developmental changes are similar to changes in biological structures, then no amount of "phase one" naturalistic description and mental testing is going to give us satisfactory explanations. The same can be said for "phase two" learning experimentation. The two approaches represented in these phases are essential, but not adequate. The naturalistic-descriptive will continue to sensitize psychologists to the wealth of developmental phenomena to be had merely by carefully watching and testing children. The experimental-theoretical will continue to give developmental psychologists analytic clarity, conceptual rigor and a sense of control. Nevertheless, preoccupation with them as sole strategies for attacking developmental problems seems to me to be on the way out because the third phase of developmental psychology is slowly coming into being.

What this third phase will consist of is not clear, but a vague shape of it has made itself felt in the massive effort put forth in the new version of *Carmichael's Manual of Child Psychology* (Mussen, 1970). A provision of this shape is partly captured in Stella Chase's review of the book when she notes: "There is a keen awareness of the need for more longitudinal studies, interdisciplinary projects, and cross-cultural explorations." It is also partly captured in concerns for ontogenetic phenomena generated in other disciplines such as ethology and behavioral genetics. This will be explained later. Right now a parting note on very recent history is relevant.

Recent Stresses that Exacerbate the Problem

The need for a third phase, in my estimation, has arisen because of a number of relatively recent stresses within the field as well as outside of it. It is too early to detect the total source of this stress, but we can be fairly certain that future historians will trace a good part of it to the efforts of those working in psycholinguistics and language development—Chomsky, Lenneberg, McNeil—and those at the nexus of language and cognition—Flavell, Furth, and others. These investigators have raised questions concerning the genetic and environmental origins of language and thought as well as questions concerning distinctions between competence and performance. As a consequence, existing learning and developmental theories of ontogenetic change are being challenged. Piaget's formulations of stages and transitions between stages, for example, will most probably need revision, as Flavell and Hill (1969) point out. Abilities such as attention, language, and memory which are subsidiary to the complexes of cognitive operations described by Piaget may do much more than traditionally thought to account for differences between transitional and fully operational children. If this is so, it will seriously challenge the Genevan notion of more or less discrete stages of cognitive development.

An additional stress is evident in the growing discontent amongst developmental psychologists themselves. Flavell and Hill (1969) point out that many current learning studies are very crude instruments for measuring and producing "organismic" or cognitive changes and are consequently not yielding much new information. They also point out that even in the relatively new areas of research opened by Piaget, areas which tend to oppose traditional learning research, there are increasing signs of a "doctrinaire" and "unimaginative" reliance upon untested methods which have failed to live up to expectations. It seems, in short, that the field is faced with a more basic problem than that posed by the traditional controversy between the learning and cognitive approaches. A redirecting of our thinking appears necessary and this, in my estimation, is what the third phase will be about.

The "Third Phase"

As noted earlier, the current problem for developmental psychology, as I see it, is its almost total ignorance of maturational processes, i.e., those subsurface structures and functions of the organism which are correlated with and most probably causally related to the behavioral changes which historically distinguish developmental psychology from general psychology. To identify these processes and their functional connections with behavior, developmental psychologists will have to become much more familiar with certain subdisciplines within the biological sciences. This has already been urged on numerous occasions in the history of the field, and it is also apparent to anyone who reads the literature in the biological sciences that before the content of such sciences can be appreciated as being relevant for understanding developmental problems. Nevertheless, the gaps separating the two disciplines will have to be bridged; developmental psychology will have a limited future if they are not.

As I see it, at least four biological disciplines will ultimately become indispensable for anyone dealing with developmental problems: genetics, neurophysiology, endocrinology, and ethology. Links between these disciplines and developmental psychology already exist. Fuller and Thompson (1960), McClearn (1968), and Vandenberg (1966) have already introduced many developmental psychologists to behavioral genetics. Eichorn (1968, 1970) and Tanner (1963, 1970) have been continually making important contributions to the developmentalist's knowledge of the physical processes associated with physical growth and the work and ideas of Bowlby (1951, 1969) Lorenz, (1937, 1965), and Tinbergen (1951, 1953) are already familiar to many in the field. However, it should be stressed that while the importance of their work has been recognized, the implications of their efforts, especially those of the geneticists and ethologists, have hardly been exploited by most developmentalists. There are at least two reasons for this—one having to do with misconceptions, the other with practical problems of methodology.

Misconceptions Delaying the "Third Phase"

Let us start with misconceptions having to do with genetics. There is, for example, a widespread belief (Kantor, 1933, 1958, for example) that man is more biological when he is an infant and more social and cultural when he is older. In other words, the older he gets, the weaker the role hereditary factors play in his development and the stronger becomes the role of social and cultural factors. Textbooks frequently get this idea across indirectly by starting off the first chapter with heredity and then never mentioning it again. This idea can be viewed as a simple stage theory of development which sums up observable differences between young and old organisms and in the process emphasizes the relative rigidity of the former and the greater plasticity of the latter. It tends, however, to ignore the similarities between them, and it also tends to violate everyday observations. It is obvious, for example, that in order to survive, the human infant must be a social creature par excellence. He must continually relate to persons stronger and more mature than himself. To avoid being mistreated or perishing, he must regulate the behavior of others by crying and smiling and it seems to help somewhat if he looks as cute and as helpless as possible. Infants have been engaging in such social behavior more or less successfully for tens of thousands of years—long before man developed the idea that it is morally wrong, as well as generally disadvantageous for the species, to kill or mistreat infants. It is only when the child is ten to twelve years older that he can be as anti-social as he desires and still survive. He could even live to old age without people, provided he is relatively healthy and adequate food sources are available.

Misconceptions about genetic factors also run contrary to the fact that such obvious phenotypes as secondary sex characteristics develop some 11 to 13 years after birth and that some pathological genetic conditions such as Huntington's Chorea make no appearance until around 35 to 40 years of age and sometimes much later. Geneticists have already good leads on how such changes take place (McClearn, 1968). There appear to be genes with different functions—some operate immediately at conception, some regulate the operations of other genes so that they do not begin to function until much later. If this is truly the case with other phenotypes which interest behavioral scientists, then the first-biological-then-cultural stage theory becomes meaningless.

The same misconception operates in the other direction as well, as most good misconceptions do. It has, for example, led many psychologists into believing that infants are so helplessly immature—because of biological limitations—that they perceive very little of their environment and are incapable of learning. In reality, however, this is not the case as research efforts in the past decade have demonstrated (Charlesworth, 1968; Fitzgerald & Porges, 1970). To make matters worse for those who believe that the infant is inflexible or inadequate in the face of cultural impositions because of biological constraints, most conditioning studies have been successful in highly artificial laboratory conditions which are culturally arbitrary and which have no representation whatsoever in the normal life experiences of human infants. Such plasticity in the face of environmental conditions novel to the whole history of the species does not support the notion that the infant's adaptation is greatly limited by innate biological constraints. In fact, natural selection seems to have favored a relatively high degree of plasticity in the human infant.

The point should be clear. Biology and culture do not work separately at separate times during ontogenesis. The two sources of influence on human behavior are inextricably merged together throughout the organism's whole existence. And a major job of the developmentalist is to disentangle such influences with the best analytic techniques available.

Behavioral geneticists are beginning to provide some of these techniques, but in the process of doing so they run into another misconception tied in with the first. There is much opposition to the notion that it is meaningful to partition phenotypic variation in behavior into genetic and environmental sources. Part of this opposition stems from the controversies over the problem of nature versus nurture. There have been numerous attempts to expunge this controversy from the psychology or at least rid psychology of the notion that one can quantify the relative effects of heredity and environment on behavior. Anastasi's (1958) paper is most quoted on this and, as I see it, she is correct when she insists that we should study the "how" of the nature-nurture issue, but wrong in arguing that knowing "how much" is misleading if not of any great utility. Her argument for focusing on interaction to answer the "how" question is well-taken, but for reasons to be spelled out later it is easier to emphasize interaction than to attack it operationally. The "how much" question cannot be disposed of so easily. What scientists do out of necessity to understand a phenomenon and how the phenomenon appears to them in terms of their private experiences and common sense are usually separate issues. Breaking down a complex behavioral and phenomenological event into stimuli and responses is a classic example. No responsible psychometrist or geneticist would claim that the statistical partitioning of behavior into hereditary and acquired components is the terminal point of understanding such behavior. Such statistical achievements, nevertheless, are worthy of our attention because they help point out the direction future research efforts should take. Phenotypes that have an important genetic component, for example, would seem to me to be most profitably studied developmentally by a team sensitive to neuro-endocrinological changes, nonspecific learning experiences, and the possible evolutionary significance of the phenotype for the species. A team sensitive to local stimulus conditions and specific learning situations would be reserved for phenotypes with less of a genetic component.

As for the problem of interaction mentioned earlier, it too can add resistance to establishing better contact between developmental psychology and biology. Frequently used as the main objection to the nature-nurture issue, interactionism has made much progress at the verbal level, but relatively little elsewhere. Asserting that interaction takes place (and who would assert that it does not?) does not necessarily imply that efforts to analyze sources of behavioral control into innate and acquired influences are essentially worthless. The main proponents of interaction such as Hunt (1961), Piaget (1960, 1968), and Schneirla (1966), who as part of their argument object to the nature-nurture dichotomy, are not clear as to what one should do after rejecting the dichotomy and its supposedly misleading implications for our understanding of developmental phenomena. No matter how accurate it may be as an explanatory construct, interactionism as represented in their proposals tends to be non-analytic, nonquantitative, and consequently operationally empty. Those who reject separating influences into innate and learned factors as meaningless fail to offer an adequate methodology to compensate for what is rejected.

An example of this can be found in Piaget's concept of equilibration which is a conceptual derivative of the interaction notion. Equilibration is postulated as a mechanism which insures cognitive growth by carefully balancing assimilative and accommodative functions regulating the organism's commerce with the environment. Nowhere associated with this construct, however, is there a hint as to how one should operationally attack it. Those, in contrast, who assume that it is worthwhile separating variances statistically into inherited and noninherited categories at least have analytic tools with which to pry further into the phenomenon at hand.

It is not the tools per se, but the ability to provide the wherewithal to pry further that is the crucial test of a theoretical position. The ideal approach, it should be emphasized, however, does not lie solely with consanguinity studies and psychometric analysis. It lies with the application of the longitudinal method along with many other methods converging on a single problem area.

The nature-nurture issue, of course, exists in ethology and is probably a main reason why American developmentalists have not been attracted strongly to ethology. The issue of what is innate and what is learned is saturated with problems of semantics and much controversy. What proponents of the ethological view, such as Lorenz (1965), want to make explicit is that many ethologists find it useful to view an organism's behavior (including the human's) as having two sources of control. One source is in the particular gene constellation within the genome, the composition of which is a result of phylogenetic (i.e., evolutionary) processes which have taken place over long periods of geological time. To the extent that such a constellation conditions behavior it can be considered an important source of variables relevant to understanding such behavior. That the formation of this constellation antedates the existence of the organism itself justifies considering it separate from the second source. The second source of variables controlling behavior develops during the organism's ontogenesis as a result of the interaction of the first source with the environment.

Most ethologists view the phylogenetic control of behavior in terms of evolutionary theory which holds that most important behaviors in lower animals, at least, can be viewed as representing the latest adaptive step by the species which aids the species' survival. As a consequence of this theory, the ethologist examines each behavior carefully to determine what its adaptive function is for the individual as well as for the species. Species- universal behaviors that occur with a high frequency in many different environments not in communication with each other are generally considered as having an innate basis until proven otherwise.

The application of ethological ideas in research has been mostly limited to animals. However, within the last five years such research has begun to be directed toward humans (Freedman, 1965; Eibl-Eibesfeldt, 1970; Hess, 1970). At the moment a bandwagon effect seems to have developed; numerous behavior scientists and practitioners working with humans are beginning to look for assistance toward ethologists whose work has been almost exclusively with other animals. As a consequence, comparative psychologists and zoologists have become uncomfortable about false analogies, cross-species comparisons, and generalizations which may be misleading. To some extent their discomfort is justified, especially in light of some popularized attempts to ethologize human behavior. Hopefully such concern will not crystallize into a rejection of the role ethology can play in human behavior. Transitional periods in science are frequently characterized by extremes of caution and recklessness which sooner or later cancel each other out and gradually give way to sensible research and conceptual strategies.

Methodological Problems Delaying the "Third Phase"

The problems of extending developmental psychology more into the directions of biology are not limited solely to misconceptions. Methodological issues also play a role. Many developmentalists interested in outside disciplines are willing to respond to them, but are faced with the question of how to put the knowledge and methods of such disciplines to appropriate use. Ethical considerations forbid assigning human mates to each other and then assigning environments to their offspring to meet certain basic methodological criteria in genetics. The deprivation experiment considered so important in ethology (Lorenz, 1965) is out of the question when dealing with most early experiences thought of as crucial for normal development in the human child. Approximations of these criteria (e.g., studying twins, foster children, half sibs, children of divorced parents, deprived children in institutions, etc.), of course, can be achieved, but for many psychologists such approximations can at best offer only hypotheses and never confident conclusions. Furthermore, a more general methodological problem exists which can be illustrated with an example.

Much good thinking and observation have gone into studying the preschool child's free play behavior. The crucial question is whether a well-trained ethologist could do better; i.e., could he come up with fresh observational data or a new insight into relationships between data, and if so, how would he do it? Is it mainly a question of how to observe or a question of what to look for or a mixture of both? There are no immediate satisfying answers to these questions.

From this perspective the third phase does not appear immediately promising, but the question still exists as to whether persisting solely with the best methodologies of the first two phases appears anymore promising. The answer is no if there are any reasons to believe that the third phase can conceivably add to, rather than subtract from, the first two phases. Such reasons exist. For example, applying the contributions of behavioral genetics and ethology to understanding problems in developmental psychology will make it easier (1) to determine the probability that certain classes of behavior showing significant changes with time are more under maturational (neurophysiological, endocrinological) or environmental control, (2) to specify, roughly at least, when the various mechanisms involved should begin and cease functioning; and (3) to offer hypotheses as to why such behaviors should develop in the first place, i.e., whether they have significant adaptive value and if so in what sense. Such information will consequently facilitate research into areas of neurophysiology and endocrinology which are currently much too esoteric or remote to have any direct bearing on the kinds of behavior of interest to developmentalists.

One example of what is meant here is in order. Piaget's work on developmental changes in cognitively controlled behavior is now well known. It is also known that the theoretical basis of this work is heavily biological. But it is not completely biological nor is the methodology that it supports. The contributions of Darwin and Mendel play almost no role in it. Nevertheless, it would seem worthwhile to attempt to apply genetics and ethology to the phenomena he has identified. The cognitive structures he describes seem to have a strong biological basis—they begin to operate at roughly the same time in most normal children; they follow each other in the same order; they appear to be species-universal, at least up to the level of concrete operations; they are hard to accelerate before a certain time and no one yet has been able to extinguish them; they seem highly relevant for survival; and some of them, especially those in the sensory motor stage, also seem to be present (although I know of no data) in primate species closely related to Homo sapiens. With such clues it seems appropriate to expand our knowledge of them by: (1) studying the developmental history of such cognitive structures in mono- and dizygotic twins; (2) testing for the same (or clearly analogous) structures in lower animals, especially primates; (3) expanding cross-cultural studies to the few remaining isolated societies that have little or virtually no contact with the majority of the world's present societies; (4) performing motor deprivation and enrichment studies in primates to determine if the motor basis of such structures is as strong as Piaget claims; and (5) attempting to locate the neural structures in primates that mediate such behaviors.

If the "Third Phase" Materializes

What will the third phase be like if the predicted collaboration between developmental psychology and biology takes place. The answer is partly provided by Sperry (1958) who so perspicaciously noted that whatever the organism inherits has to be "funneled through developmental mechanisms." This for the developmentally imperialistic mind means that behavioral genetics and ethology and the other biological disciplines having to do with behavior will ultimately be coordinated under a developmental framework, a vision that Heinz Werner had for many years. The vast chain of inside and outside events that link genotype with phenotype will lie within the aegis of the developmental approach. Behavioral constructs will be attacked longitudinally from the many sides of each participating discipline. In addition, major theoretical biases will be essentially ignored. For example, it will seem irrelevant to argue, as Skinner (1966) has, that it is more important to know the probability that an organism will behave in a given way than it is to know (as ethologists would desire) that it does behave in that particular way and not in another. Both kinds of knowledge will be considered important. To have all possible knowledge about a behavior seems much more satisfying than having knowledge prescribed arbitrarily by some "ism" or current methodological convenience.

A concrete example of this ideal convergence of efforts will make it more credible. White (1965) has identified the five-to-seven age period as one during which at least 21 response patterns occur, ranging from changes in responsiveness to classical conditioning to changes in the ability to draw largest and smallest squares. This age period has also been identified by Piaget (1960) as one during which significant changes in cognitive competence take place. The causes for these changes at this time, however, are nowhere immediately visible in the data both investigators collected. As Roose and Lipsitt (1970) point out, White's theory accounting for the changes attempts to explain why the transition occurs, but it does not explain "why the transition occurs *when* it does."

Now it seems to me that the most satisfying answer to this question will have to come from many sources—from developmental neurophysiologists and endocrinologists working across this age range to ferret out the physical and biochemical mechanisms mediating such changes; from ecologists, psychologists, and anthropologists scrutinizing environments and persons cross-culturally to determine whether such a phenomenon is universal across differential cultural treatments of children at this age; from comparative psychologists and ethologists working with other animals, especially primates, to determine whether they too undergo a similar change at the corresponding period of their development; from learning psychologists attempting to synthesize such changes in younger children and locating those discriminative and reinforcing stimulus conditions that accelerate or decelerate the appearance of such abilities; from clinicians and specialists in mental retardation diagnosing children whose problems and deficits may have influenced their status vis-a-vis these cognitive achievements; and from behavioral geneticists carrying out longitudinal consanguinity studies to determine the extent to which such abilities are influenced by hereditary and by environmental factors.

This dream of convergence is not new. Warren Bennis (1970), who seems to prophesize better than most people, sees a time in the not too distant future when scientific, technical, and economic enterprises will undergo significant changes in their structures and the way they operate. These changes will consist mainly of a shift toward building research efforts around problems rather than around existing disciplinary lines. The problems will invite a multi-disciplinary team attack by researchers with diverse professional skills.

Translating such a prediction into what has been said about developmental psychology is not difficult. The problems Bennis talks about can be envisaged as psychological and behavioral constructs, connotatively overburdened and operationally unwieldy, constructs about which we know much intuitively but about which we need to know much more scientifically—language learning, imprinting, aggression, effects of early experience, conscience development, attachment, sex-role identification, logical operations development, fear, and whatever else interests groups of persons with diverse talents who are willing to do research together within a developmental framework.

What constitutes proof and what constitutes a satisfactory explanation of such a behavioral construct would change as a result. Instead of relying solely upon the results of a limited set of critical experiments or a certain number of significant correlation coefficients to provide a satisfactory account of a particular behavior, a multitude of collateral bits of proof and evidence culled from many different but related sources would be relied upon. No single source of information from a particular discipline would be considered satisfactory or definitive because no one methodology could do justice to all aspects of a particular construct. The methodological monism characterizing much of today's research would give way to a methodological pluralism because researchers (as well as students, as will be pointed out later) would find it more useful to yield to the manifold properties of the construct than to the demands of a single theory or to a preordained commitment to a single methodology.

The expansion of developmental psychology into the biological sciences which compels recognition of different methodologies and theories would consequently not be as implausible as it may first seem. The unwieldiness of bringing together child psychologists, geneticists, ethologists, endocrinologists, etc., would be offset by their common interest in a defined set of common problems surrounding a particular phenomenon. The distinctiveness of developmental psychology would become apparent because such an enterprise would not settle for less than a total picture (as far as it is possible) of the origins and emergence of the behavior in question. This, then, would be the achievement of the third phase—the actualization of those characteristics mentioned above which distinguish developmental psychology from the rest of psychology.

And What about the Students?

If the third phase materializes the training of students would obviously be affected. Apart from changes involving restructuring universities more toward accommodating temporary training and research teams and less toward permanent departments, and organizing curricula and textbooks less in terms of superordinate categories such as perception, motivation, learning, etc., and more in terms of topics such as the constructs mentioned above, there would also be a deep change in attitudes and perspectives. Students committed to developmental psychology would learn that historical variables, ontogenetic as well as phylogenetic, are truly relevant for the understanding of contemporary behavior. Getting this across to them would be no easy task. The current activistic rejection of the past has a profound hold on the contemporary mind. On the one hand, it has positive motivational value because it continually keeps alive concerns for improving human existence. On the other hand, it has limited informational value because it encourages active neglect of, or at best indifference to, a large source of variables that have a profound effect on such existence. Human behavior takes place in a context of historical givens as well as in a context of what is currently being given. To ignore one or the other, or to ignore the more important fact of their interaction is to cut oneself off from the possibility of approaching a complete understanding of human behavior.

If the totality of the developmental viewpoint is made clear to the student and he is receptive to it, he will be sobered by the immensity of factors and preparation that go into the final production of a single behavior, a personality trait, or a developmental change. As a result, he will be less quick to judge which are the important determinants of behavior and which are not. And he will be more cautious in applying simplistic solutions to behavioral and social problems because he will be aware of the vast matrix of events from which they developed.

The emphasis on the naturalistic-descriptive aspect of research which continually reveals the visible problems of development (some of which he himself must certainly have experienced) will point out the relevance of developmental psychology for his own everyday existence and for that of others. The emphasis on the need for a systematic approach to sample selection, to observation and measurement, to experimental control and data analysis will make him aware of how hard it is to obtain reliable scientific information on those things considered important for human existence.

An immediate pedagogical criticism of this third phase offering to the student is that it demands far more accommodation on his part than he can tolerate. But this may not necessarily be the case. Historically, developmental psychology has had a number of interesting facets to it. It has one facet that Harris (1956) and others have often pointed out which is self-aggrandizing but which also makes for difficulties at times. Developmental psychology is especially open to the problems of the real world. Many students know this and take their first course in developmental to get practical knowledge about how and why people become what they are. They see the connection between many of society's problems today and the way its members are brought up, educated, and treated when they are sick and in trouble. The socially and politically active student is beginning to see knowledge as a useful instrument for social reform, but, as we all know, much of the knowledge he gets in the classroom is not in an immediately usable form. As a result, trends are developing to rectify this. Important issues courses are being created to deal with problems of growth, health, and education of children, especially of physically and socially disadvantaged children. As such trends increase, so will a demand for more facts. However, only certain kinds of facts will be demanded. Facts that emanate in what seems to be an endless flow of minutiae from the dominating theories or methodological "isms" of the day may have to yield to facts that disclose information essential for questions concerning the quality of everyday human life. These "relevant" facts will be concerned with something beyond themselves, rather than just with other facts which are viewed as having been collected "simply" to fulfill research grant requirements and to get tenure. The construct- or problem-oriented approach to research mentioned earlier will make the collection of such facts more feasible.

The instructor will, nevertheless, be required to walk a thin line. The emphasis on relevant facts represents an urgent need for greater certainty and stability in guiding courses of social action. This need, as pressing as it may be, will have to be counterbalanced by continual disclosure in the classroom and laboratory that the research enterprise cannot go on unless it is free to exercise its methodology as objectively as possible. Student zeal will have to be tempered by the recognition that reliable information is painful and slow to come by and that specialization will have to continue as well as the routines and countless details of research.

But the student stress on practical application is a good sign. It represents a quest for meaning and relevance and a request that science help make for a better connectedness between knowledge and action for individual and public good. Modern science itself has seemed to have lost its ability to achieve such unity. Its technical genius has outstripped its capacity for conceptualizing and anticipating the nature of its vast and variegated effect on the lives of people. But it is still inherent in the operations of science to seek coherence and completeness in the knowledge it acquires through the application of rational methods. Hence, it should be the duty of the instructor of science to make this clear to a generation who for obvious reasons has come to question the role of rational methods in producing a tolerable existence for everyone.

Conclusion

In conclusion, then, it seems to me that the answer to the question of whether developmental psychology has anything distinctive to offer to our understanding and teaching of the behavioral sciences is "yes." It can offer what it originally set out to offer, namely, the realization that all behavior takes place within a historical context, a realization that implies, in no trivial way, a serious recognition of the continuing interaction between past and present factors, between inherited dispositions embodied in neurophysiological structures and functions, learning histories, and contemporary stimulation. Furthermore, developmental psychology can set out to re-emphasize the heuristic value of Darwin's and Mendel's contribution to the behavioral sciences, of the relevance of phylogenesis and genetic factors for our understanding of developmental phenomena. By increasing its collaboration with such subdisciplines of biology as genetics, ethology, neurophysiology and maintaining its present reliance on environmentalist positions which emphasize the importance of learning, developmental psychology can serve as a well-needed coordinator of the many sides of general psychology and the biological sciences. Furthermore, if it maintains its traditional concern for practical problems, it can also serve as a catalyst for a problem-solving construct approach to behavior analysis that will bring together the efforts of both pure and applied disciplines.

As for the students, what they will hopefully get out of courses in developmental is the badly needed insight that the past is truly prologue to the present and that nothing lastingly significant nor beneficial can be done to, with, or for human behavior without understanding the role that earlier adaptations (whether they be biologically or culturally transmitted) play in man's behavior. What students also need is the assurance that the science of behavior is truly a life science, one that recognizes the complexity of all aspects of human existence before it sets itself to the unromantic but necessary task of dealing with it in terms of abstractions, machinelike analogies, and mathematics. It is not contrascientific (it is just overwhelming) to demand that our explanations of behavior ultimately include all the factors—phylogentic, genetic, cultural, situational as well as developmental—that make humans what they are. It is the totality of human existence which concerns many students today, and they are concerned about the extent to which man's major educational and research institutions are contributing to this totality. Science has to acknowledge the significance of their concern both for itself, as an epistemic enterprise, as well as for a society which is in serious need of its achievements.

References

Anastasi, A. Heredity, environment, and the question "How?" *Psychological Review,* 1958, **65**, 197-208.

Ausubel, D. *Theories and problems of child development.* New York: Grune & Stratton, 1958.

Bennis, W. G. A funny thing happened on the way to the future. *American Psychologist,* 1970, **25**, 595-608.

Bijou, S. W. Ages, stages, and the naturalization of human development. *American Psychologist,* 1968, **23**, 419-427.

Bowlby, J. *Maternal care and mental health.* Geneva: World Helth Organization Monograph 1951, Ser. No. 2.

Bowlby, J. *Attachment and loss.* New York: Basic Books, 1969.

Charlesworth, W. R. Cognition in infancy: Where do we stand in the mid-sixties? *Merrill-Palmer Quarterly,* 1968, **14**, 25-46.

Eibl-Eibesfeldt, I. *Ethology: The biology of behavior.* New York: Holt, Rinehart and Winston, 1970.

Eihorn, D. H. Biology of gestation and infancy: Fatherland and frontier. *Merrill-Palmer Quarterly,* 1968, **14**, 47-81.

Eichorn, D. H. Physiological development. In P. H. Mussen (Ed.), *Carmichael's manual of child psychology.* (3rd ed.) New York: Wiley, 1970, Pp. 157-283.

Elkind, D. Developmental and experimental approaches to child study. In J. H. Hellmuth (Ed.), *Cognitive studies.* New York: Brunner/Mazel, 1970. Pp. 44-54.

Fitzgerald, H. E., & Porges, S. W. A decade of infant conditioning and learning research. *Merrill-Palmer Quarterly,* 1970, **17**, 79-118.

Flavell, J. H., & Hill, J. P. Developmental psychology. *Annual Review of Psychology,* 1969, **20**, 1-56.
Freedman, D. G. An ethological approach to the genetical study of human behaviour. In S. G. Vandenberg (Ed.), *Methods and goals in human behavior genetics.* New York: Academic Press, 1965, Pp. 141-162.
Fuller, J. L., & Thompson, W. *Behavior genetics.* New York: Wiley, 1960.
Harris, D. The concept of development and child psychology. Presidential address read before Division 7 of APA, Chicago, September 1956.
Harris, D. B. *The concept of development.* Minneapolis: University of Minnesota Press, 1957. (a).
Harris, D. B. Problems in formulating a scientific concept of development. In D. B. Harris (Ed.), *The concept of development.* Minneapolis: University of Minnesota Press, 1957. Pp. 3-14. (b).
Hess, E. H. Ethology and developmental psychology. In P. H. Mussen (Ed.), *Carmichael's manual of child psychology.* (3rd ed.) New York: Wiley, 1970. Pp. 1-38.
Hunt, J. McV. *Intelligence and experience.* New York: Ronald Press, 1961.
Kantor, J. R. *A survey of the science of psychology.* Bloomington, Ind.: Principia Press, 1933.
Kantor, J. R. *Interbehavioral psychology.* Bloomington, Ind.: Principia Press, 1958.
Lewin, K. The conflict between Aristotelian and Galileian modes of thought in contemporary psychology. *Journal of General Psychology,* 1931, **5**, 141-162.
Lorenz, K. Z. The companion in the bird's world. *Auk,* 1937, **54**, 245-273.
Lorenz, K. Z. *Evolution and modification of behavior.* Chicago: University of Chicago Press, 1965.
Lorenz, K. Z. *On aggression.* New York: Harcourt, 1966.
McClearn, G. E. Behavioral genetics: An overview. *Merrill-Palmer Quarterly,* 1968, **14**, 9-24.
Munn, N. L. *The evolution and growth of Human behavior.* Boston: Houghton Mifflin, 1955.
Mussen, P. H. (Ed.). *Carmichael's manual of child psychology.* (3rd ed.) New York: Wiley, 1970.
Palermo, D. S., & Lipsitt, L. P. (Eds.) *Research readings in child psychology.* New York: Holt, Rinehart and Winston, 1963.
Piaget, J. *Psychology of intelligence.* Paterson, N. J.: Littlefield, Adams & Co., 1960.
Piaget, J. Quantification, conservation, and nativism. *Science,* 1968, **162**, 976-979.
Reese, H. W., & Lipsitt, L. P. *Experimental child psychology.* New York: Academic Press, 1970.
Russell, W. A. An experimental psychology of development: Pipe dream or possibility? In D. B. Harris (Ed.), *The concept of development.* Minneapolis: University of Minnesota Press, 1957. Pp. 162-176.

Schneirla, T. C. Behavioral development and comparative psychology. *Quarterly Review of Biology,* 1966, **41**, 283-302.
Skinner, B. F. The phylogeny and ontogeny of behavior. *Science,* 1966, **153**, 1205-1213.
Sperry, R. W. Developmental basis of behavior. In A. Roe & G. G. Simpson (Eds.), *Behavior and evolution.* New Haven: Yale University Press, 1958. Pp. 128-139.
Spiker, C. C. The concept of development: Relevant and irrelevant issues. In H. W. Stevenson (Ed.), Concept of development. *Monographs of the Society for Research in Child Development,* 1966, **31**, (5, Whole No. 107) 40-54.
Stevenson, H. W. (Ed.) Concept of development. *Monographs of the Society for Research in Child Development,* 1966, **31** (5, Whole No. 107).
Tanner, J. M. The regulation of human growth. *Child Development,* 1963, **34**, 817-847.
Tanner J. M. Physical growth. In P. H. Mussen (Ed.), *Carmichael's manual of child psychology.* (3rd ed.) New York: Wiley, 1970. Pp. 77-166.
Tinbergen, N. *The study of instinct.* London: Oxford University Press, 1951.
Tinbergen, N. *Social behavior in animals.* London: Methuen, 1953.
Vandenberg, S. G. Contributions of twin research to psychology. *Psychological Bulletin,* 1966, **66**. 327-352.
White, S. H. Evidence for a hierachical arrangement of learning processes. In L. P. Lipsitt & C. C. Spiker (Eds.), *Advances in child development and behavior.* Vol. 2. New York: Academic Press, 1965. Pp. 187-220.
Zigler, E. Metatheoretical issues in developmental psychology. In M. H. Marx (Ed.), *Theories in contemporary psychology.* New York: Crowell-Collier-Macmillan, 1963. Pp. 341-368.

Recommended Additional Readings

Bijou, S. W. Ages, stages, and the naturalization of human development. *American Psychologist,* 1968, **23**, 419-427.
Harris, D. B. (Ed.) *The concept of development.* Minneapolis: University of Minnesota Press, 1957.
Stevenson, H. W. (Ed.) Concept of development. *Monographs of the Society for Research in Child Development,* 1966, **31**, (5).

"Why don't you make just one cookie, and then Xerox it fifty times?"

Reprinted by permission of Henry R. Martin and *Saturday Review*. © 1969 *Saturday Review* Inc.

2 Conceptual Emphasis in the History of Developmental Psychology: Evolutionary Theory, Teleology, and the Nature-Nurture Issue

S. Anandalakshmy
Robert E. Grinder

Orientation

This paper continues our emphasis on adopting a historical perspective for the study of developmental psychology. Anandalakshmy and Grinder examine three conceptual issues that were of major concern in the early years of "genetic" (or developmental) psychology. Although Darwinian theory and teleology are not major areas of concern in contemporary developmental psychology, it is clear that aspects of these early issues still reside in present conceptualizations of human development.

The third issue, nature-nurture, is still very much a contemporary one. Psychologists persist in their efforts to locate the genesis of behavior and development in one or both of two camps—heredity and environmental experience. Of course, almost all scientists now claim that development is a result of interactive processes between these two elements; nevertheless, implicit in the work of many researchers is a desire to answer the antiquated question of "how much" of a given behavior can be attributed to genetics and how much to environmental history. It is of interest to note that the nature-nurture issue is a matter of concern primarily only to psychologists in England and the United States. Continental European psychologists (e.g., Piaget, Vygotsky), who tend to be more existential and/or dialectic in orientation, generally do not attempt to reduce development to these two components. A later article in this book by Hirsch (II.7) will discuss the nature-nurture controversy in detail.

Some conceptual issues within a scientific discipline eventually cease to be of importance, some issues persist in importance, and new issues ascend into importance. Like the growing and developing individual, the field of developmental psychology is growing and developing. Today's developmental psychology is the outgrowth of yesterday's developmental psychology.

Reprinted from *Child Development*, 1970, 41, 1113-1123, by permission of the authors and The Society for Research in Child Development © The Society for Research in Child Development. All rights reserved.

An earlier version of this paper was presented at the 1969 meeting of the Society for Research in Child Development, Santa Monica, California.

S. Anandalakshmy received her graduate degrees from the University of Wisconsin and now is associated with Lady Irwin College, New Delhi, India. Robert E. Grinder, former professor of educational psychology at the University of Wisconsin, is now the dean of education at the University of Maine.

2

The contemporary emphases in developmental psychology are derived from the inspirations of turn-of-the-century biology, philosophy, and pedagogy. During the first half of the nineteenth century, Western thought was dominated by the belief in special creation—that the world was populated with species whose essence was fixed permanently at creation—and developmental viewpoints were regarded as irrelevant. The Cartesian dichotomy of body and mind still persisted, and psychology was conceptualized as the study of the mind. In the second half of the century, the formulation of laws of species interrelations in evolutionary theory and the subsequent advance of genetic psychology bridged the gap of dualism and provided the first conceptual links between biology and psychology (Grinder, 1967). During the seven decades of this century, major conceptual shifts have occurred, and the present paper traces briefly the historical record of three of these: evolutionary theory, teleology, and the nature-nurture issue.

Evolutionary Theory

Darwin's theory of evolution gave rise to several speculative assumptions that influenced early developmental psychology. The first important application was Haeckel's focus on the theory of recapitulation, the belief that each individual traces the historical record (phylogeny) of his species development in his growth (ontogeny), as an explanation of individual growth. Human development, for example, presumably involved a fixed sequence of growth patterns, each of which paralleled a period of evolutionary history.

Two complementary, developmental corollaries

were derived from the theory of recapitulation. First, it was said that nature was "right." If individual growth is considered a retracing of species evolution, nothing should prevent children from realizing the fullest potential of each stage in the developmental process. In the psychological literature of the time, children's activities were catalogued as norms predetermined by nature. Linked closely to the belief in nature-is-right was the concept of catharsis—the idea that exercise of "undesirable" traits in childhood would prevent their occurence in adult years. Hall (1904), the father of child study in America, suggested that children would be immunized against asocial behavior if, for example, toddlers were permitted to play as savages, children as barbarians, and adolescents as nomadic wanderers.

The principles of nature-is-right and catharsis have survived in various forms. For example, Erikson (1965, p. 1) indicates his views on "fidelity" during adolescence as follows: "I believe it to be part of the human equipment involved with socio-genetic evolution.... Fidelity could not mature earlier in life and must not, in the crises of youth, fail its time of ascendance if human adaptation is to remain intact." And Havighurst (1948, p. 7) suggests that if a developmental task "is not achieved at the proper time it will not be achieved well, and failure in this task will cause partial or complete failure in the achievement of other tasks yet to come." Recently, other statements endorsing the significance of the invariant sequence of developmental stages have emerged in concepts about critical periods in development (Scott, 1968).

The recapitulatory implications of evolution gave rise to Social Darwinism, which dominated Western thought in the later years of the nineteenth century. War, imperialism, and racism were justified and defended in accordance with the dictum of survival of the fittest. This view was especially appealing to European scholars, who, by implication, had been placed at the pinnacle of evolutionary progress (Barzun, 1958, p. 124). In the second volume of *Adolescence,* Hall described Africans, Indians, and Chinese as members of "adolescent races," implying that persons of these ethnic groups represented examples of arrested or incomplete growth. It was commonly assumed that a scale of cultural evolution could be constructed which would parallel the pattern of children's development. Frazer (1922, p. 586) felt that "a careful study of the growth of intelligence and of the moral sense in children promises to throw much light on the intellectual and moral evolution of the race." Freud, too, perceived a relationship between individual development and evolution. He endorsed the concept of recapitulation and believed that "each individual repeats in some abbreviated fashion during childhood the whole course of the development of the human race" (Freud, 1953, p. 209). Like Hall, Freud illustrated immature ego development and growth by drawing examples from the behavior of "savages."

Such elaborate attempts to trace the origin of individual development in evolution are vulnerable to the "genetic fallacy"—"the fallacy of reducing all experiences to one condition of their origin and so killing meanings by explanations" (Barzun, 1958, p. 91). Psychologists who were imbued with the idea that genetic origins explained behavior seemed able to fit into their frame of reference all observations of children. Truancy and delinquency and any other seemingly undesirable behavior, for example, were described as genetic aberrations or as instances of regressive evolution.

Murphy (1949) has observed that Hall's approach to adolescent development represented both the strongest and weakest of what the evolutionary science of the era had to offer.

Among the alternative viewpoints to recapitulation, Thorndike's "utility theory" emerged as the most viable alternative. A trait comes into being, he said, because it is useful to the organism. Thorndike (1913) assumed that the characteristics possessed by contemporary species had once endowed their ancestors with superior advantages; thereby, on the basis of their function and utility, they became permanent traits of progency in the order of their original appearance. Thorndike advanced his views on human nature while Mendelian hereditary theory was being assimilated into scientific thought. Thorndike's functionalistic views led him to reject the catharsis doctrine as contrary to all that was known about habit formation; the older theory had to capitulate in direct confrontation with the Law of Effect. Subsequently, developmental theorists have subscribed to Darwin's early belief that natural selection working on chance variations is the sole mechanism of hereditary change. Thus, those concerned with development are no longer asking the type of questions that would evoke explanations involving a Grand System (Zigler, 1963, p. 350).

Teleology

Teleology, the belief that ends are immanent in nature and that natural phenomena are determined not only by mechanical causes but by an overall design, has been a prominent concept in the history of developmental psychology. As Nagel (1957, p. 16) points out, the concept of development connotes a strong teleological flavor, since it implies both a retrospective and a prospective reference in the change sequence. To be considered developmental, a change, he suggests, must not only have directionality but must also be cumulative and irreversible.

Teleology has been regarded as an obstacle to scientific progress. Titchener held, for example, that teleological explanations were fatal to scientific advance (Dennis, 1948), and Hull (1943) repudiated the assumption that the terminal stage of a sequence could be an antecedent cause. Scientists bound by logical circularity, he warned, would forever be helpless to predict future events.

Such reactions to teleology may have arisen from confusion of biological determinism with transcendental teleology. Biological predictability is a scientific fact; however, the extent to which human behavior is made a function of solely biological determinism varies with each theory. Teleology became an issue in developmental theory when early genetic psychologists replaced faculty psychology with functionalism (Baldwin, 1895). Consciousness, cognition, and volition were viewed as much the outcome of evolution as physical characteristics.

"The biological point of view commits us to the conviction that the mind, whatever else it may be, is at least an organ of service for the control of environment in relation to the ends of the life process" (Dewey, 1902, p. 219). Hall's philosophical assumptions included the concept of purpose. Recapitulation theory led him to percieve a grand design in nature and to exhort his colleagues to link human effort with divine purpose. Even McDougall (1908) recognized purposiveness in animal behavior; however, he took an intermediate position between functionalism and behaviorism. He saw instinctual animal behavior as serving an overall design but did not accept functionalism as a complete explanation for human nature.

On the other hand, Thorndike lauded Darwin for making psychology a natural science and giving a death blow to teleology, and he recognized that to believe in immanent purposes and nature-is-right was to forfeit rational control over events. Thorndike endorsed, instead, both eugenics and social welfare measures as means of improving human nature. Watson (1925) manifested even greater confidence in man's ability to control human behavior and the outcome of social situations.

Currently, those theorists who use near-teleological terminology define their terms with clarity. The ethologists, who appear to be genetic determinists, clearly distinguish between the causes of behavior and the biological significance of behavior. Tinbergen (1951) states that the predictability of life and life processes cannot be treated as equivalent to causal factors. A point of view at variance with the ethologists comes from organismic theorists who see the behavior of organisms as teleologically determined, that is, as comprised of functions and functionally regulated structure and processes (Kaplan, 1967, p. 716); however, Kaplan cautions that teleological determinism refers not to awareness of motives by individual organisms but to pattern and order in nature.

Perhaps a fairly representative statement of the contemporary status of teleology is that offered by Spiker (1966, p. 47). "The behavior theorists' reservation about the term 'purpose' does not have to do with purpose *in* nature but purpose *of* nature. That organisms behave purposefully is an incontrovertible fact. That nature behaves purposefully is a religious dictum." Teleology does not constitute a major dimension of developmental psychology today. With the increasing accumulation of scientific knowledge, there has been in psychology, as in other related social sciences, a decline of statements of ultimate purpose as explanatory concepts.

The Nature-Nurture Issue

Whereas evolutionary theory and teleology are primarily of historical interest, the nature-nurture issue continues to be a contemporary concern in developmental psychology. The question has been discussed from so many standpoints that a veritable word-association list could be generated, for example, nature/

nurture, heredity/environment, maturation/learning, nativistic/empiristic, constitutional/experiential.

To Hall, innate forces so dominated developmental processes that the role of nurture was relatively insignificant. Thorndike (1913) was one of the earliest to highlight the nature-nurture issue when he distinguished original nature, reflexes, instincts, and inner capacities as separate from acquired nature, learned behavior, habits, and traits. Thorndike believed in the improvement of human species through social engineering and attempted to catalogue every facet of human nature. He conveyed his conviction as follows: "The most fundamental question for human education asks precisely that we assign separate shares in the causation of human behavior to man's original nature on the one hand, and his environment or nurture on the other" (Thorndike, 1913, p. 2). Thorndike, in his own research, emphasized the learning processes but seems to have become increasingly convinced that genetic factors were of more critical importance to species improvement. Toward the end of his career, he observed: "In the *Original Nature of Man,* published in 1913, I made a rough inventory of tendencies attributable to the genes. On re-examining it, I find some cases of too great generosity to the genes, but more cases where the progress of research has transferred causation from environment to genetic forces" (Thorndike, 1949, p. 179). The early theorists associated with the nature-nurture issue promoted either nature or nurture as being the main explanation of human development. Gesell, the major proponent of the primacy of innate endowment, stressed the priority of neural ripening over learning fuctions. The unfolding of morphological features became the norms of development. "All things considered, the inevitableness and surety of maturation are the most impressive characteristic of early development. It is the hereditary ballast which conserves and stabilizes the growth of each individual infant. It is indigenous in its impulsion, but we may well be grateful for this degree of determinism" (Gesell, 1928, p. 378). As Hunt (1968, p. 298) points out, Gesell took over Hall's faith in predetermined development and suggested that development was governed by "intrinsic growth."

Watson, an advocate of egalitarianism, rejected the mystique of innate endowment. He argued instead that: "The behaviorists believe that there is nothing from within to develop. If you start with a healthy body, the right number of fingers or toes, eyes and a few elementary movements that are present at birth, you do not need anything else in the way of raw materials to make a man, be that man a genius, a cultured gentleman, a rowdy, or a thug" (Watson, 1928, p. 41).

Watson (1925, p. 91) guaranteed that he could develop a specialist in any of the professions or arts from a healthy infant, "regardless of his talents, peculiarities, tendencies, abilities, ... or race of his ancestors." Watson's optimistic approach, compounded with that of Thorndike, liberated developmental theory which had long been dominated by the genetic emphases of Hall and Gesell. Watson's famous "give-me-a-dozen-infants" statement assured parents that the potentialities for human achievement were immense, being dependent only on appropriate guidance. Woodworth (1948, p. 92) suggested that Watson sought mainly "to shake people out of the complacent acceptance of traditional views"; he appears to have succeeded to a remarkable degree. During this period, enthusiastic behaviorists dominated developmental psychology. Norsworthy and Whitley (1920, p. 203) in a widely adopted textbook captured the spirit of the new freedom: "There is no subject of child psychology more important than this one of habit formation.... Having too few habits results in insufficiency of control, and the lack of material, in narrowness of conduct and thought. The greater the number of good habits the individual possesses in all fields... thought, feeling, conduct, the more efficient he will be, especially if among them is found the habit of forming new habits."

The Gesell-Watson controversy suggests a greater polarization than actually existed. The dichotomization, however, helped in identifying critical variables in human development. In the course of time, the issue of whether heredity or environment contributed more to human development was supplanted by that of estimating their relative influence, which in turn led to the issue of interaction.

Recent analyses have revived the ancient concept of epigenesis. Aristotle described development as a process of continual integration and differentiation and insisted that the more complex growth patterns were irreducible to simpler elements or atoms. From the epigenetic viewpoint, maturity is not a simple summation of early structures but a novel synthesis of them. The organismic theory, "a variant of the doctrine of emergent evolution" (Nagel, 1957, p. 19), whose major spokesman has been Werner (1957), is derived from the epigenetic model, that is, "traits exhibited by a hierarchically organized system cannot be reduced to, or explained by, the properties of parts of the system whose mode of organization occurs on a lower rung of the hierarchy" (Nagel, 1957, p. 19).

Several contemporary theorists offer variations on the theme of epigenesis. Piaget (1963, p. 16), for example, advances the idea of "biological relativity." Structures, rather than having preformed properties, arise in accordance with needs and situations. Development is a function of internal processes—equilibration, accommodation, and assimilation—all of which are dependent upon activity and experience. Piaget characterizes structures as "far from being static and given from the start" (Piaget, 1963, p. 17). His emphasis is on the continued and progressive changes a maturing organism undergoes in response to experiential conditions, which result in a complex network of interactions. And Kuo (1967, p. 11) emphasizes that epigenesis is bidirectionalistic rather than environmentalistic. He defines epigenesis as: "A continuous developmental process from fertilization through birth to death involving proliferation, diversification, and modification of behavior patterns both in space and in time, as a result of the continuous dynamic exchange of energy between the developing organism and its environment, endogenous and exogenous." Clearly, the crux of the epigenetic viewpoint underscores environmental influences and the mutuality of the developmental impact of organism and environment.

An alternative approach to estimating the relative effects of nature and nurture has been that of finding out *how* they interact. When Thorndike distinguished between original and acquired characteristics, the analogous statistics were limited to the computation of percentages, means, and deviations. Subsequent studies have used the analysis-of-variance model to isolate the main effects of heredity and environment and their interaction, especially in reference to human intelligence. Anastasi (1958), in one of the more influential American Psychological Association addresses of recent years, pointed out that an assessment of the proportional contributions of nature and nurture using an additive model had been demonstrated as untenable. While the interaction approach was widely accepted, the evaluation of the component sources of variance presented formidable problems. Anastasi suggested, therefore, focusing the mechanisms of interaction to arrive at an estimate of predictable relationships. Both heredity and environment were considered to be distributed along continua of direct-indirect effects. For instance, sex, skin color, or body type, known to be functions of genetic factors, evoke differential environmental responses across cultures, and the resultant behavior presumably can be traced to a specific combination of factors. As the child matures, the range of possible outcomes to hereditary factors expands, and a veritable chain of interaction effects seems to be set in operation.

Working out the equation of the interaction of genetic and experiential factors is like deriving an equation without constants but with several variables of unknown, changing values. The recognition of such problems led Hunt (1961) to question the appropriateness of the analysis-of-variance model for this purpose. In a well-argued case for the importance of environmental influences upon intelligence, Hunt culled experimental evidence from a wide-ranging survey of studies to attack beliefs in predetermined development and in fixed intelligence. He suggested that for organisms on the higher levels of the phylogenetic scale, that is, as the control of the central nervous system becomes more important and the A/S ratio (based on Hebb's earlier conceptualization) becomes higher, all early experience has an impact upon further development. In a pithy metaphor, Hunt (1968, p. 329) described experience as the programmer of the human brain computer and recommended that the human being be treated as an open system.

As Endler, Boulter, and Osser (1968, p. 5) indicate in a recent editorial comment, neither heredity nor environment yield behavior; each yields structures which appear to be dependent on the influence of the other. Thus, the contemporary form of the epigenetic theory goes beyond interaction, as it is commonly understood, and is analogous to Deutsch's (1968, p. 61) delineation of the "interpenetration" of the variables of nature and nurture. The flexibility of the interacting factors, their mutual impact, and the continuous progressive mobility of the human organism, challenge the application of any type of static analysis. In this context, it should be observed that the categories and labels selected here for conceptual analysis are somewhat arbitrary and serve mainly a heuristic purpose.

There is currently a resurgence of interest in the nature-nurture issue, the incentive for which comes from a pragmatic rather than a theoretical perspective. Jensen (1969) initiates his analysis of the issue with an expressed concern about the failure of contemporary attempts to improve IQ and scholastic achievement. He argues that 80 percent of the variance in measurable intelligence can be attributed to genetic factors and that attempts to raise IQ are futile. He advocates the use of *heritability estimates* to account for differences in intelligence. Heritability is the proportion of phenotypic (measurable) variation attributed to genetic variation. Presumably, the heritability of an IQ can be derived from a formula based on quantitative genetics. For Jensen, the environment operates like a threshold; above minimum requirements it exerts little influence but below such needs, it has great influence. He indicates that heritability estimates increase as environmental sources of variation decrease. He suggests, for example, that the heritability of a trait such as stature would be very high in a population in which all of the members have excellent nutrition and hygiene, so that the measured differences could be attributed to individual genetic factors.

However, several problems arise in application of this principle to data. As Cavalli-Sforza (1969, p. 8) indicates "Heritability measurements are somewhat arbitrary and can be given in a number of different ways." He suggests that alternative methods of computing the fraction of genetic variance for intelligence might be "less striking and might be between 40 and 60 percent *from the same data.*" Jensen estimates heritability on the assumption of uncorrelated environments for twins reared apart; however, given the usual practice of selective placement of adoption agencies, it is unlikely that the environments of the twins are uncorrelated. Therefore, some of the similarity of twins' performance on an intelligence test might result from environmental similarity as well.

Although, by definition, a specific heritability estimate is made in reference to a homogenous group, Jensen consistently infers genetic differences between heterogeneous groups because they register substantially varying IQ scores. Jensen (1969), therefore, is not convincing in his strong support of genetic factors in intelligence.

Jensen also introduces race as a hereditary variable that influences intelligence. Since, in the United States, race also encompasses social and cultural factors, he has included several extraneous issues. As Gottesman (1968, p. 21) points out: "From a biological point of view, it is nonsensical to lable someone with the remotest trace of Negro ancestry as a member of the Negro American Race." Gottesman (1968) cautions agains the use of race as a variable, stating unequivocally that unless all of the relevant variables are matched except race, conclusions about the differences between the races on given traits are unwarranted.

Today there is consensus that nature, nurture, and their interaction in a complex society cannot be reduced to a simple formula. To assume, as Galton did exactly a century ago, that men who achieve eminence and those who are naturally capable are to a large extent identical seems unrealistic. Who would argue that perfect vertical social mobility exists in any society?

Both the Chinese and the Indian calendars run for a period of 60 years, and all things are said to come full circle and be repeated. Although we have accepted Mendelian genetic theory and eliminated teleological confusion from our thinking, we seem to be posing the same kind of questions about the nature-nurture issue that Thorndike asked 60 years ago.

References

Anastasi, A. Heredity, environment and the question "how?" *Psychological Review,* 1958, **65**, 197-208.
Baldwin, J. M. *Mental development in the child and the race.* New York: Macmillan, 1895.
Barzun, J. *Darwin, Marx, Wagner.* New York: Doubleday, 1958.
Cavalli-Sforza, L. L. Problems and prospects of genetic analysis of intelligence at the intra- and inter-racial level. Paper read at American Educational Research Association Meeting, San Francisco, February 1969.
Dennis, W. (Ed.) *Readings in the history of psychology.* New York: Appleton, 1948.
Deutsch, C. P. Environment and perception. In M. Deutsch, I. Katz, & A. R. Jensen (Eds.), *Social class, race and psychological development.* New York: Holt, Rinehart and Winston, 1968.
Dewey, J. The interpretation of the savage mind. *Psychological Review,* 1902, **9**, 217-230.
Endler, S., Boulter, L. R., & Osser, H. (Eds.) *Contemporary issues in developmental psychology.* New York: Holt, Rinehart and Winston, 1968.
Erikson, E. Youth: Fidelity and diversity. In E. Erikson (Ed.), *The challenge of youth.* New York: Anchor, 1965.
Frazer, J. G. The scope and method of mental anthropology. *Science Progress,* 1922, **16**, 580-594.
Freud, S. *A general introduction to psychoanalysis.* New York: Permabooks, 1953.
Gesell, A. *Infancy and human growth.* New York: Macmillan, 1928.
Gottesman, I. I. Biogenetics of race and class. In M. Deutsch, I. Katz, & A. R. Jensen (Eds), *Social class, race, and psychological development.* New York: Holt, Rinehart and Winston, 1968.
Grinder, R. E. *A history of genetic psychology.* New York: Wiley, 1967.
Hall, G. S. *Adolescence.* New York: Appleton, 1904.
Havighurst, R. J. *Developmental tasks and education.* Chicago: University of Chicago Press, 1948.

Hull, C. L. *Principles of behavior.* New York: Appleton, 1943.
Hunt, J. M. *Intelligence and experience.* New York: Ronald Press, 1961.
Hunt, J. M. Environment, development, and scholastic achievement. In M. Deutsch, I. Katz & A. R. Jensen (Eds.), *Social class, race, and psychological development.* New York: Holt, Rinehart and Winston, 1968.
Jensen, A. R. How much can we boost IQ and achievement? *Harvard Educational Review,* 1969, **39**, 1-123.
Kaplan, B. Meditations on genesis. *Human Development,* 1967, **10**, 65-87.
Kuo, Z-Y. *The dynamics of behavior development.* New York: Random House, 1967.
McDougall, W. *Introduction to social psychology.* London: Methuen, 1908.
Murphy, G. *Historical introduction to modern psychology.* New York: Harcourt, 1949.
Nagel, E. Determinism and development. In D. B. Harris (Ed.), *The concept of development.* Minneapolis: University of Minnesota Press, 1957.
Norsworthy, N., & Whitley, M. T. *The psychology of childhood,* New York: Macmillan, 1920.
Piaget, J. *The origins of intelligence in children.* New York: Norton, 1963.
Scott, J. P. *Early experience and organization of behavior.* Belmont, Calif: Brooks-Cole, 1968.
Spiker, C. C. The concept of development: relevant and irrelevant issues. In H. W. Stevensen (Ed.), Concept of development. *Monographs of the Society Research in Child Development,* 1966, **31**, (5, Serial No. 107), 40-54.
Thorndike, E. L. *The original nature of man.* New York: Teacher's College, 1913.
Thorndike, E. L. *Selected writings from a connectionist's psychology.* New York: Appleton, 1949.
Tinbergen, N. *The study of instinct.* London: Oxford University Press, 1951.
Watson, J. B. *Behaviorism.* New York: Norton, 1925.
Watson, J. B. *Psychological care of infant and child.* New York: Norton, 1928.
Werner, H. The concept of development from a comparative and organismic point of view. In D. B. Harris (Ed.), *The concept of development.* Minneapolis: University of Minnesota Press, 1957.
Woodworth, R. S. *Contemporary schools of psychology.* New York: Ronald Press, 1948.
Zigler, E. Metatheoretical issues in development psychology. In M. Marx (Ed.), *Theories in contemporary psychology.* New York: Crowell-Collier-Macmillan, 1963.

Recommended Additional Readings

Aries, P. *Centuries of childhood: A social history of family life.* New York: Knopf, 1962.

Bremner, R. H. (Ed.) *Children and youth: A documentary history.* Cambridge, Mass.: Harvard University Press, 1970.
Charles, D. C. Historical antecedents of life-span developmental psychology. In L. R. Goulet & P. B. Baltes (Eds.), *Lifespan developmental psychology: Research and theory.* New York: Academic Press, 1970. Pp. 23-52.
Grinder, R. E. *A history of genetic psychology.* New York: Wiley, 1967.
Riegel, K. F. Influence of economic and political ideologies upon the development of developmental psychology. *Psychological Bulletin,* 1972, **78**, in press.

3 Some Representative Theoretical Orientations in Developmental Psychology

L. Joseph Stone
Joseph Church

Orientation

The focus of this selection is the role of theory in our understanding of human behavior and development. There are a great number of theories about behavior and development in psychology, and all of these are presented by their creators and advocates in as convincing a manner as possible. All of them also have at least a minimal amount of empirical support, i.e., each theory has been supported with evidence gathered as a result of scientific research. To complicate matters even more, all existing theories, or at least certain aspects of them, have been refuted or in some way unsupported by other scientific research. Why must things be so complicated and confusing? Why can't psychologists sit down and mutually agree upon a theory and then be done with it? Certainly this would simplify our business enormously, but there is an overriding reason why one general theory of behavior has not been accepted by all psychologists (and why this is unlikely to happen, at least for a very long time to come): The human organism is an incredibly complex creature, and his functioning is not necessarily in accordance with our theorizations about him.

Why, then, should we bother with theories at all?

From *Childhood and Adolescence*, 2nd Edition, by L. Joseph Stone and Joseph Church. Copyright © 1957, 1968 by Random House, Inc. Reprinted by permission of the authors and the publisher.

Why not just forget about such apparently futile intellectual exercises and get down to the business of what we have to do? Such a position, though easy to arrive at, is a dangerous one. While it is unfortunately true that theories of behavior are complex and many, we must nevertheless attend to them, understand what they are saying about the organism, and try to refine them whenever possible. A critical point to understand is that not only psychologists deal with theories about people; everyone *holds or believes in a particular theory about people. Each person, whether educated or uneducated, possesses some assumptions, some beliefs, some attitudes about people and how they work and how they think. Perhaps these notions are not easy for one to spell out in great detail, or maybe the individual is only dimly aware of what his beliefs are, but they are there. Perhaps these might be called one's assumptions about "human nature," or perhaps it could be called one's "model of man." Whatever these beliefs might be, the individual governs his personal conduct toward other people according to the dictates of his personal theory. For example, if one believes that people are acquisitive, competitive creatures, he will act accordingly; he will do all he can to acquire more possessions, to crowd ahead of others in line, to scramble to get the highest grades in the class, to compete with his neighbor in any way perceived as necessary. Or if another individual happens to hold these same attitudes but is personally passive and has feelings of inferiority, he may give up the fight and allow others to scramble for the rewards. Perhaps a teacher believes that people are generally lazy and therefore will do nothing unless provided with some sort of external motivation; this teacher's behavior toward his students is thus easy to predict (grades, gold stars, verbal praise, or admonishment, etc.).*

The point is, each one *holds some personal theory about "human nature," and guides his actions accordingly. Most likely the theory is not well formulated, such as a textbook description of Freudian psychodynamic theory, but it is there and it influences one's relations with other people.*

It is common to hear someone say "Well, it exists in theory but not in practice." If a particular theory postulates that something exists, then that something should be in evidence in the "real world." If it is not, then the theory is no good and therefore should be appropriately modified or perhaps abandoned altogether in favor of a more promising theory. This is true for our own personal theories as well as for those formulated by scientists. (It is a quirk of "human nature" that some people insist upon stubbornly clinging to their preferred theories, even long after they have been disproven. Apparently, bad theories never die; they live on in the hearts of their admirers.)

As you will discover in the present article, many conceptions of the human individual have been formulated. It is important to examine these theories closely: What we believe about people determines what we will do to, for, and about them.

L. Joseph Stone is professor of psychology at Vassar College. Joseph Church is associated with the Graduate Center of the City University of New York. They have coauthored the popular textbook, Childhood and Adolescence: A Psychology of the Growing Person, *now in its second edition (New York: Random House, 1968).*

3 Psychoanalytic Theories

Let us begin by taking a somewhat closer look at a few of the more salient views of human nature, starting with the views of Sigmund Freud, who was a major modern contributor to the discovery of human childhood as an area of serious investigation. The reader should remember that Freud had very little direct contact with children and constructed his account of development from self-observations, some observations of his own children, and the reconstructions and recollections of his adult patients—his one famous child patient, Little Hans, was treated indirectly, through Hans' father. Freud's was primarily a dynamic, that is, motivational, theory of behavior, although, as we shall see, it always had cognitive and intellectual components. These were later elaborated on by some of Freud's followers, such as Ernest Kris, Heinz Hartmann, David Rapaport, Erik Erikson, and Freud's own daughter, Anna, all of whom became known as psychoanalytic ego psychologists. Freud acknowledged a wide variety of human motives, or drives, but in the end he concerned himself mainly with two: libido, or sexual energy, and the death wish, or drive to destruction of self or others (these two motivational forces have been dubbed Eros and Thanatos, from the Greek words for [eternal] love and death).

Freud proposed a three-way conceptual partition of the person into *id,* the blind strivings of Eros and Thanatos and the source of all the person's motives and energies; *ego,* the rational, reality-oriented part of the person (whence ego psychology); and *superego,* corresponding approximately to the conscience, the moral and ethical part of the person's make-up. Developmentally, the baby at birth is considered to be pure id, a seething mass of passions governed exclusively by what Freud called the *pleasure principle,* unbridled seeking for immediate gratifications. Needless to say, the baby's greedy id-strivings collide almost at once with an only partially yielding reality, and out of this collision of forces and frustrations there develops a layer of ego, constituting the practical strategies and the capacity for delay and detour by which the developing person can take account of reality in the quest for id gratification. The ego, that is, is governed by the *reality principle* taking account of what is already possible or of the means of making gratification possible. The kinds of thinking that accompany the id-based, pleasure principle in its search for immediate gratification, whether in the young baby, in adult dreams, in daydreams or fantasies, or in psychosis, are called *primary-process* thought, whereas the reality-oriented thinking associated with the ego is called *secondary process.* Finally, not in infancy but in early childhood, and in ways that we shall discuss shortly, a portion of the ego becomes further differentiated as superego, imparting an ethical-moral, socially responsible dimension, in addition to the purely pragmatic reality principle, to the search for gratification of the id-impulses.

The differentiation of some of the id into ego and of ego into superego can be seen as a process of cognitive development. Note that the motivations remain the same throughout life; it is the manner of their expression that changes. Psychosexual development refers to shifts in the channels or zones of the body through which id gratification is sought and obtained, and in the objects which serve as gratifiers.

The first of the psychosexual stages is called the *oral phase,* during infancy, when the main channel of gratification is the mouth and upper digestive tract, and the gratifying objects are the nipple and the mother who provides it—plus, to a lesser extent, the thumb, pacifier, or whatever else the baby puts in his mouth. In the Freudian view, the infant's eating is not merely a matter of appeasing hunger and assuring organic survival, but also a source of proto-erotic satisfaction. Furthermore, since the baby destroys what he has consumed, he is simultaneously satisfying his death wish. Early infancy, before the baby's teeth have sprouted, is called the *oral-passive* period. Later, when he has acquired enough teeth to begin to bite, he enters the *oral-sadistic* period in which the aggressive (Thanatos) component becomes even more conspicuous.

In the period we call toddlerhood, according to Freud, the main channel of gratification, subject to some restrictions we shall touch on later, ceases to be the mouth and becomes the lower digestive tract and anus, and the baby enters the *anal stage* of psychosexual development. This period, in which the baby gains control over the anal sphincter and the process of holding on to and expelling bowel movements, is supposedly marked by preoccupation with anal functions, to the point where having his bottom spanked may be pleasurable to the baby. The anal period was thought of by Freud as a time of inevitable conflict between parents and child over who shall regulate the time and place of defecation, and the baby gets ample exercise of his destructive tendencies in withholding bowel movements when his parents want him to let go, by expelling feces when his parents want him to hold on, and by smearing his feces in an act of defiance.

If all goes normally during the anal period, the source of gratification shifts to the genitals, and the preschool years are described as the *phallic stage*—a designation betraying Freud's male-centered view of behavior. The phallic stage, in which gratification is sought without concern for the feelings of other people, is distinguished from the *genital stage* of adolescence and adulthood, characterized not only by mature sexuality in the usual sense by also by love and a concern for others' feelings.

The phallic stage evolves into and culminates in the *Oedipus complex,* named for the tragic hero of Greek legend, King Oedipus, who unwittingly but in fulfillment of a prophecy made at his birth slew his father and married his mother. According to Freud, the boy (we shall talk in a moment about the girl's Oedipus—or, as it is sometimes called, Electra-complex) directs his phallic strivings toward the mother, putting himself in direct rivalry with the father, whom he ambivalently both hates and loves. The father is seen as retaliating with an open or implied threat to injure or cut off the boy's penis—a threat of castration. This threat is usually viewed as being expressed in the context of punishing the boy's masturbation: In actuality such threats are made by some parents, and were doubtless more prevalent in Freud's day.

The child, by now about age five, moved both by fear of castration and by love for his father, capitulates, renouncing his claim on the mother's favors and at the same time *repressing* (forcing out of consciousness) all sexual cravings. This renunciation of sexuality brings the child into the stage of psychosexual *latency,* corresponding to the middle years of childhood, which ends only when the physiological developments of adolescence bring a new upsurge of more mature sexual feelings, destroying the quiescence of latency. Furthermore, in capitulating to the father, the son is said to *identify* with him, to take on his styles of action, and to *internalize,* to make a part of himself, his father's moral values, which form the basis of the superego. That is, the parental prescriptions and proscriptions that formerly came to him as "thou shalt" and "thou shalt not" now speak to him in his own voice, saying, "*I* must" and "*I* must not," and arouse feelings of *guilt* when violated. To keep sexuality (and excessive aggression) at bay, the ego brings into play several of the *mechanisms of defense,* one of which is repression, already mentioned, and other stratagems by which one reinforces the repression of dangerous cravings and impulses.

The superego is initially a rigid, tyrannical sort of conscience, which only later becomes more flexible and tolerant (corresponding accurately to empirical findings that children shift from literal to relative concepts of morality), and the ego is hard put to mediate among the insistent clamor of the id, the harsh moralism of the superego, and the realities of the outside world. For Freud, a psychiatrist oriented to disturbed—particularly neurotic—forms of behavior, the defense mechanisms were always defenses against oneself, not the world outside. For instance, the child engages in *reaction formation,* a going to the opposite extreme, as when he rejects not only sexuality but the opposite sex in general— middle-years boys notoriously will have nothing to do with girls or "sissy stuff." By the mechanism of *sublimation,* the id-energies are transformed into socially acceptable drives, as, for example, the school-age child's new concern with intellectual, objective learning—Freudians would say that it is the resolution of the Oedipus complex that makes children ready for school, and that some of those children who do badly in school are suffering from an unresolved Oedipus complex.

The *obsessive-compulsive* mechanisms found in adult neurotics (extremes of meticulousness, repetitive hand-washing, and so on) have their counterparts in the *repetition* and *ritual* of the normal child as seen in such activities as counting all the cars of a freight train, having to touch every lamppost, holding one's breath until one reaches the corner, or avoiding the cracks in the sidewalk to the tune of "Step on a crack, break your mother's back"—which Freudians would interpret as a disguised expression of ambivalent hostility (and associated guilt and fear) against the now unattainable mother. Other well-known defense mechanisms are *projection,* whereby one ascribes one's own unconscious motivations to other people, and *rationalization,* whereby one justifies evilly motivated behavior in terms of some virtuous or neutral purpose.

Freud never formulated to his own satisfaction the course of the Oedipus complex in girls, but offered the following sketch. Initially, the girl's Oedipus complex is identical with the boy's: The girl, considered to assume that she has a penis, addresses her attentions to the mother in competition with the father. For the girl, however, the crisis comes not with the threat of castration at the hands of the father but with the dreadful discovery that she has been castrated (and amputated)—presumably her interpretation of the female genitalia. At this juncture, the girl can react in either of two ways. Ordinarily she is seen as now identifying with the mother, and competing for the father—seeking a baby from him in fantasy as a substitute counterpart of the lost penis. In the face of reality she then will shift from actual striving for the father's favors to substitute father-figures, repressing her impulses as she reaches latency. The girl's (less common) alternative reaction to castration is called the *masculine protest,* a persistent if symbolic denial of the loss of the penis and the pursuit of an essentially masculine style of life (whether or not involving the practice of

Lesbianism), with a more or less permanent inability to play a feminine role in love or marriage. Such women are called by Freudians "castrating females," possessed by *penis envy,* as though seeking to take away from men what rightfully belongs to women.

Both boys and girls, reaching maturity, are seen as seeking mates with whom (in part) to reenact their earlier Oedipal attachments—"I want a girl just like the girl who married dear old Dad." Indeed, all close attachments, including that formed to the psychoanalyst in therapy (transference), are viewed as partial reenactments of the ambivalent relations to the parents formed in infancy and early childhood.

It is worth mentioning parenthetically that Freudians invoke the repression that climaxes the Oedipal drama to explain *infantile amnesia,* the fact that we find it hard or impossible to remember anything more than disconnected fragments of the events happening before age five or six. It is as though the repressed sexual feelings took with them all the associated events, to be recaptured only in the special condition of hypnosis or the *abreaction*—the reliving process—of psychoanalytic psychotherapy.

We should note that normal psychosexual development requires the enactment by child and parents alike of strictly fixed roles in the psychosexual drama. These roles are thought of as instinctive, and so are considered *biological imperatives.* Any departure from the biological imperatives is considered to lead to more or less serious disturbances of functioning.

Before discussing some of the more common departures from the normal course, we must introduce another Freudian concept, that of *cathexis.* At each psychosexual stage, one's supply of libidinal energy is said to be invested, or *cathected,* in the relevant channel and the important gratifying objects associated with that stage. Thus, in the oral stage of infancy, the child's mouth is said to be libidinized (that is, it has acquired erotic meaning), and the bottle or breast—and, by extension, the mother—cathected (has erotic meaning attached to it). When the needs associated with a given stage are gratified, the child is said to withdraw his cathexes and reinvest his energy in the new zones and objects of the next stage. If, however, inadequate gratification is given, all or part of the child's cathexes may remain *fixated* at that stage, that is, the child may still go on seeking the gratifications appropriate to a stage which he is supposed to have outgrown. Paradoxically, fixation can be produced both by insufficient gratification, which leaves him ever hungering, or by overgratification, which leaves the child so well satisfied with his present mode of existence that he is, so to speak, reluctant to move on. But assuming that a child has found satisfaction at one level and moved on, if he then encounters excessive frustration at a later stage, he may *regress*—that is, go back—to the last earlier stage at which he found satisfaction, so that the baby who finds the demands of toilet training in toddlerhood, say, too taxing, may revert to the clinging helplessness and orality of infancy.

Now it is clear that total fixation at or permanent regression to childhood patterns of behavior would be a rare and profoundly pathological style of life. However, the Freudians have a system of adult character types said to express partial fixations at early levels of psycho-sexual development. Thus, the adult "oral character" is marked by infantile, dependent, demanding behavior, and by a preoccupation with oral gratifications: eating, gum-chewing, smoking, nail-biting, drinking, pencil-chewing, garrulous talking, and whatever. Oral sadism, deriving from the second oral stage, would be shown not only in a literal tendency to bite but in the symbolic biting expressed in harsh words. Fixation at the anal stage might be expressed in either the "anal-retentive character," who seems to have learned too well the lessons of control, orderliness, and regularity associated with bowel training, and whose behavior is marked by stinginess, accumulating and collecting of money or other valuables, rigid adherence to forms and routines, suspicion, legalistic ways of thinking, and what has been described as "emotional constipation"; or the "anal-aggressive character," who seems still to be fighting the battle of bowel control and who expresses his rebellion against authority by extreme messiness in personal habits and grooming, by indifference to routines and schedules, and even by the tendency to "dirty" language, which can be seen as a symbolic befouling or "soiling" of the environment or those around him. The "phallic character" would be one who selfishly exploits others sexually, without regard to their needs, concerns, or feelings—in males, the Don Juan or the sexual psychopath or the rapist, in females the self-centered nymphomaniac.

A few other features of the Freudian system should be mentioned. For Freud, significant motivations and their alterations by defense mechanisms were primarily unconscious, and the true mainsprings of behavior lay deep in organic nature. In the course of ego development, some features of experience enter consciousness, including a certain limited knowledge of oneself and one's motives. The knowledge that is not conscious at the moment but is available to consciousness is said to reside in the *preconscious,* as in the case of all those things we can easily remember. The *unconscious* is thought to contain both those strivings that have never been conscious and those that have started forth and then been repressed, for instance, the Oedipal impulses. The presence of unconscious strivings is inferred from evidence of the operation of defense mechanisms, from self-betrayals in such forms as selective forgetting, making meaningful mistakes and slips of the tongue, and from the content of dreams, which are supposed to be symbolic, more or less disguised representations of all the unfulfilled cravings that come out of hiding during sleep, when the person's defenses are down. In short, people constantly do things for reasons that they do not recognize and may even deny.

Having sketched the formal elements of the Freudian system, let us say that Freud made a monumental contribution to our understanding of human behavior. Certainly we can see something like oral preoccupations in infancy-people sometimes describe the baby as being "all mouth." Certainly we can see Oedipus-like or Electra-like behavior in the preschool years, as when the child lavishes affection on the opposite-sex parent, even inviting that parent to share the child's bed, meanwhile telling the same-sex parent to go away, or turning that parent's picture to the wall. (In adults, we can see persistent Oedipal tendencies in people who marry mother-surrogates or father-surrogates.) Certainly we can see people living at odds with their own motives, practicing self-deception (which may deceive no one else) on a staggering scale, and in general struggling to get loose from leftover problems of childhood, including overinvolvement with parents. Certainly we can see ambivalence, strong hostility as well as strong affection to parents and to others who are very close—and certainly the hostility is sometimes veiled and indirect though indubitably there. We can see Freudian symbolism not only in our dreams (one six-year-old girl dreamed that a raging bull trampled down a fence and gored her in the bottom, but, strangely enough, it was "sort of pleasant"), but in myths and fairy tales. Consider the symbolism of the ogre in "Jack and the Beanstalk," of the oven in which the ogre's wife hides Jack, of Jack's cutting down the beanstalk, and of his using the gold stolen from the slain ogre to live alone with his mother. Irrationality and absurdity are everywhere around us, and Freud has taught us to look beyond face values.

Nevertheless, we feel that Freud's brilliant contribution has serious shortcomings, some of them acknowledged by his less orthodox followers. It accounts much better for pathological behavior than for normality, and almost not at all for what we consider superior styles of life. It is possible that babies born in nineteenth- and early twentieth-century Vienna were turbulent bundles of id—or were made so by parental practices—but the babies we know are not only intent on immediate gratifications and sometimes perverse but also cheerful, curious, playful, often affectionate without detectable undertones of salacious desire, and given to joyously pursuing and practicing mastery of skills and objects.

In general, like most drive theories, Freud's deals with unsatisfied (and even insatiable) instincts, but, as Robert W. White[1] has pointed out, a complete psychology has to deal also with the behavior of organisms which are awake and in reasonable equilibrium, free from the goading of deficiency drives. Indeed, one may say that human beings are at their most human when their basic biological needs are least involved. The behavior of a human being in the throes of asphyxiating, choking, starving, or parching is not much different from that of any other animal. And when biological needs are satisfied, they cannot be invoked to account for love, curiosity, the higher learning, imagination, humor, foresight, symbolic activity, and all else that is distinctively human. Just as the old economics of scarcity, in which there is not enough to go around, must yield, in an age of mechanized production and population control, to an economics of abundance, so in the behavioral sciences must deficiency theories yield to accounts of the behavior of organisms whose basic needs have been satisfied. For the replete organism, as White says, does not invariably drift off into the blissful Nirvana of semicoma, but looks around to see what there is to do next. (This Nirvana concept of motivation is the psychological extension of the physiological concept of motivation is the psychological extension of the physiological concept of homeostasis—seeking the resting-state.) In White's view, boredom becomes as great a spur to action as lust or hunger.

In terms of development, Freud laid more stress on the determining power of biology (in his own version), and less on the influence of experience with people and things (except for the preordained acting out of the biological imperatives), than present evidence on the effects of early experience makes tenable. But whatever the flaws in Freud's formulations, we can see them only because Freud himself opened our eyes (in Erich Fromm's phrase, people stand on Freud's shoulders and, because they can see so much farther than he, consider him a pigmy; this is a variant of the usual figure of speech, which has the modern pigmy standing on the shoulders of giants[2]), and he showed us phenomena that any adequate theory of behavior is going to have to include. While it may be that Freud's theory was too intricate and complex to lend itself easily to empirical verification or disproof, he at least tried to deal with the full richness of human behavior as it is known to the dramatist, to the novelist, and to the painter.

Freud was the father of psychoanalysis, but his was not the only analytic school. There are several varieties, most of them founded by former colleagues of Freud's as schismatic movements which Freud himself considered heretical. Carl Jung, of Switzerland, presided over a religiously oriented school which lays much stress on the *racial* or *collective unconscious* (an idea found in Freud, too), the basic patterns or structures which guide ideation. These supposedly universal patterns are called *archetypes,* and, as Jung has shown, are found in many cultures. In this view, life is more a search for understanding (as of the way archetypes shape behavior) than for direct or indirect carnal gratifications.

Otto Rank originally built his sytem around the *trauma of birth,* a lifelong attempt to return to the warm, safe nothingness of intrauterine existence, from which the newborn babe is cast into a cold, hostile, and unmanageable world. As Rank's thinking evolved, he became more concerned with *will therapy,* designed to help people know what they really wanted, instead of living with the agony of formless dissatisfactions, and to feel free to want it. Rank's influence is still visible in several current versions, including the *non-directive therapy* of Carl Rogers, which places responsibility for his own psychological welfare on the client rather than the therapist, and, more explicitly, in the teachings of many schools of social work.

Alfred Adler, a Viennese socialist, was far more concerned than Freud with social influences on development, in keeping with his Marxian orientation. To Adler we owe such notions as *sibling rivalry,* the warfare among brothers and sisters for the affection of the parents, with much emphasis on the dethronement of the firstborn child from his place of unique eminence by the arrival of younger children. Adler also gave us *inferiority,* the sense of weakness or inadequacy (beginning with every child's smallness and helplessness) that keeps people striving and drives them to extremes of *compensation* to make up for their real or imagined insufficiencies in stature or competence. Organ inferiority was the basis of Adler's concept of "inferiority complex." Basically, Adler's is a more optimistic, problem-solving view of man than is Freud's.

In the United States, as we have said, a school of anthropologically and sociologically oriented psychoanalysis has developed alongside the orthodox one. The roster of this group includes such names as Harry Stack Sullivan, Karen Horney, and Erich Fromm. Horney and Fromm have been extremely critical of modern civilization and the social and personal ills it begets. Where Freud saw civilization as a painfully and partially attained modification of the violence of the id, in their view, modern society is itself inhuman and dehumanizing, and those who try to live lives of love and virtue, representing man's basic strivings, will inevitably find themselves in torment. We have already mentioned the work of Erik Erikson, who perhaps comes closest to bridging the Freudian and post-Freudian approaches. Erikson evolved a scheme of considerable use in the study of development, describing the eight stages of man, which are a more generalized statement of the Freudian psychosexual stages. Thus, as we have seen, the oral stage of infancy becomes for Erikson the period for the formation of basic trust or mistrust, which are seen as closely implicated with the oral experience of hunger and feeding but which are also related to the dependability with which other basic needs are met. As we shall see in the next chapter, Freud's anal period of toddlerhood is elaborated by Erikson as the time for the development of either autonomy or attitudes of shame and doubt.

Learning Theories of Development

The learning-theory tradition, which lays heavy emphasis on the role of the environment in shaping modes of behavior, has two main antecedents. The first of these is British associationism, characterized by John Locke's doctrine of the newborn as a *tabula rasa,* a clean slate (literally, a wax tablet shaved clean) on which the world writes or impresses its message. The message the world writes on the individual consists of associations, a set of connections between events that occur together in time and space. Associationism found its scientific expression in the second major forebear of learning theories, Ivan Pavlov's conditioned reflex theory.

The basic tenet of learning theories is that the elaboration of behavior observable in the course of development can be explained, apart from a few concessions to physical growth and maturation, as the continuous formation of connections among stimuli and responses. Pavlov, as is widely known, demonstrated that one can link arbitrary, experimentally manipulated stimuli (called *conditioned* or *conditional* stimuli) to already existing "natural," *unconditioned* stimuli that reliably elicit a fixed, *unconditioned* response. This is accomplished introducing the conditioned stimulus just prior to the natural, unconditioned one. Thus, if food is an unconditioned stimulus for salivation, sounding a bell immediately before giving a dog food causes the bell alone to become the stimulus, after a few pairings, for salivation, even when no food is given. One can, in the laboratory, establish a great many conditioned associations, limited only by the number of unconditioned stimulus-response sequences one can identify. One can make some arbitrary event cause the pupil to contract as though a bright light were about to flash, or the eye to blink as though it were about to be subjected to a puff of air.

Pavlov and his contemporary, Bekhterev, for the most part, studied the conditioning of simple reflexes, the way the organism reacts to having something done to it—secretory responses, muscle twitch or limb flexion to vibration or electric shock, and so forth—many of them describable as vegetative responses mediated through the autonomic nervous system. In contrast to reflex reactions, there are also instrumental acts, the things the organism does in response to its own inner states, such as the hungry rat's looking for food in a maze. American behaviorists such as Watson extended Pavlov's *classical conditioning* model to include instrumental, goal-directed behavior. As we shall see shortly, another branch of American behaviorism provides separate accounts for classical and instrumental conditioning. The classical conditioning technique has proved very useful in the study of perceptual discrimination and generalization. For instance, one can sound one particular tone as a signal for food, and introduce a different tone unrelated to feeding, and the animal learns to react to the relevant tone and to disregard the irrelevant one. As the difference between the tones is reduced, the animal's responses allow us to infer the limits of discrimination. On the other hand, if only a single tone is used in a conditioning experiment, the animal will "generalize" his response to a great variety of tones introduced later on.

The fundamental form of classical conditioning is between a stimulus and a response, but repeated associations of stimuli can lead to stimulus-stimulus conditioning, as when Tristram Shandy's mother formed a connection between the periodic conjunction of her husband's winding the clock and then having intercourse with her, causing her to cry out at a delicate moment, "Pray, my Dear, have you not forgot to wind up the clock?" One can also form response-response connections, as in the habitual sequence of actions by which one gets dressed— here we have the phenomenon of "equivalent acts" which may be substituted in a response-response sequence, so that if the child asks his father to tie his shoes just as the father is about to tie his own shoes, the father may go forth with his shoelaces dangling. Response-response conditioning can be invoked to account for behavior when no external stimulus can be found to explain it. Thus, in the language of associationism, states of organism such as hunger, thirst, sexual arousal, attitudes, or expectations can be described as responses which in turn stimulate yet other responses.

Associationism has been a potent force in American psychology. In the Pavlovian tradition, Clark Hull and his students, notably Kenneth Spence, have greatly elaborated on the classical conditioning model to take account of more subtle learning such as the formation of concepts and latent learning.

In this view, *concept formation* is seen as establishing an equivalence among a group of stimuli on the basis of the number of "identical elements" the stimuli have in common. Thus, the concept "man" includes all those beings who share such identical elements as deep voices, facial hair, masculine styles of dress and haircuts, and so forth.

Latent (incidental) *learning* takes place in the absence of any demonstrable motivation or reinforcement (reward), as when the hungry white rat, exploring the maze for food, also learns incidentally where water is to be found in the maze, so that he goes directly to the water when he is placed in the maze thirsty instead of hungry. A great deal of our learning, such as our learning about space and time, about values, and, indeed, the cultural learning we have already described, seems to be of this latent, incidental sort. The fact of latent learning was long used to counter association theory, and for a time associationists denied its very existence. Now the Hull-Spence position seems to be that latent learning follows the same course as overt stimulus-response learning, except that the motivation, response, and reinforcement are fractional rather than full-blown, so that the somewhat thirsty animal, even if he does not actually drink the water, is nevertheless partially reinforced by the mere sight of it. Hull evolved a whole set of axioms to describe the learning process, including a set of propositions about the strength of drives and inhibitions, habit-family hierarchies (the likelihood that any one of a group of competing responses would be released in given circumstances), and secondary or substitute reinforcers (the tokens that can appease the animal without actually satisfying his physical needs, as a pacifier lessens hunger pangs). The Hull-Spence model (the axioms and the predictions derived from them) gives a fairly good account of some cases of learning, but its predictions have been disappointed in numerous other cases, even after much tinkering with the basic axioms and the adding of new assumptions (such as those used to account for latent learning). The reader who wishes a detailed discussion is referred to Hilgard.[3] In general, the system has become more cumbersome than the facts it sets out to explain, and so has little utility. Shortly before his death, however, Spence published papers assigning much greater weight to cognitive factors than a strict behaviorism would allow, which may signal a decided change in the learning-theory camp.[4]

A number of Hull's collaborators, including John Dollard, Neal Miller (whose more recent work has been in brain biochemistry), and Robert Sears, have sought to fuse Hullian learning theory and Freudian psychodynamics with a particular view to showing the learned (as opposed to instinctive) basis of motives. Their best-known contribution has been the hypothesis that aggression is always produced by frustration, and, conversely, that frustration always leads to aggression. The hypothesis was fruitful in inspiring research, and many of the predictions derived from it were verified. However, while it can be shown that in certain circumstances, frustration and aggression do go together, the generality of the hypothesis has been mooted by numerous observations. A famous study by Barker, Dembo, and Lewin,[5] for instance, showed that one possible outcome of frustration is regression. Behavior at a less mature level than before frustration was introduced. Rosenzweig[6] has shown the variety of reactions frustration may produce. Common-sense observation suggests that one can sometimes react to frustration constructively, as by seeking new solutions to a problem.

Another way of looking at instrumental acts, in addition to the stimuli with which they are associated, is in terms of their outcomes, the practical consequences they produce. In classical instrumental conditioning, the experimenter provides rewards if the animal behaves properly, for example, there is food waiting if the animal goes to the correct arm of the maze. A second kind of instrumental conditioning, made prominent initially by E. L. Thorndike, arranges things so that the animal itself makes things happen and is rewarded as an intrinsic result of doing the right thing. Thus, if the animal can learn to work the latch on the cage door, he is rewarded by being liberated from the cage. Thorndike phrased this relationship between the animal's own activity and the rewarding results it produces as the *law of effect,* which says simply that those behaviors that lead to satisfying consequences tend to be repeated under like circumstances. Initially, the law of effect went parallel with another that said that annoyance (equivalent to punishment) as a consequence of behavior tended to weaken and, eventually, to *extinguish* the behavior, but later observations led Thorndike to drop this part of his formulation. Instead, the extinction of behavior is said to occur simply because of the lace of satisfying consequences, the lack of reinforcement.

As we have said, Solomon[7] has collected evidence suggesting that Thorndike and his followers may have been too hasty in abandoning the principle of annoyance, since it seems to be the timing of punishment that determines whether it inhibits behavior. Thus, punishment given at the very inception of an act thoroughly inhibits its repetition, while punishment given immediately on completion of an act is a somewhat less effective deterrent. The longer punishment is delayed after completion of the act, the less effective it becomes. This formulation might serve to explain why punishment rarely deters criminals. In fact, the criminal act itself, such as committing a burglary, may be successful and thus get reinforced, whereas it is behavior later in the total sequence, such as trying to dispose of the stolen goods or confiding to a girl friend, that gets punished. Strictly by theory, the criminal would be encouraged by the reinforcement of the act of burglary to try again, whereas the punishment would teach him to modify his post-burglary behavior. In Thorndike's view, behavior was essentially shapeless and random, being produced by a diffuse state of disequilibrium, such as that engendered by hunger or fear. Those components of random behavior that happened to produce satisfying effects were retained, whereas those that did not were not. In this way, amorphous behavior could be given shape and made to appear as though it were directed from within, whereas in fact it was the pattern of effects that was responsible. Reinforcement can be thought of as effective either because it is rewarding (that is, because it satisfies some need and reduces tension) or because it supplies information, akin to feedback, that guides the organism in intelligent action. Estes and his collaborators[8] have devised techniques to assess separately the relative contributions of reward and information to the learning process, and it appears, as in Spence's finding mentioned earlier, that information is at least as important as need reduction.

Work on the law of effect has been carried forward by B. F. Skinner, whose prescriptions for child-rearing we have already outlined, and his collaborators. Skinner prefers to speak of *operant* learning or conditioning instead of instrumental, an operant being a whole class of instrumental acts which may differ in the way they are performed but still produce the same effect. Thus, the somewhat different instrumental acts of saying at the dinner table, "Please pass the salt," "Give me the salt, please," and "May I please have the salt?" all produce the same result and so belong to the operant class of asking-for-the-salt. Skinner's radical behaviorism goes beyond Thorndike also in its doctrine of the empty organism (the ultimate *tabula rasa*), according to which the newborn baby can be described entirely in terms of the capacities for action built into his physical constitution, the reflexes that enter into classical conditioning and the motivational states that will set the baby in random motion. From there on, the law of effect takes over to give shape to the baby's behavior and can be exploited by the parent or educator to make of the child what he will—likewise the trainer of animals.

Among the specific contributions of Skinner and his co-workers is the analysis of *schedules of reinforcement*. Instead of reinforcing the animal (usually a pigeon in Skinner's laboratory) every time he gives the correct response, one can reinforce some fixed percentage of occurrences of a particular response, or space reinforcements according to some average time interval. Some interesting findings emerge from this sort of analysis. For instance, if reinforcement is given consistently and then discontinued, the reinforced response extinguishes very rapidly; but if reinforcement has been given only intermittently, extinction is greatly attenuated. However, this principle does not seem to hold, as we have seen, when it comes to extinguishing bedtime crying in young infants. Skinner has also demonstrated a phenomenon he calls "superstition"—a completely irrelevant act occurring just prior to a reinforced relevant one may also be reinforced and thus be incorporated into the total response, so that the pigeon invariably stretches its neck or pirouettes before pecking the key that brings it food. Skinner has also demonstrated the possibility of teaching an animal behavioral *chains,* elaborate sequences of actions which bring the animal reinforcement.

Skinner shares with S. L. Pressey the paternity of the teaching-machine and programed-instruction movement in the United States. In programed instruction (whether or not presented via a machine) an area of knowledge is broken down into small steps or units which are presented one at a time in logical order, each presentation being followed by one or more opportunities for the learner to apply his learning, as by answering a question, with immediate knowledge of results (usually, knowing that one is right is sufficient reinforcement). Programed instruction has proved fairly useful with some subject matters, such as arithmetic, and with some subjects, particularly children who find it hard to pay attention and to learn in regular classroom settings. A learning program can be attacked in solitude, at one's own pace, away from distractions, and the manipulation of the materials, especially on a machine, provides an additional incentive.

As a technique for total education, however, programed instruction has disappointed the original high hopes of its proponents, at least partly because some of its assumptions about how people learn seem to be in error. Some forward-looking technicians, let it be said, anticipate the day when a whole field of learning, including original documents, pictures, tables, charts, diagrams, and movie footage, can be put on a reel of videotape, or, more flexibly, entered in the memory of a computer together with a highly flexible program that permits the learner to move about through the material according to his own interests, to be used on the individual's private viewing machine.[9]

The principles of operant conditioning have also had some application in psychotherapy for a few conditions and in the treatment of minor problems such as shyness or timidity.[10] The technique here is to discover what will serve as a reinforcer for the individual whose behavior is to be modified—such things as expressions of approval, or pieces of candy, or money—and then to use these to reinforce successively closer approximations to the desired behavior. Zeaman and House[11] have used operant techniques to good effect in experiments with the mentally retarded, suggesting applications to their education.

In spite of its contributions, the learning-theory tradition, in all its variations and expressions, fails in our view as a general account of human behavior. This is not because of its environmentalism—if there is any hope of improving human-kind, it lies in the manipulation of the environment, and learning theory is pushing hard to test the limits of environmental manipulation. Learning theory is weak because it ignores so many areas of learning. It is concerned hardly at all, or at most programmatically, with knowledge, with the broad interconnected system of facts and ideas and meanings and values that school learning is supposed to deal with. It has very little patience with the learning of concepts and relations and principles, with thinking, reasoning, and insight, or with the consequences of learning—for instance, how learning contributes to the development of intelligence or honesty or sympathy or friendliness, or how it can serve to make a person less egocentric or ethnocentric. Learning theorists sometimes go to extravagant lengths to avoid coming to grips with phenomena that do not fit their conceptions. One device is expressed in *Maier's Law*,[12] if you can't explain it, call it something else. The child's acquisition of his native tongue does not happen in ways that fit the usual conceptions of learning, so it has come to be called what we have called it—"acquisition."

Another area in which learning theory is unsatisfactory is that of unlearning, which is something different from simple forgetting. The child must unlearn, in due time, such beliefs as fairies or Santa Claus, or that babies come by stork, not to mention the many superstitions he invents or accumulates. Indeed, one of the chief problems of higher education is to discover what misinformation or misconceptions the student brings to college so that they can be untaught. Learning theory remains silent on how, in the first place, one can learn things that are not so, such as a belief in Santa Claus, and all the other phantom entities and forces with which people populate the natural order. Nor does learning theory tell us any more about learning through language, by having things described to us, than it does about learning language itself. In general, learning theorists concentrate on the externals of learning, on the easily measurable atomistic indices, such as changes in skin conductivity, salivation, muscle twitches, bar-pressing, or moving from place to place. When they look beneath the surface, it is usually for mediating processes in the central nervous system, bypassing the subjective realm of knowing, wanting, and feeling, as in the accounts of thinking by Osgood[13] and by Miller, Galanter, and Pribram,[14] or in the work of Pavlov and his followers. This criticism is not to say that learning is independent of the nervous system, but that the facts of learning transcend our knowledge of nervous system functioning, and even where correspondences can be found, learning still exists on its own level and can be studied in its own right, regardless of what the neural processes may be.

Let us point out that there are other learning theorists who work outside the Pavlovian and Thorndikean mainstreams, and in some cases even contrary to them. Howard and Tracy Kendler[15] study the role of cognitive or verbal mediation, rather than neural mediation, in learning. Their favored technique is to compare the learning of *reversal shifts* and *nonreversal shifts,* by which the animal is taught a discrimination and then has to learn a new discrimination. A reversal shift is learning that the same dimension of discrimination continues to be the relevant one, but it is now the previously unrewarded stimulus that gets reinforced; thus if color is the relevant dimension, the animal is first rewarded for responding, say, to the blue stimulus and not the red, and then has to learn to respond to the red and not the blue. A nonreversal shift involves changing the basis of reinforcement to a completely different, previously irrelevant dimension. Thus, if the animal first learns to respond on the basis of color, he then must learn to respond on the basis of size or shape or whatever. Interestingly enough, white rats, young children, and young monkeys, but not older monkeys and people, find reversal shifts much more difficult than nonreversal.[16] That is, it is harder to unlearn to respond to one end of a dimension than to unlearn the whole dimension and shift to another. The Kendlers interpret such findings to mean that linguistic mediation is the key to reversal shifts. While it seems likely that verbal analysis facilitates reversal shifts, perhaps by making possible the framing of hypotheses about what the rules of the game have now become, it is not the whole story, since nonlinguistic organisms like rats and monkeys do, with however much difficulty, eventually learn to make reversal shifts.

Another student of learning who escapes the objections we have entered against learning theorists in general is Harry Harlow,[17] who studies *learning set* (also called learning to learn), the fact that the animal generalizes principles to new learning situations, instead of learning a specific response to a specific stimulus. For instance, in so-called oddity learning, the child or animal has to learn to respond to the stimulus that is different from the rest of the set. Having done so, he can then transfer the general principle of oddity to a new oddities problem, as when he first learns to pick the block that is a different color from the rest, and then has to learn to select the block that is different in shape from the others.

Harlow has also stressed that curiosity (without other reward) is a sufficient motivation for learning, and that solving problems is intrinsically satisfying, and not merely a means to satisfying some physical need. We have already mentioned Harlow's research with surrogate mothers, which had as its original aim to disprove the learning-theory hypothesis that the mother comes to be valued as a conditioned stimulus for the satisfaction of physical needs: The monkeys, it will be recalled, preferred the terrycloth mother, who did not feed them, to the bare wire mother who did.

It is an assumption of learning theory that the basic processes of learning are the same in all animal species, but M. E. Bitterman and associates[18] have shown that there are qualitative differences among animals of different species in their performance on standard learning tasks such as reversal shifts and probability learning, lending support to the discontinuity hypothesis (to be discussed later in this chapter) as it applies to evolutionary differences, if not directly to individual development.

A number of students of learning have addressed themselves to the question of learning by imitation, with findings that go counter to the usual learning-theoretical formulation that imitation itself is learned as an operant out of an accidental coincidence of doing something just after somebody else has done it and then being rewarded for it. One incidental finding, by Darby and Riopelle,[19] is that monkeys who learn by watching the behavior of other monkeys learn better from another monkey's errors than from his correct responses, indicating that nonimitation is an important part of observational learning.

Maturational Theories

The usual textbook formulation of developmental change is that psychological development is a product of learning and *maturation.* Maturation includes but is not the same as growth, which refers primarily to changes in size. Maturation is the continuation of the prenatal developmental processes which produce qualitative changes in tissues or in anatomical and physiological organization. Thus, cartilage ossifies into true bone, myelin sheaths come to cover the peripheral nerves, new enzymes arise in the digestive tract, the body changes in proportions, the endocrine glands change in relative size and activity, yielding new organizations. The thymus gland, for instance, atrophies during childhood, and the decrease in thymus activity, long a riddle, is now seen as important to the development of somatic individuality, the idiosyncratic tissue composition that makes tissue grafts between individuals impossible (although progress has been made in circumventing the body's immune response to permit organ transplants, and even in utilizing the immune response to combat cancer). The pituitary gland has periods of greater and lesser activity during development, and the gonads do not mature until adolescence.

There is no doubt that maturation of the somatic equipment takes place and that it affects our capacities for acting and reacting. A whole group of theorists has seen in maturation the chief explanation for behavioral changes in the course of development. While such theorists grant that learning takes place, learning is thought of primarily as the activation of structures that have already taken shape in the given organism. It is as though the nervous system becomes a coded or pictorial replica of the outside world, so that perceiving is a matching of neural structure with object, and thinking a running off in miniature of events in the real world. Since it is the organization of the nervous system that structures thought, language is not a mechanism of thinking but only its accompaniment.

Some of the chief proponents of this viewpoint have been Gesell, Mantagu, Carmichael, the great Swiss psychologist Jean Piaget, whose thinking has become influential in this country only in the last decade but who began his researches forty-odd years ago. Kurt Goldstein, Martin Scheerer, and Heinz Werner. Characteristically, these writers have been more concerned with cognitive processes than with emotion or motivation, and have postulated sets of polar opposites defining the differences between immature and mature behavior. The progression from immature to mature is usually described in terms of stages, epochs in the child's life characterized by particular themes or modes of operation, as in Freud's psychosexual stages, although the exact age periods differ from authority to authority. We shall give only a sampling of the principles and polarities set forth by these theorists.

Gesell proposed a principle of *reciprocal interweaving* which seems to mean a shuttlelike pattern of movement into the future, as though the child takes two steps forward to gain new experience, and then one step back while he consolidates his gains and integrates them with the past. For Gesell, the child's overt behavior is an expression of reciprocal interweaving in nervous system maturation. Gesell's principle of *epigenesis* was simply his statement of the principle of discontinuity, the idea that the individual undergoes qualitative reorganization as he develops.

Goldstein is best known for the polarity of *concrete* and *abstract* behavior, which he proposed first in connection with his studies of the effects of brain injury. Concreteness means being bound to the present situation and its sometimes irrelevant dynamics, as when the patient sorts a collection of blocks according to color and then is unable to make a new sorting on some other basis such as shape, or when he cannot gratuitously go through the motions of knocking on an imaginary door, or is unable to name a comb until he actually uses it to comb his hair. Abstract behavior requires detachment from the immediacies, consideration of alternative modes of action, and thinking and acting in terms of general principles and future outcomes. Goldstein's formulation is also relevant to normal human development which can be viewed as a progression from concrete to abstract modes of functioning.

Piaget has contributed such notions as egocentrism, already discussed, and its polar opposite, *relativism,* the ability to shift perspectives and see how a situation appears to someone else. He has given the name *primary adualism*—that is lack, of two-ness, or nondifferentiatedness—to the baby's initial inability to distinguish clearly between events happening inside himself and those originating in the outside environment, so that he experiences dreams as happening in the outside world and acts as though he could control external events by wishing, by sympathetic magic. Two major notions in Piaget's system are *assimilation* and *accommodation.* Assimilation refers to the way the child reshapes his experience to fit his own level of functioning. Accommodation, by contrast, refers to explicitly, as when the child says that the broken plaything hurts. The next

Piaget has given us a sequence of ways in which, the child is supposed to perceive causal relations. The first of these is *animism,* whereby material objects, living or not, are regarded as having an animal spirit that makes them behave as they do. Animism may be implicit, as when the child simply acts as though something were alive, as when he shows sympathy to a broken plaything, or explicity, as when the child says that the broken plaything hurts. The next *artificialism,* according to which all events are regulated by some humanlike entity, as when the child asks, "Why do they have thunder?" At a more advanced level comes *naturalism,* the acceptance of impersonal natural forces as the governing agent in many events, including some portion of human behavior, as in scientific accounts. We have already mentioned Piaget's notions of the nonconservation and conservation of objects. Piaget has also defined a number of stages in the development of logical thinking, such as the sensori-motor stage of babyhood, when all behavior is assumed to be stimulated reflex-fashion from without, and the impulsive *pre-operational* stage of early childhood, which yields to the more systematic, analytical approach of the *operational* stage. Piaget has embellished his stages with descriptive terms drawn from symbolic, logic and esoteric branches of mathematics. In addition, he has supplied a model of brain organization to account for the operations of thought and how they change with age.

Piaget's thinking, even after many decades of study, is still evolving and changing, notably in the direction of giving more weight to learning and of making the learning and use of language and symbolic forms a major component in the development of thinking.[20] His astute observation and ingenious tasksetting have worked a revolution in the study of psychological development and made this a significant segment of general psychology. The authors, over the years, have assimilated and accommodated to much of Piaget's thinking, and his ideas are implicit in many parts of this book.

Heinz Werner was dedicated to making the principles of individual psychological development the key to a general philosophy embracing animal evolution, historical change, cultural differences, and psycho-pathology. Werner incorporated Goldstein's abstract-concrete and Piaget's egocentric-relativistic polarities into his view in addition to several developmental polarities of his own. One such dimension is development from syncretic (global), undifferentiated functioning, in which perception and thought resemble what William James called a "blooming, buzzing confusion," to differentiated and articulated functioning. Whenever the child or adult meets a new situation he may have to recapitulate this sequence. Many of us can remember the first time we ever saw the engine of a car. At first encounter, it appears as a syncretic, poorly differentiated, burbling, throbbing, ticking, whirring, faintly ominous metal mass enmeshed in a tangle of wires and pipes and hoses and protuberances—although this description is already too well articulated to convey our first global impression. With experience, we become able to sort out—to differentiate—the visible components: the engine block, the manifolds, the carburetor and air cleaner, the fuel pump, the distributor, the generator, the water pump, the radiator and connections, and so forth. And as we are instructed in the engine's functioning, including the movement of parts hidden from view—the rotor, the pistons and rods, the crankshaft, the camshaft, the valves, the timing gear, and so on—we become able to see the engine as an intelligible, orderly, stable, integrally functioning unit: but his now differentiated and articulated whole is quite different from the global, undifferentiated whole with which we began. In the chapters that follow we will point out a number of phenomena that seem to follow the pattern of differentiation and higher-order, emergent integration expressed in the *syncretic-articulated* polarity.

Two other polarities of development are closely linked in Werner's system. These are development from *rigidity* to *flexibility* of thought and action, and from living in a *labile* world to inhabiting a *stable* one. In the unstable world of the young child, where feelings mingle with objects and the spirit world coexists with the material, where meanings and identities change without warning, and where there are no limits on what is possible, stability can be achieved only by adhering to rigid, habit-bound, and ritualistic patterns of behavior. We find similar behavior in adults who, faced with the flux and complexities and perplexities of life, take refuge in tradition and simplistic slogans and dogmas and formulas. As and if the world becomes more stable, reliable, orderly, and predictable, more possibilities for effective action become apparent and behavior can be increasingly flexible and varied.

Werner also had a special interest in language, in both its factual, descriptive, propositional functions and its evocative, affective, expressive ones, as in poetry and metaphor. Some of his ideas in this area are expressed in the account of language learning in the next chapter.

It might be supposed that psychoanalytic theories, which are concerned primarily with motivation and emotion, and maturational theories, which focus on cognition, could simply be combined to give a reasonably complete picture of the individual and his development. The thought has occurred to quite a few people, for the most part originally of a psychoanalytical persuasion, including Silvano Arieti, E. J. Anthony, René Spitz, Peter Wolff, Ernest Schachtel, and David Rapaport, all of whom have sought to integrate the two approaches, largely in terms of parallels between Freud's primary-process thinking and the primitive (egocentric, concrete, rigid, syncretic) thinking described by maturational theorists. It seems to us that these attempts have been abortive, perhaps because psychoanalysts find it hard to cope with the emergentism of maturational theories, and because neither psychoanalysis nor maturational theory makes sufficient provision for learning and the role of context on behavior. Werner, for instance, talked a great deal about cultural differences, but had almost nothing to say about how one comes to assimilate a particular cultural outlook.

Kurt Lewin's Field Theory

Maturational theories came out of one branch of *Gestalt psychology.* Whereas other psychologies tried to analyze psychological functions in terms of their elements, Gestalt psychology was concerned with the qualities of organized things-as-a-whole. The Gestalt psychologist would point out that a square constructed of either dots or lines is perceived in terms of its squareness, which is designated a *whole-quality,* rather than in terms of the particular elements that go to make it up.

It is hard to summarize the Gestalt position because it contains a number of contradictions. On the one hand, the Gestaltists insisted upon the distinctive human qualities of human beings, complaining that associationism dealt with "mere facts" to the neglect of values. On the other hand, Wolfgang Kohler (who, along with Max Wertheimer and Kurt Koffka, was one of the fathers of the Berlin school of Gestalt, the one best known in the United States) vigorously denies the possibility of qualitative transformations in the course of either normal evolution or individual human development. At the same time, he emphasizes the role of *insight* (understanding, often rapid, as expressed in "Eureka!" or the "Aha experience") in learning and the discontinuous learning curves that are produced when the organism arrives at insightful understanding of the task. Kohler is a reductionist, and seeks to explain behavior and perception in terms of maplike representations of situations in the brain. Although much of his work was concerned with perception and the effects of different organizations of stimuli on perception, Kohler staunchly maintains that the real world is in principle unknowable, holding that what we take to be reality is only something that we infer or reconstruct out of discharges in the nervous system.

Although both Wertheimer and Koffka were interested in work with children, the Berlin school was nondevelopmental in its orientation, but it produced one psychologist, Kurt Lewin, who was at least partly developmental in his outlook and who did considerable research with children. Lewin's point of view is called *topological field theory,* and is based on his concept of *life space.* The life space consists of those aspects of the environment which stand out for the individual as psychologically relevant. Lewin represented the life space diagrammatically as a potato-shaped region of space marked off with routes and barriers representing types of activity allowed and available, or prohibited and blocked. The routes in this life space lead to goal objects with positive and negative valences which indicate whether the objects are attractive or repulsive. The person in the field (which may itself be something thought about as well as the person's perceived world) is moved or immobilized by the action of these valences along the pathways that are open to him, representing courses of action. In this way one can represent a number of classical situations. Frustration, for instance, can be shown as an attractive goal across an insurmountable barrier. Various states of conflict can be represented, as when the person is torn between two competing but incompatible attractions, or when a negatively valenced object lies across the path to an attractive one, or the condition of ambivalence, when a goal has both positive and negative valences. Lewin was avowedly "ahistorical" in his approach; that is, he was less interested in how a life space comes to be constituted as it is than in its structure at the moment of action. He did say, however, that the life space becomes progressively more differentiated in the course of development, and, as suggested earlier, he conceived of regression as a de-differentiation of the life space. And it does not seem incompatible with his views to assume that the life space also undergoes qualitative reorganizations and reintegrations as development proceeds.

Lewin represented the person in the life space's field of forces as an empty circle, reminiscent of Skinner's empty organism, so that people are no longer described in terms of their own capacities but in terms of their environments. Thus, instead of talking about stupid and intelligent people, we can talk about stupid or intelligent life spaces. Or, if we want to drop Lewin's terminology, we can talk about more or less intelligent environments. Thus, intelligence is not a property of nervous systems but of how the world appears to people, its qualities and flavors and meanings and possibilities for action. If we adopt this view of intellectual differences, then the goal of education becomes that of making people more intelligent by teaching them what the environment consists of and how it operates (and how they can interact with it), including the action of forces and agents lying beyond perception, whether x-rays or infrared or molecular structure or ultrasound or viruses or RNA or human psysiology or the law of supply and demand. For the life space changes as we come to understand the principles by which we and the world work, including the fact that different cultures shape the life space of people in very different ways, including the moral, esthetic, and emotional valences it assigns to things.

References

1. White, R. W. Motivation reconsidered: The concept of competence. *Psychological Review,* 1959, **66**, 297-333 and ego and reality in psychoanalytic

theory. *Psychological Issues,* 1963, Monograph No. 11; Smith, M. B. Socialization for competence. *Social Science Research Council Items,* 1965, **19**, 17-23; see also Chodoff, P. A. Critique of Freud's theory of infantile sexuality. *American Journal of Psychiatry,* 1966, **123**, 507-518.
2. Merton, R. K. *On the shoulders of giants.* New York: Free Press, 1965.
3. Hilgard, E. R., and Bower, G. H. *Theories of learning.* (3rd ed.) New York: Appleton, 1966.
4. Spence, K. W. Cognitive factors in the extinction of the conditioned eyelid response in humans. *Science,* 1963, **140**, 1224-1225.
5. Barker, R. G., Dembo, T., & Lewin, K. Frustration and regression: An experiment with children. *University of Iowa Studies in Child Welfare,* 1941, **18**, no. 1.
6. Rosenzweig, S., & Kogan, K. L. *Psychodiagnostics.* New York: Grune & Stratton, 1949.
7. Solomon, R. L. Punishment. *American Psychologist,* 1963, **18**, 239-253.
8. Keller, L., Cole, M., Burke, C. J., & Estes, W. K. Reward and information values of trial outcomes in paired-associate learning. *Psychological Monographs,* 1965, **79**, no. 12.
9. Nelson T. H. The hypertext. International Federation for Documentation: *Abstracts, 1965 Congress,* p. 80.
10. Allen, K. E., Hart, B., Buell, J. S., Harris, F. R., & Wolf, M. M. Effects of social reinforcement on isolate behavior of a nursery school child. *Child Development,* 1964, **35**, 511-518.
11. House, B. J., & Zeaman, D. Reward and nonreward in the discrimination learning of imbeciles. *Journal of Comparative and Physiological Psychology,* 1958, **51**, 614-618.
12. Maier, N. R. F. Maier's law. *American Psychologist,* 1960, **15**, 208-212.
13. Osgood, C. E. On understanding and creating sentences. *American Psychologist,* 1963, **18**, 735-751.
14. Miller, G. A., Galanter, E., & Pribram, K. H. *Plans and the structure of behavior.* New York: Holt, Rinehart and Winston, 1960.
15. Kendler, T. S., & Kendler, H. H. Reversal and nonreversal shifts in kindergarten children. *Journal of Experimental Psychology,* 1959, **58**, 56-60.
16. Tighe, T. J. Reversal and nonreversal shifts in monkeys. *Journal of Comparative and Physiological Psychology,* 1964, **58**, 324-326.
17. Harlow, H. F., & Kuenne, M. Learning to think. *Scientific American,* 1949, **181**, 36-39.
18. Bitterman, M. E., Wodinsky, J., & Candland, D. K. Some comparative psychology, *American Journal of Psychology,* 1958, **71**, 94-110.
19. Darby, C. L. & Riopelle, A. J. Observational learning in the rhesus monkey. *Journal of Comparative and Physiological Psychology,* 1959, **52**, 94-98; Riopelle, A. J. Observational learning of a position habit by monkeys. *Journal of Comparative and Physiological Psychology,* 1960, **53**, 426-428.

20. Inhelder, B., & Piaget, J. *The early growth of logic in the child.* New York: Harper & Row, 1964.

Recommended Additional Readings

Baldwin, A. L. *Theories of child development.* New York: Wiley, 1967.
Bromley, D. B. An approach to theory construction in the psychology of development and aging. In L. R. Goulet & P. B. Baltes (Eds.), *Life-span developmental psychology: Research and theory.* New York: Academic Press, 1970. Pp. 71-114.
Bronfenbrenner, U. Developmental theory in transition. In H. W. Stevenson (Ed.), *Child psychology: The 62nd yearbook of the National Society for the Study of Education.* Chicago: National Society for the Study of Education, 1963. Pp. 517-542.
Langer, J. *Theories of development.* New York: Holt, Rinehart and Winston, 1969.
Reese, H. W., & Overton, W. F. Models of development and theories of development. In L. R. Goulet & P. B. Baltes (Eds.), *Life-span developmental psychology: Research and theory.* New York: Academic Press, 1970. Pp. 115-145.

4 Age Changes and Age Differences

K. Warner Schaie

Orientation

The following article by K. Warner Schaie presents some especially provocative and important ideas. Developmental psychologists have been myopic in their approach to the study of changes within the individual. They have studied developmental phenomena primarily through the use of two methods: the cross-sectional *method, which compares several samples of people of different ages, and the* longitudinal *method, which looks at only one sample of people over a period of time. Both of these methods have their merits, but unfortunately they both produce distorted pictures of change over time. The fundamental problem with the cross-sectional and longitudinal approaches is that, while they focus upon the changing nature of the growing and developing individual, they ignore the fact that the individual exists within an environment that is also changing. Thus, the generation (or cohort) of people born into the world in 1920 experienced a set of conditions that is far different from those conditions encountered by the generation born in 1960. An intensive study of the 1920 cohort, while informative about that particular group, tells us very little about the processes of development within the 1960 cohort.*

Perhaps Schaie's primary lesson is that we must broaden our understanding of time and change. In addition to our interest in the changes within the individual, we must also concern ourselves with the changes occurring within the cultural environment in which the individual lives.

K. Warner Schaie is a professor and chairman of the department of psychology at West Virginia University. This department offers one of the very few programs in developmental psychology that has an explicit life-span focus.

Reprinted from *The Gerontologist*, 1967, 7, 128-132, by permission of the author and The Gerontological Society. A more extended discussion of the issues presented in this paper is given in Schaie, K. W., "A general model for the study of developmental problems." *Psychological Bulletin*, 1965, 64, 92-107.

Preparation of this paper was supported in part by Public Health Services Research Grant HD-00367-02 from the National Institute of Child Health and Human Development.

4

Almost as soon as objective measures were defined which could be used to index intellectual abilities and other cognitive functions, researchers began to express interest in individual differences on such measures. One of the most persistent of such interests has been the investigation of developmental changes in cognitive behavior. Most treatments covering age changes in cognitive behavior have closely followed the prevalent approaches in the description of developmental theories. Although great attention has always been paid to early development, and maturation during childhood and adolescence is fully described, very little is said about the further development of intelligence and other cognitive variables during adulthood or senescence. In fact, the concern with age changes in cognitive behavior during adulthood did not come to be of serious interest to psychologists until it became clear that the I.Q. concept used in age scales was inapplicable for the measurement of intelligence in adults. As a consequence of the work of Wechsler (1944) in developing special measures for the description of the intelligence of adults but also due to the earlier descriptive works of Jones and Conrad (1933) with the Army Alpha and that of the Stanford group working with Miles (1933), it soon became clear that somewhat different conceptual models would be required for the proper understanding of adult cognitive development.

It will be noted that emphasis has been placed upon the term "age changes." The literature on the psychological studies of aging has long been haunted by a grand confusion between the terms "age change" and "age difference." This confusion has beclouded the results of studies involving age as a principal variable and has loaded the textbook literature with contradictory findings and what will be shown to be spurious age gradients. This presentation intends to clarify in detail the relationship between age changes and age differences and to show why past methodologies for the study of age-related changes have been inadequate.

Much of the literature on aging and cognitive behavior has been concerned with describing how older individuals differ from their younger peers at a given point in time. Such a descriptive attempt is quite worth while and is necessary in the standardization of measurements. This approach, however, is restricted to a description of the very real differences between organisms of various lengths of life experience at a given point in time. Unless some very strong assumptions are made, these attempts beg the issue and fail to produce relevant experiments on the question of how the behavior of the organism changes over age. This is a strong statement, and it is not made rashly since it clearly questions much of the work in the current literature. But it is required since we find ourselves increasingly puzzled about the results of our own and others' studies of age differences. Let us be explicit in clarifying the basis of our concerns and in tracing the resulting implications for the interpretation of much of the data in the developmental literature.

A general model has been developed which shows how the previously used methods of developmental analysis are simply special cases which require frequently untenable assumptions. This model has been described elsewhere in more detail (Schaie, 1965). At this time, however, it would be useful to state the most important characteristics of a general model required for the explanation of aging phenomena as they pertain to the relationship between age changes and age differences.

Let us begin then by clearly distinguishing between the concepts of age change and age difference. Before we can do so effectively, we must also introduce some new concepts and redefine various familiar concepts. The concept of *age* is, of course, central to our discussion. It needs to be carefully delineated, however, and whenever used will be taken to denote the age of the organism at the time of occurence of whatever response is to be measured. Even more precisely, age will refer to the number of time-units elapsed between the entrance into the environment (birth) of the organism and the point in time at which the response is recorded.

In addition, it is necessary to introduce two concepts which are relatively unfamiliar in their relevance to developmental study. The first of these concepts is the term *cohort*. This term has frequently been used in population and genetic researches and is useful for our purpose. The term implies the total population of organisms born at the same point or interval in time. Restrictions as to the nature of the population and the latitude in defining the interval in time designated as being common to a given cohort or generation must be determined by the special assumptions appropriate to any given investigation.

The second concept to be introduced is that of *time of measurement*. It will take on special significance for us as it denotes that state of the environment within which a given set of data were obtained. In any study of aging it is incumbent upon the investigator to take pains to index precisely the temporal point at which his measurements occur. Such concern is most pertinent since changes in the state of the environment may contribute to the effects noted in an aging study.

With these definitions in mind let us now examine Figure 1 which will help us in understanding the distinction between age changes and age differences. Figure 1 contains a set of six independent random samples, three of which have been given some measure of cognitive behavior at the same point in time, and three of which have been drawn from the same cohort; i.e., whose date of birth is identical. If we compare the performance of samples 1, 2, and 3 we are concerned with *age differences*. Discrepancies in the mean scores obtained by the samples may be due to the difference in age for samples measured at the same point in time. But note that an equally parsimonious interpretation would attribute such discrepancies to the differences in previous life experiences of the three different cohorts (generations) represented by these samples.

If, on the other hand, comparisons were made between scores for samples 3, 5, and 6, we are concerned with *age changes*. Here the performance of the same cohort or generation is measured at three different points in time. Discrepancies between the mean scores for the three samples may represent age changes, or they may represent environmental treatment effects which are quite independent of the age of the organism under investigation. The two comparisons made represent, of course, examples of the traditional cross-sectional and longitudinal methods and illustrate the confounds resulting therefrom.

Lest it be thought that there is no way to separate the effects of cohort and time differences from that of aging, we shall now consider a further set of differences which may be called *time lag*. If we compare samples 1, 4, and 6, it may be noted that the resulting differences will be independent of the organism's age, but can be attributed either to differences among generations or to differences in environmental treatment effects or both.

Any definitive study of age changes or age differences must recognize the three components of maturational change, cohort differences, and environmental effects as components of developmental change; otherwise, as in the past,

Figure 1.
Examples of a Set of Samples Permitting All Comparisons Deducible from the General Developmental Model.

Time of Testing	1955	1960	1965
1910	Sample 3 Age 45 $A^1C^3T^1$	Sample 5 Age 50 $A^2C^3T^2$	Sample 6 Age 55 $A^3C^3T^3$
1905	Sample 2 Age 50 $A^2C^2T^1$	Sample 4 Age 55 $A^3C^2T^2$	
1900	Sample 1 Age 55 $A^3C^1T^1$		

Time of Birth (Cohort)

A — Age level at time of testing.
C — Cohort level being examined.
T — Number of test in series.

we shall continue to confuse age changes with age differences and both with time lag. Hence, it may be argued that studies of age differences can bear upon the topic of age changes only in the special case where there are no differences in genetically or environmentally determined ability levels among generations and where there are no effects due to differential environmental impact. It follows, therefore, that findings of significant age differences will bear no necessary relationship to maturational deficit, nor does the absence of age differences guarantee that no maturational changes have indeed occurred.

As a further complication, it is now necessary to add the notion that differences in the direction of change for the confounded developmental components may lead to a suppression or an exaggeration of actual age differences or changes. As an example, let us suppose that perceptual speed declines at the rate of one half sigma over a five-year interval. Let us suppose further that the average level of perceptual speed for successive five-year cohorts declines by one-half sigma also. Such decrement may be due to systematic changes in experience or to some unexplained genetic drift. Whatever their cause, if these suppositions were true, then a cross-sectional study would find no age differences whatsoever because the maturational decrement would be completely concealed by the loss of ability due to some unfavorable changes in successive generations.

As another example, let us suppose that there is no maturational age decrement but that there is systematic improvement in the species. In such a case successive cohorts would do better than earlier ones, and cross-sectional studies would show spurious decrement curves, very much like those reported in the literature for many intelligence tests.

One of the most confusing facets of aging studies therefore is the fact that experimental data may reveal or fail to reveal a number of different combinations of underlying phenomena. Yet the understanding of the proper conceptual model which applies to a given set of data is essential before generalizations can be drawn. Let us illustrate the problem by considering some of the alternative models that might explain the behavior most typically represented in the literature on developmental change. Reference here is made to cross-sectional gradients such as those reported by Wechsler (1944) or by Jones and Conrad (1933). These gradients typically record a steep increment in childhood with an adult plateau and steep decrement thereafter.

When we address ourselves to the question of what developmental changes are represented by such data, we face relatively little difficulty in determining whether maturational changes are contained in the age differences noted during childhood and adolescence. Our own children provide us with at least anecdotal evidence of the longitudinal nature of such change. Whether this portion of the developmental curve, however, is a straight line or a positive asymptotic curve is still in doubt. Also, it should be remembered that even if we agree upon the validity of evidence for maturational changes, we must still consider that such changes will be overestimated by cross-sectional data if there are positive cohort differences and/or negative environmental experience effects. Similarly, maturational growth will be underestimated in the event of cohort decrement or the effect of positive environmental influences.

For the adult and old-age portions of the developmental span, matters are much more complicated. While we can readily accept the fact of psychological maturational growth during childhood, similar evidence of maturational decline on psychological variables by means of longitudinal study remains to be demonstrated. As a consequence, we must at least entertain models which would account for age differences in the absence of maturational age changes.

The detailed analysis of the general developmental model (Schaie & Strother, 1964a) shows that it is possible to differentiate as many as 729 models to account for developmental change if one considers the direction and slope as well as the three components involved in developmental change gradients. Of the many possible models, three will be considered now which seem to be high probability alternatives for the classical textbook age-gradients. Our three examples are models which not only would fit these textbook gradients but would furthermore predict that the cross-sectional data depicted by the gradients could not possibly be replicated by longitudinal studies.

The first of these models might be called an "improvement of the species" model. It holds that the form of the maturational gradient underlying the typical representatives of the textbook gradients is positive asymptotic, i.e., that there is systematic increment in performance during childhood, slowing down during early adulthood, and that there is no further maturational change after maturity. The model further holds that the cohort gradient, or the differences between generations, should also be positive asymptotic.

Successive generations are deemed to show improved performance for some unspecified genetic or prior experience reason, but it is also assumed that improvement has reached a plateau for recent generations. The effect of the environment is furthermore assumed to be constant or positive asymptotic also. When these components are combined they are seen to provide a cross-sectional age gradient which shows steady increment during childhood, a plateau in midlife, and accelerating decrement in old age.

The same model, however, when applied to longitudinal data will predict steady increment during childhood, but slight improvement in midlife, and no decrement thereafter. The only reason the cross-sectional gradient will show decrement is that the younger generations start out at a higher level of ability and thus in the cross-sectional study the older samples will show lower performance. Of course, this means no more than that the older samples started out at a lower level of ability even though they showed no decrement over their life-span.

A second no less plausible alternative to account for the textbook age gradients might be called the "environmental compensation" model. This model also specifies a concave maturational gradient with increment in youth and decrement in old age, much as the cross-sectional gradient. In addition, however, this alternative calls for a positive environmental experience gradient. Here the effect of an environmental experience increases systematically due to a progressively more favorable environment. The effects of cohort differences in this model are assumed to be neutral or positive asymptotic.

If the second model were correct, then our prediction of longitudinal age changes would result in a gradient with steep increment in childhood but no decrement thereafter, since maturational changes would be systematically compensated for by a favorable environment. Since the environmental component of change over time is not measured in the cross-sectional study, assessment would be made only of the maturational decrement yielding information on the state of a population sample of different ages at a given point of time. But it would provide misleading information as to what is going to happen to the behavior of this population sample as time passes.

Third, let us propose a more extreme alternative which we might label the "great society" model. This model specifies a positive asymptotic maturational gradient; i.e., increment during childhood and a plateau thereafter. The model further specifies a positive asymptotic cohort gradient; i.e., successively smaller increments in performance for successive generations. Finally the model specifies increasingly favorable environmental impact. The reason for calling this alternative the "great society" model should be readily apparent. The model implies (a) that maturity is an irreversible condition of the organism, (b) that the rapid development of our people is reaching the plateau of a mature society, and (c) that any further advance would now be a function of continually enriching the environment for us all. Note that the cross-sectional study of groups of different age at this time in our history will still conform to the textbook cross-sectional gradients. Their longitudinal replication, however, would result in a gradient which would be steep during childhood, which would level off during adulthood, but which would show continued growth until the demise of the organism.

Obviously, it is still possible that the straightforward decrement model might hold equally well for the classical gradients. The information we have on longitudinal studies such as those of Owens (1953) and Bayley and Oden (1955) and the more recent sequential studies by Schaie and Strother (1964b; c) let it appear that any one of the above alternatives may be a more plausible one.

It is hoped that the examples just given have alerted the reader to some of the flaws in the traditional designs used for the studies of aging phenomena. Caution is in order at this time lest the premature conclusion be reached that the increase in sophistication of our methods has indeed led to a better understanding of how and why organisms age. Thus far it seems just as likely that all which has been investigated refers to differences among generations and thus in a changing society to differences which may be as transient as any phase of that society. Only when we have been successful in differentiating between age changes and age differences can we hope therefore that the exciting advances and methods in the more appropriate studies now in progress will truly assist us in understanding the nature of the aging process.

Summary

The concepts of age change and age difference were differentiated by introducing a three-dimensional model for the study of developmental change involving the notions of differences in maturational level (age), differences among generations (cohorts), and differential environmental impact (time of measurement). It was shown that age differences as measured by cross-sectional methods confound age and cohort differences while age changes as measured by the longitudinal method confound age and time of measurement differences. Conceptual unconfounding permits specification of alternate models for the prediction of age changes from age differences and resolution of the meaning of discrepancies in the finding yielded by cross-sectional and longitudinal studies. Examples of alternative models for aging phenomena were provided.

References

Bayley, N., & Oden, M. H. The maintenance of intellectual ability in gifted adults. *Journal of Gerontology*, 1955, **10**, 91-107.
Jones, H. E., & Conrad, H. S. The growth and decline of intelligence: A study of a homogeneous group between the ages of 10 and 60. *Genetic Psychology Monographs*, 1933, **13**, 223-298.
Miles, W. R. Age and human ability. *Psychological Review*, 1933, **40**, 99-123.
Owens, W. A. Age and mental abilities: A longitudinal study. *Genetic Psychology Monographs*, 1953, **48**, 2-54.
Schaie, K. W. A general model for the study of developmental problems. *Psychological Bulletin*, 1965, **64**, 92-107.
Schaie, K. W., & Strother, C. R. Models for the prediction of age changes in cognitive behavior. *Gerontologist*, 1964, **4**, 14. (Abstract) (a)
Schaie, K. W., & Strother, C. R. A cross-sequential study of age changes in cognitive behavior. Paper presented at the meetings of the Midwestern Psychological Association, St. Louis, 1964. (b)
Schaie, K. W., & Strother, C. R. The effect of time and cohort differences upon age changes in cognitive behavior. *American Psychologist*, 1964, **19**, 546. (Abstract) (c)
Wechsler, D. *The measurement of adult intelligence.* (2nd ed.) Baltimore: Williams & Wilkins, 1944.

Recommended Additional Readings

Baltes, P. B. Longitudinal and cross-sectional sequences in the study of age and generation effects. *Human Development*, 1968, **11**, 145-171.
Kuhlen, R. G. Age and intelligence: The significance of cultural change in longitudinal vs. cross-sectional findings. *Vita Humana*, 1963, **6**, 113-124.
Ryder, N. The cohort as a concept in the study of social change. *American Sociological Review*, 1965, **30**, 843-861.
Schaie, K. W. A general model for the study of developmental problems. *Psychological Bulletin*, 1965, **64**, 92-107.

5 The Course of Human Life as a Psychological Problem

Charlotte Buhler

Orientation

The following article presents a view that is clearly not in the mainstream of American psychology. For several decades Charlotte Buhler has worked as a clinical psychologist, engaging in intensive, in-depth psychotherapeutic interviews with her patients. Out of this lifelong work has emerged a framework within which to study the course of human development. Though her view of human nature may be a unique one in current psychology, it is a very rich and sensitive approach to the study of people and their lives, and for that reason it is included in this book.

Buhler's approach has been to study individual persons and to note their uniquenesses as well as their similarities with other individuals. This is in sharp contrast to the more typical approach of studying groups of people (as you will note later in this book in articles that are reports on empirical studies) in order to obtain normative, general-tendency information. Her approach toward theory construction also has been taken by certain other psychotherapists notably Sigmund Freud and Erik Erikson.

Buhler places heavy emphasis on the individual's spontaneous activities: A person does what he does chiefly because of his own initiative. This aspect of human nature is referred to as "intentionality." Human behavior is not to be viewed as "solely a reaction to environmental stimuli." This perspective conceives life as being primarily concerned with the achievement of self-fulfillment; unhappy people— those displaying some sort of "pathology"—are those whose lives are in some way unfulfilled. As in many other theories of human development, Buhler's theory uses a stage approach to describe the life span. Each of the five phases she proposes are characterized by strivings toward the attainment of a personal developmental goal.

Charlotte Buhler conducts her practice in clinical psychology in Los Angeles, California; she has also been associated with the University of Southern California. She is a leading figure in the development of the field known as "humanistic psychology."

Reprinted from *Human Development*, 1968, 11, 184-200, by permission of the author and S. Karger, Basel.

5

1. A model of human development through the life cycle. This paper restates a theory of human development set forth in a much earlier publication (Buhler, 1933) and evaluates the applicability of this theory to more recent data and thinking.

The model was originally developed from analysis of 202 biographical studies and some additional statistical and interview material. The biographies were assembled under the direction of E. Frenkel and E. Brunswik in Vienna and were selected for completeness and reliability of the sources, particularly statements regarding motives, attitudes and other inner experiences.

The majority of published biographies concern themselves with the lives of personalities who for one reason or another have become outstanding. An impressive fact was that most of these lives seemed to have an inner coherence which appeared due to some unifying or integrating principle. This integrating principle seemed to evolve from certain expectations which permeated these people's lives; it suggested that human life was lived under certain directives. I called this principle intentionality.

2. Evidence of intentionality. In the majority of the biographies, the individual's intentionality became repeatedly apparent. Only those lives in which failure predominated seemed to lack this unifying principle.

Some amnestic interview studies of lives of average middle class and working class people, as carried out by Marie Jahoda and Gertrud Wagner in Vienna, seemed to confirm the view that healthy persons experience their lives as a unit. A more recent study of the life of Bill Roberts (Buhler, 1961) also confirmed this view.

Yet studies of a great many younger people do not seem to justify this assumption. When asked about life goals, younger people often cannot point them out clearly and there is no evidence of consistent pursuits.

Intentionality becomes quite apparent in interview and clinical studies of individuals over 50 or 55 years old. When asked to tell the story of their lives, they usually end up with a summarizing statement: "All in all it was a good life," or "There were so many disappointments," or "It all came to nothing."

Such summary statements do not necessarily always indicate complete or appropriate evaluations. Some neurotic individuals evaluate their lives unrealistically. Thus a woman who died at age 66 after a period of depression, said a short time before her death: "I have lived as good a life as anyone." This was actually untrue, because her main interests in life, her marriage and her relationship with her children, had been mostly unhappy and unsuccessful.

While in many cases no distinct intents could be established for the earlier periods of life, study of later phases revealed that practically always there had been expectations which had either been fulfilled or disappointed.

Sometimes the expectations a person has held all through the years become clear only late in life. Thus a man of age 55 discovered after two years of psychotherapy, what it was he always had hoped for.

"All I ever wanted was recognition and attention...." "If I had the assurance of getting it I would make a supreme effort." Then he said: "I feel I am through. In a few weeks I am 55—I have no incentive—just hope to die." And some time later: "I feel like I am dead. I don't seem to want to come to life any more...." "I am unhappy about my whole life....."

The fact that toward their end, if not before, people feel their expectations were fulfilled or disappointed points to the occurrence of earlier, even if unconscious, directives regarding the whole of their lives. Thus my original hypothesis seemed confirmed that a person's life is permeated by some kind of intentionality, an intentionality directed toward fulfillment. Fulfillment is defined as a closure experience of an overall feeling of satisfaction, accomplishment and success, which in different individuals is anticipated and visualized differently. This anticipation helps in varying degrees to direct and to unify a person's endeavors.

The next question was to see in what way intentionality was put into effect. In the detailed study of 53 of the 202 available biographies, a general structure of the lives became apparent. This structure applied to the great majority of the biographies, especially to outstanding persons. Although it did not apply in the same manner to disrupted and disorganized lives, this normal structure could be thought of as a normative developmental structure.

The overall picture seemed to reveal five phases of self-determination to certain results of life, results which promised some kind of personal fulfillment. These results were to be achieved by means of the successful pursuit of certain goals. The latter were called life goals, in distinction to other intermediate goals. According to this theory and to the findings, the unification and integration of a life takes place by means of the fulfillment of certain long-range goals.

The phases were seen as: (1) before self-determination of life goals sets in (to about age 15); (2) a tentative and preparatory self-determination of life goals (from age 15 to about age 25); (3) self-determination of life goals becomes more specified and definitive (from age 25 to about age 45 or 50); (4) assessment of the foregoing life and the attained or failed fulfillment (from age 45 or 50 to age 60 or 65); (5) a phase in which either more or less complete fulfillment is acknowledged, and a post-self-determination life sets in with rest and memories, sometimes illness and decline; or elsewhere partial fulfillment and partial failure motivate the individual to return to previous forms of striving, sometimes also to resignation; or elsewhere the feeling of more or less complete failure ends in depression or despair.

In all these phases there are, in addition to the normative patterns, various deviations of problematic and neurotic personalities.

3. Areas of fulfillment. When this theory of intentionality and self-determination was conceived, studies of goals were only sporadic. Ach's "determining tendencies" and Lewin's "aspiration level" were early predecessors of a line of inquiry that is only now coming to the fore in psychology. Neither then or now have there been any lists of goals available, nor any empirical studies of the hierarchy of goals. The question of how far goals represent *values* is only a recent subject of discussion. Allport (1937) was a forerunner in this area.

My hypothesis of what kind of goals were designated as life goals, and the model of the progressive phases in the pursuit of life goals, were both based on an interpretative study of the aforementioned biographies. In the biographies, three areas of life seem to stand out as significant: One is the area of activities. In the fully-developed adult life, the range of activities seem to yield fulfillment under a number of varying conditions. Fulfillment seems to depend on the appropriateness of activities in terms of an individual's activity needs and preferences, his aptitudes and potentialities, the achievement and acknowledgement he attained, and his beliefs and values. Thus overall fulfillment depends, to varying degrees, on fulfillments in the direction of four basic tendencies, *Need-Satisfaction, Creative Expansion, Self-Limiting Adapatation* and *Upholding the Internal Order* (Buhler, 1959).

A person wants to enjoy activities functionally, wants to achieve and find acknowledgement of his achievements, wants to use his best potentialities and aptitudes and wants to believe that what he is doing is valuable. This can be seen with respect to occupational as well as avocational activities. For most people, however, the occupational area seems to be the most decisive.

A second area of life goals is personal relationships. Here, too, fulfillment depends on conditions which can be analyzed in terms of the four basic tendencies. The most important are the finding of love, the satisfactory development of a marital relationship, and the relationship with children who are developing well. Important also are relationships with other people—friends and acquaintances, co-workers, neighbors and others.

A third area of life goals, that of the development of the self, comes less often into awareness. Often without being conscious of it, people have certain expectations regarding their own selves. Again corresponding to the four basic tendencies, they want to feel good about themselves and to like themselves (Need

Satisfaction); they want to be able to accept themselves and feel they belong (Self-limiting Adaptation); they want to feel they grow and develop as persons (Creative Expansion); and they want to believe in themselves and to feel they have worth (Upholding the Internal Order).

Self-fulfillment is usually experienced as incidental to fulfillment in work or in personal relationships, but even simple people are able to abstract the notion that they "let themselves down" or that they are "content with themselves" with respect to how they have lived their lives.*

4. The beginnings of self-determination in the first two phases of life. As already mentioned, the first phase of childhood (until approximately 15 years of age) was originally conceived as a period in which there is not yet an awareness of life as a whole. This observation must however be qualified. In more recent clinical studies (Buhler, 1962), as well as in nursery-school studies to which Nancy Dilworth (1968) contributed, it was found that sometimes even three to five-year old children concern themselves with long-range plans regarding what they want to do or to be. Of course, the future into which they project these ideas is only vaguely perceived, but even so, the thoughtful child's reflections about himself must not be considered altogether futureless.

Thus it is only generally speaking that we may call childhood the period before self-determination of life goals. Childhood is primarily a period in which different types of behavior develop which are preparatory to long-range goal-setting. These categories, described in detail in another context (Buhler, 1966), are: (1) "spontaneous" activity with which the newborn takes hold of the world around him; (2) selective perception, with which even the young infant begins to build his own world; (3) contact which the individual establishes, from infancy on, with those who take care of him; (4) the beginning of will, identity formation and conscience with which the two- to four-year old child begins to find direction of his own; (5) experiences of "I can" and "I cannot," which give even the baby his first inklings of mastery; (6) more complex reactions to the experiences with the outside world, an attitude of predominant constructiveness or destructiveness which the older school child decides upon in going along with or going against his environment; (7) a gradually consistent attitude toward achievement as a goal; (8) an increasingly established set of beliefs and values which allows the older school child to handle matters on the basis of opinions and convictions; (9) the adolescent's first commitments in love, friendship and other more permanent relationships; (10) the adolescent's struggles to integrate all he knows and wants in a first *"Weltanschauung"*; (11) the adolescent's first attempts to find *direction, purpose and meaning* for his life as a whole; (12) the

---

*Development throughout the life cycle is not sufficiently characterized by the development of the self alone, as Erikson (1959) suggests. Self-development seems to this author to be only one aspect of an individual's maturation.

adolescent's first concepts of the different fulfillments life might offer. These last three categories mark the arrival of the second phase of life.

This list includes all those behavioral categories which indicate an inner-determined directiveness. Early child psychologists, from Preyer (1882) on, spoke in this connection of "spontaneous" behavior. American developmental psychologists, those oriented toward behaviorism as well as those oriented toward psychoanalysis, have only recently accepted the idea that activity is not solely a reaction to environmental stimuli. Only very recently has the individual's own initiative been acknowledged as a factor that determines a selective and directive thrust into the world, as it were.

From the present author's point of view, the briefly enumerated behavioral advances of the first phase of life are in part an expression of the underlying intentionality, the self's active contribution to the individual's developing self-determination.

In the *"second phase,"* the period from about 15 to 25 years, life goals are conceived of tentatively and experimentally. There is a first grasp of the idea that one's own life belongs to oneself and represents a time unit with a beginning and an end. He becomes interested in how it all began. A thoughtful person may write his first autobiography during this phase.

The production of autobiographies are just as phase-characteristic of adolescents as the autobiographies of older people who see their lives in retrospect. Often the adolescent autobiography represents the beginning of a diary and gives evidence that this youth has begun to see himself in a historical perspective. After World War I, in the years between 1921 and 1934, the author analyzed and published a considerable number of diaries. Several other authors, among them particularly Bernfeld, Kupsky and Stern, followed with publications of further diary material which they considered eye-openers to the inner life of adolescents.

While admittedly most of these diaries brought out the thinking and feeling of rather introverted youths, they were the first documentary material to reveal certain general adolescent trends. Most important, they revealed that kind of longing with which the young person starts out his life, and begins to search for partly undefinable goals.

Among these goals may be the search for fulfilling encounters with a partner to love or with a God to believe in. But beyond that, vaguer concerns are opened up: the questions regarding the purpose and meaning of life and the "quests," as Darion (1966) calls them in "The Man of La Mancha," this man who "dreams the impossible dream" and "reaches for the unreachable stars."

While a great many youths are too realistic and even too materialistic to go out for causes or to "right the unrightable wrongs," I have found in clinical work that these sober attitudes reflect often more a manufactured detachment than the reality of these adolescents' feelings. Often all this hardness proves to be a facade, which eventually is relinquished.

Kuppers (1964) who recently published a post-World War II collection of diaries of German adolescents, expresses a similar view. She denies that present-day youth are skeptical and without ideals. They may be more reticent in the expression of feelings, but she does not find a great difference between them and the previous generation.

This may of course be different in American culture. One new trend is the emphasis on the *immediate experience,* through which adolescents try to find themselves. They have a problem with the future in that they feel rebellious regarding the goals and values of their elders. Some choose to turn away even from thinking about the future. Others, however, concern themselves with values and attempts at re-evaluations.

5. Healthy, problematic and neurotic patterns of self-determination. Unfortunately there is not as yet any comprehensive study of people's procedures in life-goal planning nor of how they handle success and failure at various stages of their lives.

In my original life cycle model, the adolescent period was seen as one in which life goals are set tentatively and experimentally, while in the following adult period (about 23 to 45 or 50 years) life goals were seen as set in a more specified and definitive way. Between 45-50 and 60-65 years, people were assumed to concern themselves with assessments of their fulfillments and failures. These *normative* motivational stages occur in different patterns. The different patterns appear to be more or less favorable for progress to the next phase, depending on how problematic or even neurotic the development of a person. I would like to forward a further hypothesis about these developments and patterns.

The hypothesis is two-fold. One part concerns how the present is experienced; the other concerns how past and future are handled. The two problems are interrelated. The manner in which a person experiences the present depends greatly on what type of past he had. His outlook on the future, on the other hand, is greatly determined by the way he experiences the present.

As we know from psychoanalysis the past is apt to intrude on a person's present life to the degree that it was emotionally disturbing and damaging. Certain neurotics are so preoccupied with their past that they handle their present lives poorly and are prevented from a healthy self-realizing existence.

The healthy, self-realizing person lives predominantly in the present. He sees his future as tied to his present life in a meaningful continuity (Shostrom, 1964).

On close inspection, one might distinguish an immediate, an intermediate and a distant future. The immediate future is closely intertwined with a person's present in that it requires immediate preparation. The intermediate and the distant futures may be planned more or less tentatively, clearly, definitely, flexibly or rigidly. Hope or fear may prevail depending on successes and failures of the present and past.

The manner in which a person experiences the present is hypothesized to occur in five different directions:

(1) A person may experience a phase maturely, which means that his self-determination is phase-characteristic. If the person is immature, his goal behavior will be that of the foregoing phase. If he is premature he will rush to the goal behavior of the next phase.

(2) A person may experience satisfaction or disappointment in the self-determination he gave himself, because he feels he pursued the right or the wrong goal, right or wrong in the sense of appropriateness to his tastes and talents, his intentionality and his potentialities.

(3) A person may experience himself as capable or incapable of handling the problems and conflicts which he encounters in pursuing his goals.

(4) A person may register successes or failures with respect to the results of his pursuits, which give him feelings of victory or defeat.

(5) A person may experience fulfillment or unfulfillment, even despair, in an existential awareness of the totality of his life. This may be a momentary feeling of fulfillment or despair, or a feeling about one's life as a whole, as related to one's intentionality and to the world-at-large and the universe.

The following cases illustrate these five modes of experience in the different phases of life.

David, 22, is an example of a person who lives his second phase adequately. He is a science student who plans to become a college professor. His occupation is preparatory as is his engagement to a girl he intends to marry. He is satisfied with his plans and feels he will be developing his best potentials: he is on the whole quite capable of coping with various problems and conflicts; he can register more successes than failures, and he looks hopefully to a fulfillment of his goals and his life as a whole.

David's past has not been happy. His early-divorced and enormously-harrassed mother was not up to this brilliant boy, her only child, and she tried to control him with excessive demands and harsh, punitive discipline. David grew up defending himself against his mother. He conceived early in life of a future in which he would live on his own and work toward a scholarship in a distant university. He managed to fulfill this hope and to succeed relatively well in not letting memories of his unhappy childhood detract him. Inasmuch as he was disturbed, he sought out psychotherapy to work them through.

David has a schematically outlined program for the future. He is more definite about it than many other young people. He might even become quite disturbed if the program did not work out; that is to say, he is probably not quite flexible enough about it. But it seems a reasonable enough plan to become realized.

His program for his immediate future is to study for his college degree and to earn enough money to support himself. In the intermediate future he plans to become a college teacher and to marry. His fiancee agrees to this plan. On his distant future, David visualizes himself ambitiously in an outstanding college position. He also expects to be a happy husband and father and a truly satisfied person.

This tri-partite division of expectations and planning would apply to almost any age; however, in adolescence the more immediate plans are likely to be more predominant than at a later age. And often they are not as yet successfully tied to further plans. David has an unusually specified program as compared with many young people. He is helped in his self-determination by his early awareness of his potentials as well as his needs.

Emotional immaturity prevents a great many young people from committing themselves even tentatively in occupations or personal relationships. As Otto (1964) established in questionnaire studies, astonishingly few people analyze and consider their own potentialities in choosing a direction for themselves. Many young people also fail to clarify the values they believe in and want to live for. They make tentative choices of certain life patterns and are disappointed because they do not seem to yield what they really want.

A third point in the development of self-determination is connected with the individual's ability or inability to handle intervening problems and conflicts. A 28 year-old woman, for example, overwhelmed by guilt feelings she experienced after leaving a convent was subsequently unable to achieve a new self-determination.

A fourth point in the development of self-determination is the experience of success and failure which may validate or invalidate the choice. One adolescent, whose father had nourished in him from childhood the idea that he would become a doctor, came close to a breakdown when he found his undergraduate courses to be far beyond his capacities.

A fifth point is found in the experience of fulfillment or unfulfillment, even despair, which a person may feel with regard to his whole existence. Fulfillment or unfulfillment, as used here, refers to an inclusive feeling about one's life, or occasionally an episode or a period related as a whole to one's intentionality. It is what leads people to say that their "life was good," or "it all came to nothing," or "there were so many disappointments" or similar summarizing statements. David represents the feeling of a young adult on the way to fulfillment.

> Wayne, age 21, whose ambition was to become an actor, represents complete relinquishment and withdrawal. Wayne had been a great success as an actor in high school and he assumed that the world was open and waiting for him. He experienced severe disappointment when he did not find any access to any theater. Refusing to compromise his plans in any way, he withdrew and returned to his divorced mother who doted on her only son. Wayne excused himself with various psychosomatic illnesses requiring care and rest and left the question of his future completely unresolved. He felt hopeless and as if he were at the end.

Wayne's relinquishing of effort toward his self-determination resulted not only from the failure of his attempts, but also from his inability to handle the problems he encountered on the way. Refusing detour solutions, he insisted on his goal inflexibly.

His rigid self-determination in the area of a career was, furthermore, *premature,* since he could not ascertain his chances of realizing it.

All premature self-determinations in the second phase create definitive situations without sufficient preparation. Frequent examples are the many very young marriages of our time, on which young girls more often than boys insist, fearing they might not find again a partner who would want them or simply impatient for a life of their own. These premature self-determinations often end in failure.

A healthy person moving from a second to a third phase self-determination proceeds gradually from tentative to more specified and more definite goals. This process takes place from about 25-30 years of age to about 45-50 years of age. The case of Bill Roberts is a good example.*

Bill had worked at a variety of unskilled jobs since he was 14 years old. At age 26 he married. A year in the army at the end of World War I interrupted his career plans—upon his return however, he obtained a job as an employee of a big trucking company. After three years of special training, he advanced to an excellent position in this firm, which he held until the depression when he was forced to switch to other jobs. Most of the time, however, he was able to stick to the occupation he was trained for. At age 46, a severe illness as well as his firm's bankruptcy stopped his established way of life.

Similar to the many variations of preliminary planning in adolescence, a variety of normative patterns of self-determination are found in the third phase of life. In the ideal case, this is the time of life in which a person's potentials have evolved more or less fully and in which many aptitudes are culminating. Also an individual's personal life normally reaches a certain peak during this period, a marital bond is established, sexual capacity is at its best, family life is developing and increasing the scope of the individual's life and a circle of friends is being acquired within the settled circumstances of a stable career and an established domicile.

The middle of life might then be expected to be the period in which an individual could settle down to live within the frame of reference of the present in pursuing many immediate objectives; he also might be expected to acquire a clearer outlook on the intermediate and even distant future. But the many conditions of a settled third phase are of course seldom fulfilled. A satisfactory

---

*Case presented originally in Buhler, *Values in psychotherapy* (1962).

third phase pattern is probably more the exception than the rule. Clinical material reveal the many hindrances people encounter in their effort to settle down. A few of the most frequent problem patterns are indicated here.

Again we find first an *immaturity* pattern, that is the inability to settle down into definite circumstances.

> At 34 years old, Myron has not as yet been able to establish himself in any career or to marry. Extremely dependent on his domineering mother, he allowed himself to be employed in the family business, although he disliked it intensely. He had not married because his mother disapproved of whatever girl he brought home.
>
> Sex fears prevented Diane, age 33, from accepting as marriage partners any suitors she had had. She was also unable to place herself appropriately in her career.

While immature middle-age persons persist in tentative and preparatory pursuits, others are handicapped by disappointments and conflicts or are forced to regress because of failures.

Corresponding to the disappointed group of adolescents, individuals in the third phase also experience disappointments over what seem to be wrong choices of goals.

> Bob, who in his second phase experienced failure in his self-determination to become a doctor, suffered a breakdown in the third phase over the disappointment he felt in the career he finally chose.
>
> When Bob found himself unable to meet the requirements of medical studies, he chose to leave college and to become a salesman. This activity distressed him to such a degree that he took to alcohol and later even to drugs to escape his feelings of failure and disappointment.
>
> In psychotherapy he gradually realized that in both his occupational choices he had failed to inquire into his own potentialities and that both his choices were not related to what he could do best and would have liked to do. What he really felt he could do well was to advise, guide and help people. While this was what appealed to him in the doctor's role, he could have planned for this kind of functioning just as well in any kind of counselor's role. He decided that while it was too late for him to start over at age 33, he could use his free time to study for a degree in business and to get managerial training for an executive position.

Very frequently, disappointments of this third phase occur from the wrong choice of marital partners.

> Ethel, age 41, had married a man with whom she was deeply in love. While he soon proved to be unreliable and irresponsible in many respects, Ethel

could for years not reconcile herself to a divorce. She went on hoping things would change for the better and when, during psychotherapy, she finally decided to free herself from her bondage, she suffered a breakdown.

Emotionally-disabling conflicts unsettle the third phase for some. This is parallel to the disablement in self-determination found in the previous phase.

Georgia, age 35,* needed psychotherapy to resolve her conflict over caring for her demanding mother at the risk of dissolving her otherwise happy marriage.
Frances, age 29, was also emotionally disabled. After an unsatisfactory marriage she had started to live in a homosexual relationship and was in inner conflict about her way of life.

Success or failure experiences may also modify the third phase. Both Bob and Ethel, who were disappointed over their wrong choices, perceived themselves as failures. Both regressed to indetermination. Bob escaped from his unresolved conflicts into alcohol and later to drugs. Ethel was unable to hold a job. She had to fall back on accepting help from her parents.

The third phase may also generate premature self-determinations in anticipation of the critical self-assessment of the fourth phase. As mentioned earlier, self-assessments take place repeatedly. To different degrees people are motivated repeatedly to evaluate their lives and themselves. But they are particularly stringent in the fourth phase which, like the second phase, represents a crisis period of life.

The fourth phase self-assessment, occurring in the climacteric years from about 45-50 to about 60-65, is normally a more comprehensive survey of a person's whole life in retrospect. It comes in a critical period in which many people have to re-orient themselves for a number of reasons. In many occupations a person in this age group is no longer employable; retirement looms ahead for the majority. This means for the most the potentiality of a drastic reduction of income. It also means for many the end of a career that offers an effective means to interact with the world outside themselves. In many cases, reduced abilities or the beginning of severe illnesses testify to the beginning of decline.

In the healthy person, the fourth phase self-assessment includes a stock-taking of the past and leads to revised planning for the future in light of necessary limitations. There may also be some anticipatory thinking about the last years of life.

Bill Roberts is again a good example. At 46, Bill had to accept a settlement from his firm, which had gone bankrupt. He decided to take his family on a three-month tour of the United States. When interviewed at 67, Bill

*Case described originally in Bühler, *Values in psychotherapy* (1962).

spoke of this trip as the highlight of his life, a "peak experience" (Maslow, 1962).

Upon return, he found a good job, but ill health soon forced him to look for a new occupation. He bought a grocery store which he ran with the help of his wife until age 64, when he retired from this occupation also.

Bill is an example of an average healthy person, who in his fourth phase twice assesses his situation and makes two important decisions on the strength of the evaluation. Forced to interrupt a successful career, he considers this a good moment at which to rest and to reach out for a unique treat. On returning, he feels lucky to find a good job. When severe illness forces him to seek reduced activity, he chooses a new occupation with appropriate consideration of what might happen when later he is forced to retire more and more completely from work. His decision to retire at 64 marks his retreat from competitive gainful occupations as he enters the fifth phase of life.

As in the foregoing phases, different patterns are observable in the fourth phase in the attitude toward self-assessment. The first is again an immaturity pattern seen in those who try to avoid the issue of self-assessment, even though circumstances almost enforce it on them.

One example is Richard, age 50, a contractor whose business went bankrupt. If he had assessed himself and his past life sincerely, he would have had to admit to himself that the same personality defects which were destroying his family life were responsible for his business failure. In both situations he was dishonest and exploited people, and a summarizing self-appraisal would at this point have been very much in order. But Richard refused to face himself and insisted that he knew remedies for all his troubles.

Gary, age 48, whose third marriage failed because of his repeated extramarital involvements, also demonstrates immaturity. Deeply hurt because he truly loved his third wife, he was nevertheless unable to see himself and the repetitious pattern in his life.

A second group reflects *disappointment* over hoped-for fulfillments in life.

George, age 50, felt that he should have accomplished more in life than he had. This was true of his marriage to a woman who had never quite responded to his love. It was also true of his career as a teacher, where he never had been as successful as he had wanted to be. He found it difficult to resign himself to it being unlikely that he would get a full measure of satisfaction out of life.

A third group encounters insoluble *problems* in their chosen careers or with their partners. In the fourth phase, the insoluble problems enforce job changes, often with losses or marital separations and divorces.

> Randolph, now 63, had to give up a political career which was highly meaningful to him because he lost the confidence of his supporters. Forced to switch to a business career, he experienced considerable bitterness.

As in the earlier phase of life, the experience of failure may be the mode of an individual's fourth phase.

> Flora, a woman of 65, had divorced her husband some twelve years before. Her attitude toward her marriage had been ambivalent, and when after prolonged suffering over her conflict, she decided on divorce, her action did not bring her the hoped-for relief. She mourned for what she had lost and resented her fate. Flora also had problems with her only daughter and in her career as a public school teacher. She experienced what many others would interpret as almost all-round failure. She tried to sustain herself with what she could save out of a tenuous relationship with her daughter and later her grandchildren. She ended her teaching career early and took up church activities in its place. Defensively, she thought of the outcome of her life rather in terms of unfortunate developments than of failure on her part.

Finally, feelings of fulfillment or unfulfillment are generated in the fourth phase. As mentioned earlier, this phase represents a crisis period in which fulfillment becomes often problematic.

> Ben, age 51, a twice-divorced and childless salesman, entered psychotherapy at fifty because of impotence, various psychosomatic complaints and depression. He explained his depression as owing to his guilt over his empty, meaningless life. He felt he had wasted his life in the pursuit of goals in which he did not believe.
>
> He had started this way in opposition to a father he hated because he had tried to push the gifted boy in a harsh and unloving manner. Ben quit school, left home and got a clerical job in order to be independent. Only after his father's death, when he was in his twenties, did he finish high school, go to college and get a degree in law. He practiced law for some time, but he was restless and was unsucessful in this and in later managerial positions.
>
> At heart, Ben was a writer and a scholar, and he grieved forever that he had never gotten beyond haphazard attempts at pursuing his real interests and his true potentialities. Now he was too restless and too disturbed to settle down to anything like that. He did not believe in anything he did.

His bad feelings about himself were reinforced by the failure of two meaningless marriages which ended after short durations.

Psychotherapy enabled Ben to experience meaningful love for the first time in his life, but his attempts to write failed tragically. He realized that his life history had led to a tragic reduction of his original potentialities, but he could not resign himself to this insight. His early death in an accident was perhaps not quite as accidental as it seemed.

When convinced he was a hopeless failure, Ben hastened prematurely into his fifth phase of life in resigning himself to the end of his life.

The *fifth phase* of life, beginning around 65 to 70 years, is one in which most people rest and retire from the self-determination to goals of life. In doing so, they may regress to the need-satisfaction or self-limiting living patterns of the pre-self-determined childhood period. If in good physical and mental condition, they may enjoy leisure activities like travel, games or collecting, or they may put themselves at the disposal of some welfare, church or political cause.

If ill, they may of course have to resign themselves to an inactive end, almost a waiting for death to come. If unusually strong and productive, they may go on with previous expansive pursuits.

Important, however, in the fifth phase experience is the gradually-evolving awareness of the past life as a whole resulting essentially in fulfillment or unfulfillment and even despair. Not infrequent also, is a kind of closure experience of resignation. Case material indicates that people who hardly ever before thought of their life as a whole, do so in the fifth phase.

A healthy fifth phase attitude can be exemplified again with the case of Bill Roberts.

Bill Roberts, interviewed at age 67, volunteered some statements summarizing his life as a whole. He says he lived what he calls a "decent and religious life without being a fanatic." He says he had a good wife, and normal children who gave them wonderful grandchildren. He has always been able to provide for his family. His life was worthwhile to him throughout, and he is lucky to have no major regrets. His one big decision to move to California proved to be a very good one. He described his present activities in terms of "interests," emphasizing that he always had many interests and knew what to do with his free time.

In comparison with this mature and fulfilled final phase, other fifth phase structures are more problematic or even desperate.

Lowell, age 62, looks at his life immaturely. A man who worked himself up from the poorest beginnings to moderate riches, he does not find anything worthwhile he can do any more. Having reached his goal he feels empty. In continuous work, he has lost the ability to enjoy himself. He has never thought about his life as a whole, nor its meaning. Uncertain about what to do with himself, his attitude is one of disgust and boredom.

The same case could also be cited with reference to the experience of disappointment, instead of fulfillment, when the goal was reached and there seemed nothing more to do.

Other people take unresolved problems and conflicts into the last stages of their lives.

> Norman, age 70, an immigrant, had been a German military officer whose partly Jewish ancestry cost him his career during the Nazi regime. With only a military education and no interest or aptitude in business, he had been unable to establish himself in the new world and ended up playing small military roles in the movies. He responded with great bitterness and hostility which he expressed even to those who tried to help him. He was sarcastic and hardly close to anyone. He never married and lived alone.

This man's life ended in inner rebellion against the world and his own fate to which he never realized he contributed. Thus it must also be looked at as a failure, resulting in unfulfillment if not despair.

While there are not as yet any statistical data available, the impression gained from a survey of cases is that relatively few people enjoy the feeling of essential or even relative fulfillment in the last stages of their lives. On the other hand the incidence of actual despair also seems small. The majority of aging persons whose cases were studied, showed a closure pattern that might best be termed as resignation. The resigned ending may be caused by very different kinds and degrees of disappointments, which either prevent the feeling of fulfillment or come close to unfulfillment and despair.

The pattern found in the case of Amelia seems to appear frequently.

> Amelia, age 72, had lived alone for about twenty years, after a late divorce from a husband who had shown little interest in her or their two children. The family atmosphere was one of unhappiness and both children left home early in life.
>
> As time went on, Amelia felt that she had failed herself and her children. After her divorce, she tried to get closer to her children. Psychotherapy enabled her to see her own lack of understanding which she tried to remedy to a degree.
>
> In talking about her life, she expressed feelings of regret regarding her past life, but had resigned herself to having found a somewhat belated satisfaction in her recent relationship with her children and grandchildren.

Summary

The process of human life is discussed in relation to life goals. Human life is seen as a process characterized by an intentionality, generated in the core-self system, and directed toward a fulfillment at the end. The intentionality materializes in phases of self-determination towards goals which in the ideal case bring the individual's best potentials to realization. The five phases are one in which self-determination is prepared and built up; one in which life goals are set experimentally and programmatically; one in which life goals become specific and definite; one in which success and failure are being assessed; and one in which rest or continuation of striving occur with a closure experience of fulfillment or unfulfillment, resignation, sometimes despair.

References

Allport, G. *Personality.* (2nd ed.), New York: Holt, Rinehart and Winston, 1961.
Buhler, C. Der menschliche Lebenslauf als phsychologisches Problem (The course of human life as a psychological problem). Gottingen: Verlag fur Psychologie, 2nd. ed., 1959; Leipzig: S. Hirzel, 1933.
Buhler, C. Theoretical observations about life's basic tendencies. *American Journal of Psychotherapy,* 1959, **13**, 3:561-581.
Buhler, C. Meaningful living in the mature years. In Kleemeier, R. W. (ed.), *Aging and leisure.* London: Oxford University Press, 1961.
Buhler, C. *Values in psychotherapy.* New York: Free Press, 1962.
Buhler, C. Human life goals in the humanistic perspective. AAHP Presidential Address, New York 1966. *Journal of Humanistic Psychology,* 1967, **6**.
Buhler, C. & Massarik, F. (Eds.) *The course of life: A study of goals in the humanistic perspective.* New York: B. Springer, 1968.
Darion, J. In Wasserman, D., Darion, J., & Lehigh, M. *Man of la Mancha.* New York: Random House, 1966.
Erikson, E. *Identity and the life cycle.* New York: International Press, 1959.
Kuppers, W. Madchentagebucher der Nachkriegszeit (Girls diaries of the post-war period). Stuttgart: E. Klett, 1964.
Maslow, A. Lessons from the peak-experiences. *American Journal of Humanistic Psychology,* 1962, **2**, 9-18.
Otto, H. The personal resources development research—the multiple strength perception effect. The Proceedings of the Utah Academy of Science, Arts and Letters, 1961-1962, **38**, 182-186.
Preyer, W. *Die Seele des Kindes* (The soul of the child). Leipzig: Grieben, 1882.
Shorstrom, E. L. Personal orientation inventory, San Diego. Educational and Industrial Testing Service, 1962.

Recommended Additional Readings

Buhler, C. Basic theoretical concepts of humanistic psychology. *American Psychologist,* 1971, **26**, 378-386.
Buhler, C., & Massarik, F. (Eds.) *The course of human life: A study of goals in the humanistic perspective.* New York: Springer, 1968.
Erikson, E. H. *Childhood and society.* (2nd ed.) New York: Norton, 1963.

6 Behavior Problems of Normal Children: A Comparison between the Lay Literature and Developmental Research

Hermine H. Marshall

Orientation

The next article is of significance in that it deals directly with the relationship between the scientific research carried out by developmental psychologists and the "advice-to-parents" literature disseminated through popular magazines and books. The phenomenal sale of such books as Benjamin Spock's Baby and Child Care *and Haim Ginott's* Between Parent and Child *and* Between Parent and Teenager *attests to the apparent desire on the part of at least a great number of parents for professional advice about the "proper" way to rear their children. It is difficult to ascertain, of course, just what impact these books and other popular magazines have on actual child-rearing practices, but it seems likely that this literature has some eventual effect.*

Hermine Marshall has empirically compared what the "experts" in the lay literature are saying and what research investigations have actually found. Five behavior problems—food finickiness, physical timidity, specific fears, jealousy, and temper tantrums—were identified as occurring in 90 percent of children, according to the research. Marshall found that while current lay literature is much more sophisticated than that produced in earlier periods, there is nevertheless only limited attention paid to the findings of contemporary research. She recommends that more and different channels of communication be opened between developmental psychologists and parents.

Marshall's criterion (that it must occur in 90 percent of children) for the inclusion of a behavior problem in this study was quite high. There are other problems and concerns of parents that occur less frequently but still merit our attention. These might include thumb sucking, childhood masturbation, enuresis, and encopresis, among others. There is a genuine need for well-written articles or books on these matters that the nonprofessional (i.e., the parent) can read and easily comprehend; this litera-

Reprinted from *Child Development*, 1964, 35, 469-478, by permission of the author and The Society for Research in Child Development.
© 1964 The Society for Research in Child Development.

ture should summarize the available research findings, and should draw conclusions that would aid parents in their dealings with these concerns and hopefully would calm their anxieties.

Hermine H. Marshall is a lecturer in the school of education at the University of California, Berkeley.

6

This paper focuses on the manner in which those behavior problems that are most common among normal children are treated in the literature that is written specifically for parents. No attempt is made towards an exhaustive review of the research. Rather, concern centers both on the frequency with which articles on these behavior problems appear in the lay literature and on the question of whether such articles transmit to the parents the research findings from the fields of child development and child psychology. For example, does this literature take cognizance of the commonness of these problems, of their transitory nature, or of cultural determinants?

In the present study, five of the behavior problems most frequently occurring in normal children are investigated to determine the extent to which the literature for parents corresponds to certain of the research findings on these problems.

Method

Selection of Problem Behavior

In order to have a basis for selecting those problems which occur most frequently among normal children, the *Developmental Study of Behavior Problems of Normal Children between 21 months and 14 Years* by Macfarlane, Allen, and Honzik was used.[10] It should be noted that this study of physical, mental, and personality development from birth to maturity was published in 1954, although the collection of data began in 1929.

For purposes of the present investigation, those behavior problems reported by Macfarlane, Allen, and Honzik as occurring in 90 per cent of the boys or 90 per cent of the girls at one or more age levels between

21 months and 14 years were chosen for further study. The problems that occurred in 90 per cent of the boys were "specific fears," "jealousy," and "temper tantrums." Those occurring in 90 per cent of the girls were "food finickiness," "physical timidity," and "specific fears." Not only were these five problems found to be the most common, but they were also transitory in nature.

As it is not the purpose of this paper to review the Guidance Study monograph, the interested reader is referred to the original publication [10] for a more detailed analysis of these problems. Results of this Study will be presented only when they are pertinent to an analysis of the lay literature.

There are certain questions in this method of selecting common behavior problems which must be considered. First is the question of cross-validation. Would other comprehensive studies of normal children arrive at these same five problems as occurring most frequently? Second is the Guidance Study sample. Are the findings from these parents, who have had more formal education but a lower average income, applicable to the nation at large? Thirdly, is behavior which was considered to be a problem in the decade of the thirties—when the Study was conducted—still considered so today? This question becomes pertinent when one considers Martha Wolfenstein's analysis of the U. S. Children's Bureau *Infant Care* bulletins between 1914 and 1951 regarding the changing concept of the nature of children's impulses and of methods of handling them. [22] She noted that those impulses which were thought of as dangerous in 1929 appeared benign by 1942. Therefore, one might hypothesize that attitudes regarding those particular behavior constellations which constitute a problem might likewise undergo a change between the 1930's and the late 1950's. The final question regards the possible pitfalls in the interviewing procedure which was used for the Control Group of the Guidance Study: what the mother conceives problem behavior to be and what the mother is ready to reveal, as well as distortions of memory and perception, may not reflect the psychological reality of the matter. Nevertheless, the data from the Control Group of the Guidance Study remains the most suitable basis for selecting common behavior problems of normal children available at this time.

Sources of Parent Literature

Two criteria were established for selecting sources of parent literature. The first of these is that it be addressed specifically to parents—as opposed to that addressed to professional or semiprofessional readers, e.g., *Today's Health.*

The second criterion concerns the particular years during which this literature was most likely to be in the hands of the parents. As noted above, there is a possibility that the literature written for parents at the time the data for the Guidance Study was collected, i.e., from 1929 to 1944, might differ in some way from that being written currently, i.e., since the publication of the monograph in 1954. Therefore, two time intervals were differentiated in the present study. The first time interval considered is that between 1929 and 1940. Because all the children in the Guidance Study were born in 1928 and 1929 and because as a consequence the parents in this Study would probably be more aware of literature addressed to parents from this time on—if they had not already been so—1929 was chosen as the beginning of this earlier period. The second or current time interval was established as 1955 to 1961, because the data from the Control Group of the Guidance Study were not published until 1954. Consequently, the second criterion for selecting sources was that it be available to parents either between 1929 and 1940 or between 1955 and 1961.

The following comprised all the sources of parent literature into which investigation was made: (a) All the articles on the five common problems which were printed in *Parents' Magazine** during these two periods were reviewed. (b) Those relevant publications from the U. S. Children's Bureau that were available to parents during these time intervals were used. The one published in 1927[18] was available to readers in the earlier period of 1929 to 1940, and the equivalent ones—now in two sections according to age—dated 1945[23] and 1949[24] are the publications that are currently available to parents. (c) One of Dr. Benjamin Spock's books was taken as typical of the present period[16] as well as a comparable one of the earlier period by Fisher and Gruenberg.[5] (d) Pamphlets by Ross Laboratories,[25, 26] which might be found in a pediatrician's office, and one put out by the Metropolitan Life Insurance Company[19] were also added for the contemporary period.

A total of 37 books and articles have been searched in order to determine if and how the problems of specific fears, jealousy, temper tantrums, food finickiness, and physical timidity are treated for the parent population. It should be noted, however, that some of the books and pamphlets deal with more than one problem.

Results

Specific Fears

Table I indicates the number and percentage of sources on each behavior problem in the two time intervals. As can be seen by this table, the sheer number of articles in the literature for parents suggests general public recognition of the problem of specific fears. Besides being mentioned most frequently of any of the problems investigated in the parent literature, four of these sources on children's fears specially stated that fears were common or natural in children.[2, 13, 19, 24] The transitory nature of these fears was mentioned in three sources,[2, 7, 19] all of which were published in the contemporary period of 1955 to 1961. Two of these latter sources[7, 19] also took note of the decrease in fears with increased age, but not of the subsequent preadolescent increase found in the Control Group of the Guidance Study. Three articles considered the frequency of the fears of dogs and the dark, though without referring to the particular age level at which they are most commonly found. The parent literature contained no dis-

---

* As indexed in *Readers' guide to periodic literature.*

cussions concerning the Control Group results regarding sex differences or birth order differences as affecting susceptibility to fears. The remainder of the sources concentrated on the possible causes of fears, e.g., threats, conditioning, and suggestions for overcoming these fears. Although the value for parents of articles concerning causes and coping devices is not to be taken lightly, one can see that only a relatively small percentage of the sources dealing with children's fears brought out any of the findings of the Control Group survey. It is interesting to note, however, that, with one exception,[13] all of the articles which corresponded in some way with the Control Group findings were those written in the later period. Perhaps this can be taken as an indication of some degree of recognition of research findings in the current parents' literature. Since the

Table 1
Frequency of Consideration of Each Behavior Problem
in the Parent Literature

Behavior Problem	1929-1940 %	1929-1940 N	1955-1961 %	1955-1961 N	Total %	Total N
Specific Fears	21	6	36	9	28	15
Jealousy	29	8	24	6	26	14
Food Finickiness	25	7	20	5	23	12
Temper Tantrums	18	5	12	3	15	8
Physical Timidity	7	2	8	2	8	4
Total	100	28	100	25	100	53

Note:
Although there are 37 different sources, some of these sources deal with more than one problem. Consequently, totaling each separate account of a behavior problem yields 53 scorable considerations of the five behavior problems.

Control Group data were not published until 1954, the lack of recognition in the earlier period may have been due to lack of research rather than to an inadequate review of the professional literature by the authors.

Besides the difference between the two periods in the amount of correspondence to the control Group findings, one other interesting contrast was noted: the prevalence in the earlier period of conditioning as a cause or cure. One might hypothesize that this orientation was due to the impact of the work of John Watson, particularly regarding what he considered the cause of "fear" in infants.[20] More sophisticated research which superseded Watson's findings probably contributed to the diminution of the mention of conditioning and the augmentation of the stress on other approaches. That in the decade of the 1930's the literature showed the influence of Watson, and current literature gives recognition to more recent findings, may be an indication that literature for parents does take some cognizance of contemporary research.

Jealousy

Table I shows that the problem of jealousy was treated relatively frequently in the literature for parents. In addition, five of the articles on this topic implied that this problem was a common one.[11, 14, 16, 21, 24] Only one source specifically stated that this problem was of a transitory nature[15] —though this may be related to the Control Group finding that jealousy was among the more persistent problems. One source[15] specifically mentioned its higher frequency among first-born children, although this was implicit in the majority of the articles on this problem for they showed the cause of jealousy to be the birth of a sibling. However, the important exception at age 12—the time of the dependence-independence struggle—when second-born boys were found to be more jealous was completely neglected. One source[23] took cognizance of its more frequent occurence in girls at the preschool level, but no mention was made of its being more common in school-age boys, nor of this sex difference being caused by a difference in cultural expectations between boys and girls. In fact, the literature for parents noticeably neglected the relation of jealousy to the child in the older age group. Thus, only in the frequency of the problem was there more than occasional correspondence between the lay literature and the data presented by the Control Group Study.

Almost all of the literature on the subject of jealousy was concerned with suggestions as to how to prevent or overcome this problem. An interesting differentiation between the literature of the two time periods was the comparative naiveté of that written during the earlier time interval of 1929 to 1940. Illustrative of this was the statement that jealousy would be overcome "invariably" if the child were confided in and told of the impending birth of a sibling.[17] This was in contrast to the more current feeling that jealousy cannot be completely prevented, although its more extreme forms can be avoided.[4, 15, 16] Similarly, the earlier literature was more likely to consider jealousy a dangerous emotion or a bad trait;[5, 9, 14, 18] whereas later literature emphasized the possibility of converting jealousy into a constructive force. Perhaps this is an indication that, with a deeper understanding of the problem, a more sophisticated approach to it can be expounded in the literature for parents.

Temper Tantrums

The literature for parents does not give the impression that tantrums are as common a concern as the Control Group Study showed them to be. Only 8 of the 37 sources dealt with this problem (see Table 1), and only three of them mentioned that this problem is a common one.[1, 16, 18] Two articles pointed out the decrease with age,[6, 8] one of which also noted the more rapid decline in girls found in the Control Group Study.[8] Although the Control Group found several other factors related to temper tantrums, none of these were mentioned in the articles for parents. These factors were the influence of birth order, the mother's lack of verbal facility, and cultural expectations contributing to both the higher degree of acceptance of temper tantrums in boys and the tapering of tantrums into oversensitiveness among girls. The finding that tantrums are likely to be a sign of disturbance among girls likewise went unnoticed. Most articles were concerned with causes and methods for handling the problem behavior.

One of the most interesting articles was the report of a study on this problem by Florence Goodenough—one of the few psychologists among the authors of parent literature.[8] She studied 45 children to discern, not only the cause, frequency, and time of the tantrum, but also the most successful methods of coping with this behavior. Reasoning and ignoring were used more by parents whose children had fewer tantrums, whereas indulgence was used more by parents whose children had more frequent outbursts. Although her report in *Parents' Magazine* was not written as forcefully as it might have been for this type of literature, it was notable in that it was the only source whose suggestions for overcoming the problem were based on research.

While Goodenough's rather sophisticated article was written in the earlier period of 1929 to 1940, certain of the literature of the same period contained such apparently naive notions as a child with frequent trantrums being "emotionally unstable by nature."[18, 1, 6] Whether this concept has been completely rejected or not, it was not brought out at all in the more current literature. As with specific fears and jealousy, perhaps this change may be considered an indication of improvement in publications for parents as understanding of the problem progresses—despite the minimal amount of correspondence between the parent literature and the Control Group findings.

Food Finickiness

The commonness of food finickiness would seem to be indicated by the number of sources on the subject (see Table 1), but the commonness of this problem was mentioned specifically in only one of these [16] and its transitoriness in two.[3, 16] Sex and birth order differences as well as cultural factors, e.g., the association of this problem with the dependence-independence struggle in boys, were completely omitted, thus evidencing minimal correspondence with the Control Group data. The articles concentrated rather on the causes and cures of the problem.

As with specific fears, there was a greater emphasis in the literature of the earlier period on conditioning as a cause and a cure. This was exemplified by the report of an experiment by Marguerite Gauger in which she experimentally conditioned preschool children to like vinegar, salt solution, and egg white by offering chocolate after each of these foods, then gradually withdrawing the chocolate. She implied that parents should try this type of conditioning at home.[12]

Although the concept of a good appetite indicating a healthy child and a healthy child reflecting an adequate mother was implicit as a cause of the problem in much of the literature of both periods, it was perhaps given an unfortunate impetus in the early period by the public school's practice of sending home health reports stating flatly whether the child was under-, over-, or of normal weight.[5] Too frequently these reports were false or did not take into account other factors, e.g., body build. Hopefully, this practice has been altered, thereby eliminating one cause of the mother's anxiety about feeding her child. Nevertheless, as can be seen in Table 1, it does not appear that concern over this problem has decreased very much over time, if the number of articles written on the subject can be taken as an indication.

Physical Timidity

Despite the fact that 90 per cent of the girls in the Control Group Study were physically timid at one age level or another, Table 1 shows that only four of the sources surveyed were concerned with this problem, and one of these[24] included only a brief mention. No mention was made of the cultural expectations that caused fewer boys to exhibit this problem, nor of birth order differences, nor of the relation the Control Group Study found physical timidity to have to other withdrawal patterns rather than to aggressive overt behavior. For some reason, this behavior which was commonly considered a problem in data gathered in the thirties, was not treated as such in the literature of the period. Nor is it so treated currently—since the publication of the Control Group findings. One wonders whether this discrepancy is due to the Guidance Study's definition of timidity at the behavior problem level: "more cautious than average—watches others first before participating; always tense, so that efficiency is interfered with" (p. 42);[10] It may be that, while this type of behavior is considered to be a problem by the investigators, it is not generally thought of as a behavior problem by perhaps a wider cross-section of our culture, or at least by those who write for and edit the literature for parents. Whether this degree of physical timidity was conceived by the parents in the survey to be a problem or whether it was rather a function of the investigators' code and psychological orientation is not clear. One might hypothesize the latter. This difference in orientation, then, would account for the lack of correspondence between the literature for parents and the Control Group findings. Indeed, the Study does point out that previous research has shown a zero correlation between what is considered significant behavior by teachers—who were found to agree highly with parents in this regard—and that considered significant by mental hygienists.

Conclusions

In determining the extent to which the literature for parents corresponds to the research findings in the fields of child development and child psychology, one notes that all five behavior problems found to be most common by the report of the Control Group of the Guidance Study occurred in 90 per cent of the boys or the girls. However, concern shown in the parent literature was not at the same high for all five problems. Referring again to Table 1, it can be seen that specific fears, jealousy, and food finickiness, in that order, were the problems most frequently written about. Writers for parents seemed least interested in physical timidity. Neither was the commonness of these problems specified on an equal level. Table 2 summarizes the frequency—or lack of frequency—with which the factors of commonness, transitoriness, sex differences, birth order differences, and cultural expectations were found in the sources canvassed.

If one accepts the findings of the Control Group Study as valid, then only a small percentage of them seem to have seeped into the literature for parents. Nevertheless, certain indications of a somewhat more sophisticated approach to these problems have been noted in the later literature as contrasted to that written two decades ago. But one may justly wonder what is preventing their

Table 2
Number of Times Each Contributing Factor Was Noted in Parent Literature on Each Behavior Problem

	Behavior Problem				
Contributing factor	Specific fears	Food finickiness	Jealousy	Temper tantrums	Physical timidity
Commonness	4	1	5	3	0
Transitoriness	3	2	1	1	0
Sex differences	0	0	1	1	0
Birth order	0	0	1	0	0
Cultural expectations	0	0	0	0	0

Note:
The contributing factors were drawn from those factors which the Control Group Study found related to these five behavior problems. Both time intervals, 1929 to 1940 and 1955 to 1961, are combined in this table.

freer flow into the hands of parents. Is it due to some reluctance on the part of the editors of literature for parents? Or do research psychologists hesitate for some reason to share their findings with the lay public—which stands to gain a great deal from such knowledge?

Only three experiments on the five behavior problems were reported to the parents: Clara Davis's famous experiment on self-selection of food,[3] Marguerite Gauger's conditioning of food tastes,[12] and Florence Goodenough's survey of temper tantrums.[8] Goodenough's method of studying the most successful coping measures used by parents and then making suggestions based on the results of this research should be stressed as a fruitful approach.

One easily arrives at two conclusions. First, since some research appears to result in improved advice for parents, as noted in the discussions of specific fears, jealousy, and temper tantrums, how much more adequate it would be with research of increased scope and depth. Perhaps some combination of the methods used by the Guidance Study—which yielded the frequency and persistence of the problem as well as the influence of other factors—with those used by Goodenough—which indicated the best solutions used by parents who had overcome the problem—would be most productive of pertinent information.

The second conclusion serves to complete the first: the channels between the research laboratories and the publications for parents need to be widened. In this way, research findings can be funneled into the homes in order to help parents alleviate the problems which hamper their children.

References

1. Bacmeister, R. How to avoid temper tantrums. *Parents' Magazine*, 1935, **10**, 18-19.
2. Bloomquist, E. R. When a child is afraid of the dark. *Parents' Magazine*, 1959, **34**, 36-37.
3. Davis, C. M. Can babies choose their food? *Parents' Magazine*, 1930, **5**, 22-23.
4. Finnigan, J. How we dealt with jealousy. *Parents' Magazine*, 1930, **30**, 42-43.
5. Fisher, D., & Gruenberg, S. M. *Our Children: A handbook for parents.* New York: Viking, 1932.
6. Fisher, M. S. Common faults and what they mean. *Parents' Magazine*, 1934, **9**, 20-21.
7. Fraiberg, S. H. Big fears of little children. *Parents' Magazine*, 1960, **35**, 40-41.
8. Goodenough, F. L. How to handle temper tantrums. *Parents' Magazine*, 1933, **8**, 20-21.
9. Hill, A. B. How to deal with jealousy: The child and the new baby. *Parents' Magazine*, 1940, **15**, 22-23.
10. Macfarlane, J. W., Allen, L., & Honzik, M. P. *A developmental study of the behavior problems of normal children between 21 months and 14 years.* Berkeley: University of California Press, 1954.
11. McMaster, M. How to handle jealousy. *Parents' Magazine*, 1938, **13**, 26-27.
12. Pierce, A. How to get your children to eat what they should and like it. *Parents' Magazine*, 1929, **4**, 25ff.
13. Sapin, R. Helping children to overcome fear. *Parents' Magazine*, 1933, **8**, 14-16.
14. Schultz, G. D. When little brothers are pests. *Parents' Magazine*, 1937, **12**, 32-33.
15. Spector, R. Jealousy came with our new baby. *Parents' Magazine*, 1960, **35**, 38-39.

16. Spock, B. *Baby and child care.* New York: Pocket Books, 1945, 1957.
17. Strait, S. H. What tantrums taught us. *Parents' Magazine,* 1956, **31**, 48-49.
18. Thom, D. A. *Child management.* (Rev.) U. S. Children's Bureau, Publ. No. 143. U. S. Government Printing Office, 1927.
19. *Understanding your child.* Metropolitan Life Insurance Company, 1951.
20. Watson, J. B. *Psychology from the standpoint of a behaviorist.* Philadelphia: Lippincott, 1919.
21. Weill, B. G. When children quarrel. *Parents' Magazine,* 1930, **5**, 22-23.
22. Wolfenstein, M. Fun morality: an analysis of recent American child-training literature. in M. Mead & M. Wolfenstein (Eds.), *Childhood in contemporary cultures.* Chicago: University of Chicago Press, 1955. Pp. 168-178.
23. *Your child from one to six.* (Rev.) U. S. Children's Bureau, Publ. No. 30. U. S. Government Printing Office, 1945.
24. *Your child from six to twelve.* U. S. Children's Bureau, Publ. No. 324. U. S. Government Printing Office, 1945.
25. *Your child's appetite.* Ross Laboratories, 1960.
26. *Your child's fears.* Ross Laboratories, 1960.

Recommended Additional Readings

Ginott, H. *Between parent and child.* New York: Crowell-Collier-Macmillan, 1965.

Ginott, H. *Between parent and teenager.* New York: Crowell-Collier-Macmillan, 1969.

Thomas A., S., & Birch, H. G. *Your child is a person.* New York: Viking, 1965.

Part Two   Influences on Development

7 Behavior-Genetic Analysis and Its Biosocial Consequences

Jerry Hirsch

Orientation

The first article in this section, which is broadly concerned with "Influences on Development," focuses upon the important question of genetic determinants and/or influences on behavior. Eventually, any discussion of this question reverts to the troublesome issue of "nature versus nurture." This misleading phrase has caused the scientific enterprise—and ultimately society itself—untold difficulties. The difficulties arise immediately because of the insistence by so many persons to insert the word "versus" between nature and nurture, thereby implying that the behavior or characteristic in question was brought about by one of two components: biological inheritance or environmental experience. Actually, almost all contemporary scientists would say that behavior and development are the products of complex interactional processes between these two components; nevertheless, because of certain historical developments that occurred within biological and psychological disciplines, the nemesis of the "versus" fallacy lingers on in much of our scientific research and in the assumptions underlying our social planning.

Reprinted from *Seminars in Psychiatry*, 1970, 2, 89-105, by permission of the author and Henry M. Stratton, Inc.

Invited address presented to the XIXth International Congress of Psychology, London, England, July 30, 1969, and dedicated to Prof. Th. Dobzhansky on his 70th birthday.

This work was prepared with the support of Mental Health Training Grant 1 TO1 10715-04 BLS for Research Training in the Biological Sciences.

In this article Professor Hirsch first presents a historical overview of the nature-nurture issue and identifies significant events and persons who greatly influenced the scientific study of individual differences. As Hirsch so ably illustrates, it is clear that predominant views of human nature (or "models of man"), which themselves are influenced by or derived from prevailing economic and political ideologies, influence what scientists study and how they study it. Even "objective" scientists are not uninfluenced by subjective forces in society.

Hirsch then attempts to put the nature-nurture issue into what he believes to be the proper perspective; he calls his approach the "behavior genetic analysis." The explanatory concept he heavily emphasizes is the norm of reaction.

The reader may find many of the ideas and concepts presented in this article to be rather difficult to grasp upon first reading. The editor feels, nevertheless, that the material therein is so important that it merits intensive study.

Jerry Hirsch is a professor of psychology at the University of Illinois.

7

As a psychology student I was taught that a science was founded on the discovery of lawful relations between variables. During my student days at Berkeley, the true psychological scientist was preoccupied with the major learning theories. We read, studied, and designed experiments to test the theories of Thorndike, Guthrie, Hull and Tolman. Many of their verbally formulated laws of behavior were replaced by the mathematical models that have since come into vogue.

Afterwards I learned empirically the truth of what might be the most general of all behavioral laws, the Harvard law of animal behavior: "Under the most carefully controlled experimental conditions the animals do as they damn please." Still later I discovered the low esteem in which post World War II psychology was held by two of the best minds this century has seen. In 1947, John Dewey, eighth president of the American Psychological Association, wrote to discourage young Robert V. Daniels from studying psychology at Harvard:

> Psychology ... is on the whole, in my opinion, the most inept and backwards a tool ... as there is. It is much of it actually harmful because of wrong basic postulates—maybe not all stated, but actually there when one judges from what they do—the kind of problems attacked and the way they attack them.[3]

On the final page of the last book written before his death in 1951 Ludwig Wittgenstein, perhaps the most influential of the founders of modern philosophical analysis, observed:

> The confusion and barrenness of psychology

is not to be explained by calling it a "young science"; its state is not comparable with that of physics, for instance, in its beginning. (Rather with that of certain branches of mathematics. Set theory.) For in psychology there are experimental methods and *conceptual confusion*. (As in the other case conceptual confusion and methods of proof.)

The existence of the experimental method makes us think we have the means of solving the problems which trouble us; though problem and method pass one another by.[11]

Laws of Genetics

It was then while overcome by feelings of disenchantment (obviously without laws behavior study could never be science) that I embraced genetics. There was true science. My passion became even more intense when I realized that, like thermodynamics, genetics had three laws: segregation, independent assortment and the Hardy-Weinberg law of population equilibria. What a foundation they provided for my beloved individual differences.

Since both my teaching and research involved considerable work with *Drosophila*, I knew and would recount to my classes in somewhat elaborate detail the story of Calvin Bridge's classic experiments on sex determination as a function of a ratio between the sex chromosomes and the autosomes. As the important discoveries in human cytogenetics were made throughout the 1950's and 60's and "abnormalities" like Klinefelter's, Turner's and Down's syndromes and the violence-prone males with an extra Y chromosome became genetically comprehensible, I began to realize that the so-called laws of genetics were no more universal than the so-called laws of behavior. Every one of the above-mentioned clinical conditions involved, at the very least, a violation of Mendel's law of segregation. Of course, so did Bridge's experiments, but it had been to easy to rationalize them as clever laboratory tricks.

Behaviorism

Over the past two decades the case against behaviorist extremism has been spelled out in incontrovertible detail. The behaviorists committed many sins: they accepted the mind at birth as Locke's *tabula rasa*, they advocated an empty-organism psychology, they asserted the uniformity postulate of no prenatal individual differences; in short they epitomized typological thinking. Many times we have heard quoted the famous boast by the first high priest of behaviorism, John B. Watson:

> Give me a dozen healthy infants, well-formed, and my own specified world to bring them up in, and I'll guarantee to take any one at random and train him to become any type of specialist I might select—doctor, lawyer, artist, merchant-chief and yes, even beggar-man and thief, regardless of his talents, penchants, tendencies, abilities, vocations, race of his ancestors.

However, it is only when we read the next sentence, which is rarely, if ever, quoted, that we begin to understand how so many people might have embraced something intellectually so shallow as radical behaviorism. In that all important next sentence Watson explains:

> I am going beyond my facts and I admit it, but so have the advocates of the contrary and they have been doing it for many thousands of years (p. 104)[33]

Racism

Who were the advocates of the contrary and what had they been saying? It is difficult to establish the origins of racist thinking, but certainly one of its most influential advocates was Joseph Arthur de Gobineau, who published a four-volume *Essay on the inequality of the human races*[10] in the mid-1850s. De Gobineau preached the superiority of the white race, and among white it was the Aryans who carried civilization to its highest point. In fact, they were responsible for civilization wherever it appeared. Unfortunately, de Gobineau's essay proved to be the major seminal work that inspired some of the most perverse developments in the intellectual and political history of our civilization. Later in his life, de Gobineau became an intimate of the celebrated German composer, Richard Wagner. The English-born Houston Stewart Chamberlain, who emigrated to the Continent, became a devoted admirer of both de Gobineau and Wagner. In 1908, after Wagner's death, he married Wagner's daughter, Eva, settled in and supported Germany against England during World War I, becoming a naturalized German citizen in 1916.

In the summer of 1923, an admirer who had read Chamberlain's writings, Adolf Hitler, visited Wahnfried, the Wagner family home in Bayreuth where Chamberlain lived. After their meeting, Chamberlain wrote to Hitler: "My faith in the Germans had never wavered for a moment, but my hope ... had sunk to a low ebb. At one stroke you have transformed the state of my soul!"[11] We all know the sequel to that unfortunate tale. I find that our modern scientific colleagues, whether they be biological or social scientists, for the most part, do not know the sad parallel that exists for the essentially political tale I have so far recounted. The same theme can be traced down the main stream of biosocial science.

Today not many people know the complete title of Darwin's most famous book: *On the Origin of Species by Means of Natural Selection or the Preservation of Favoured Races in the Struggle for Life*. I find no evidence that Darwin had the attitudes we now call racist. Unfortunately many of his admirers, his contemporaries, and his successors were not as circumspect as he. In Paris in 1838, J. E. D. Esquirol first described a form of mental deficiency later to become well known by two inappropriate names unrelated to his work. Unhappily one of these names, through textbook adoption and clinical jargon, puts into wide circulation a term loaded with race prejudice. Somewhat later (1846 and 1866), E. Seguin described the same condition under the name "furfuraceous cretinism" and his account has only recently been recognized as "the most ingenious description of physical characteristics . . ."[2]

Unhappily that most promising scientific beginning was ignored. Instead the following unfortunate events occurred: In 1866, John Langdon Haydon Down published the paper entitled "Observations on an ethnic classification of idiots."[6]

> . . . making a classification of the feeble-minded, by arranging them around various ethnic standards—in other words, framing a natural system to supplement the information to be derived by an inquiry into the history of the case.
>
> "I have been able to find among the large number of idiots and imbeciles which comes under my observation, both at Earlswood and the out-patient department of the Hospital, that a considerable portion can be fairly referred to one of the great divisions of the human family other than the class from which they have sprung. Of course, there are numerous representatives of the great Caucasian family. Several well-marked examples of the Ethiopian variety have come under my notice, presenting the characteristic malar bones, the prominent eyes, the puffy lips, and retreating chin. The woolly hair has also been present, although not always black nor has the skin acquired pigmentary deposit. They have been specimens of white negroes, although of European descent.
>
> Some arrange themselves around the Malay variety, and present in their soft, black, curly, hair, their prominent upper jaws and capacious mouths, types of the family which people the South Sea Islands.
>
> Nor have there been wanting the analogues of the people who with shortened foreheads, prominent cheeks, deep-set eyes, and slightly apish nose, originally inhabited the American Continent.
>
> The great Mongolian family has numerous representatives, and it is to this division, I wish, in this paper, to call special attention. A very large number of congenital idiots are typical Mongols. So marked is this, that when placed side by side, it is difficult to believe that the specimens compared are not children of the same parents. The number of idiots who arrange themselves around the Mongolian type is so great, and they present such a close resemblace to one another in mental power, that I shall describe an idiot member of this racial division, selected from the large number that have fallen under my observation.
>
> The hair is not black, as in the real Mongol, but of a brownish colour, straight and scanty. The face is flat and broad, and destitute of prominence. The cheeks are roundish, and extended laterally. The eyes are obliquely placed, and the internal canthi more than normally distant from one another. The palpebral fissure is very narrow. The forehead is wrinkled

transversely from the constant assistance which the levatores palpebrarum derive from the occipito-frontalis muscle in the opening of the eyes. The lips are large and thick with transverse fissures. The tongue is long, thick, and is much roughened. The nose is small. The skin has a slightly dirty yellowish tinge and is deficient in elasticity giving the appearance of being too large for the body.

The boy's aspect is such that it is difficult to realize that he is the child of Europeans, but so frequently are these characters presented, that there can be no doubt that these ethnic features are the result of degeneration.

And he means degeneration from a higher to a lower race. The foregoing represents a distasteful but excellent example of the racial hierarchy theory and its misleadingly dangerous implications. That was how the widely-used terms Mongolism and Mongolian idiocy entered our "technical" vocabulary. For the next century, this pattern of thought is going to persist and occupy an important place in the minds of many leading scientists.

Alleged Jewish Genetic Inferiority

In 1884, Francis Galton, Darwin's half cousin, founder of the Eugenics movement and respected contributor to many fields of science, wrote to the distinguished Swiss botanist, Alphonse de Candolle: "It strikes me that the Jews are specialized for a parasitical existence upon other nations, and that there is need of evidence that they are capable of fulfilling the varied duties of a civilized nation by themselves" (p. 209).[27] Karl Pearson, Galton's disciple and biographer, echoed this opinion 40 years later during his attempt to prove the undesirability of Jewish immigration into Britain: ". . . for such men as religion, social habits, or language keep as a caste apart, there should be no place. They will not be absorbed by, and at the same time strengthen the existing population; they will develop into a parasitic race . . ." (p. 125).[28]

Beginning in 1908 and continuing at least until 1928, Karl Pearson collected and analyzed data in order to assess "the quality of the racial stock immigrating into Great Britain . . ." (p. 33).[25] He was particularly disturbed by the large numbers of East European Jews, who near the turn of the century began coming from Poland and Russia to escape the pogroms. Pearson's philosophy was quite explicitly spelled out:

> Let us admit . . . that the mind of man is for the most part a congenital product, and the factors which determine it are racial and familial: we are not dealing with a mutable characteristic capable of being moulded by the doctor, the teacher, the parent or the home environment" (p. 124).[28]

The ancestors of the men who pride themselves on being English today were all at one time immigrants; it is not for us to cast the first stone against newcomers, solely because they are newcomers. But the test for immigrants in the old days was a severe one; it was power, physical and mental, to retain their hold on the land they seized. So came Celts, Saxons, Norsemen, Danes and Normans in succession and built up the nation of which we are proud. Nor do we criticize the alien Jewish immigration simply because it is Jewish; we took the alien Jews to study, because they were the chief immigrants of that day and material was readily available" (p. 127).[28]

His observations led him to conclude: "Taken *on the average,* and regarding both sexes, this alien Jewish population is somewhat inferior physically and mentally to the native population" (p. 126).[28]

Alleged Black Genetic Inferiority

Quite recently there has appeared a series of papers disputing whether or not black Americans are, in fact, genetically inferior to white Americans in intellectual capacity. The claims and counterclaims have been given enormous publicity in the popular press in America. Some of those papers contain most of the fallacies that can conceivably be associated with this widely misunderstood problem.

The steps toward the intellectual cul-de-sac into which this dispute leads and the fallacious assumptions on which such "progress" is based are the following: (1) A trait called intelligence, or anything else, is defined and a testing instrument for the measurement of trait expression is used; (2) the heritability of that trait is estimated; (3) races (populations) are compared with respect to their performance on the test of trait expression; (4) when the races (populations) differ on the test whose heritability has now been measured, the one with the lower score is genetically inferior, Q.E.D.

The foregoing argument can be applied to any single trait or to as many traits as one might choose to consider. Therefore, analysis of this general problem does *not* depend upon the particular definition and test used for this or that trait. For my analysis I shall pretend that an acceptable test exists for some trait, be it height, weight, intelligence, or anything else. (Without an acceptable test, discussion of the "trait" remains unscientific.)

Even to consider comparisons between races, the following concepts must be recognized: (1) the genome as a mosaic, (2) development as the expression of one out of many alternatives in the genotype's norm of reaction, (3) a population as a gene pool, (4) heritability is not instinct, (5) traits as distributions of scores, and (6) distributions as moments.

Since inheritance is particulate and not integral, the genome, genotype or hereditary endowment of each individual is a unique mosaic—an assemblage of factors many of which are independent. Because of the lottery-like nature of both gamete formation and fertilization, other than monozygotes no two individuals share the same genotypic mosaic.

Norm of Reactions

The ontogeny of an individual's phenotype (observable outcome of development) has a norm or range of reaction not predictable in advance. In most cases the norm of reaction remains largely unknown; but the concept is nevertheless of fundamental importance, because it saves us from being taken in by glib and misleading textbook clichés such as "heredity sets the limits but environment determines the extent of development within those limits." Even in the most favorable materials only an approximate estimate can be obtained for the norm of reaction, when, as in plants and some animals, an individual genotype can be replicated many times and its development studied over a range of environmental conditions. The more varied the conditions, the more diverse might be the phenotypes developed from any one genotype. Of course, different genotypes should not be expected to have the same norm of reaction; unfortunately psychology's attention was diverted from appreciating this basic fact of biology by a half century of misguided environmentalism. Just as we see that, except for monozygotes, no two human faces are alike, so we must expect norms of reaction to show genotypic uniqueness. That is one reason why the heroic but ill-fated attempts of experimental learning psychology to write the "laws of environmental influence" were grasping at shadows. Therefore, those limits set by heredity in the textbook cliché can never be specified. They are plastic within each individual but differ between individuals. Extreme environmentalists were wrong to hope that one law or set of laws described universal features of modifiability. Extreme hereditarians were wrong to ignore the norm of reaction.

Individuals occur in populations and then only as temporary attachments, so to speak, each to particular combinations of genes. The population, on the other hand, can endure indefinitely as a pool of genes, maybe forever recombining to generate new individuals.

Instincts, Genes and Heritability

What is heritability? How is heritability estimated for intelligence or any other trait? Is heritability related to instinct? In 1872, Douglas Spalding demonstrated that the ontogeny of a bird's ability to fly is simply maturation and not the result of practice, imitation or any demonstrable kind of learning. He confined immature birds and deprived them of the opportunity either to practice flapping their wings or to observe and imitate the flight of older birds; in spite of this, they developed the ability to fly. For some ethologists this deprivation experiment became the paradigm for proving the innateness or instinctive nature of a behavior by demonstrating that it appears despite the absence of any opportunity for it to be learned. Remember two things about this approach: (1) the observation involves experimental manipulation of the conditions of experience during development, and (2) such observation can be made on the development of one individual. For some people the results of a deprivation experiment now constitute the operational demonstration of the existence (or non-existence) of an instinct (in a particular species).

Are instincts heritable? That is, are they determined by genes? But what is a gene? A gene is an inference from a breeding experiment. It is recognized by the measurement of individual differences—the recognition of the segregation of distinguishable forms of the expression of some trait among the progeny of appropriate matings. For example, when an individual of blood type AA mates with one of type BB, their offspring a uniformly AB. If two of the AB offspring mate, it found that the A and B gene forms have segregated during reproduction and recombined in their progeny to produce all combinations of A and B: AA, AB, and BB. Note that the only operation involved in such a study is *breeding* of one or more generations and then at an appropriate time of life, observation of the separate individuals born in each generation—controlled breeding with experimental material or pedigree analysis of the appropriate families with human subjects. In principle, only one (usually brief) observation is required. Thus we see that genetics is a science of *differences,* and the breeding experiment is its fundamental operation. The operational definition of the gene, therefore, involves observation in a breeding experiment of the segregation among several individuals of distinguishable differences in the expression of some trait from which the gene can be inferred. Genetics does not work with a single subject, whose development is studied. (The foregoing, the following, and all discussions of genetic analysis presuppose sufficiently adequate control of environmental conditions so that all observed individual differences have developed under the same, homogeneous environmental conditions, conditions never achieved in any human studies.)

How does heritability enter the picture? At the present stage of knowledge, many features (traits) of animals and plants have not yet been related to genes that can be recognized individually. But the role of large numbers of genes, often called polygenes and in most organisms still indistinguishable one from the other, has been demonstrated easily (and often) by selective breeding or by appropriate comparisons between different strains of animals or plants. Selection and strain crossing have provided the basis for many advances in agriculture and among the new generation of research workers are becoming standard tools for the experimental behaviorist. Heritability often summarizes the extent to which a particular population has responded to a regimen of being bred selectively on the basis of the expression of some trait. Heritability values vary between zero and plus one. If the distribution of trait expression among progeny remains the same no matter how their parents might be selected, then heritability has zero value. If parental selection does make a difference, heritability exceeds zero, its exact value reflecting the parent-offspring correlation. Or more generally, as Jensen says: "The basic data from which . . . heritability coefficients are estimated are correlations among individuals of different degrees of kinship" (p. 48).[18] Though, many of the heritabilities Jensen discusses have been obtained by comparing mono-and di-zygotic twins.[17]

A heritability estimate, however, is a far more limited piece of information than most people realize. As was so well stated by Fuller and Thompson: "heritability is a property of populations and not of traits."[9] In its strictest sense, a heritability measure provides for a given population an estimate of the proportion of the variance it shows in trait (phenotype) expression which is correlated with the segregation of the alleles of independently acting genes. There are other more broadly conceived heritability measures, which estimate this correlation and also include the combined effected of genes that are independent and of those that interact. Therefore, heritability estimates the proportion of the total phenotypic variance (individual differences) shown by a trait that can be attributed to genetic variation (narrowly or broadly interpreted) in some particular population at a single generation under one set of conditions.

The foregoing description contains three fundamentally important limitations which have rarely been accorded sufficient attention: (1) The importance of limiting any heritability statement to a specific population is evident when we realize that a gene, which shows variation in one population because it is represented there by two or more segregating alleles, might show no variation in some other population because it is uniformly represented there by only a single allele. Remember that initially such a gene could never have been detected by genetic methods in the second population. Once it has been detected in some population carrying two or more of its segregating alleles, the information thus obtained might permit us to recognize it in populations carrying only a single allele. Note how this is related to heritability: the trait will show a greater-than-zero heritability in the segregating population but zero heritability in the non-segregating population. This does *not* mean that the trait is determined genetically in the first population and environmentally in the second!

Up to now my discussion has been limited to a single gene. The very same argument applies for every gene of the polygenic complexes involved in continuously varying traits like height, weight and intelligence. Also, only *genetic* variation has been considered—the presence or absence of segregating alleles at one or more loci in different populations.

(2) Next let us consider the ever-present environmental sources of variation. Usually from the Mendelian point of view, except for the genes on the segregating chromosomes, everything inside the cell and outside the organism is lumped together and can be called environmental variation: cytoplasmic constituents, the maternal effects now known to be so important, the early experience effects studied in so many psychological laboratories, and so on. None of these can be considered unimportant or trivial. They are ever present. Let us now perform what physicists call a Gedanken, or thought, experiment. Imagine Aldous Huxley's *Brave New World* or Skinner's *Walden II* organized in such a way that every individual is exposed to precisely the same environmental conditions. In other words, consider the extreme, but *un*realistic, case of complete environmental homogeneity. Under those circumstances the heritability value would approach unity, because only genetic variation would be present. Don't

forget that even under the most simplifying assumptions, there are over 70 trillion potential human genotypes—no two of us share the same genotype no matter how many ancestors we happen to have in common.[16] Since mitosis projects our unique genotype into the nucleus, or executive, of every cell in our bodies, the individuality that is so obvious in the human faces we see around us must also characterize the unseen components. Let the same experiment be imagined for any number of environments. In each environment heritability will approximate unity but each genotype *may* develop a different phenotype in every environment and the distribution (hierarchy) of genotypes (in terms of their phenotypes) must not be expected to remain invariant of environments.

(3) The third limitation refers to the fact that because gene frequencies can and do change from one generation to the next, so will heritability values or the magnitude of the genetic variance.

Now let us shift our focus to the entire genotype or at least to those of its components that might co-vary at least partially with the phenotypic expression of a particular trait. Early in this century Woltereck[7] called to our attention the norm-of-reaction concept: the same genotype can give rise to a wide array of phenotypes depending upon the environment in which it develops. This is most conveniently studied in plants where genotypes are easily replicated. Later Goldschmidt[11] was to show in *Drosophila* that, by careful selection of the environmental conditions at critical periods in development, various phenotypes ordinarily associated with specific gene mutations could be produced from genotypes that did not include the mutant form of those genes. Descriptively, Goldschmidt called these events *phenocopies*—environmentally produced imitations of gene mutants of phenotypic expressions only manifested by the "inappropriate" genotype if unusual environmental influences impinge during critical periods in development, but regularly manifested by the "appropriate" genotype under the usual environmental conditions.

In 1946, the brilliant British geneticist J.B.S. Haldane[12] analyzed the interaction concept and gave quantitative meaning to the foregoing. For the simplest case but one, that of two genotypes in three environments or, for its mathematical equivalent, that of three genotypes in two environments, he showed that there are 60 possible kinds of interaction. Ten genotypes in 10 environments generate 10^{144} possible kinds of interaction. In general m genotypes in n environments generate $\frac{(mn)!}{m!n!}$ kinds of interaction. Since the characterization of genotype-environment interaction can only be ad hoc and the number of possible interactions is effectively unlimited, it is no wonder that the long search for general laws has been so unfruitful.

For genetically different lines of rats showing the Tryon-type "bright-dull" difference in performance on a learning task, by so simple a change in environmental conditions as replacing massed-practice trials by distributed-practice trials, McGaugh, Jennings and Thompson[23] found that the so-called dulls moved right up to the scoring level of the so-called brights. In a recent study of the open-field behavior of mice, Hegmann and DeFries[13] found that heritabilities measured repeatedly in the same individuals were unstable over two successive days. In surveying earlier work they commented: Heritability estimates for repeated measurements of behavioral characters have been found to increase (Broadhurst & Jinks, 1961), decrease (Broadhurst & Jinks, 1966), and fluctuate randomly (Fuller & Thompson, 1960) as a function of repeated testing" (p. 27). Therefore, to the limitations on heritability due to population, situation and breeding generation, we must now add developmental stage, or, many people might say, just plain unreliability! The late and brilliant Sir Ronald Fisher, whose authority Jensen cites (p. 34)[18] indicated how fully he had appreciated such limitations when he commented: "the so-called coefficient of heritability, which I regard as one of those unfortunate short-cuts which have emerged in biometry for lack of a more thorough analysis of the data" (p. 217).[8] The plain facts are that in the study of man a heritability estimate turns out to be a piece of "knowledge" that is both deceptive and trivial.

The Roots of One Misuse of Statistics

The other two concepts to be taken into account when racial comparisons are considered involve the representation of traits in populations by distributions of scores and characterization of distributions by moment-derived statistics. Populations should be compared only with respect to one trait at a time and comparisons should be made in terms of the moment statistics of their trait distributions. Therefore, for any two populations, on each trait of interest, a separate comparison should be made for every moment of their score distributions. If we consider only the first four moments, from which are derived the familiar statistics for mean, variance, skewness, and kurtosis, then there are four ways in which populations or races may differ with respect to any single trait. Since we possess 23 independently assorting pairs of chromosomes, certainly there are at least 23 uncorrelated traits with respect to which populations can be compared. Since comparisons will be made in terms of four (usually independent) statistics, there are $4 \times 23 = 92$ ways in which races can differ. Since the integrity of chromosomes is *not* preserved over the generations, because they often break apart at meiosis and exchange constituent genes, there are far more than 23 independent hereditary units. If instead of 23 chromosomes we take the 100,000 genes man is now estimated to possess (p. IX)[24] and we think in terms of their phenotypic trait correlates, then there may be as many as 400,000 comparisons to be made between any two populations or races.

A priori, at this time we know enough to expect no two populations to be the same with respect to most or all of the constituents of their gene pools. "Mutations and recombinations will occur at different places, at different times, and with differing frequencies. Furthermore, selection pressures will also vary" (p. 1441).[16] So the number and kinds of differences between populations now waiting to be revealed in "the more thorough analysis" recommended by Fisher literally staggers the imagination. It does not suggest a linear hierarchy of inferior and superior races.

Why has so much stress been placed on comparing distributions only with respect to their central tendencies by testing the significance of mean differences? There is much evidence that many observations are not normally distributed and that the distributions from many populations do not share homogeneity of variance. The source of our difficulty traces back to the very inception of our statistical tradition.

There is an unbroken line of intellectual influence from Quetelet through Galton and Pearson to modern psychometrics and biometrics. Adolphe Quetelet (1796-1874), the Belgian astronomer-statistician, introduced the concept of "the average man"; he also applied the normal distribution, so widely used in astronomy for error variation, to human data, biological and social. The great Francis Galton followed Quetelet's lead and then Karl Pearson elaborated and perfected their methods. I know of nothing that has contributed more to impose the typological way of thought on, and perpetuates it in, present-day psychology than the feedback from these methods for describing observations in terms of group averages.

There is a technique called composite photography to the perfection of which Sir Francis Galton contributed in an important way. Some of Galton's best work in this field was done by combining—literally averaging— the separate physiognomic features of many different Jewish individuals into his composite photograph of "the Jewish type." Karl Pearson, his disciple and biographer, wrote: "There is little doubt that Galton's Jewish type formed a landmark in composite photography . . ." (p. 293).[27] The part played by typological thinking in the development of modern statistics and the way in which such typological thinking has been feeding back into our conceptual framework through our continued careless use of these statistics is illuminated by Galton's following remarks: "The word generic presupposes a genus, that is to say, a collection of individuals who have much in common, and among whom medium characteristics are very much more frequent than extreme ones. The same idea is sometimes expressed by the word typical, which was much used by Quetelet, who was the first to give it a rigorous interpretation, and whose idea of a type lies at the basis of his statistical views. No statistician dreams of combining objects into the same generic group that do not cluster towards a common centre; no more can we compose generic portraits out of heterogeneous elements, for if the attempt be made to do so the result is monstrous and meaningless" (p. 295).[27] The basic assumption of a type, or typical individual, is clear and explicit. They used the normal curve and they permitted distributions to be represented by an average because, even though at times they knew better, far too often they tended to think of races as discrete, even homogeneous, groups and individual variation as error.

It is important to realize that these developments began before 1900, when Mendel's work was still unknown. Thus at the inception of biosocial science there was no substantive basis for understanding individual differences. After

1900, when Mendel's work became available, its incorporation into biosocial science was bitterly opposed by the biometricians under Pearson's leadership. Galton had promulgated two "laws": his Law of Ancestral Heredity (1864)[27] and his Law of Regression (1877).[27] When Yule[35] and Castle[4] pointed out how the Law of Ancestral Heredity could be explained in Mendelian terms, Pearson[26] stubbornly denied it. Mendel had chosen for experimental observation seven traits, each of which, in his pea-plant material, turned out to be a phenotypic correlate of a single gene with two segregating alleles. For all seven traits one allele was dominant. Unfortunately Pearson assumed the universality of dominance and based his disdain for Mendelism on this assumption. Yule[36] then showed that without the assumption of dominance, Mendelism becomes perfectly consistent with the kind of quantitative data on the basis of which it was being rejected by Pearson. It is sad to realize that Pearson never appreciated the generality of Mendelism and seems to have gone on for the next 32 years without doing so.

Two Fallacies

Now we can consider the recent debate about the meaning of comparisons between the "intelligence" of different human races. We are told that intelligence has a high heritability and that one race performs better than another on intelligence tests. In essence we are presented with a racial hierarchy reminiscent of that pernicious "system" which John Haydon Langdon Down used when he misnamed a disease entity "mongolism."

The people who are so committed to answering the nature-nurture pseudo-question (Is heredity or environment more important in determining intelligence?) make two conceptual blunders. (1) Like Spalding's question about the instinctive nature of bird flight, which introduced the ethologist's deprivation experiment, their question about intelligence is, in fact, being asked about the development of a single individual. Unlike Spalding and the ethologists, however, they do not study development in single individuals. Usually they test groups of individuals at a single time of life. The proportions being assigned to heredity and to environment refer to the relative amounts of the variance between individuals comprising a population, not how much of whatever enters into the development of the observed expression of a trait in a particular individual has been contributed by heredity and by environment respectively. They want to know how instinctive is intelligence in the development of a certain individual, but instead they measure differences between large numbers of fully, or partially, developed individuals. If we now take into consideration the norm-of-reaction concept and combine it with the facts of genotypic individuality, then there is no general statement that can be made about the assignment of fixed proportions to the contributions of heredity and environment either to the development of a single individual, because we have not even begun to assess his norm of reaction, or to the differences that might be measured among members of a population, because we have hardly begun to assess the range of environmental conditions under which its constituent members might develop!

(2) Their second mistake, an egregious error, is related to the first one. They assume an inverse relationship between heritability magnitude and improvability by training and teaching. If heritability is high, little room is left for improvement by environmental modification. If heritability is low, much more improvement is possible. Note how this basic fallacy is incorporated directly into the title of Jensen's article "How much can we boost IQ and scholastic achievement?"[18] That question received a straightforward, but fallacious, answer on his page 59: "The fact that scholastic achievement is considerably less heritable than intelligence . . . means there is potentially much more we can do to improve school performance through environmental means than we can do to change intelligence. . . ." Commenting on the heritability of intelligence and "the old nature-nurture controversy" one of Jensen's respondents makes the same mistake in his rebuttal: "This is an old estimate which many of us have used, but we have used it to determine what could be done with the variance left for the environment." He then goes on "to further emphasize some of the implications of environmental variance for education and child rearing" (p. 419).[3]

High or low heritability tells us absolutely nothing about how a given individual might have developed under conditions different from those in which he actually did develop. Heritability provides no information about norm of reaction. Since the characterization of genotype-environment interaction can only be ad hoc and the number of possible interactions is effectively unlimited, no wonder the search for general laws of behavior has been so unfruitful, and *the* heritability of intelligence or of any other trait must be recognized as still another of those will-o-the-wisp general laws. And no magic words about an interaction component in a linear analysis-of-variance model will make disappear the reality of each genotype's unique norm of reaction. Such claims by Jensen or anyone else are false. Interaction is an abstraction of mathematics. Norm of reaction is a developmental reality of biology in plants, animals and people.

In Israel, the descendants of those Jews Pearson feared would contaminate Britain are manifesting some interesting properties of the norm of reaction. Children of European origin have an average IQ of 105 when they are brought up in individual homes. Those brought up in a Kibbutz on the nursery rearing schedule of 22 hours per day for 4 or more years have an average IQ of 115. In contrast, the mid-Eastern Jewish children brought up in individual homes have an average IQ of only 85, Jensen's danger point. However, when brought up in a Kibbutz, they also have an average IQ of 115. That is, they perform the same as the European children with whom they were matched for education, the occupational level of parents and the Kibbutz group in which they were raised (p. 420).[3] There is no basis for expecting different overall results for any population in our species.

Some Promising Recent Developments

The power of the approach that begins by thinking first in terms of the genetic system and only later in terms of the phenotype (or behavior) to be analyzed is now being demonstrated by an accumulating and impressive body of evidence. The rationale of that approach derives directly from the particulate nature of the gene, the mosaic nature of the genotype and the manner in which heredity breaks apart and gets reassembled in being passed on from one generation to the next. We now have a well-articulated picture of the way heredity is shared among biological relatives.

That madness runs in families has been known for centuries. The controversy has been over whether it was the heredity or the environment supplied by the family that was responsible for the madness. Franz Kallmann and some others collected large amounts of data in the 1940s and 1950s showing that monozygotic twins were much more concordant than dizygotic twins. Since David Rosenthal of NIMH has provided some of the best criticism of the incompleteness, and therefore inconclusiveness, of the twin-study evidence for the role of heredity in schizophrenia, Rosenthal's own recent findings become especially noteworthy.

He has divided foster-reared children from adoptive homes into two groups: those with a biological parent who is schizophrenic and those without a schizophrenic biological parent. If was found by Rosenthal,[31] and by Heston[15] in a completely independent but similar study, that the incidence of schizophrenia was much greater among the biological children of schizophrenics. Most significantly, combining the two studies, the risk of schizophrenia in offspring is four to five times greater if a biological parent is schizophrenic. Still other recent studies support the Rosenthal and the Heston findings. Both Karlsson[19] and Wender[30] found a high incidence of schizophrenia in the foster-reared relatives of schizophrenics.

Thinking genetically first in terms of biological relationship has already paid off in the analytical detail revealed as well as in the mere demonstration of concordance with respect to diagnostic category. Lidz and co-workers[20] reported marked distortions in communicating among many of the non-hospitalized parents of schizophrenic hospital patients. McConaghy,[22] using an objective test of thought disorder, assessed the parents of 10 schizophrenic patients and compared them to a series of control subjects. Sixty per cent of the patients' parents, including at least one parent in every pair, registered test scores in the range indicative of thought disturbance. In contrast, less than 10 per cent of the controls had such scores.

The major features of McConaghy's findings have since been replicated by Lidz and co-workers.[21] More recently Phillips and co-workers[29] studied 48 relatives of adult schizophrenics and 45 control subjects using a battery of tests to assess thought disorder. They found cognitive disorders to be much more frequent among the relatives of schizophrenics; 17 or 18 parents registered "pathological" scores, even though their social behavior had never been diagnosed as pathological.

In 1962, Anastasopoulos and Photiades[1] assessed susceptibility to LSD-induced "pathological reactions" in the relatives of schizophrenic patients. After studying 21 families of patients and 9 members of two control families, they reported "... it was almost invariable to find reactions to LSD in one of the parents, and often in one or more of the siblings and uncles and aunts, which were neither constant nor even common during the LSD-intoxication of healthy persons."

Analogous work has been done studying the responses of the relatives of patients with depressive disorders using anti-depressant drugs like imipramine (Tofranil) or an MAO inhibitor. Relatives tend to show a response pattern similar to that of their hospitalized relations.

Some very interesting human behavior-genetic analyses are currently being done on these affective disorders by George Winokur and his colleagues in St. Louis.[30] Out of 1075 consecutive admissions to a psychiatric hospital, 426 were diagnosed as primary affective disorders. So far, these appear to fall into two subtypes, the first of which shows manic episodes; some first-degree relatives show similar manifestations. The other subtype is characterized by depressive episodes and lack of concordance among close relatives. Furthermore, evidence is now accumulating implicating a dominant factor or factors on the X-chromosome in the manic subtype: (1) the condition is considerably more prevalent in females than in males; (2) the morbid risk among siblings of male probands is the same for males and females, but the morbid risk among siblings of female probands is quite different—sisters of female probands are at a 21 per cent risk while their brothers are only at a 7.4 per cent risk. More detailed study in several appropriately chosen family pedigrees suggests that there is a dominant gene on the short arm of the X-chromosome. The condition has so far shown linkage with color-blindness and the Xg blood groups, both of which are loosely linked on the short arm of the X-chromosome.

To examine the structure of the phenotypic variation in a trait whose development is in no obvious way influenced by environment and which, though ostensibly a simple trait, has been sufficiently well-analyzed phenotypically to reveal its interesting complexity, we have chosen to study dermatoglyphics, or fingerprints, in my laboratory. For his doctoral dissertation, R. Peter Johnson is making these observations on both parents and offspring in individual families. His preparatory survey of the previous literature revealed one study which reported data on a cross-sectional sample of 2000 males.[32] Scoring them on all ten fingers with respect to four distinguishable pattern types, the following data reveal the interesting but sobering complexity that exists in such a "simple" trait: the same type of pattern was shown on all 10 fingers by 12 per cent, on 9 of 10 fingers by 16 per cent, and on 8 of 10 fingers by 10 per cent of the men.

In addition, 5 per cent of the men showed all four pattern types. This included 1 per cent of the individuals who had all four pattern types on a single hand.

While probably everybody has heard that there are some unusual hospitalized males who carry two Y chromosomes, are rather tall, and prone to commit crimes of violence, few people know that when a comparison was made between the first-order relatives of both the Y-Y chromosome males and control males hospitalized for similar reasons (but not carrying two Y chromosomes), there was a far greater incidence of a family history of crime among the controls. In this control group there were over six times as many individual first-order relatives convicted and many, many times the number of convictions.

In summary, the relationship between heredity and behavior has turned out to be one of neither isomorphism nor independence. Isomorphism might justify an approach like naive reductionism, independence a naive behaviorism. Neither one turns out to be adequate. I believe that in order to study behavior, we must understand genetics quite thoroughly. Then, and only then, can we as psychologists forget about it intelligently.

References

1. Anastasopoulos, G., & Photiades, H. Effects of LSD-25 on relatives of schizophrenic patients. *Journal of Mental Science,* 1962, **108**, 95-98.
2. Benda, C. E. "Mongolism" or "Down's syndrome." *Lancet,* 1962, **1**, 163.
3. Bloom, B. S. Letter to the editor. *Harvard Educational Review,* 1969, **39**, 419-421.
4. Castle, W. E. The laws of heredity of Galton and Mendel, and some laws governing race improvement by selection. *Proceedings of the American Academy of Arts and Sciences,* 1903, **39**, 223-242.
5. Dewey, J. Correspondence with Robert V. Daniels, 15 February, 1947. *Journal of Historical Ideas,* 1959, **20**, 570.
6. Down, J. L. H. Observations on an ethnic classification of idiots. London Hospital Reports, 1866. In V. A. McKusick (Ed.), *Medical genetics,* 1961. *Journal of Chronic Diseases,* 1962, **15**, 417-572.
7. Dunn, L. C. *A short history of genetics.* New York: McGraw-Hill, 1965.
8. Fisher, R. A. Limits to intensive production in animals. *British Agricultural Bulletin,* 1951, **4**, 217-218.
9. Fuller, J. L. & Thompson, W. R. *Behavior Genetics.* New York: Wiley, 1960.
10. de Gobineau, J. A. Essai sur l' inégalité des races humaines. Rééd. intégrale en 1 vol., avec une préface de Hubert Juin. Paris: P. Belfond, 1967.
11. Goldschmidt, R. B. *Theoretical genetics.* Berkeley: University of California Press, 1955. P. 257.
12. Haldane, J. B. S. The interaction of nature and nurture. *Annals of Eugenics,* 1946, **13**, 197-205.
13. Hegmann, J. P. & DeFries, J. C. Open-field behavior in mice: Genetic analysis of repeated measures. *Psychonomic Science,* 1968, **13**, 27-28.

14. Heiden, K. *Der Führer.* London: Houghton, 1944. P. 198. Cited in A. Bullock: *Hitler: A study in tyranny.* Harmondsworth: Penguin, 1962. P. 80.
15. Heston, L. L. Psychiatric disorders in foster home reared children of schizophrenic mothers. *British Journal of Psychiatry,* 1966, **112**, 819-825.
16. Hirsch, J. Behavior genetics and individuality understood: Behaviorism's counterfactual dogma blinded the behavioral sciences to the significance of meiosis. *Science,* 1963, **142**, 1436-1442.
17. Jensen, A. R. Estimation of the limits of heritability of traits by comparison of monozygotic and dizygotic twins. *Proceedings of the National Academy of Science,* 1967, **58**, 149-156.
18. Jensen, A. R. How much can we boost IQ and scholastic achievement? *Harvard Educational Review,* 1969, **39**, 1-123.
19. Karlsson, J. L. *The biologic basis of schizophrenia.* Springfield, Ill., Charles C. Thomas, 1966.
20. Lidz, T. Cornelison, A., Terry, D., & Fleck, S. Intrafamilial environment of the schizophrenic patient: VI. The transmission of irrationality. *AMA Archives of Neurological Psychiatry,* 1958, **79**, 305-316.
21. Lidz, T., Wild, C., Schafer, S., Rosman, B., & Fleck, S. Thought disorders in the parents of schizophrenic patients: A study utilizing the Object Sorting Test. *Psychiatric Research,* 1962, **1**, 193-200.
22. McConaghy, N. The use of an object sorting test in elucidating the hereditary factor in schizophrenia. *Journal of Neurology, Neurosurgery and Psychiatry,* 1959, **22**, 243-246.
23. McGaugh, J. L., Jennings, R. D., & Thomson, C. W. Effect of distribution of practice on the maze learning of descendants of the Tryon maze bright and maze dull strains. *Psychological Reports,* 1962, **10**, 147-150.
24. McKusick, V. A. *Mendelian inheritance in man: Catalogs of autosomal dominant, recessive, and X-linked phenotypes.* Baltimore: Johns Hopkins Press, 1966.
25. Pastore, N. *The nature-nurture controversy.* New York: King's Crown Press (Columbia University), 1949.
26. Pearson, K. On a generalized theory of alternative inheritance, with special reference to Mendel's laws. *Philosophical Transactions of the Royal Society of London,* 1904, **A203**, 53-86.
27. Pearson, K. *The life, letters and labours of Francis Galton.* Vol. II. *Researches of middle life.* Cambridge: Cambridge University Press, 1924.
28. Pearson, K., & Moul, M. The problem of alien immigration into Great Britain, illustrated by an examination of Russian and Polish Jewish children. *Annals of Eugenics,* 1925, **1**, 5-127.
29. Phillips, J. E., Jacobson, N., & Turner, W. J. Conceptual thinking in schizophrenics and their relatives. *British Journal of Psychiatry,* 1965, **111**, 823-839.
30. Rose, R. J., Department of Psychology, University of Indiana, private communication, 1969.

31. Rosenthal, D., Wender, P. H., Kety, S. S., Schulsinger, F., Welner, J., & Ostergaard, L. Schizophrenics' offspring reared in adoptive homes. *Journal of Psychiatric Research,* 1968, **6**, 377-391.
32. Waite, H. Association of fingerprints. *Biometrika* 1915, **10**, 421-478.
33. Watson, J. B. *Behaviorism.* Chicago: University of Chicago Press, 1959.
34. Wittgenstein, L. *Philosophical investigations.* (2nd ed.). Translated by Anscombe, G. E. Oxford: Blackwell, 1963. P. 232.
35. Yule, G. U. Mendel's laws and their probable relation to intra-racial heredity. *New Phytologist,* 1902, **1**, 193-207, 222-238.
36. Yule, G. U. On the theory of inheritance of quantitative compound characters on the basis of Mendel's laws—A preliminary note. Report of the third International Conference on Genetics, 1906, 140-142.

Recommended Additional Readings

Dobzhansky, T. Of flies and men. *American Psychologist,* 1967, **22**, 41-48.
Hirsch, J. Behavior-genetic, or "experimental," analysis: The challenge of science versus the lure of technology. *American Psychologist,* 1967, **22**, 118-130.
McClearn, G. E.s Genetic influences on behavior and development. In P. H. Mussen (Ed.), *Carmichael's manual of child psychology.* (3rd ed.) Vol. 1. New York: Wiley, 1970. Pp. 39-76.

8   **Intelligence, Intelligence Tests, Intellectual Differences, and Intellectual Development**

L. Joseph Stone
Joseph Church

Orientation

From *Childhood and Adolescence,* 2nd Edition, by L. Joseph Stone and Joseph Church, Copyright ©1957, 1968 by Random House, Inc. Reprinted by permission of the authors and the publisher.

In the next article Stone and Church present a perceptive—and very personal—analysis of the concept of intelligence and other issues related to it, including the persistent nature-nurture question, individual differences, and intellectual development. It will be plainly evident to the reader that these authors do not view the psychometric (i.e., test-oriented) view of intelligence very kindly; in fact, they decry the negative effects that the widespread deployment of this concept has had upon schools in particular and society in general.

Notions of "intelligence" and the use of tests to "measure" whatever intelligence might be have become incredibly well-entrenched in the American public school system (and also that of Great Britain). Paradoxically, France, the native country of the creator of intelligence tests, Alfred Binet, has been most reluctant to adopt these tests into its educational system. Binet's tests received the most enthusiastic reception in the United States, mostly through the work and influence of Lewis M. Terman of Stanford University.

Stone and Church propose the novel suggestion that we abandon the use of the noun, "intelligence"; instead, they recommend the use of the adjective, "intelligent," and apply it to person-environment relations. Thus, we can speak of intelligent environments as well as (or rather, in place of) intelligent persons. From this point of view, education is redefined to be the process of giving a child an intelligent environment.

The reader is strongly encouraged to consider carefully the critique of the concept of intelligence and related educational practices that Stone and Church have so eloquently presented here. The American populace, as well as the American system of education, has uncritically accepted the notion of intelligence, almost as if it were a concrete reality residing somewhere inside one's body (not unlike the notions of "soul" and "conscience"). Perhaps someday soon we will adopt different and more constructive ways to interpret the functioning of the human organism in substitution for the traditional approach of explaining away his behavior according to how much "intelligence" he possesses.

L. Joseph Stone is a professor of psychology at Vassar College. Joseph Church is associated with the graduate center of the City University of New York. They have coauthored the popular textbook, Childhood and Adolescence: A Psychology of the Growing Person, *now in its second edition (New York: Random House, 1968).*

8

Now, when the child has moved out of the natural environment of the family into the seminatural habitat of the preschool and is on the threshold of formal school, seems an appropriate time to talk about intellectual diversity, the fact that some people are smarter or dumber than others. The fact of intellectual diversity is not to be disputed. The debate centers around the ever persistent nature-nurture controversy, where intellectual differences come from. This is an important issue in terms of practical action. One's willingness to invest large sums of money in school depends on how much faith one has that education can bring about important changes in cognitive functioning. One's attitudes toward racial integration depend on the extent to which one accepts the debased social and intellectual status of Negroes—or of the poor in general—as a product of inferior biological endowment or of the degrading conditions in which they have been forced to live. A proper scholarly discussion of the nature-nurture issue would require a book in itself, and indeed has been subject of several recent volumes, such as Hunt's.[1] We can only outline the course such an argument would take, referring the reader to the relevant literature, and put forth a few assertions which the reader can accept or reject— but, we hope, on the basis of careful consideration.

Let us say that we are in general agreement with the view that intellectual development represents an interaction between genetic and biological factors and experience, the physical and social settings and events with which a person grows up. But let us caution that we know very little about the biological side of this interaction. Apart from the organic pathologies that affect some 10 per cent of the population, very little is known about the constitutional correlates of intelligence. The only reasonably clear evidence about differences in brain function associated with differences in learning ability comes from studies of rats and mice, and here the most impressive evidence is on central nervous system differences that have been produced by differences in experience.[2] We are also coming to know something about differences in brain structure and function produced by dietary factors, as in the case of protein deficiency, and by the action of hormones. What is missing from our picture of the constitutional correlates of intelligence is the role of the genes in controlling intellectual development. All the evidence makes clear that there is no single gene for intelligence. Whatever the constitutional differences may be that make for differences in intellectual ability, they must be produced by the action of a multiplicity of genes. We must also be aware that intelligence expresses itself in a variety of ways, and the various forms of intelligence may represent the action of very different genes and gene combinations.

A full discussion of the matter would require an analysis of the instruments by which intelligence is assessed, the so-called intelligence tests. Most of the common instruments, like the Wechsler Intelligence Scale for Children, the Stanford-Binet, the Otis Self-Administering, or the Kuhlmann-Anderson, are described in textbooks on tests, measurements, and individual differences. The tests have been around for so long and have been used so widely that people have come to take them for granted as valid measuring instruments.

When one looks closely, however, as writers such as Hunt, Anastasi, Church, Fowler, Sigel, Spiker and McCandless, Stott and Ball, and Tyler[3] have done, the tests are almost pathetically flimsy tools with which to study an important topic like intellectual ability. They are open to both empirical and logical criticism on many grounds: lack of independent validation, inadequate standardization, cultural bias, conventionality of content, errors of mathematical logic, lack of any theory or even a good definition of what it is that they are supposed to measure, instability of individual performance from one testing to another, and untested assumptions about how intelligence is distributed. It is a curious bit of intellectual and cultural history that the tests have become as solidly entrenched as they have without the test-makers ever responding in any serious way to the damning criticisms that have been made. Particularly objectionable has been the indiscriminate use of tests for school placement, so that young children who score at different levels on the test are assigned to different educational tracks, which raises suspicions of self-fulfilling prophecies. This is particularly true of the use of simple, overall, summated tests scores, like the IQ. Individual tests are nonetheless valuable instruments for trained observations of the cognitive processes, if only because each trained clinician has seen hundreds of children perform in this setting and can therefore readily recognize what is individual and diagnostic for understanding a child's ways of learning or note disturbances in his development.

For a complete discussion of intelligence and intellectual diversity, we should also have to examine the indirect evidence for a biological view of behavior. This evidence would include Tryon's selective breeding of maze-bright and maze-dull strains of rats;[4] comparisons of the intelligence of identical twins raised apart with the intelligence of identical twins raised together, of identical twins with fraternal twins and nontwin siblings; and comparisons of the correlations of intelligence in adopted children with that of their true parents and adoptive parents.[5]

The findings from all such studies support a hereditarian view of intelligence, but all have met with environmentalist criticism. For instance, Searle[6] has concluded that the difference between Tryon's two rat strains is a difference of temperament rather than intelligence (although the measurement of animal intelligence is as confused an undertaking as the measurement of human intelligence). We must note also that most animal studies have omitted the vital environmental control of cross-fostering. This technique swaps newborn litters between mothers of different genetic strains to insure that the behavior of the young is due to genetic causes rather than behavior learned through being raised by their own mothers. When the data from the most famous of the twin studies, which compared identical twins raised apart, by Newman, Freeman, and Holzinger,[7] are regrouped, as we have done in Table 1, to take account of psychological dissimilarity between the physically separate environments in which the members of twin pairs were raised, it can be shown that environment did in fact exert an influence on the twins' intellectual development.

Decreasing similarity in IQ's as we go from identical twins raised together to identical twins reared apart to fraternal twins to siblings in general can be seen to be associated with decreasingly similar environments. Even fraternal twins, because of their differing stimulus characteristics, can provoke quite different reactions from people around them and so in principle can inhabit very different psychological environments even while living in the same physical one. The critic has no ready answer to the finding that foster children often (although by no means invariably) are more like their real parents than their adoptive parents in measured intelligence, except to say that nobody has yet sufficiently studied the psychology of adoption and the possible problems of identification involved. In addition, those who are interested in possible environmental influences on intelligence point to positive evidence of the kind discussed in the next paragraph.

Most of our knowledge of environmental effects on intelligence comes from studies of infrahuman species, although Wellman conducted a still-controversial series of studies which seemed to indicate that nursery-school experience of the right sort could markedly raise a child's IQ, and we are beginning to get highly

Table 1
Comparison of Identical Twins in Terms of Differences in
IQ and Differences in Educational and Social Advantages (DESA)

	Large DESA	Small DESA
IQ differences < 10	1, 4, 7	1, 1, 1, 2, 2, 5, 6, 8, 9
IQ differences > 10	10, 12, 12, 15, 17, 19, 24	—

Source:
Adapted from Newman, H. H., Freeman, F. N., & Holzinger, K. J. *Twins.* Chicago: University of Chicago, 1937; and Johnson, R. C., Similarity in IQ of separated identical twins as related to length of time spent in same environment. *Child Development,* 1963, 34, 745-749.

optimistic returns from studies of the effects of Head Start programs.[8] At the animal level, we have such studies as that by Forgays and Forgays[9] on the behavioral consequences of growing up in rich or dull environments, and analogous studies by C. J. Smith[10] on brain localization and by Bennett, Krech, Rosenzweig, et al.,[11] on brain biochemistry. The impaired intellectual functioning of institution-reared children could also be used to show the effects of environment, except that we have no possible way of knowing that such children would have been brighter had they grown up in more favorable circumstances. Skeels,[12] however, in a thirty-year follow-up study, found dramatic differences in the life histories of thirteen "retarded" children given enriched experience, by contrast with twelve others raised in a standard institutional environment.

Sayegh and Dennis [13] found differences in institution children given only three weeks of supplementary experience. Those concerned with environmental influences would say further that we must also take account of the prenatal environment and its possible role in developing the person's intellectual ability. Wellman was widely attacked on statistical and logical grounds, but in general few psychologists nowadays uphold a strongly hereditarian view of intelligence (at least in public), and those who do, like Hirsch, [14] insist that they are interested only in population distributions rather than individuals.

It seems safe to conclude from the available evidence that we do not know much about the relative contributions of constitution and environment to intellectual development or how, concretely, they interact to produce a spread of intelligence. Pragmatically, our ignorance is not important, since we cannot do much to manipulate heredity anyway (although we do know how to overcome such hereditary defects as phenylketonuria and diabetes, and although there are those who claim that the hereditary strings are almost in our hands), and we do know that environment can have a large impact on intellectual development. There are two logical possibilities: One could, by particular educational arrangements, either squeeze almost everybody into a narrow range of high intelligence; or simply displace the present distribution curve upward as a whole, so that the new average IQ of 100 would correspond to a higher level than it now does. Either of these outcomes, it seems to the authors, would be socially desirable.

Nevertheless, it is unsatisfactory, if only because we want to advance our understanding, to be purely atheoretical and pragmatic in our thinking about intelligence. We believe that some of the confusion on this subject comes about because people have tended to localize intelligence in the organism, or even more narrowly in the central nervous system, as though it were some kind of mechanism or structure. Instead, following Stoddard, [15] we would like to propose abandoning the noun "intelligence," which invites reification, the translation of a convenient hypothetical construct into an entity, and substitute the adjective "intelligent" as a characteristic of organism-environment relations. In this view, it makes just as much sense to talk about more or less intelligent environments as about more or less intelligent people. Intellectual development in this view becomes the progressive discovery of reality, and the task of education to give the child an intelligent—differentiable, thinkable, manipulable—environment.

Our transactions with the environment—which includes, let us remember, the natural and man-made physical world, animals, people, things, institutions, and symbols, and all the meanings, sounds, odors, and demand qualities that emanate from them—are obviously conditioned on the way our environment appears to us, the way our personal life space is constituted. Out of our cultural learning and our personal experience we come to perceive the world in diverse ways, so that different individual life spaces have different organizations of what stands out as figure and what blends into the background, different possibilities for action and manipulation, different causal dynamics, and in sum, different patterns of meaning. The *unintelligent life space* is the sphere of primitive experience into which we are all born and which is perceived egocentrically, dynamistically, phenomenalistically, and magicalistically, activated by psychic and spiritual forces of a kind unknown to science. The psychologically primitive environment permits full play to superstition and is populated with imagined beings and entities, from poltergeists to the miasma to "intelligence." Such an environment is phrased in tradition-bound formulas, platitudes, axioms, precepts, taboos, myths, and slogans. All action has a moral component and must be accompanied by the proper ritual observances lest it misfire. The primitive environment invites acceptance and submission and discourages invention and innovation. By contrast, the *more intelligent life space* is one in which primitive forces are subordinate to more orderly and manageable ones, as generally known in the educated parts of our society, so that magic gives way to science, we look beneath the surface of things to see what makes them tick, and we evolve sets of principles applicable to various domains of reality. We learn to distinguish between moral and morality-free spheres of action and to emphasize the pragmatic rather than the ritual. We invent new—and sometimes better—ways of doing things, we learn to exploit natural forces, we let machines do our work for us (including a certain amount of our thinking). In the arts, we experiment with novel combinations of light and sounds and words and meanings. We search for new, largely secular formulations of the meaning of it all. We invent new social doctrines. Most radically, we look for ways of altering and improving human nature, through child-rearing practices, education, drugs, surgery, and psychotherapy.

Let us note that even the most emancipated of us are still hobbled by superstitions, dogmas, irrational convictions and intuitions, rigidities, and some imperviousness to evidence, but we have moved away from the Stone Age and the Middle Ages, from witch trials and public executions, from mass manias and holy crusades. We still have a considerable way to go, as one can read in the headlines, but we have made a beginning. In this interplay of person and environment, the part that seems to remain most stubbornly as in the person is language and symbols, and the environment can be no more intelligent than the symbolic system with which the person can manipulate facts and relationships intellectually, in thought.

If, as we propose, we now *consider education to be a process of giving the child an intelligent environment,* one whose workings and possibilities and impossibilities he understands, we must take account of our society's profound ambivalence about intellectual development and education. For it is only some fraction of the community that has worked its way into the light of modern understanding, and that fraction is none too well represented in the seats of power. On the whole, many parents are more interested in whether the child is "doing well," in the sense of getting good marks, than in what and whether he is learning and whether *he finds learning exciting.* This duality is also expressed in the contradictory attitudes communicated implicitly to boys and to girls. Boys are expected to do well, but it is assumed that they will find intellectual activity unmanly and will dislike school. Girls are expected to be more docile and to accept the schooling process, but they are assumed not to be capable of serious intellectual achievement, which in any case is viewed as irrelevant and perhaps even inimical to their eventual feminine role. In fact, girls do learn better in school than boys, partly at least because the schools are run by women and offer an effeminate, prettified curriculum.

Equally important is the distinction that can be drawn between safe and unsafe areas of learning. We can see the various school subject matters as a spectrum running from the skill subjects (reading, writing, and arithmetic) through the physical sciences, the biological sciences, the social sciences, the arts and humanities, to philosophy and metaphysics. Branching off from this array are the applied fields: engineering, agriculture, medicine, social work and so on. We encourage the child in mathematics and the natural sciences, particularly if he is a boy, just as we welcome new knowledge in these fields. We smile upon the learned technologies such as engineering and medicine, but as we move away from this region of the spectrum, learning and logic begin to clash both with established social doctrines and with our emphasis on what is practical, manly, and uncontroversial. Even the generally esteemed, not to say revered, field of medicine becomes controversial in such areas as birth control and surgical abortion. The applications of chemistry and biology in agriculture become controversial when they raise economically important questions about pollution of the natural environment. In biology itself, the hypothesis of evolution still arouses distrust and antagonism and its teaching is still outlawed in several of the United States. *Sex education* is taught in many school systems, but most courses stop short of the psychology, sociology, and ethics of sex. Conservatism and conventionality restrict thinking in many fields, and it is extremely easy for the child to acquire a conventional outlook, beginning in the preschool years. Even when young people realize that the adult world has not dealt with them honestly—as when adolescent disillusionment sets in—they lack the knowledge necessary to arrive at reasonable formulations of their own.

As educators, we have to bear in mind that intellect is not purely intellectual: We sometimes become so preoccupied with the symbolic tools of thought that we neglect the factual substance of knowledge, including the reaches of human virtue and evil, reason and stupidity, nobility and pettiness, plasticity and fixity, sensibility and opacity, passion and prejudice. As a rule, schools do not discuss the ethical demands of modern life. Since most children, prior to entering school, have grown up among adults who live in a twilight zone between primitive and sophisticated thought, some part of an intellectual education has to be devoted to dis-educating children of errors they have learned, to divesting them of the clichés, the slogans, the platitudes, the dogmas, and the unspoken assumptions that dominate much of conventional thinking. A free and rich preschool education—whether in school or at home—which encourages questioning and deliberately avoids dogmatism seems to be a vital part of the total span of intellectual development.

Notice, however, that giving the child an intelligent life space also means giving him an intelligent view of himself. For every differentiation and higher-order integration of external reality implies a corresponding differentiation and integration of the person himself, a greater freedom to think, to act, to know himself, to become involved, to care, to be committed, and to wonder. This, in sum, is what we mean by being intelligent.

References

1. Hunt, J. McV. *Intelligence and experience.* New York: Ronald Press, 1961.
2. Bennett, E. L. Diamond, M. C., Krech, D., & Rosenzweig, M. R. Chemical and anatomical plasticity of brain. *Science,* 1964, **146**, 610-618; Levine, S., Mullins, R. F. Hormonal influences on brain organization in infant rats. *Science,* 1966, **152**, 1585-1592; Smith, C. J. Mass action and early environment in the rat. *Journal of Comparative and Physiological Psychology,* 1959, **52**, 154-156.
3. Hunt, *op. cit.,* and Anastasi, A. Heredity, environment, and the question "How?" *Psychological Review,* 1958, **65**, 197-208; Church, J. *Language and the discovery of reality.* New York: Random House, 1961. Chapter 7; Fowler, W. Cognitive learning in infancy and early childhood. *Psychological Bulletin,* 1962, **59**, 116-152; Sigel, I. E. How intelligence tests limit understanding of intelligence. *Merrill-Palmer Quarterly of Behavior and Development,* 1963, **9**, 39-56; Spiker, C. C., & McCandless, B. R. The concept of intelligence and the philosophy of science, *Psychological Review,* 1954, **61**, 255-266; Stott, L. H., & Ball, R. S. Infant and preschool mental tests. *Monographs of the Society for Research in Child Development,* 1965, **30**, no. 3; Tyler, L. E. *Tests and Measurements.* Englewood Cliffs, N. J.: Prentice-Hall, 1963, Chapter 4. For a striking recent study of self-fulfilling prophecies, see Rosenthal, R. *Pygmalion in the classroom.* New York: Holt, Rinehart and Winston, 1968.
4. Tryon, R. C. Genetic differences in maze-learning ability in rats. *Thirty-ninth Yearbook of the National Society for the Study of Education,* 1940, 111-119.
5. These studies are reviewed in Tyler, L. E. *The Psychology of Human Differences.* New York: Appleton, 1965. Chapters 17 and 18.
6. Searle, L. V. The organization of hereditary maze-brightness and maze-dullness. *Genetic Psychology Monographs,* 1949, **39**, 279-325.
7. Newman, H. H., Freeman, F. N., & Holzinger, K. J. *Twins.* Chicago: University of Chicago Press, 1937; Johnson, R. C. Similarity in IQ of separated identical twins as related to length of time spent in same environment. *Child Development,* 1963, **34**, 745-749.
8. Crowell, D. C. Shiro, L. K., Cade, T. M., Landau, B., & Bennett, H. L. Progress report: Preschool readiness project. Honolulu: University of Hawaii, 1966. (Mimeographed); Kamii, C. K., Radin, N. L., & Weikart, D. P.

A. A two-year preschool program for culturally disadvantaged children. Paper read at meetings of American Psychological Association, 1966; Gray, S. W., & Klaus, R. A. An experimental preschool program for culturally deprived children. *Child Development,* 1965, **36**, 887-898.
9. Forgays, D. G., & Forgays, J. W. The nature of the effect of free environmental experience in the rat. *Journal of Comparative and Physiological Psychology,* 1952, **45**, 322-328.
10. Smith, *op. cit.* [See ref. 2.]
11. Bennet, *et al., op. cit.* [See ref. 2.]
12. Skeels, H. M. Adult status of children with contrasting early life experiences. *Monographs of the Society for Research in Child Development,* 1966, **31**, no. 3.
13. Sayegh, H., Dennis, W. The effect of supplementary experiences upon the behavioral development of infants in institutions. *Child Development,* 1965, **36**, 81-90.
14. Hirsch, J. Behavior genetics and individuality understood. *Science,* 1963, **142**, 1436-1442.
15. Stoddard, G. D. *The meaning of intelligence.* New York: Crowell-Collier-Macmillan, 1943.

"Don't you understand? This is _life_, this is what is happening. We _can't_ switch to another channel."

Drawing by Robt. Day; ©1970 The New Yorker Magazine, Inc.

9 Social Structure and the Socialization of Competence

Alex Inkeles

Orientation

Reprinted from *Harvard Educational Review*, 36, Summer 1966, 265-283, by permission of the author and the publisher. Copyright © 1966 by President and Fellows of Harvard College. Prepared for a Conference on the Socialization of Competence sponsored by the Committee on Socialization, Social Science Research Council, meeting in Puerto Rico April 29—May 1, 1965. M. Brewster Smith summarized the results of the Conference in *Items*, published by the Social Science Research Council, Vol. XIX, (June 1965). The work of the Committee on Socialization is supported by the National Institutes of Mental Health (Grant #MH 4 160), whose aid is gratefully acknowledged.

In the following article the focus of attention is upon the concept of competence, *which refers to one's ability to acquire and perform successfully important kinds of behaviors. In the previous article Stone and Church (II.8) strongly urged the adoption of terms other than "intelligence" in our attempts to describe and understand the functioning of individuals. The concept of competence represents an especially useful alternative.*

Almost everyone—regardless of age, socioeconomic status, physical or mental condition—possesses a tremendous variety of skills and aptitudes that enable him to function quite satisfactorily within the environment in which he lives. For the young child, these might include the attainment of the ability to manipulate such "behavioral objects" as doorknobs, shoestrings, and spoons. And as the individual gets older, of course, his range of specific competences enlarges enormously. A problem for many adults, however, is that the tremendously complex American society demands that, to be "successful," in the traditional sense, one must possess a great number of specific skills; as Inkeles points out, these include the telling and management of time, language skills, arithmetic skills, and possession of all kinds of specific information. To this list might be added the necessity of being competent in filling out forms of all kinds. Even a small amount of reflection on the part of the reader will reveal the incredible importance of form-filling-out behaviors in our culture; an individual is confronted with forms and blanks—often rather formidable ones—for everything from applying for admission to college to requesting food stamps. The situation is complicated even more by the fact that our technological society is changing at an amazing pace; skills developed now may be disfunctional tomorrow, and new situations continually demand new skills. Perhaps we shall have to develop "Head Start" type programs for adults to assist them in keeping pace with the dizzying rate of change.

The Stone and Church article and the Inkeles

article, taken together, present us with a view that there are better, more beneficial ways to view the abilities of people than the traditional method of applying tests that measure that nebulous notion of "intelligence." The competence approach takes a more optimistic perspective: It looks at how successfully—and even creatively—the individual has fit himself into the realities of his environmental situation.

Alex Inkeles has been associated with the center for international affairs at Harvard University and with the school of education at Stanford University.

9

I will define competence as the ability effectively to attain and perform in three sets of statuses: those which one's society will normally assign one, those in the repertoire of one's social system one may appropriately aspire to, and those which one might reasonably invent or elaborate for oneself. In contrast to socialization, then, the concept of competence stresses the end-product, the person as he is *after* he has been socialized, rather than the formative process itself. This conception is also broader than that of socialization in that the latter usually is defined with reference to a fixed repertoire of roles provided by a given socio-cultural system, whereas competence is here defined to include an individual's capacity to move to *new* statuses and to *elaborate* new roles. Despite these differences, however, the concepts socialization is to produce competent people, as competence is defined in any given society. It aims to develop a person who can take care of himself, support others, conceive and raise children, hunt boar or grow vegetables, vote, fill out an application form, drive an auto, and what have you.

As soon as we specify some of these qualities, it becomes evident that the research on socialization in our scientific literature has little to say about these matters. Research on socialization addresses itself predominantly to understanding how the child learns to manage his own body and his primary needs. It inquires mainly how the child is guided in learning to manage the intake of food, the discharge of waste, and the control of sexual and aggressive impulses. Except for the rather isolated, even if highly interesting, forays in the direction of studying modes of moral functioning,[1] little is done in socialization research to study the acquisition of a broad array of qualities, skills, habits, and motives, which are essential to the adequate social functioning of any man or woman and in fact occupy the great bulk of the time of all socializing agents. A discussion of competence provides an opportunity to correct that imbalance in some small degree. I propose, therefore, to emphasize some of the qualities of individuals which are of most interest to society (and, incidentally, the focus of a good part of its socialization effort after infancy) but which seem largely to have escaped systematic study by students of socialization.

Two paths are open to me. One would be to list demands on the individual typically made by society, followed by specification of the requisite personal qualities these demands assume and of the socialization patterns presumed to engender these "socially demanded" personality dispositions. While this might be manageable if I were dealing with a particular stratum of a single society, it is otherwise too large and diffuse a perspective for the limited space at my disposal. I have therefore chosen to follow a second path, that of presenting a model of the personal system, essentially an accounting scheme of the elements of personality, broadly conceived, which I have found highly serviceable in all my efforts systematically to relate personality to social structure. The elements of this scheme will then serve me as indicators, pointing to more specific personal attributes I consider relevant to a discussion of competence.

A Model of Personality and Its Implications for Research on the Socialization of Competence

In this section, I will present a model of the personality, one which represents essentially an accounting scheme rather than a theory.[2] I make no claim for originality in this scheme, but rather emphasize its practical usefulness as a framework for the discussion of personality in those researches in which interest is centered not in intra-personal dynamics but rather on interaction between the person and the socio-cultural system. I have used the scheme here, as elsewhere, mainly as a way of organizing the discussion of a social issue which we assume cannot be dealt with adequately unless we take systematic account of personality.[3] In this case, the theme is competence to perform social roles. Each element in the scheme will serve as an opportunity to point up an aspect on the "social demand" side of the equation which requires certain personal system attributes in the incumbents of social statuses. Further, to keep the discussion focussed, I approach competence mainly as a requirement for participation in contemporary and "modern" urban industrial settings. And rather than attempt any thorough coverage of the relevant issues, I have concentrated my energies on pointing to what seem to me important neglected opportunities in socialization research.

One of the virtues of the model is that it encompasses, in a decidedly limited set of some twelve to fifteen major headings, a great deal of what is ordinarily included in "personality." Even within that restricted frame, space limitations require that I forgo altogether discussion of most elements of the model. My objective here is not to be exhaustive, but rather, by suggestion and illustration, to open up a discussion. The following list presents the scheme in full, and asterisks indicate those elements of the model which are actually taken up in the subsequent presentation.

The scheme, I repeat, is arbitrary, meant to serve as an accounting device. This applies not only to the main entries but also to the sub-systems into which I

An Accounting Scheme for Personality Study

Psychomotor System	Temperament *Aptitudes *Skills
Idea System	*Information Opinions and Attitudes
Motivational System	Values *Motives and Needs
Relational System	Orientation to Authority Figures Orientation to Intimates and Peers Orientation to Collectivities
Self System	Conceptions of Self Modes of Defense Modes of Moral Functioning
Modes of Functioning	*Cognitive Modes Affective Modes Conative Modes

* Indicates an element discussed in the text.

have suggested they may be grouped. But I hope that in the discussion which follows, the utility of the scheme will nevertheless be manifest. As already indicated, because of space limitations I have in the section which follows undertaken to discuss only five of the sixteen main entries.

Aptitudes

By an aptitude I mean an innate capacity or potential capacity to perform exceptionally well specific and difficult acts of sensing, muscular coordination, or the like. I include this heading more for the sake of completeness than out of a conviction that it has a definite relevance to the study of social competence. At the present time we are not at all sure which, if any, qualities meet the test of biological distinctiveness as special rather than general aptitudes characterizing all men more or less equally. Musical ability very likely does, manual dexterity and coordination may. Intelligence seems to be the clearest case, although some will question all the evidence yet available. Of course, at the extremes of the distribution, there may be nothing very problematical here. A child with impaired brain functioning is clearly not competent to learn more than the most rudimentary forms of social behavior and can never become a full participant in his society. Such a person may be defined legally as incompetent and therefore may be barred from exercising any of the formal rights allowed to most individuals in our society, such as rights to hold property, to vote, to conduct vehicles on the public thoroughfares, or to choose one's place of residence or occupation. Yet an adult male moron or imbecile may be perfectly competent to impregnate some female. So the *physical* capacity, drive, and coordination aspect of competence—the aptitude if you wish—must be clearly discriminated from the social and legal definition of competence.

Large numbers of our citizens have been and are defined, by the powers that be, as legally incompetent to participate in society on the basis of their alleged lack of aptitude. I have in mind not only the 600,000 who occupy our mental hospitals, but also much of the Negro population of the American South and a great many of the American Indians. Not unlike many tribal and colonially dominated people in other parts of the world, they have often been deemed by authorities as incompetent to exercise the rights of citizenship and in other ways conduct their own affairs. This legal definition of incompetence has frequently been justified on the grounds that mentally and psychologically, the Negroes affected are incapable of managing their own affairs, presumably because endowed by our maker with insufficient innate intelligence or social maturity. But in our Northern cities, the argument takes a rather different form. There we can encounter many social workers, often motivated by the highest ideals, who believe that the disadvantaged minorities are basically equipped by nature as are other men. Yet large numbers of these social workers in Harlem and Brooklyn have come to the conclusion that many of their clients are not able to manage their own lives and will likely be more or less permanent wards of the State or of private welfare agencies. If we hold fast to the belief that this is not due to *innate* defects, i.e., to lack of aptitude, then we must ask: what is lacking in their clients and what produced this condition? Presumably we may find the clue in some other aspect of the personal system.

Skills

Skills are aptitudes which have been trained or developed in accord with some cultural pattern. In other words, a skill is a socialized aptitude. Skills may therefore be based on the special aptitudes such as musical or mathematical ability, or on the more general aptitudes for muscular coordination, or for learning and using language, which virtually everyone possesses. The essential condition of the social groups and individuals who perform at a disadvantage in our society is their lack of the primary social skills. Our disadvantaged minorities are disadvantaged not only in winning so small a share of the available goods, services, and psychic rewards, but precisely because they so often lack the specific skills which could enable them to win a larger, or at least more adequate, share.

The main facts are today so well known that they have become almost clichés. While the end results, as reflected in low IQ scores, abysmal performance on aptitude tests, and consistent failure in the classroom, are well known, we must admit that students of socialization have done little to study the process which presumably yields this low capacity, especially as the process unfolds in the home, on the street, and in the primary grades of the classroom.

Of course, as we begin to explore these antecedent conditions, we may be led to re-examine our impressions of the actual pattern of skill deficiencies. The Harlem boy and girl may have an extremely meager vocabulary and very little ability to manipulate concepts—but are they also less well coordinated muscularly? Is the little Indian girl less able to cook or care for a younger sibling? And is the little Puerto Rican boy less able to bat a ball, to put it through a basket, or to sew on its torn cover? These questions, which unhappily we cannot pursue here, point to the important distinction between the social desirability of skills and their intrinsic difficulty, rarity, or aesthetic value. Some of the skills one must possess in minimum degree to participate in one society might be totally irrelevant or even outright disadvantageous in another.

Yet it does not do much good to mourn the fact that the other skills men *do* have are fine, even exquisite. The problem is painfully evident in the developing countries. There, the demands of the industrial order and urban living insistently undermine the relevance of venerable and often exquisitely developed skills which were highly important in the past; and they elevate to great importance *new* skills which, to the citizens of more developed countries, seem most elementary, yet which may seem difficult to master, even occult, to those in the developing nations.[4] It is notable, at least to me, that many of these skills are precisely those which *also* seem to be seriously underdeveloped in those groups in American society which are most disadvantaged, i.e., our ethnic minorities, including those in our Indian population and in our pockets of Protestant white poverty such as Appalachia and the South. Among the more obvious skills which are relevant to adaptation to modern life and insufficiently mastered by these disadvantaged groups I note, for illustrative purposes:

The telling and management of time: I know of no simpler or better indicator of a man's desire to show himself as modern than the acquiring and *demonstrative* wearing of a wrist watch. Of course, what is involved here is a value as well as a skill. However we may *feel* about the coercion of the clock, we must recognize the ability to tell time and to order one's affairs in relation to the clock as a critical skill for participation in the modern world. From social workers in the more deteriorated slums, and researchers who have worked with juvenile delinquents, one hears the frequent complaint that "they *never* keep appointments." Although my search is probably not exhaustive, I have not come upon a single article comparing children of different backgrounds in their ability to tell time and to meet the exigencies of the clock. This is all the more striking since the literature on developing countries gives the problem so much attention, and long ago Lewis Mumford argued that the industrial revolution and the modern world were ushered in not by the steam engine but by the elaborate organization of time in Christian monasteries.

The command of language, especially in its written form. The point certainly needs no elaboration, but perhaps I may be permitted to inquire what light our numerous studies of socialization can throw on the persistent failure of our Negro slum children to master language in a way appropriate to adequate performance in our schools? We cannot be satisfied with the usual reference to "poverty," not only because it is so global but also because it seems hardly to explain the extraordinary richness of the language and the elaborateness of the spoken style of Oscar Lewis' poor Mexican families.[5] And poverty can hardly explain the late readers and language cripples who bloom so profusely in some of our most favored suburban communities. Here again, then, we would like to see fuller study of those practices of child rearing, those patterns of interaction between parent, environment, and child, which lead to greater or lesser degree of command over spoken and written language.

The research cupboard is not so bare in the case of studies of the acquisition of language skills as it is of those involving management of time, but neither are we presented with a bursting granary. We may eagerly await the second volume of *The Review of Child Development,* which is to give us a chapter on "Language Development and its Social Context." If we are to judge from advance reports on the studies of Martin Deutsch and Irving Taylor at the Institute for Developmental Studies, we may yet meet some surprises in discovering that it is not the number but the use of words that distinguishes the underprivileged child.[6] For them, apparently, words stand for objects and objects for action. One word does not elicit another word, but rather elicits the *image* of action. This would seem to be highly congruent with the findings of Miller and Swanson.[7] It could also serve somewhat to explain not only the disadvantaged minority groups' performances on our typical word tests, but also the problems of later incorporating these groups into environments where one word leads to another word and not to an action.

Arithmetic: I suppose I cannot mention the language of words without also mentioning the language of numerical and mathematical symbols. On a four item arithmetic test we gave to thirty-seven miners and ex-miners in Appalachia whose education averaged three to four years, we found that two "occupational" groups—the physically disabled and those not employed on relief jobs—could collectively answer only 6 per cent and 17 per cent of the problems, respectively. In the physically disabled group, for example, only three of the twenty men could correctly subtract 9 from 23!. Two of these three could also multiply 8 x 9. So two men got half the problems right, one man got a quarter right, and the remaining seventeen men missed all four. We must note not only the appalling average lack of skill, but the individual variation. The group employed on relief did better, collectively, but the difference is due largely to individual variation. Two of the seventeen could answer all four questions, one could do three, another man one, and the remaining thirteen could answer none!

We had a very similar experience with tests of verbal ability. Using a larger sample from the same region of Appalachia, now augmented by groups with an average of seven or eight years of school, we asked the respondent to give the opposite of each of fifteen words. Before the fifteen words were presented, the interviewee was given an example such as "black" and "white." Many of the men simply could not grasp the task, that is, they could not dominate the *concept* of a word's opposite. By contrast, there were some who could give twelve correct answers, including words as difficult as "intelligent," "modest," "corpulent," and "affluent." Although education played a large role in this outcome, the variation within groups of the same education was very wide.[8] Again we are led to wonder what are the precise qualities of intelligence, of home environment, or of later experience which yield these differences.

Getzels and Jackson have presented some highly suggestive research relevant to these issues as they apply at a quite different educational and cultural level.[9] In their research they distinguished between adolescents with a high IQ on standard tests and those with special skill to use ideas and information creatively. Perhaps this is matter not merely of basic skill, but rather of cognitive style, which will be discussed below.

Information

The cognitive element of the personality may be divided into two broad categories—levels of information and styles of thinking. I refer to the former here. Almost every public opinion poll ever conducted has shown great differences in the sheer quantity of fact known by middle and upper class individuals as against those in the lower classes. Such differences are not restricted to news events and public figures, but include as well all sorts of practical and useful information, such as how to get your gas turned on or your garbage collected, how to get permission to conduct publicly controlled activities, how to organize a meeting, and so on. We may assume that this kind of knowledge about getting along in the everyday world is an important pre-condition to effective and independent participation in the modern social order.

The difference in information among various social class groups is generally attributed to the differences in their average education. While acknowledging the importance of education, I am rather of the opinion that these differences are already quite marked by the time children enter grade school, and that a great deal of the variance is to be explained as a result of the different early socialization experiences which the children of the several groups receive.

I am further strengthened in this belief by the clear evidence of great individual variation in information levels among persons with equal education. Thus, when we asked former Soviet citizens what they could do if some bureaucrat were taking an action injurious to them, the most common response among workers was "nothing," whereas most members of the intelligentsia cited at least two sources to which they might turn for help.[10] But there were ordinary workers who named four, five, six, and more agencies they would write to—the Communist party, the trade union committee, the factory manager, the newspapers, and finally Stalin himself! These different levels of information undoubtedly reflect the influence of motivation to know and perhaps also of a sense of efficacy—but that does not make it any less interesting to inquire what are the mechanisms of socialization which lead to these differences both in the desire to know and the resultant knowing.

The topic of language skills and information provides an opportunity to highlight the conceptual problems of relating major social-structural factors to a socialization problem such as "the development of competence." Again I emphasize the necessarily schematic form which must characterize my discussion. The points I can make within these limits are rather obvious, but perhaps they will serve to provide a starting point for future discussion.

Let us consider the command of language, including the size and content of vocabulary and the capacity to form sentences and larger units of speech in grammatical and culturally acceptable ways, thus enabling one to bring his language capacity effectively to bear on situations of action.[11] Assume we deal with a Negro boy in Harlem with normal intelligence and no more than the average number of situational disabilities which affect residents of that area. How would the hypothetical "self-sustaining vicious circle" we so often hear about be apparent in his case? In the home, we may identify the following inputs:

Low capacity models: From a sheer learning point of view, the total vocabulary available to be learned in the home is likely to be quite small because of the limited education and experience of the parents or others in the household. Each incumbent of the home probably adds very little that is not already in the vocabulary of others, so that the total word pool is likely to be small and restricted. Since many homes are broken and one parent absent, the available pool of models is further reduced. Inevitably, fewer words will be learned than actually are available in the limited total potential vocabulary of even this group. The prognosis is for a very limited vocabulary.

Interaction effects: Whatever the pool of potential words to be learned, they cannot be learned if those who know them do not use them in situations in which the child can learn them. If mother and father are gone most of the day and/or leave the child alone a great deal, effective learning is greatly reduced. Mother and father may be there, but not communicate much with each other. Thus, in my current cross-national research I found in all countries marked social group and individual differences in the frequency with which men indicate any interest in communicating with their wives on a variety of themes, such as work, educating the children, running the house, and sex.

Content and tone: Interaction effects are felt not only in that they make words "available" for learning. The quality of the interaction clearly will affect the relative frequency of words and, beyond that, the emotional tone associated with verbal exchange in general. If the parents talk to each other mainly to complain, grumble, or quarrel, the words in that realm will obviously be learned sooner and more fully. Perhaps more important, the unconscious conception of what language is mainly "about" or "for" will be affected. With parental communication mostly demanding or quarrelling, the likelihood is greater that, unconsciously, language and verbal expression will be associated mainly with unpleasant experiences and hence be something more or less avoided.

Social Valuation: The interest or motivation for involvement in language will be affected not only by unconscious association with situation and content, but also by the more explicit cultural valuation put on relative degrees of skill and interest in the use of language. If the most common evaluations are generally negatively toned, as I believe they are in Harlem, the effect will most likely be to induce the child to view language skills as relatively undesirable qualities to cultivate in oneself. I cannot establish the point as a matter of fact, but I feel fairly certain that in Negro Harlem, the most common evaluations of language facility are negative: "big talker" and "talking big" are clearly negative. "Loudmouth," "shooting off the mouth" speak for themselves. "Preaching" is not too good. "Nagging" is, after all, mainly verbal behavior, and "talking foolish like a woman" reflects a similar feeling. There may be some positive associations to the facile use of language, but they seem fewer and less strongly toned than the negative associations.

The brief sketch of the Harlem boy's start in language in the home could be extended, but I trust I have made my point. The same mode of analysis could be applied as well to later stages in his language training, but I had better forbear. Let me merely note a few salient points. In the peer group, again, the pool of words collectively shared will be small and each boy will add little new. New vocabulary may be amply introduced, but mainly in areas relating to sex, aggression, and the law, and such nature that the words cannot be carried over for use in polite society. On the streets, the skills valued and encouraged will be mainly physical, and indeed, the verbal may be actively disvalued.

When the boy later arrives at school, interesting new elements are introduced. In contrast to the rewards the school offers the middle class child for what he already *knows,* it is likely to greet our Harlem boy with horror for what he does *not* know and *cannot* do with language. The result, on his part, will be more avoidance of words and language. Much of the language he does know will be unacceptable, and if expressed will produce more negative reactions. The mode of expression of which he is most capable, the physical, will likely find no valued outlet at school, or may even be punished.

This analysis could be carried further in the life cycle, to the first job and beyond, but I suspect everyone can tell it for himself. Whether the story is accurate or not is unfortunately not as well documented as it might be, but I doubt that it qualifies as a "just-so" story.

Motives

The preceding discussion points to the importance of motives as an underpinning for the acquisition of information, and thus alerts us to the relevance of motivation for a discussion of competence. Certainly at first blush, motivation would seem to have little to do with competence, since competence refers to the respected social roles. I am rather of the opinion that this question cannot be it. But if we recognize skills and information as contributing to competence to perform social roles, then we must recognize the *motive* to attain socially valued status positions as a necessary, even if not sufficient, requirement for competence in social action.

It might be argued that the point is trivial, on the grounds that little if anything socially desirable, yet relatively scarce, will be acquired without motivation to do so. The more challenging question is whether there are any motives which in and of themselves can be seen as more adaptive, and in this sense contributing more to one's competence to attain and perform in available and respected social roles. I am rather of the opinon that this question cannot be answered if it is put in general terms, as applying to any sociocultural system. Every sort of motivation, including aggression and hostility or extreme dependency and passivity, has been found to be relevant and adaptive in some culture somewhere. If we specify performance in a particular type of social system, however, then we can identify more or less precisely motives which are differentially adaptive. For example, cross-cultural anthropological research indicates that hunting and gathering societies are more likely to train youngsters for autonomy and independence, while pastoral societies give more emphasis to inculcating compliance and dependence.[12] In the context of modern industrial society, we might expect the need for achievement to be more adaptive than the need for affiliation, the need for autonomy more productive than the need for dependence, at least for those competing for middle class positions. Studies of the adult population of the United States indicate that there is some such pattern in the distribution of motives,[13] and studies of child rearing in the different class and ethnic groups suggest that these adult differences most likely rest on differences in socialization practices.[14]

Cognitive Modes of Functioning

When we considered the theme of information, we were concerned mainly with the amount and type of knowledge possessed by the individual. The cognitive modes refer to the forms of thinking and to the "style" characterizing the individual's mental processes. We ask: Is thinking abstract, concrete, or both? Is it slow and deliberate, or quick and mercurial? Is interest focussed or diffuse? Is the language of emotion more elaborated than is the conceptual apparatus for dealing with objects or material relations?

Cognitive functioning, at least in the realm of concept formation, has certainly been of interest to students of child socialization, but the interest has been mainly the usual developmental one of fixing the ages at which different conceptual skills emerge, with little systematic attention paid to individual, and even less to group, differences.[15] The work of Miller and Swanson and their students in establishing the stronger tendency to conceptual expression in the middle class child and of motoric expression in the working class child points the way. Unhappily, this path has been little followed by other workers in the field.

The modes of cognitive functioning clearly influence the child's initial performance in meeting the demands of the school and other social agencies; his later preferences for academic work, trade school, or practical apprenticeship; and his eventual choice and performance in his occupational and other adult roles. The boy who feels inadequate or is made uncomfortable in an environment which gives much emphasis to manipulating symbols instead of things will soon be drifting out of school to a world in which experience is more immediate and concrete. And this applies not only at the level of primary school and in disadvantaged neighborhoods. The problem is also very real for those who are already in college and choosing their professional careers. Thus, Stern, Stein, and Bloom report how great a role cognitive styles played in the adjustment of freshmen at the College of the University of Chicago. The College program stressed and rewarded "abstract analysis and relativity of values and judgment rather than fixed standards." (p. 191) Teachers introduced a good deal of ambiguity and often departed from conventional standards of judgment. It was precisely those students whose cognitive style inclined them to concrete thinking, to an insistence on one "correct" answer, who made up the bulk of the academic casualties at the end of the year. And this was true despite close matching of the students on measures of intelligence and scholastic aptitude.[16]

Unfortunately, Stern, Stein, and Bloom did not systematically explore the home environments which produced these different types. Indeed, we have very little knowledge about the socialization experiences from which stem one or another style of cognitive functioning. Rokeach, for example, tells us almost nothing of the home environment of those with "open" and "closed" minds. He does suggest, however, that ambivalence toward parents which is not permitted expression generates anxiety and narrowed possibilities for identification with persons outside the family. Rokeach sees both of these conditions, in turn, leading to the development of closed belief systems.[17] The point seems closely paralleled by the observation of Getzels and Jackson with regard to the home environments producing more creative adolescents.[18] The creative family, they concluded, "is one in which individual divergence is permitted and risks accepted." Those which produced a mere high IQ without "creativity" seemed more conventional, with the mothers stressing cleanliness, good manners, and studiousness. Miller and Swanson help us to see that physical punishment is more intimately tied to motoric than to conceptual expression, whereas psychological discipline, such as the threatened withdrawal of love, more often yields conceptual expression.[19] Hoffman's research also contributes to isolating the style of training he calls "inductive discipline," referring to efforts to explain to the child the effects of his action on others. The outcome which interests Hoffman, however, is less a mode of cognitive and more a mode of moral functioning.[20] We should, of course, here recall the explorations into the origins of the authoritarian personality, one component—some would argue the main component—of which is a certain cognitive style. Adorno *et al.* suggest a series of antecedents in the home environment of those who display prejudice and authoritarianism, but their evidence is mainly clinical and has not been more systematically tested on large samples.[21]

Cognitive styles emerge as an extremely important component of the individual's equipment for coping with the demands of society and a critical element in determining what kinds of roles he may seek out and successfully play. Cognitive style will evidently play an early role in school performance; it will channel—and limit—the choice of occupations, and will affect the nature of one's political participation. The evidence seems unmistakable that the observed adult differences in cognitive style have their origins in childhood experience. Unfortunately, very little has been done by specialists on socialization to follow these insights in programs of systematic research. The implications for future research seem clear.

Conclusions

This presentation of a comprehensive conception of the elements of personality may have led the reader astray, and may leave me exposed to the charge of neglecting my announced subject. Admittedly, I have not discussed as fully as I perhaps should have social structure and the socialization of competence. If this is so, I beg indulgence on the ground that it seemed to me the main topic could not be properly understood unless we first dealt with matters more fundamental. In my discussion of the elements of the personal system, I have in effect sought to establish the basis on which—and in a sense the language *in* which—a more meaningful discussion can be presented. But it has taken so long to compile the vocabulary and explain the grammar that little space remains for telling the tale I meant to recite.

The message is very simple. Like so many obvious things, it is not only fundamental but also much overlooked. The main business of socialization is the training of infants, children, adolescents (and sometimes adults) so that they can ultimately fulfill the social obligations that their society and culture will place on them. Implicit in this statement is the expectation that, in meeting these societal demands, the individual will not be placed under so much strain as to fall apart psychologically. And not excluded is the thought that the term "social obligations" includes elaborating and acting effectively in roles not commonly assigned by the given sociocultural system. Indeed, we do not by any means exclude the possibility that the most creative way of meeting the demands of a given social situation may be to reject that situation as it presents itself, to insist on a new deal, and to forge new roles and new styles of life.

I am firmly convinced that concern about the ultimate playing of social roles is the decisive element in the child-rearing behavior of most parents or chief parent-surrogates. This is not meant to deny that at some periods in a person's life—especially in early infancy and perhaps again at puberty—the problems of sheer management of physical need or the facts of physiological change may not come briefly to dominate the concerns of the socializers. Yet, I believe that even in dealing with such ultimately physiological needs as the hunger drive, socializers never lose sight of the long-run adaptive significance, both of the sheer mastery of this drive and of the *way* in which it is mastered. Evidence for this can be found in the common speech of every mother. Nor do I mean to deny that there may be periods in the child's development when what is done to him more expresses the psychic needs and desires of the parent for giving or withholding, for restraining or indulging, than it represents any conscious or subconscious thoughts about the social roles the child will ultimately play in society. The issue is clearly one of relative emphasis. My chief point is that the degree to which, and the ways in which, socialization is a relatively conscious process of training in anticipation of future social roles, have been neglected relative to the conception of socialization as mainly a process in which adults cope with the challenge of the infant and child as *organism*. The same criticism applies to those who approach the study of socialization as mainly expressive of the parents' needs and dispositions.

So far as competence is concerned, some children face a situation in which almost everything conspires to insure that most of the more favored positions in society will be closed to them. They will grow up ill-equipped to compete for entrance into the more advantaged roles, and those desirable positions they may acquire they will be unable to hold successfully. This is what we mean by competence—the ability to attain and perform in valued social roles. In our society this means, above all, the ability to work at gainful and reasonably remunerative employment, to meet the competition of those who would undo us while yet observing the rules for such competition set down by society, to manage one's own affairs, to achieve some significant and effective participation in community and political life, and to establish and maintain a reasonably stable home and family life. We should not for a moment forget the massive and cruel formal obstacles our society has devised to prevent the disadvantaged minorities from sharing equally in the opportunities inherent in the level of wealth and civilization we have attained. But we must also recognize that these obstacles—such as overt discrimination, segregated schools and communities, color-bar hiring practices, and even legal disfranchisement—are not the *only* barriers to effective functioning on the part of disadvantaged minorities. The most cruel aspect of discrimination and disadvantage lies in its ability to deprive the individual of that competence which is essential to effective functioning once the formal barriers to free competition have been breached. Lack of competence effectively to take advantage of new opportunities in a competitive system can make the attainment of nominal legal equality a hollow victory, and make a self-fulfilling prophesy of the bigots' claim that minority members are unable to perform effectively even when not formally discriminated against. To deny people the means for attaining competence while yet granting them technical equality under the law is the contemporary equivalent of saying that the majesty of the law confers on the rich as on the poor alike the right to sleep under bridges.

To perform effectively in contemporary society, one must acquire a series of qualities I believe to be developed mainly in the socialization process. Effective participation in a modern industrial and urban society requires certain levels of skill in the manipulation of language and other symbol systems, such as arithmetic and time; the ability to comprehend and complete forms; information as to when and where to go for what; skills in interpersonal relations which permit negotiation, insure protection of one's interests, and provide maintenance of stable and satisfying relations with intimates, peers, and authorities; motives to achieve, to master, to persevere; defenses to control and channel acceptably the impulses to aggression, to sexual expression, to extreme dependency; a cognitive style which permits thinking in concrete terms while still permitting reasonable handling of abstractions and general concepts; a mind which does not insist on excessively premature closure, is tolerant of diversity, and has some components of flexibility; a conative style which facilitates reasonably regular, steady, and persistant effort, relieved by rest and relaxation but not requiring long periods of total withdrawal or depressive psychic slump; and a style of expressing affect which encourages stable and enduring relationships without excessive narcissistic dependence or explosive aggression in the face of petty frustration.

This is already a long list, and surely much more could be added. My purpose here is not to strive for an exhaustive list. I want simply to indicate the *kinds* of personal attributes which I feel a modern industrial society requires in significant quantity of substantial numbers of its citizens. Without most of this array, one is not competently prepared for life in our society, and must sink into some form of dependency or deviance. There is no great difficulty in demonstrating that these qualities are very unevenly distributed in the several strata of our society— educational, occupational, ethnic, and regional. The challenge for the students of child-rearing is to show whether, and explain how, these differences came about as a result of differential socialization practices and experiences. My assessment of recent work in the field of socialization research is that very few of these issues have been the object of much systematic study on a significant scale. More than that, I incline to the conclusion that this situation is not now rapidly changing. The cause, I believe, lies in the scientific "culture" of those doing socialization research; they are beginning at the wrong end.

The master key to understanding socialization, in my opinion, lies not in further deepening our involvement in the innate propensities of the child and the situation of action this defines for the parent. The key lies rather in a redefinition of the problems of socialization research which starts with a clear statement of what are the massively evident observed *differences* among adults which appear socially important enough to be worth the trouble of explaining. I have tried to show through a brief and limited discussion of competence what a few such differences may be. But any other social issue—mobility, political participation, delinquency and crime, or occupational performance—could have served the same purpose. I applaud the fact that the six culture study sponsored by the Laboratory of Human Development at Harvard included, in the nine standard behavior systems to be observed, not only the old and tried themes of nurturance, succorance, aggression, and obedience, but also responsibility, sociability, achievement, and self-reliance. [22] Even in this case one may ask: If you have not first defined what is the quality of the adult you wish to understand, how do you know what to look for in the disciplining of the child when you study "responsibility" and "achievement"? And beyond responsibility, social ability and achievement, we still want to know about information, values, motives, skills, moral functioning, self-conceptions, cognitive, conative, and affective modes; about the ability to trust others and enter into enduring relationships of cooperation or undestructive competition; about images of and relations to authority figures, and the sense of membership in, and feeling of obligation to, the community.

Before concluding, I should clarify some issues not necessarily important to the student of socialization, but nevertheless fundamental to the functionalist perspective on social structure, a perspective which my analysis in this paper represents.

The first issue concerns the appropriateness of a *general* model of competence, such as I have presented, for the analysis of performance in what is inevitably a highly differentiated social structure. Another, and blunter, way of making the point is to claim that I have presented a model of competence as defined by the middle class in American society. The charge is correct. The aspects of competence I have sketched above are precisely those which one requires either to continue as part of, or to attain to a position in, middle class America. Every model of competence is, in large measure, specific to some culture, and often even to some stratum of a particular society. The elements of competence, as I have sketched them above, would not necessarily loom equally important for a man who was hoping to be the world's heavyweight champion. And they might be quite beside the point for a Trobriand Island fisherman or an Arctic Eskimo. The point I mean to make in this way, again, is that socialization research generally begins from the wrong end. In my opinion, the starting point of every socialization study should be a set of qualities "required" by, i.e., maximally adaptive in, a given sociocultural system and/or manifested in a given population. The task of students of socialization should be to explain how these qualities came to be manifested by individuals, thus rendering them competent, or why individuals failed to manifest these same qualities, thus being rendered less competent to perform in the given social setting.

The very form of my last sentence raises the second and last issue on which I wish to touch briefly—the issue of "competence for what?" or "adjustment for whom?" A functional perspective always runs the risk of leading one to assume that what is good for society is good for the individual, and vice versa. Those interested in encouraging competence, or excellence, or whatever desirable quality, run the same risk. If we define competence as the capacity to organize one's life and to strive so as to achieve some degree of social stability or desired mobility, it means that many individuals, in seeking to meet the competence requirements of their society, may in that very act also be inviting more or less certain frustration. The Negro in Harlem who is quite comfortably able to accept his dependence on welfare authorities, to be passive in the face of middle class society's expectation of constant effort and striving, and to find release from his tensions through extreme physical and vocal expression in his store-front church, may be making a more appropriate adjustment to the realities of his situation, and to that degree be more *competently* managing his life than is his neighbor who has all the white middle class virtues, which will in turn increase the probability that he will run up against a solid wall of frustration and futility.

Everywhere today—by continent, by nation, by region, by class—there is a vast process of social change exerting its force. To manage their lives in a satisfying way, men need new information, skills, motives. New problems and situations everywhere constantly challenge their competence. Tragically, men find that the skills and talents which formerly made them models of competence in their community are of no value or are even demeaned and degraded in the new scheme of things. In this turbulent sea, we often glimpse some remarkable people who seem especially equipped to navigate freely and easily through conditions which are tumbling most people overboard. Are there then some qualities of man which give him a general competence useful in all places and times, qualities especially suited to adapt a man to all waters no matter how fast the current or sudden its changes? What are these qualities and what are the special forms of socialization which bring them into being? Here is a challenge to the student of socialization worthy of *his* competence.

References and Notes

1. See the review by L. Kohlberg, Development of moral character and moral ideology. In M. L. and L. W. Hoffman, *Review of child development research,* Vol. I. New York: Russell Sage Foundation, 1964.
2. The model is as much Daniel J. Levinson's as it is mine. Although we have

used it extensively in our work, we have never published a full and systematic account of the scheme. We sketched some of its elements in the article "National character" in G. Lindzey (Ed.), *Handbook of social psychology*, Vol. II, 1954, and later in "The personal system and the sociocultural system in large scale organizations," *Sociometry*, 1963, **36**. I used the scheme systematically in empirical research reported in A. Inkeles, E. Hanfmann, and H. Beier, "Modal personality and adjustment to the Soviet sociopolitical system," *Human Relations* 1958, **11**. The conception is outlined in schematic form most fully in A. Inkeles, "Sociology and psychology" in S. Koch (Ed.), *Psychology: A study of a science,* Vol. VI. New York: McGraw-Hill, 1963.
3. For a fuller statement of this issue see: A. Inkeles, Personality and social structure. In R. Merton, *et al.* (Eds.), *Sociology today.* New York: Basic Books, 1959.
4. One of the most sensitive accounts is offered by E. Erikson in his remarks on the Dakota Indians, in *Childhood and society* (New York: Norton, 1950), whom he describes as no longer having a socially satisfying mode of using either their skill as riders and hunters or the character traits of cruelty and generosity, which were apparently meaningful, rewarding, and encouraged when their culture was whole.
5. The richness of the vocabulary and the fluency of expression of his Mexican subjects has so struck readers of Lewis' accounts that many wonder to what extent these qualities are the product of translation and editing. But those who have heard the tapes in Spanish say they are often quite poetic. And Lewis says: "Despite their lack of formal training, these young people express themselves remarkably well. . . ." (O. Lewis, *The children of Sanchez.* New York: Random House, 1961. P. xii.) Of course, one might well retort: "So do many Harlem Negroes." Has Lewis been unusually selective? In any case, we may ask how two environments so much alike in their poverty and related conditions can produce groups of individuals so different in their command of language. Surely socialization practices played some role.
6. These have been briefly summarized by F. Riessman in *The culturally deprived child.* New York: Harper & Row, 1962.
7. I have in mind their finding, reported in *Inner conflict and defense* (New York: Holt, Rinehart and Winston, 1960), that middle class boys more often adopted a conceptual mode of expression, whereas the motoric mode was more typical for working class boys.
8. These results are presented in an unpublished senior honors thesis by William W. Lawrence, Department of Social Relations, Harvard College, April, 1965.
9. Getzels, W., & Jackson, P. W. Family environment and cognitive style: A study of the sources of highly intelligent and of highly creative adolescents. *American Sociological Review,* 1961, **26**, 351-59.

10. See Inkeles, A., & Bauer, R. *The Soviet citizen.* Cambridge, Mass. Harvard University Press, 1959.
11. Any discussant of this topic is necessarily heavily indebted to Basil Bernstein's pioneering work in the development of socio-linguistics. See Some sociological determinants of perception: An inquiry into sub-cultural differences. *British Journal of Sociology,* 1958, **9**; Language and Social class. *British Journal of Sociology,* 1960, **11**.
12. Herbert, B., Child, A. I., & Bacon, M. Relations of child training to subsistence economy. *American Anthropologist,* 1959, **61**, 51-63. Also see D. Aberle, Culture and socialization. In F. S. Hsu (Ed.), *Psychological anthropology.* Homewood, Ill.: Dorsey, 1961.
13. See Veroff, J., *et al.* The use of the thematic apperception test to assess motivation in a nationwide interview study. *Psychological Monographs,* 1960, **74**.
14. See Rosen, B. C. The achievement syndrome. *American Sociological Review,* 1956, **21**, 203-211. Strodtbeck, F. L. Family interaction, values and achievement. In D. C. McClelland, *et al* (Eds.), *Talent and society.* Princeton, N. J.: Van Nostrand, 1958. Miller, D. R., & Swanson, G. E., *op. cit.*
15. This should be readily apparent to anyone who consults the review by I. I. Sigel, The attainment of concepts. In M. L. Hoffman & L. W. Hoffman (Eds.), *Review of child development research,* Vol. I New York: Russell Sage Foundation, 1964.
16. Stern, G., *et al.* Methods in personality assessment, New York: Free Press, 1956.
17. Rokeach, M. *The open and closed mind.* New York: Basic Books, 1960.
18. Getzels & Jackson, *op. cit.*
19. Miller & Swanson, *op. cit.*
20. Hoffman, M. L. Report of research sponsored by N.I.M.H. Merrill-Palmer Institute, October 1964. (Mimeographed.)
21. Adorno, *et al. The Authoritarian Personality.* New York: Harper & Row, 1950. Also see Frenkel-Brunswik, E., & Havel, J. Authoritarianism in the interviews of children. *Journal of General Psychology,* 1953, **83**, 91-136.
22. Whiting, B. B. (Ed.) *Six cultures.* New York: Wiley, 1963.

Recommended Additional Readings

Mehler, J., & Bever, T. G. The study of competence in cognitive psychology. *International Journal of Psychology,* 1968, **3**, 273-280.
White, R. W. Motivation reconsidered: The concept of competence. *Psychological Review,* 1959, **66**, 297-333.

10 Nutrition and Learning

Heinz F. Eichenwald
Peggy Crooke Fry

Orientation

Within the past decade great interest has emerged in the effects of various kinds of early experiences upon later intellectual development. The impetus for this concern arose with the increasing social agitation regarding the impoverished and disadvantaged portions of the human community. That disadvantaged children did poorly in school has long been known, but this phenomenon was traditionally attributed to the "inferior" genetic inheritance of these people. More recently attention has been given to the effects of an impoverished, "deprived" early environment on later performance. An article later in this book by Caldwell (#11) explores these environmental considerations. The present article by Eichenwald and Fry, however, examines a different aspect of early experience—the effects of nutritional deprivation on later development.

While the recent emphasis on psychological deprivation is exceedingly important and long overdue, there has been insufficient attention given to the effects of biological deprivation. In this article the authors review the existing research on inadequate nutrition in animals, and then they examine the available data on human subjects. (Obviously, a scientist cannot conduct an experiment with human infants in which food intake is deliberately withheld.) The data leave little room for doubt: Early malnutrition produces long-lasting, deleterious, and generally irreversible effects on many physical structures, including the brain and other parts of the central nervous system. These structural deficiencies then, of course, bring about impaired psychological functioning in both the intellectual and emotional spheres.

Eichenwald and Fry point to the relation between inadequate nutrition and societal imperfections; nutritional deprivation is found most frequently among the underclass and the financially impoverished. Thus, this article provides a compelling example of the dual sets of emphases operating in the course of human development: Both biological and societal elements influence psychological functioning.

Reprinted from *Science*, 1969, 163 (3868), 644-648, by permission of the authors and the publisher. Copyright© 1969 by the American Association for the Advancement of Science.

Supported in part by research grants CC-00131 and CC-00132 of The National Communicable Diseases Center, Atlanta, Georgia, and by the Eugene and Margaret McDermott Foundation.

It would seem that how society sets out to resolve this profound social problem will be a testimony to its commitment to provide the best possible life for all its citizens.

Heinz F. Eichenwald is William Buchanan Professor and chairman of the department of pediatrics, and Peggy Crooke Fry is an assistant professor of pediatrics at the University of Texas Southwestern Medical School at Dallas.

10

It has long been scientifically acceptable and, in some circles, even fashionable to ascribe many behavioral characteristics of the older child and adult to conditioning experiences received during infancy and perhaps prenatally. More recent data have indicated that some of these conditioning factors, rather than being psychosocial in nature, have a biochemical basis. Alterations of the biological and physical environment produce profound and lasting disturbances of the anatomical, chemical, and thus developmental and behavioral pattern of the organism. This course of events has been aptly termed "biological freudianism" by Dubos et al.[1]

The best studied aspect of biochemical conditioning is nutrition. In most areas of the world malnutrition in early life is directly or indirectly responsible for more deaths among children than all other causes combined; recent evidence has indicated that deficiencies in nutrition not only affect physical growth but may produce irreversible mental and emotional changes. Many aspects of these long-term effects of malnutrition have been studied in animals;[2] from these experimental data a series of hypotheses relating to human development has been proposed.

Nutrition and Physical Growth

Numerous observations in animals dating back many years indicate that malnutrition retards physical growth;[3] if growth is suppressed for a sufficiently long period during a critical phase in the early weeks or days of life, subsequent supplementation of the diet, while initially enhancing the rate of growth, usually does not permit the animal to attain its normal mature size. The same sequence of events has been demonstrated to occur in children who had suffered from severe malnutrition during early life.

To a considerable extent, biochemical development parallels this suppressed physical development in both children and experimental animals. Maturation of a variety of biochemical processes is delayed; thus, malnutrition results in inhibition of the biochemical maturation of the organism, and may, under certain conditions, produce retrogressions to earlier functional patterns. In malnourished children, such widely different measurements as water distribution, fat absorption, concentrations of plasma lipids and cholesterol, and excretion of creatinine approximate corresponding values observed in well-nourished younger children of the same height and weight.[4] Furthermore, the metabolism of phenylalanine to tyrosine is depressed in malnourished older infants. Such patients show an abnormally high ratio of phenylalanine to tyrosine in the blood plasma and excrete excessive quantities of phenylalanine in the urine; this suggests that the enzyme system responsible for the conversion of the one amino acid to the other has either been depressed or has not fully developed. This result strongly parallels the status of the metabolic system in younger infants. In older children, malnutrition affects the ability to metabolize aromatic amino acids; thus these patients show defects closely resembling those associated with certain inborn errors of metabolism.[5] Of particular interest is the manner in which these defects affect the development and function of the central nervous system.

Nutrition and Neural Growth

Recent investigations[6] on the physical and biochemical growth of the central nervous system in animals have provided information helpful to an understanding of the effects of malnutrition on the nervous tissue. Between the 50th day prior to birth and the 40th day postpartum, the brain of the pig grows intensively, gaining an average of 5 to 6 percent of its eventual mature weight every 2 weeks and increasing in weight considerably faster than the spinal cord. In spite of this difference in weight, the biochemical development of the brain is quite similar to that of the spinal cord. During maturation and growth, the amount of water in both organs gradually diminishes, whereas the amount of cholesterol steadily increases. In the pig, the adult concentration of cholesterol is not reached until after the 3rd year, although the cholesterol concentration of DNA-phosphorus, which parallels cell concentration, rises to a peak and begins to fall in the brain and cerebellum prior to birth, while reaching a plateau in the cord prior to birth and then slowly declining. These data have been interpreted as defining two separate phases in the development of the central nervous system. One of these, maturation, consists of a rapid increase in cells, as indicated by the increased concentrations of DNA-phosphorus. The other phase, growth, parallels myelination and is represented by increased amounts of cholesterol. Thus, in the pig, the rate of maturation of the central nervous system peaks prior to birth, whereas the rate of growth of the central nervous system peaks shortly after that time.[6]

It is therefore not surprising that experiments on various animals have indicated that nutrition inadequate in calories and protein, coinciding with the period in life in which the brain is growing most rapidly, produces a brain which is not only smaller at maturity than in control animals but also one which matures biochemically and functionally at a slower rate.[2,7-11] Even if rats are underfed for only a few weeks after weaning and are then placed on an adequate diet, this same result is produced. From a biochemical standpoint, malnutrition in young rats is associated with decreased synthesis of sulfatide, a lipid which is a component of myelin; thus the time of most active myelin formation is the most vulnerable period for the developing central nervous system.[7] The brain of the mature rat malnourished during infancy is not only physically smaller but histologically may show degenerative changes of neurons and neuroglial cells. In the adult rat, starvation does not result in significant changes in brain weight or in similar damage to neural tissue.

Results obtained from simple caloric deprivation differ from those based on diets restricted not only in amount but also in quality. For example, simple nutritional deprivation of rats during the weaning period results in lower brain weights as well as in reduction in total brain lipids, phospholipids, and cholesterol.[8] These three substances are reduced to about the same extent as brain size, with the relative concentrations unchanged. However, the amounts of cerebroside and proteolipid proteins were considerably lower in the brains of undernourished weanlings as compared to normal controls. When the animals were subsequently fed adequate diets, the defects were largely overcome. Similarly, the process of myelination, while being depressed during the period in which the newborn rat was undernourished, proceeded at an increased rate when an adequate diet was fed, so that after several weeks the degree of myelination was the same in the test animals as in the control groups. Thus, in the rat, an insufficient supply of food during the time of most rapid growth results in biochemically immature brains; the rat is capable of recovery if an adequate diet is fed by the 21st day of life.

The interpretation of these results requires some qualification. In the first place, while short-term simple undernutrition during the weaning period of the rat appears to leave few, if any, detectable biochemical traces if the animal is subsequently adequately fed, these studies do not indicate whether the function of the brain is permanently altered. Furthermore, the timing and duration of malnutrition during infancy is probably critical in determining whether anatomic and biochemical damage can subsequently be healed. Another study indicates that in rats neither the deficit in brain size nor the DNA concentration resulting from a restricted diet in the early days of postnatal life could be repaired by subsequent feeding; if, on the other hand, the dietary restriction occurred during a later period of infancy, subsequent adequate feeding would produce a normal-sized brain and normal concentration of DNA.[9] Thus, early restriction of calories may slow the rate of cell multiplication (maturation) whereas later restriction affects only cell size (growth).

The rat may not represent a suitable model for the study of human development, nor do the experimental conditions resemble those encountered among malnourished infants. In infants there frequently is a long-term deficit in both the quantity of the diet as well as its quality, particularly with respect to protein and perhaps certain vitamins.[12]

In this regard, observations on human infants have shown that inadequate feeding of pyridoxal phosphate, which serves as a coenzyme for most enzymatic reactions of amino acids, results in a series of changes in the physiological function of the brain and in the appearance of clinical symptoms.[13] In the newborn baby, the ingestion of a diet deficient in this substance but otherwise adequate results within 6 weeks in hyperirritability, convulsive seizures, abnormalities in development, and behavioral disorders. If this deficit continues for a sufficiently long period, irreversible alterations of cerebral function will occur, resulting in severe mental retardation.[14] Since pyridoxal phosphate is important in amino acid transport and protein synthesis, it is not surprising that the administration of puromycin, which inhibits protein synthesis, produces degrees of disorientation and loss of memory in mice which parallel the duration of the inhibition of protein synthesis by this antibiotic.[15] These data suggest that inadequate protein nutrition or synthesis, or both, during brain development could result in changes in function and that, if the degree of deprivation were sufficiently severe and prolonged, the changes in function might be permanent. Other experimental observations do in fact indicate that insufficient intake of protein during early neural development affects mentation.[16]

Nutrition and Mentation

In rats and swine, simple caloric deprivation during the nursing period apparently results in behavioral changes but does not seem to affect the animal's problem-solving ability. On the other hand, protein deprivation in early life not only causes the same behavioral changes but also reduces the capacity of the experimental animal to learn at a later age. Furthermore, rats born of and suckled by malnourished mothers are similarly deficient in their learning capacity.[2, 10]

It has been suggested that these deficits do not result entirely from nutritional factors, but that infection and behavioral disturbances produced by the experimental situation itself might be the most important factors. Recent studies with pathogen-free mice and a suitably stabilized environment indicate that these indirect, nonnutritional events have little effect on experimental animals.[17] In the more complex human case, these environmental factors undoubtedly contribute significantly to the effects of malnutrition on behavior and function.

Protein-calorie malnutrition in the human infant, if severe enough, produces two clinical conditions—infantile marasmus and kwashiorkor.[12] It seems likely that these two conditions are not separate clinical entities, but that they represent different manifestations of the same basic problem. Infantile marasmus occurs most commonly among babies weaned early without receiving suitable substitutes for human milk.[12] Since mothers in preindustrialized societies increasingly imitate the cultural practices of more advanced groups, early weaning without the availability of subsequent adequate nutrition has resulted in a rapidly increasing proportion of children with severe protein-calorie deficits.[12]

The clinical condition called kwashiorkor is most commonly encountered in children who had not been weaned until the 2nd or 3rd year of life and had not been offered adequate food supplements. The diet of the child afflicted with kwashiorkor may be reasonably adequate in calories, but it is grossly deficient in protein.[18]

The early investigators of protein-calorie malnutrition in infants found that apathy was universally present in these children.[19] The patients lacked the curiosity and activity found in normal children of similar ages and responded sluggishly, if at all, to a variety of stimuli. Not all of these emotional changes are necessarily the result of malnutrition, since malnourished children often are born to parents who themselves are intellectually and educationally depressed. Furthermore, the degree of apathy shown by malnourished children compares to that observed in healthy children who are abruptly separated from their mothers during the 1st year and placed in institutional situations. Nevertheless, it seems unlikely that apathy in hospitalized malnourished children is entirely due to separation from their mothers, because exactly the same lack of emotional responsiveness is observed among children lovingly cared for at home with an inadequate diet.[19] Furthermore, one of the early signs of recovery from marasmus and kwashiorkor is the return of responsiveness to the environment.[19]

There is a parallel between electroencephalographic changes and behavior. The electrical activity of the brain in protein-malnourished children shows consistent abnormalities in the form, frequency, and amplitude of activity. After the child successfully recovers from malnutrition, the wave form tends to conform more closely to that of healthy children of similar age.[20]

Follow-up studies of children who have been treated for infantile marasmus and kwashiorkor indicate that during recovery they at first grow physically with great rapidity.[19] However, if observations are continued over sufficiently long periods, the child never completely catches up with his healthy peers; growth stops at the usual chronological age, and the height of the adult is significantly shorter than that of his healthy peers.[19] Head circumference, a useful but not absolute indicator of brain size, is also smaller, although this measurement is not necessarily related to variations in intellectual capacity. Nevertheless, the intellectual attainments of children who have recovered from a clinically severe episode of protein-calorie malnutrition are consistently lower than those of individuals with adequate nutrition during infancy.[21, 22] These findings suggest that less severe but more chronic forms of malnutrition, which do not result in dramatic and life-threatening nutritional diseases, might contribute to the small stature universally observed among the economically poorer families of pre-industrialized societies and might be correlated with a decrease in intellectual development.

Nutrition, Growth, Infection, and Environment

Several pioneering studies, on a variety of population groups in whom malnutrition is indigenous, have suggested a relation between short stature and low intellectual achievement.[19, 21, 23] Such investigations are highly complex, since malnutrition does not constitute a single, definable condition and other related factors cannot be adequately controlled. The basic condition, malnutrition, occurs primarily among underprivileged populations and thus is located in a particular physical, social, and biological environment. This environment, and probably malnutrition as well, results in an increased frequency and chronicity of infection, especially diarrhea, among the very young (the so-called "weanling diarrhea"). Repeated respiratory and gastrointestinal infections produce a cycle in which infection increases metabolic demands, decreases food intake, and inhibits the absorption of nutrients, resulting in a further deterioration of the nutritional state.[24]

The relation between famine and pestilence was recognized in ancient civilizations; conditions associated with acute famine also provide an environment in which infectious diseases can flourish. Thus, the synergism between infectious disease and malnutrition in humans is both direct and indirect.

Infections, such as measles, may result in severe diarrhea and a high incidence of pneumonia in malnourished children.[25] The reasons for greater severity of disease and increased incidence of complications remain somewhat obscure. Protein deficiency leads to decreased immunologic responses in animals as well as humans; furthermore, some of the normal components of protective serum, such as complement, are reduced. However, it appears unlikely that these changes in blood serology account for the great susceptibility to infection in the malnourished child. Rather, the major defect is probably on a cellular level; not only is it possible that the integrity of the respiratory and gastrointestinal epithelium is altered, permitting more ready invasion of the body, but normal inflammatory and healing responses may be more or less inhibited.[26] In kwashiorkor, reduced inflammatory response can be demonstrated clinically: the child enters the hospital without showing signs and symptoms of infection. After the patient has been rehydrated and his nutritional requirements have been replenished, the presence and the physical site of infection become obvious, and the signs of inflammation appear, thus illustrating the similarity between the malnourished child and the immature infant.[27] In the latter, little inflammation around the site of infection may occur, the delayed hypersensitivity response is repressed, and there is little tendency for the invading pathogen to be localized. All of these factors result in increased frequency of infection, increased severity of disease, and disease produced by organisms which under more normal circumstances might not be considered pathogenic.[27]

Furthermore, certain infections in malnourished children may produce severe and prolonged hypoglycemia, which in itself can cause brain damage.[28] Furthermore, infection tends to accentuate various biochemical deficits of children with malnutrition; this, in turn, presumably renders the patient more susceptible to further infections. Finally, many infectious diseases or the therapy used in treatment may result in damage to the central nervous system, without this effect necessarily being evident during the acute stage of illness.[28] Infection and malnutrition thus act synergistically to produce a chronically and recurrently sick child less likely to react to sensory stimuli from his already inadequate social environment. Thus, a third factor, sensory and emotional deprivation related to illness and environment, becomes significant.

Studies among different cultures have demonstrated that one of the more important factors contributing to the intellectual development of a child is the level of maternal education; this influence becomes progressively stronger as the child grows older.[29] Furthermore, studies of families of severely malnourished children have indicated that the majority of these mothers have low intelligence quotients.[30] Socioeconomic factors also intervene indirectly; children suffering from severe malnutrition generally come from homes where the immediate economic pressures are such that the parents cannot provide the child with the stimulation necessary for optimum intellectual development.[31]

Studies of human populations generally have not permitted the investigator to separate the effects of malnutrition from infection and from environmental factors, such as lack of intellectual stimulation and other socioeconomic conditions. Even tissue examination may not be sufficiently sensitive to permit discrimination between these factors; in animals pathological changes quite similar to those accompanying malnutrition can be produced by placing the animal in an environment entirely free of stimulation and by preventing him from exploring it. Furthermore, infections associated with hypoglycemia or severe electrolyte disturbances, as well as recurrent infections (particularly those affecting the central nervous system, the lungs, or the gastrointestinal tract) may act to retard physical growth and may produce tissue lesions similar to those observed in simple malnutrition.[28]

Thus, it is exceedingly difficult to design field studies which would unequivocally decide whether human malnutrition results in permanent impairment of learning and of adaptive behavior. From a purely pragmatic standpoint, this probably makes little difference. If adequate nutrition in early childhood diminishes the incidence of infection as well as the opportunity for sensory and cultural deprivation, the end result might be much the same. To compound the difficulty, observations would have to be carried on for several decades in order to provide a complete answer, since experimentation with animals suggests that poor nutrition of the infant female may affect the development of her offspring born many years later.[3]

Research Needs

Progress in this area would be greatly enhanced by comprehensive field and laboratory studies designed to determine the crucial time in development, if any, when malnutrition causes its major effects, as well as the quantitative and qualitative factors involved. No methods exist to identify and quantitate the biochemical abnormalities of mild, moderate, and severe malnutrition. Little is known of the effects of nutrition on the biochemical processes of nervous tissue, nor is there much information on the relation between mentation and the metabolism of these tissues.

Problems of Psychological Testing

Additional studies, based on psychological testing, would also be useful. It is difficult to perform an adequate evaluation of the intellectual capacity of children from preindustrial societies because more generally available tests have been standardized for other groups and cannot be applied cross-culturally. Furthermore, one cannot accept the idea that the motor development at a particular time is an index which will predict the future intellectual development of the child,[16] since accelerated motor behavior does not necessarily represent a superior intellectual potential. In this instance, intelligence is best defined as the mental adjustment to new circumstances, and it is characterized by increasing complexity in the channels through which the subject acts on objects. Some techniques, such as the Gesell tests, have proved useful, and their reliability on a cross-cultural plane may be high. Thus, examination of the adaptive sphere by the Gesell method, which is based upon the organization of stimuli, the perception of interrelations, and the separation of the whole into its component parts with subsequent resynthesis, can perhaps serve as an analog to later intelligence.[23]

The determination of the Gesell development quotient for a group of chronically undernourished children indicated that infants recovering from severe protein-calorie malnutrition persisted in low performance scores in tests of adaptive behavior during rehabilitation.[23] Similarly, a measurement of visual, haptic, and kinesthetic integration, which appears to be unaffected by cultural factors, shows that children subjected to chronic but moderate degrees of malnutrition, as reflected by their heights, exhibit major functional lags in the development of this capacity.[19] The ability to integrate visual with haptic, haptic with kinesthetic, and visual with kinesthetic stimuli is undoubtedly involved in most learning experiences which depend on the ability to integrate patterned information.

Should studies[32] indicate that the backwardness of many preindustrial societies is related to the inhibitory effects of malnutrition on physical, mental, and emotional development, one would still have to determine how this defect can best be corrected. Simply providing food is not the answer to the problem, since tradition has convinced the people of many cultures that foods we consider important to young children are harmful to them. Thus, nutritional supplements may be sold or traded for other substances thought to be more useful or of greater importance to the adult. Hence, where tradition rather than education determines food utilization, the forces behind the tradition must be understood before positive approaches toward reeducation in food utilization can be undertaken.

Political and Social Considerations

In any discussion of the relation between malnutrition and intellectual achievement and behavior, the enormous political and socioeconomic consequences of such a finding also need be considered.[33] In preindustrial societies, the socially and politically dominant classes can utilize malnutrition as a means of maintaining their control over vast numbers of economically deprived groups. If chronic malnutrition leads to decreased mental function and apathy, it represents a powerful tool of the oppressor over the oppressed. Thus, it would be to the advantage of the ruling group to maintain the majority of the population under conditions favoring widespread malnutrition.

If the socially deprived groups had access to increased amounts of food, then agitation for social, cultural, and political reforms would increase, and should political oppression continue, revolutionary movements and violence would probably ensue. Thus, it seems likely that improvements in nutrition may have the paradoxical effect of causing revolts, although classic political-economic theory would indicate the opposite. Recognition of this fact might therefore handicap any major attempt to reduce the high incidence of malnutrition in many areas of the world.

Summary

Observations on animals and human infants suggest that malnutrition during a critical period of early life results in short stature and may, in addition, permanently and profoundly affect the future intellectual and emotional development of the individual. In humans, it is not known whether these results may be caused by malnutrition alone or whether such intimately related factors as infection and an inadequate social and emotional environment contribute significantly to the problem. Field studies to test these hypotheses are, at best, difficult to design and to carry out; it seems likely that it will prove impossible to separate clearly the individual effects of malnutrition, infection, and social environment. While progress has been made in understanding the biochemical development of nervous tissue, little is known of the effects of nutrition on the physiological processes of these organs, nor is there adequate information on the relation between mentation and cellular metabolism.

References and Notes

1. Dubos, R., Savage, D. & Schaedler. *Pediatrico,* 1966, **38**, 789.
2. Cowley, J. J., & Griesel, R. D. *Journal of Genetic Psychology*, 1963, **103**, 233.
3. Platt, B. S., Heard, C. R. C., Steward, R. J. C. In H. N. Munro & J. B. Allison (Eds.), *Mammalian Protein Metabolism*, Vol. 2. New York: Academic Press, 1964.
4. Cravioto, J. *American Journal of Clinical Nutrition*, 1962, **11**, 484.
5. Dean, R. F. A., & Whitehead, R. G. *Lancet*, 1963-I, 188.
6. Dickerson, J. W. T. & Dobbing, J. *Proceedings of the Royal Society of Biology*, 1967, **166**, 384.
7. Chase, H. P., Dorsey, J., & McKhann, G. M. *Pediatrics*, 1967, **40**, 551.
8. Benton, J. W., Moser, H. W., Dodge, P. R., & Carr, S. *Pediatrics*, 1966, **38**, 801.
9. Winick, M., & Noble, A. *Journal of Nutrition*, 1966, **89**, 300.
10. Barnes, R. H., Cunnold, S. R., Zimmerman, R. R., Simmons, H., MacLeod, R. B., & Krook, L. *Journal of Nutrition*, 1966, **89**, 399.
11. Pond, W. G., Barnes, R. H., Bradfield, R. B., Kwong, E., & Krook, L. *Journal of Nutrition*, 1965, **85**, 57.
12. DeSilva, C. C. *Advances in Pediatrics*, 1964, **13**, 213.
13. Coursin, D. B. *Federal Process*, 1967, **26**, 134.
14. Snell, E. E., Fasella, P. M., Braunstein, A. E., & Fanelli, A. R. *Chemical and biological aspects of pyridoxal catalysis*. London: Pergamon Press, 1963.
15. Flexner, L. B., Stellar, E., de la Halba, G., & Roberts, R. B. *Journal of Neurochemistry*, 1962, **9**, 595.
16. Cravioto, J. Pre-school child malnutrition. *National Academy of Science, National Research Council Publication No. 1282,* 1966.
17. Dubos, R., Schaedler, R. W., & Costello, R. *Journal of Experimental Medicine*, 1968, **127**, 783.
18. Trowell, H. C., Davies, J. N. P., & Dean, R. F. A. *Kwashiorkor*. London: E. Arnold, 1954. Brock, J. F., & Autret, M. *World Health Organization Monograph,* 1952, No. 8.
19. Cravioto, J., DeLicardie, E. R., & Birgh, H. G. *Pediatrics*, 1966, **38**, 319.
20. Valenzuela, R. H., Peniche, J. H., & Macias, R. *Gaceta Med. Mex.*, 1959, **89**, 651.
21. Cabak, V., & Nydanvic, R. *Archives of Diseases in Children*, 1965, **40**, 532.
22. Nables, B., Ramos-Galvan, R., & Cravioto, J. *Bol. Med. Hosp. Infant. Mex.*, 1959, **16**, 317.
23. Cravioto, J. & Robles, B. *American Journal of Orthopsychiatry*, 1965, **35**, 449.

24. Gordon, J. E., Chitkara, I. D., & Wynon, J. B. *American Journal of Medical Science*, 1963, **245**, 345.
25. Gordon, J. E. Prevention of mental retardation through control of infectious disease. *W. S. Public Health Service Publication No. 1692*, 1968.
26. Scrimshaw, N. S. In ref. 16.
27. Gomez, F., Ramos-Galvan, R., Cravioto, J., & Frenk, S. *Advanced Pediatrics*, 1955, **7**, 131.
28. Eichenwald, H. F. In N. S. Scrimshaw & J. E. Gordon (Eds.), *Malnutrition, learning and behavior*. Cambridge, Mass.: M. I. T. Press, 1968.
29. Knobloch, H. & Pasamanick, B. *American Journal of Diseases in children*, 1963, **106**, 77.
30. Martinez, P. D., Ramos-Galvan, R., & De La Fuente, R. *Bol. Med. Hospital Infant. Mex.*, 1951, **6**, 743.
31. Cravioto, J. & Birch, H. G. In Ref. 25.
32. Canosa, D. A. In ref. 28.
33. Richardson, S. A. In ref. 28.

Recommended Additional Readings

Brockman, L. M., & Ricciuti, H. N. Severe protein-calorie malnutrition and cognitive development in infancy and early childhood. *Developmental Psychology*, 1971, **4**, 312-319.

Thompson, W. R., & Grusec, J. Studies of early experience. In P. H. Mussen (Ed.), *Carmichael's manual of child psychology*. (3rd ed.) Vol. 1. New York: Wiley, 1970. Pp. 565-654.

11 What Is the Optimal Learning Environment for the Young Child?

Bettye M. Caldwell

Orientation

The previous article by Eichenwald and Fry examined the effects of deficient early nutrition on later physical structures and cognitive performance. From another focus, but still within the framework of the vital issue of early-experience, the present article centers about the question of the early psychological, learning-inducing environment of the child. (Of course, it will be obvious that these issues of early nutritional and learning experiences are intimately related.)

Two fundamental questions underly the issue of early experiences: (1) How influential are early environmental experiences for later cognitive and emotional development? and (2) Are specific early periods of time in one's life more important than others for the optimal growth of psychological functioning? The answer to the first of these two questions seems to be "greatly influential"; the second question has not as yet been answered satisfactorily, but available data suggest that certain kinds of "critical periods" during the first few months of human life are indeed crucial for establishing important underpinnings for intellectual and personality development. It was Freud's view that early experiences ultimately are determinant for normal psychosexual development; in recent years, much evidence has also pointed to the influence of certain environmental experiences for many kinds of cognitive abilities.

Nevertheless, while developmental psychologists now are in general agreement on the effects of early environmental interactions for later beneficial or detrimental development, there yet remains much uncertainty as to what is the effective *early environment. That is, just what experiences, stimuli, social interactions, or whatever, are essential for optimal development? In this article, Caldwell takes a critical, insightful look at the available evidence—much of which is confusing and contradictory—and draws some tentative conclusions.*

Bettye M. Caldwell is presently associated with the

Reprinted from *American Journal of Orthopsychiatry*, 1967, 37, 8-21, by permission of the author. Copyright, © 1967, The American Orthopsychiatric Association, Inc. Reproduced by permission.

Presented at the 1965 annual meeting of the American Orthopsychiatric Association, New York, New York. The author's work is supported by Grant Nos. MH-07469 and MH-08542, NIMH, U. S. Public Health Service, and by Grant No. D-156(R), Children's Bureau, Social Security Administration, Department of Health Education, and Welfare.

center for early development and education at the University of Arkansas. Formerly she was editor of Child Development *and director of the children's center, department of pediatrics, Upstate Medical Center, State University of New York at Syracuse.*

11

A truism in the field of child development is that the milieu in which development occurs influences that development. As a means of validating the principle, considerable scientific effort has gone into the Linnaean task of describing and classifying milieus and examining developmental consequences associated with different types. Thus we know something about what it is like to come of age in New Guinea,[29] in a small Midwestern town,[4] in villages and cities in Mexico[25] in families of different social-class level in Chicago[12] or Boston,[27, 31] in a New York slum,[46] in Russian collectives,[9] in Israeli Kibbutzim,[23, 34, 41] in the eastern part of the United States,[3] and in a Republican community in Central New York.[10] Most of these milieu descriptions have placed great stress on the fact that they were just that and nothing more, i.e., they have expressed the customary scientific viewpoint that to describe is not to judge or criticize. However, in some of the more recent milieu descriptions which have contrasted middle- and lower-class family environments or highlighted conditions in extreme lower-class settings,[81, 46] often more than a slight suggestion has crept in that things could be better for the young child from the deprived segment of the culture. Even so, there remains a justifiable wariness about recommending or arranging any environment for the very young child other than the type regarded as its natural habitat, viz., within its own family.

Of course, optimizing environments are arranged all the time under one guise or another. For example, for disturbed children whose family environments seem effectively to reinforce rather than extinguish psychopathology, drastic alterations of milieu often are attempted. This may take the form of psychotherapy for one or both parents as well as the disturbed child, or it may involve total removal of the child from the offending environment with temporary or prolonged placement of the child in a milieu presumably more conducive to normal development. Then there is the massive milieu arrangement formalized and legalized as "education" which profoundly affects the lives of all children once they reach the age of five or six. This type of arrangement is not only tolerated but fervently endorsed by our culture as a whole. In fact, any subculture (such as the Amish) which resists the universalization of this pattern of milieu arrangement is regarded as unacceptably deviant and as justifying legal action to enforce conformity.

For very young children, however, there has been a great deal of timidity about conscious and planned arrangement of the developmental milieu, as though the implicit assumption has been made that any environment which sustains life is adequate during this period. This is analogous to suggesting that the intrauterine environment during the period of maximal cellular proliferation is less important than it is later, a suggestion that patently disregards evidence from epidemiology and experimental embryology. The rate of proliferation of new behavioral skills during the first three years of life and the increasing accumulation of data pointing to the relative permanence of deficit acquired when the environment is inadequate during this period make it mandatory that careful attention be given to the preparation of the developmental environment during the first three years of life.

Conclusions from Inadequate Environments

It is, of course, an exaggeration to imply that no one has given attention to the type of environment which can nourish early and sustained growth and development. For a good three decades now infants who are developing in different milieus have been observed and examined, and data relating to their development have made it possible to identify certain strengths and deficiencies of the different types of environments. Of all types described, the one most consistently indicred by the data is the institution. A number of years ago Goldfarb[19] published an excellent series of studies contrasting patterns of intellectual functioning shown by a group of adopted adolescents who had been reared in *institutions* up to age three and then transferred to foster homes or else placed shortly after birth in foster homes. The development of the group that had spent time in the institution was deficient in many ways compared to the group that had gone directly into foster homes. Provence and Lipton[33] recently published a revealing description of the early social and intellectual development of infants in institutions, contrasting their development with that of home-reared children. On almost every measured variable the institutional infants were found wanting—less socially alert and outgoing, less curious, less responsive, less interested in objects, and generally less advanced. The findings of this study are almost prototypic of the literature in the field, as pointed out in excellent reviews by Yarrow[47] and Ainsworth.[1]

Although there are many attributes in combination that comprise the institutional environment, the two most obvious elements are (1) absence of a mother and (2) the presence of a group. These basic characteristics have thus been identified as the major carriers of the institutional influence and have been generalized into an explicit principle guiding our recommendations for optimal environments—learning or otherwise—for young children whenever any type of milieu arrangement is necessary. This principle may be stated simply as: the optimal environment for the young child is one in which the child is cared for in his own home in the context of a warm, continuous emotional relationship with his own mother under conditions of varied sensory input. Implicit in this principle is the conviction that the child's mother is the person best qualified to provide a stable and warm interpersonal relationship as well as the necessary pattern of sensory stimulation. Implicit also is the assumption that socioemotional development has priority during the first three years and that if this occurs normally, cognitive development, which is of minor importance during this period anyway, will take care of itself. At a still deeper level lurks the assumption that attempts to foster cognitive development will interfere with socio-emotional development. Advocacy of the principle also implies endorsement of the idea that most homes are adequate during this early period and that no formal training (other than possibly some occasional supervisory support) for mothering is necessary. Such an operating principle places quite an onus on mothers and assumes that they will possess or quickly acquire all the talents necessary to create an optimal learning environment. And this author, at least, is convinced that a majority of mothers have such talents or proclivities and that they are willing to try to do all they can to create for their children the proper developmental milieu.

But there are always large numbers of children for whom family resources are not available and for whom some type of substitute milieu arrangement must be made. On the whole, such attempts have followed the entirely logical and perhaps evolutionary approach to milieu development—they have sought to create substitute families. The same is usually true when parents themselves seek to work out an alternate child-care arrangement because of less drastic conditions, such as maternal employment. The most typical maneuver is to try to obtain a motherly person who will "substitute" for her (not supplement her) during her hours away from her young child.

Our nation has become self-consciously concerned with social evolution, and in the past decade a serious attempt has been made to assimilate valid data from the behavioral and social sciences into planning for social action. In this context it would be meaningful to examine and question some of the hidden assumptions upon which our operating principle about the optimal environment for the young child rests.

Examining the Hidden Assumptions

1. *Do intermittent, short-term separations of the child from the mother impair the mother-child relationship or the development of the child?* Once having become sensitized to the consequences of institutionalization, and suspicious that the chief missing ingredient was the continued presence of the mother, the scientific and professional community went on the *qui vive* to the possibly deleterious consequences of any type of separation of an infant from its mother. Accordingly, a number of studies[10, 18, 21, 35, 39] investigated the consequences of short-term intermittent separation and were unable to demonstrate in the children the classical syndrome of the "institutional child." In reviewing the literature, Yarrow[47] stressed the point that available data do not support the tendency to assume that maternal deprivation, such as exists in the institutional environment, and maternal separation are the same thing. Apparently short cyclic interruptions culminated by reunions do not have the same effect as prolonged interruptions, even though quantitatively at the end of a designated period the amount of time spent in a mother-absent situation might be equal for the two experiences. Also in this context it is well to be reminded that in the institutional situation there is likely to be no stable mother-child relationship to interrupt. These are often never-mothered rather than ever-mothered children a fact which must be kept in mind in generalizing from data on institutional groups. Thus until we have data to indicate that such intermittent separation-reunion cycles have similar effects on young children as prolonged separations, we are probably unjustified in assuming that an "uninterrupted" relationship is an essential ingredient of the optimal environment.

2. *Is group upbringing invariably damaging?* In studies done in West European and American settings, social and cognitive deficits associated with continuous group care during infancy have been frequently demonstrated. Enough exceptions have been reported, however, to warrant an intensification of the search for the "true" ingredient in the group situation associated with the observed deficits. For example, Freud and Dann[17] described the adjustment of a group of six children reared in a concentration camp orphanage for approximately three years, where they were cared for by overworked and impersonal inmates of the camp, and then transported to a residence for children in England. The children, who had never known their own mothers but who had been together as a group for approximately three years, were intensely attached to one another. Although their adjustment to their new environment was slow and differed from the pattern one would expect from home-reared children, it was significant that they eventually did make a reasonably good adjustment. That the children were able to learn a new language while making this emotional transition was offered as evidence that many of the basic cognitive and personality attributes remained unimpaired in spite of the pattern of group upbringing. The accumulation of data showing that Kibbutz-reared children[34] do not have cognitive deficits also reinforces the premise that it is not necessarily group care

per se that produces the frequently reported deficit and that it is possible to retain the advantages of group care while systematically eliminating its negative features. Grounds for reasonable optimism also have been found in retrospective studies by Maas[26] and Beres and Obers,[6] although in both cases the authors found evidence of pathology in some members of the follow-up sample. Similarly Dennis and Najarian[14] concluded from their data that the magnitude of the deficit varied as a function of the type of instrument used to measure deficit, and Dennis[13] showed that in institutions featuring better adult-child ratios and a conscious effort to meet the psychological needs of the infants the development of the children was much less retarded than was the case in a group of children residing in institutions with limited and unsophisticated staff. It is not appropriate to go into details of limitations of methodology in any of these studies; however, from the standpoint of an examination of the validity of a principle, it is important to take note of any exceptions to the generality of that principle.

In this context it is worth considering a point made by Gula.[20] He recently has suggested that some of the apparent consistency in studies comparing institutionalized infants with those cared for in their own homes and in foster homes might disappear if it were possible to equate the comparison groups on the variable of environmental adequacy. That is, one could classify all three types of environments as good, marginal, or inadequate on a number of dimensions. Most of the studies have compared children from palpably "inadequate" institutions with children from "good" foster and own homes. He suggests that, merely because most institutions studies have been inadequate in terms of such variables as adult-child ratio, staff turn-over, and personal characteristics of some of the caretakers, etc., one is not justified in concluding *ipso facto* that group care is invariably inferior or damaging.

3. *Is healthy socio-emotional development the most important task of the first three years? Do attempts to foster cognitive growth interfere with social and emotional development?* These paired assumptions, which one finds stated in one variety or another in many pamphlets and books dealing with early child development, represent acceptance of a closed system model of human development. They seem to conceptualize development as compartmentalized and with a finite limit. If the child progresses too much in one area he automatically restricts the amount of development that can occur in another area. Thus one often encounters such expressions as "cognitive development at the expense of socio-emotional development." It is perhaps of interest to reflect that, until our children reach somewhere around high school age, we seldom seem to worry that the reverse might occur. But, of course, life is an open system, and on the whole it is accurate to suggest that development feeds upon development. Cognitive and socio-emotional advances tend on the whole to be positively, not negatively correlated.

The definition of intelligence as *adaptivity* has not been adequately stressed by modern authors. It is, of course, the essence of Piaget's definition[32] as it was earlier of Binet.[7] Unfortunately, however, for the last generation or so in

America we have been more concerned with how to measure intelligent behavior than how to interpret and understand it. Acceptance of the premise that intelligent behavior is adaptive behavior should help to break the set of many persons in the field of early child development that to encourage cognitive advance is to discourage healthy socio-emotional development. Ample data are available to suggest that quite the reverse is true either for intellectually advanced persons[42, 43] or an unselected sample. In a large sample of young adults from an urban area in Minnesota, Anderson[3] and associates found that the best single predictor of post-high school adjustment contained in a large assessment battery was a humble little group intelligence test. Prediction based on intelligence plus teacher's ratings did somewhat better, but nothing exceeded the intelligence test for single measure efficiency.

It is relevant here to mention White's[45] concept of competence or effectance as a major stabilizing force in personality development. The emotional reinforcement accompanying the old "I can do it myself" declaration should not be undervalued. In Murphy's report[30] of the coping behavior of preschool children one sees evidence of the adjustive supports gained through cognitive advances. In his excellent review of cognitive stimulation in infancy and early childhood, Fowler[16] raises the question of whether there is any justification for the modern anxiety (and, to be sure, it is a modern phenomenon) over whether cognitive stimulation may damage personality development. He suggests that in the past severe and harmful methods may have been the culprits whenever there was damage and that the generalizations have confused methods of stimulation with the process of stimulation *per se.*

4. *Do cognitive experiences of the first few months and years leave no significant residual?* Any assumption that the learnings of infancy are evanescent appears to be a fairly modern idea. In his *Emile*, first published in 1762, Rousseau[38] stressed the point that education should begin while the child is still in the cradle. Perhaps any generalization to the contrary received its major modern impetus from a rather unlikely place—from longitudinal studies of development covering the span from infancy to adulthood. From findings of poor prediction of subsequent intellectual status[5] one can legitimately infer that the infant tests measure behavior that is somewhat irrelevant to later intellectual performance. Even though these bahaviors predictive of later cognitive behavior elude most investigators, one cannot infer that the early months and years are unimportant for cognitive development.

Some support for this assumption has come from experimental studies in which an attempt has been made to produce a durable effect in human subjects by one or another type of intervention offered during infancy. One cogent example is the work of Rheingold,[36] in which she provided additional social and personal stimulation to a small group of approximately six-month-old, institutionalized infants for a total of eight weeks. At the end of the experimental period, differences in social responsiveness between her stimulated group and a control group composed of other babies in the institution could be observed.

There were also slight but nonsignificant advances in postural and motor behavior on a test of infant development. However, when the babies were followed up approximately a year later, by which time all but one were in either adoptive or boarding homes or in their own natural homes, the increased social responsiveness formerly shown by the stimulated babies was no longer observed. Nor were there differences in level of intellectual functioning. Rheingold and Bayley[37] concluded that the extra mothering provided during the experimental period was enough to produce an effect at the time but not enough to sustain this effect after such a time as the two groups were no longer differentially stimulated. However, in spite of their conservative conclusion, it is worth noting that the experimentally stimulated babies were found to vocalize more during the follow-up assessments than the control babies. Thus there may have been enough of an effect to sustain a developmental advance in at least this one extremely important area.

Some very impressive recent unpublished data obtained by Skeels, offer a profound challenge to the assumption of the unimportance of the first three years for cognitive growth. This investigator has followed up after approximately 25 years most of the subjects described in a paper by Skeels and Dye.[40] Thirteen infants had been transferred from an orphanage because of evidence of mental retardation and placed in an institution for the retarded under the care of adolescent retardates who gave them a great deal of loving care and as much cognitive stimulation as they could. The 13 subjects showed a marked acceleration in development after this transfer. In contrast a group of reasonably well matched infants left on the wards of the orphanage continued to develop poorly. In a recent follow-up of these cases, Skeels discovered that the gains made by the transferred infants were sustained into their adult years, whereas all but one of the control subjects developed the classic syndrome of mental retardation.

The fact that development and experience are cumulative makes it difficult ever to isolate any one antecedent period and assert that its influence was or was not influential in a subsequent developmental period. Thus even though it might be difficult to demonstrate an effect of some experience in an adjacent time period, delayed effects may well be of even greater developmental consequence. In a recent review of data from a number of longitudinal studies, Bloom[8] has concluded that during the first three to four years (the noncognitive years, if you will) approximately 50 per cent of the development of intelligence that is ever to occur in the life cycle takes place. During this period a particular environment may be either abundant or deprived in terms of the ingredients essential for providing opportunities for the development of intelligence and problem solving. Bloom[8] states:

> The effects of the environments, especially of the extreme environments, appear to be greatest in the early (and more rapid) periods of intelligence development and least in the later (and less rapid) periods of development. Although there is relatively little evidence of the effects of changing the

environment on the changes in intelligence, the evidence so far available suggests that marked changes in the environment in the early years can produce greater changes in intelligence than will equally marked changes in the environment at later periods of development. (pp. 88-89)

5. *Can one expect that, without formal planning, all the necessary learning experiences will occur?* There is an old legend that if you put six chimpanzees in front of six typewriters and leave them there long enough they eventually will produce all the works in the British Museum. One could paraphrase this for early childhood by suggesting that six children with good eyes and ears and hands and brains would, if left alone in nature, arrive at a number system, discover the laws of conservation of matter and energy, comprehend gravity and the motions of the planets, and perhaps arrive at the theory of relativity. All the "facts" necessary to discern these relationships are readily available. Perhaps a more realistic example would be to suggest that, if we surround a group of young children with a carefully selected set of play materials, they would eventually discover for themselves the laws of color mixture, of form and contour, of perspective, of formal rhythm and tonal relationships, and biological growth. And, to be sure, all this could occur. But whether this will necessarily occur with any frequency is quite another matter. We also assume that at a still earlier period a child will learn body control, eye-hand coordination, the rudiments of language, and styles of problem solving in an entirely incidental and unplanned way. In an article in a recent issue of a popular woman's magazine, an author[22] fervently urges parents to stop trying to teach their young children in order that the children may learn. And, to be sure, there is always something to be said for this caution; it is all too easy to have planned learning experiences become didactic and regimented rather than subtle and opportunistic.

As more people gain experience in operating nursery school programs for children with an early history deficient in many categories of experience, the conviction appears to be gaining momentum that such children often are not able to avail themselves of the educational opportunities and must be guided into meaningful learning encounters. In a recent paper dealing with the preschool behavior of a group of 21 children from multiproblem families, Malone[28] describes the inability of the children to carry out self-directed exploratory maneuvers with the toys and equipment as follows:

> When the children first came to nursery school they lacked interest in learning the names and properties of objects. Colors, numbers, sizes, shapes, locations, all seemed interchangeable. Nothing in the room seemed to have meaning for a child apart from the fact that another child had approached or handled it or that the teacher's attention was turned toward it. Even brief play depended on the teacher's involvement and support. (p. 5)

When one reflects on the number of carefully arranged reinforcement contingencies necessary to help a young child learn to decode the simple message, "No," it is difficult to support the position that in early learning, as in anything else, nature should just take its course.

6. *Is formal training for child-care during the first three years unnecessary?* This assumption is obviously quite ridiculous, and yet it is one logical derivative of the hypothesis that the only adequate place for a young child is with his mother or a permanent mother substitute. There is, perhaps unfortunately, no literacy test for motherhood. This again is one of our interesting scientific paradoxes. That is, proclaiming in one breath that mothering is essential for the healthy development of a child, we have in the very next breath implied that just any mothering will do. It is interesting in this connection that from the elementary school level forward we have rigid certification statutes in most states that regulate the training requirements for persons who would qualify as teachers of our children. (The same degree of control over the qualifications and training of a nursery school teacher has not prevailed in the past, but we are moving into an era when it will.) So again, our pattern of social action appears to support the implicit belief in the lack of importance of the first three years of life.

In 1928, John B. Watson[44] wrote a controversial little trade book called *The Psychological Care of Infant and Child.* He included one chapter heretically entitled, "The Dangers of Too Much Mother Love." In this chapter he suggested that child training was too important to be left in the hands of mothers, apparently not because he felt them intellectually inadequate but because of their sentimentality. In his typical "nondirective" style Watson[44] wrote:

> Six months' training in the actual handling of children from two to six under the eye of competent instructors should make a fairly satisfactory child's nurse. To keep them we should let the position of nurse or governess in the home be a respected one. Where the mother herself must be the nurse—which is the case in the vast majority of American homes—she must look upon herself while performing the functions of nurse as a professional woman and not as a sentimentalist masquerading under the name of "Mother." (p. 149)

At present in this country a number of training programs are currently being formulated which would attempt to give this kind of professional training called for by Watson and many others. It is perhaps not possible to advance on all fronts at the same time, and the pressing health needs of the young child demanded and received top priority in earlier decades. Perhaps it will now be possible to extend our efforts at social intervention to encompass a broader range of health, education, and welfare activities.

7. *Are most homes and most parents adequate for at least the first three years?* Enough has been presented in discussing other implicit assumptions to make it unnecessary to amplify this point at length. The clinical literature, and much of the research literature of the last decade dealing with social-class differ-

ences, has made abundantly clear that all parents are not qualified to provide even the basic essentials of physical and psychological care to their children. Such reports as those describing the incidence of battered children[15, 24] capture our attention, but reports concerned with subtler and yet perhaps more long-standing patterns of parental deficit also fill the literature. In her description of the child-rearing environments provided by low lower-class families, Pavenstedt[31] has described them as impulse determined with very little evidence of clear planfulness for activities that would benefit either parent or child. Similarly, Wortis and associates[46] have described the extent to which the problems of the low-income mother so overwhelm her with reactions of depression and inadequacy that behavior toward the child is largely determined by the needs of the moment rather than by any clear plan about how to bring up children and how to train them to engage in the kind of behavior that the parents regard as acceptable or desirable. No social class and no cultural or ethnic group has exclusive rights to the domain of inadequate parentage; all conscientious parents must strive constantly for improvement on this score. However, relatively little attention has been paid to the possibly deleterious consequences of inadequacies during the first three years of life. Parents have been blamed for so many problems of their children in later age periods that a moderate reaction formation appears to have set in. But again, judging by the type of social action taken by the responsible professional community, parental inadequacy during the first three years is seldom considered as a major menace. Perhaps, when the various alternatives are weighed, it appears by comparison to be the least of multiple evils; but parental behavior of the first three years should not be regarded as any more sacrosanct or beyond the domain of social concern than that of the later years.

Planning Alternatives

At this point the exposition of this paper must come to an abrupt halt, for insufficient data about possible alternative models are available to warrant recommendation of any major pattern of change. At present there are no completed research projects that have developed and evaluated alternative approximations of optimal learning environments for young children in our culture. One apparent limitation on ideas for alternative models appears to be the tendency to think in terms of binary choices. That is, we speak of individual care *versus* group care, foster home *versus* own home, and so on. But environments for the very young child do not need to be any more mutually exclusive than they are for the older children. After all, what is our public education system but a coordination of the efforts of home plus an institution? Most of us probably would agree that the optimal learning environment for the older child is neither of these alone but rather a combination of both. Some of this same pattern of combined effort also may represent the optimal arrangement for the very young child.

A number of programs suggesting alternatives possibly worth considering are currently in the early field trial stage. One such program is the one described by Caldwell and Richmond.[11] This program offers educationally oriented day care for culturally deprived children between six months and three years of age. The children spend the better part of five days a week in a group care setting (with an adult-child ratio never lower than 1:4) but return home each evening and maintain primary emotional relationships with their own families. Well child care, social and psychological services, and parent education activities are available for participating families. The educational program is carefully planned to try to help the child develop the personal-social and cognitive attributes conducive to learning and to provide experiences which can partially compensate for inadequacies which may have existed in the home environment. The strategy involved in offering the enrichment experience to children in this very young age group is to maximize their potential and hopefully prevent the deceleration in rate of development which seems to occur in many deprived children around the age of two to three years. It is thus an exercise in circumvention rather than remediation. Effectiveness of the endeavor is being determined by a comparison of the participating children with a control group of children from similar backgrounds who are not enrolled in the enrichment program. Unfortunately at this juncture it is too early for such projects to do more than suggest alternatives. The degree of confidence which comes only from research evidence plus replicated experience will have to wait a little longer.

Effective social action, however, can seldom await definitive data. And in the area of child care the most clamorous demand for innovative action appears to be coming from a rather unlikely source—not from any of the professional groups, not particularly from social planners who try to incorporate research data into plans for social action, but from *mothers*. From mothers themselves is coming the demand that professionals in the field look at some of the alternatives. We need not be reminded here that in America at the present time there are more than three million working mothers with children under six years of age.[2] And these mothers are looking for professional leadership to design and provide child-care facilities that help prepare their children for today's achievement-oriented culture. The challenge which has been offered is inevitable. After almost two decades of bombarding women with the importance of their mothering role, we might have predicted the weakening of their defenses and their waving the flag of truce as though to say, "I am not good enough to do all that you are saying I must do."

It is characteristic of social evolution that an increased recognition of the importance of any role leads to the professionalization of that role, and there can be no doubt but that we are currently witnessing the early stages of professionalization of the mother-substitute role—or, as I would prefer to say, the mother-supplement. It is interesting to note that no one has as yet provided a satisfactory label for this role. The term "baby-sitter" is odious, reminding us of just about all some of the "less well trained" professionals do—sit with babies. If

English were a masculine-feminine language, there is little doubt that the word would be used in the feminine gender, for we always speak of this person as a "she" (while emphasizing that young children need more contact with males). We cannot borrow any of the terms from already professionalized roles, such as "nurse" or "teacher," although such persons must be to a great extent both nurse and teacher. Awkward designations such as "child-care worker," or hybridized terms such as "nurse-teacher" do not quite seem to fill the bill; and there appears to be some reluctance to accept an untranslated foreign word like the Hebrew "metapelet" or the Russian "Nyanya." When such a word does appear, let us hope that it rhymes well and has a strong trochaic rhythm, for it will have to sustain a whole new era of poetry and song. (This author is convinced that the proper verb is *nurture*. It carries the desired connotations, but even to one who is not averse to neologisms such nominative forms as "nurturist," "nurturer," and "nurturizer" sound alien and inadequate.)*

Another basis for planning alternatives is becoming available from a less direct but potentially more persuasive source—from increasing knowledge about the process of development. The accumulation of data suggesting that the first few years of life are crucial for the priming of cognitive development call for vigorous and imaginative action programs for those early years. To say that it is premature to try to plan optimal environments because we do not fully understand how learning occurs is unacceptable. Perhaps only by the development of carefully arranged environments will we attain a complete understanding of the learning process. Already a great deal is known which enables us to specify some of the essential ingredients of a growth-fostering milieu. Such an environment must contain warm and responsive people who by their own interests invest objects with value. It must be supportive and as free of disease and pathogenic agents as possibly can be arranged. It also must trace a clear path from where the child is to where he is to go developmentally; objects and events must be similar enough to what the child has experienced to be assimilated by the child and yet novel enough to stimulate and attract. Such an environment must be exquisitely responsive, as a more consistent pattern of response is required to foster the acquisition of new forms of behavior than is required to maintain such behavior once it appears in the child's repertoire. The timing of experiences also must be carefully programmed. The time table for the scheduling of early postnatal events may well be every bit as demanding as that which obtains during the embryological period. For children whose early experiences are known to be deficient and depriving, attempts to program such environments seem mandatory if subsequent learning difficulties are to be circumvented.

---

*In a letter to the author written shortly after the meeting at which this paper was presented, Miss Rena Corman of New York City suggested that the proper term should be "nurcher," a compound of the words, "nurse" and "teacher." To be sure, a "nurcher" sounds nurturant.

Summary

Interpretations of research data and accumulated clinical experience have led over the years to a consensual approximation of an answer to the question: what is the optimal learning environment for the young child? As judged from our scientific and lay literature and from practices in health and welfare agencies, one might infer that the optimal learning environment for the young child is that which exists when (a) a young child is cared for in his own home (b) in the context of a warm and nurturant emotional relationship (c) with his mother (or a reasonable facsimile thereof) under conditions of (d) varied sensory and cognitive input. Undoubtedly until a better hypothesis comes along, this is the best one available. This paper has attempted to generate constructive thinking about whether we are justified in overly vigorous support of (a) when (b), (c) or (d), or any combination thereof, might not obtain. Support for the main hypothesis comes primarily from other hypotheses (implicit assumptions) rather than from research or experimental data. When these assumptions are carefully examined they are found to be difficult if not impossible to verify with existing data.

The conservatism inherent in our present avoidance of carefully designed social action programs for the very young child needs to be re-examined. Such a re-examination conducted in the light of research evidence available about the effects of different patterns of care forces consideration of whether formalized intervention programs should not receive more attention than they have in the past and whether attention should be given to a professional training sequence for child-care workers. The careful preparation of the learning environment calls for a degree of training and commitment and personal control not always to be found in natural caretakers and a degree of richness of experience by no means always available in natural environments.

REFERENCES

1. Ainsworth, M., Reversible and irreversible effects of maternal deprivation on intellectual development. *Child Welfare League of America*, 1962, 32-62.
2. American Women. Report of the President's Commission on the Status of Women, 1963.
3. Anderson, J. E. et al. *A survey of children's adjustment over time.* Minneapolis: University of Minnesota Press, 1959.
4. Barker, R. G., & Wright, H. F., *Midwest and its children: The psychological ecology of an American town.* New York: Harper & Row, 1955.
5. Bayley, N. Consistency and variability in the growth of intelligence from birth to eighteen years. *Journal of Genetic Psychology*, 1949, **75**, 165-196.
6. Beres, D., & Obers, S., The effects of extreme deprivation in infancy on psychic structure in adolescence. *Psychoanalytic Study of the Child*, 1950, **5**, 121-140.

7. Binet, A., & Simon, T. *The development of intelligence in children.* E. S. Kite, trans. Baltimore: Williams & Wilkins, 1916.
8. Bloom, B. S., *Stability and change in human characteristics.* New York: Wiley, 1964.
9. Bronfenbrenner, U., Soviet studies of personality development and socialization. In *Some views on Soviet psychology.* Washington: American Psychological Association, 1962. Pp. 63-85.
10. Caldwell B. M., *et al.* Mother-infant interaction in monomatric and polymatric families. *American Journal of Orthopsychiatry,* 1963, **33**, 653-64.
11. Caldwell, B. M., & Richmond, J. B. Programmed day care for the very young child—a preliminary report. *Journal of Marriage and the Family,* 1964, **26**, 481-488.
12. Davis, A., & Havighurst, R. J. Social class and color differences in child-rearing. *American Sociological Review,* 1946, **11**, 698-710.
13. Dennis, W. Causes of retardation among institutional children. *Journal of Genetic Psychology,* 1960, **96**, 47-59.
14. Dennis, W., & Najarian, P. Infant development under environmental handicap. *Psychological Monographs,* 1957, **71**, (7, Whole No. 536).
15. Elmer, E. Identification of abused children. *Children* 1963, **10**, 180-184.
16. Fowler, W. Cognitive learning in infancy and early childhood. *Psychological Bulletin,* 1963, **59**, 116-152.
17. Freud, A., & Dann, S. An experiment in group upbringing. *Psychoanalytic Study of the Child,* 1951, **6**, 127-168.
18. Gardner, D. B., Hawkes, G. R., & Burchinal, L. G. Noncontinuous mothering in infancy and development in later childhood. *Child Development,* 1961, **32**, 225-234.
19. Goldfarb, W. Rorschach test differences between family-reared, institution-reared and schizophrenic children. *American Journal of Orthopsychiatry* 1949, **19**, 624-633.
20. Gula, H. Paper given at Conference on Group Care for Children. Children's Bureau, January 1965.
21. Hoffman, L. W., Effects of maternal employment on the child. *Child Development,* 1961, **32**, 187-197.
22. Holt, J. How to help babies learn—without teaching them. *Redbook.* 1965, **126 (1)**, 54-55, 134-137.
23. Irvine, E. E. Observations on the aims and methods of child-rearing in communal settlements in Israel. *Human Relations,* 1952, **5**, 247-275.
24. Kempe, C. H. *et al.* The battered child syndrome. *Journal of the American Medical Association,* 1962, **18**, 17-24.
25. Lewis, O. *Five Families.* New York: Basic Books, 1959.
26. Maas, H. Long-term effects of early childhood separation and group care. *Vita Humana,* 1963, **6**, 34-56.
27. Maccoby, E., & Gibbs, P. K. Methods of child-rearing in two social classes.

In W. E. Martin & C. B. Stendler (Eds.), *Readings in child development.* New York: Harcourt, 1954. Pp. 380-396.
28. Malone, C. A. Safety first: comments on the influence of external danger in the lives of children of disorganized families. *American Journal of Orthopsychiatry*, 1966, **36**, 3-12.
29. Mead, M. *Growing up in New Guinea.* New York: New American Library, 1953.
30. Murphy, L. B. et al. *The widening world of childhood.* New York: Basic Books, 1962
31. Pavenstedt, E. A comparison of the child-rearing environment of upper-lower and very low-lower class families. *American Journal of Orthopsychiatry,* 1965, **35**, 89-98.
32. Piaget, J., *The origins of intelligence in children.* M. Cook, trans. New York: International Universities, 1952.
33. Provence, S. & Lipton, R. C. *Infants in institutions* New York: International Universities, 1962.
34. Rabin, A. I. Personality maturity of Kibbutz and non-Kibbutz children as reflected in Rorschach findings. *Journal of Projective Techniques*, 1957, 148-153.
35. Radke Yarrow, M. Maternal employment and child rearing. *Children* 1961, **8**, 223-228.
36. Rheingold, H. The modification of social responsiveness in institutional babies. *Monographs of the Society for Research in Child Development*, 1956, **21**, (63).
37. Rheingold, H. L., & Bayley, N. The later effects of an experimental modification of mothering. *Child Development*, 1959, **30**, 363-372.
38. Rousseau, J. J. *Emile,* (1762). Great Neck, N. Y.: Barron's Educational Series, 1950.
39. Siegel, A. E., & Hass, M. B. The working mother: a review of research. *Child Development*, 1963, **34**, 513-42.
40. Skeels, H., & Dye, H. A study of the effects of differential stimulation on mentally retarded children. *Proceedings of the American Association on Mental Deficiency,* 1939, **44**, 114-136.
41. Spiro, M. *Children of the Kibbutz.* Cambridge, Mass.: Harvard University Press, 1958.
42. Terman, L. M. et al. *Genetic studies or genius*: Vol. 1. *Mental and physical traits of a thousand gifted children.* Stanford, Calif.: Stanford University Press, 1925.
43. Terman, L. M. & Oden, M. H. *The gifted child grows up: twenty-five years' follow-up of a superior group.* Stanford, Calif.: Stanford University Press, 1947.
44. Watson, J. B. *Psychological care of infant and child.* London: G. Allen, 1928.

45. White, R. W. Motivation reconsidered: the concept of competence. *Psychological Review*, 1959, **66**, 297-333.
46. Wortis, H., *et al*. Child-rearing practices in a low socio-economic group. *Pediatrics*, 1963, **32**, 298-307.
47. Yarrow, L. J. Maternal deprivation: toward an empirical and conceptual re-evaluation. *Psychological Bulletin*, 1961, **58**, 459-490.

Recommended Additional Readings

Fowler, W. The effect of early stimulation: The problem of focus in develop-developmental stimulation. *Merrill-Palmer Quarterly*, 1969, **15**, 157-170.
Gray, S. W., & Miller, J. O. Early experience in relation to cognitive development *Review of Educational Research*, 1967, **37**, 475-493.
Hess, R. D., & Bear, R. M. (Eds.) *Early education*. Chicago: Aldine, 1968.
Jeffrey, W. E. Early stimulation and cognitive development. In J. P. Hill (Ed.), *Minnesota symposium on child psychology*. Vol. 3. Minneapolis: University of Minnesota Press, 1969. Pp. 46-67.
Kohlberg, L. Early education: A cognitive-developmental view. *Child Development*, 1968, **39**, 1013-1062.
Thompson, W. R., & Grusec, J. Studies of early experience. In P. H. Mussen (Ed.), *Carmichael's manual of child psychology*. (3rd ed.) Vol. 1. New York: Wiley, 1970. Pp. 565-654.

12   Some Socio-Cultural Factors in Adolescent-Parent Conflict: A Cross-Cultural Comparison of Selected Cultures

Lawrence B. Schiamberg

Orientation

Reprinted from *Adolescence* 1969, 4, 333-360, by permission of the author and Libra Publishers, Inc.

The significance of the growing-up experience in the American culture is perhaps best put into proper perspective by examining this early period of life as it is experienced in other cultures. The following article introduces a number of cross-cultural considerations related to the study of human development, with special emphasis on the adolescent period.

Great difficulty is encountered in the attempt to study and make generalizations about other national

cultures. We are all products of our own cultures; it is inevitable, although we endeavor to remain objective, that subjective elements will enter our analyses. Additionally, there is the problem of genuinely understanding another culture across the language barriers and unique ways of perceiving and thinking. Most cross-cultural studies of human development have been merely the application of American research designs and procedures to one or more other cultures. These studies have been plagued by linguistic differences in meaning, inadequate sampling, and variability of examiners and data-collection procedures. Thus, existing cross-cultural comparisons are probably not very accurate.

A factor which even further exacerbates our understanding of other cultures is the rapidity of technological and social change taking place over the entire world. For example, textbooks in the social sciences frequently refer to Margaret Mead's studies of the Arapesh people and to Ruth Benedict's studies of the Melanesian cultures of Southeast New Guinea. Little attention is given to the fact that these observations were made several decades ago. It seems unlikely that these societies have not been affected by the incredible technological changes that have occurred since. As the article by Schaie (1.4) so forcefully points out, significant changes take place within the culture in which one lives as well as in the individual himself. Thus, making generalizations about socialization practices, personality attributes, and other cultural attributes is hazardous. Dramatic events such as the 1967 war between the Arab countries and Israel introduce new conditions which render previous generalizations invalid.

Despite these many caveats, Schiamberg has presented an enlightening analysis of intergenerational relationships in non-American cultures. His comparison of several societies helps us understand, at least partially, the incredible richness of human diversity. It is all too easy to become locked into one's own cultural perspective; we must constantly recognize that the goggles we wear are very much a construction of our culture, and that individuals growing up in other cultures may very likely wear quite different goggles.

Lawrence B. Schiamberg is on the faculty of the department of educational psychology and guidance at the University of South Dakota.

12

Some level of adolescent-parent conflict has been virtually a constant factor in human societies. The problem of intergenerational relations has been so widespread as to have required some societal response, whether in the form of initiation rites and or rules governing intergenerational behavior.[1] It is generally during the period of adolescence that youth-parent conflicts are intensified because it is during this period that the youth must begin to make progress toward becoming an adult.

Several reasons have been suggested for the so-called "conflict of generations" in Western societies: (a) the different content of experience for youth of the present and for their parents when they were young; (b) the lack of clearly defined steps marking the recession of parental authority over children; and (c) the resulting differences between parents and youth on the psychological and sociological levels (youthful imagination versus adult experience, on the psychological level, and parental role as supervisor of child development versus child's need for independent experience on the sociological level.[2]

The brevity of these statements should not disguise the ultimate complexity of adolescent-parent conflict. For example, the idea of "different experiences" of youth and adults involves a large number of possible combinations, such as the particular style of family relationships, and lack of a sense of historical relatedness due to continual social change and particular traumatic events—to name two of the more general categories of experience. The main point of this paper is that adolescent-parent conflicts, or the so-called "conflict of generations" in the West, are not arbitrary and inscrutable but are directly related to the sociocultural background in which they occur. This point will become somewhat clearer upon examination of several different cultures—their values, norms, and their handling of adolescent-parent conflicts.

However, before examining the "generation gap" in its cross-cultural perspective, an examination of the various explanations of the adolescent-parent conflict will serve to further clarify that point.

Explanations of the conflict of adolescent and parent (father) have followed the pattern of initially treating the problem as an intrapsychic personality problem (Freud) and more recently emphasizing the equally important influence of the sociocultural milieu in helping to shape the personality.

Some Selected Theories

Perhaps the most famous explanation of the "conflict of generations" was Freud's notion of the "Oedipus Complex." Freud thought that the male youth's relationship with his father and, ultimately, with his culture, was determined by how well the son resolved the problem of identifying with his father versus his desire for sexual relations with his mother. The dilemma would hopefully be resolved when the adolescent son observed the father's dominance over the desired sexual object, the mother, and then would identify with the father as a source of power and control. These intrapsychic conflicts of the Oedipal period reappeared during the adolescent stage and then had to be permanently resolved.[3]

The weakness of the Oedipus complex theory is that it seems to underplay the fact that father-adolescent son relationships are conditioned largely by their social and cultural background. Freud's theory treats the problem of the "conflict of generations" from an intrapsychic point of view rather than as a problem of interpersonal relationships influenced by sociocultural norms and values. This is not to deny the existence of intrapsychic aspects of the problem but rather to suggest that Freud perhaps laid an undue amount of stress upon somewhat impressionistic and unverifiable constructs, such as the Oedipus complex.

Erik Erikson's notions of adolescent-parent conflict differ from Freud in that the latter places much more emphasis on the role of the Oedipus complex whereas Erikson emphasizes the social nature of the conflict.[4] Erikson's notion of adolescence involved the problem of establishing one's "ego identity." This concept primarily involved the individual's relationship to his parents and other individuals in his world, and the establishment of a general "stance" toward the world. Erikson was to some extent influenced by the work of cultural anthropology because he recognized that the method and content of adolescent ego identity would differ from culture to culture. Erikson did suggest that the achievement of adolescent ego identity had one element in common for all cultures: the adolescent must receive meaningful recognition of his achievements from his parents and from his society.[5]

Erikson recognized two major sources of conflict between adolescents and parents: (1) the failure of parents to accord recognition of adolescent achievement; and (2) adolescent revolt against the values and dominance of the parents. Erikson indicated that youths rarely identified with their parents during adolescence and often rebelled against their parents in their quest for ego identity. Ultimately, Erikson thought the adolescent must establish his ego identity by adopting and formulating a stance toward the world. Erikson seems to extend the Freudian concept of same-sex identification to include the social implications of a failure to achieve ego identity. Thus, in Erikson there is an emphasis upon the social nature of adolescent-parent conflicts, although he does not divorce himself from strict Freudian interpretation such as the Oedipus complex theory.

Perhaps the work of Kurt Lewin applies most specifically to the nature of adolescent-parent conflicts, since Lewin was concerned with both the stage of adolescence and the resolution of social conflicts. Lewin approached the problem of intergenerational conflicts from the point of view of his general theory of behavior which stated that behavior is a function of the person and the environment $[B = F(P.E)]$. The sum of all environmental and personal factors (motivation, needs, perception, etc.) in interactions is called the life space, and behavior is also a function of this construct $[B = F(L.Sp.)]$.[6] Lewinian Field theory recognizes that life spaces may differ between individuals and cultures. Field theory views adolescence as a transition from the life space of the child to the life space of the adult. Because of the rapid and somewhat abrupt shifts in the life space of the adolescent, he often becomes a "marginal man" with one foot in the world of childhood and the other in the world of the adult. The adolescent often experiences emotional tension due to the ambiguity of his social position. The resulting tension may affect adolescent-parent relationship and often produces conflicts over role expectations. Perhaps the most important contribution of Lewin to the study of intergenerational conflicts is his notion of the general cultural atmosphere or background for social situations: "In sociology, as in psychology, the state and event in any region depend upon the whole of the situation of which this region is a part."[7]

Some recent ideas on the relationship between personality and culture—and by implication the relationship between the generations—seem to have followed in the footsteps of Kurt Lewin. The background of parent-adolescent conflicts is seen to be interaction of the personality system and the sociocultural systems, which in turn are mediated by the family system.[8] The social system has certain functional prerequisites such as role differentiation, shared goals and values, and communication, while the personality system has certain requirements such as the satisfaction of needs and recognition of achievement. The family system mediates between the personality and social systems by ensuring the presence in individuals, of societal goals and values and by providing an atmosphere in which the achievement of these goals is recognized. The socialization processes of the society—of which the family system is a prominent force—serve to create a congruence between the functional prerequisites of the society and the motivational patterns of the personality system.[9] Parent-adolescent crisis, if it does develop, occurs within this complex arrangement of personality, cultural, social, and familial variables. The problem of the conflict of the father with the male adolescent—the main concern of this paper—cannot then be divorced from this complex interrelationship of variables. Whether one approaches the problem from the viewpoint of a learning theory or a cognitive theory or any other point of view, the parent-adolescent relationship is ultimately limited by the nature of the particular socio-cultural values in each particular culture.

The conflict of generations would seem to be best explained from a cultural anthropological view which recognizes that the parent-adolescent relationship is related not only to particular personalities and temperaments, and particular circumstances, but is also related to cultural and societal values and norms which influence the parent-adolescent relationship through the medium of the family. In light of this stance, it makes no sense to condemn industrialized Western societies for what appears to be a greater prevalence of such conflicts and rivalries, since many Western societies are so much more complex and more rapidly changing than some Eastern societies which apparently have less parent-adolescent conflict. Further, as Margaret Mead has indicated, the "Sturn und Drang" characteristic of American adolescence is perhaps the price we pay for our high level of technology and material welfare.[10]

In many of the cultures and societies to be discussed in this paper, Western technological innovations are beginning to reshape the society and, therefore, the parent-adolescent relationship. With increased industrialization in Eastern countries such as India and China, adolescent-parent problems begin to appear which bear a marked resemblance to those of the West. For example, in China, as more job opportunities have become available to adolescents, the traditional dependence of the young upon their elders has begun to disappear and with it has gone the centuries-old tradition of unqualified respect for one's elders and ancestors.[11] A "generation gap" has begun to appear as the experiential worlds of Chinese youth have become different from the experiential world of their elders and ancestors. These experiential differences have formed a basis for more conflict between adolescents and their parents. When this situation in China (a situation which has not, as yet, permeated all of Chinese society) is compared with the United States—in which the so-called "computer revolution" is further extending the effects of industrialization—then perhaps it is not so surprising that differences of values and intergenerational conflicts are perhaps more prevalent in the West.

The question of adolescent-parent conflict has important ramifications for society at large and for the schools in particular. In Western societies, as is the case in many non-Western societies, the family is the primary influence on the developing child. However, in static (unchanging) societies parental influence is often continued into latter childhood and often into adolescence and adulthood, as children learn virtually all they need to know from their parents. However, in American society the schools have developed as a means of training youth for adult responsibilities (especially those skills necessary in a technological era). The schools (especially those devoted to higher education) in advanced technological societies serve to provide youth with the kind of information and skills which are not the specific function of the family or other societal institutions to transmit. If the schools serve only to transmit and teach technological and social skills without confronting the problem of the adolescent-parent conflict and, more generally, the conflict of generations, then the problem can only become worse. The purpose of this paper is not to dispute whether the apparent "price" paid for social change is too high in terms of individual discontent, but rather to place the problem of adolescent-parent conflict in perspective: (1) by showing the intimate relationship between sociocultural values and methods of preventing adolescent-parent conflict in both nontechnological societies and in societies which are only beginning to be influenced by Western technology; and (2) to present some of the educational implications of adolescents and parents who are "out of step" with one another.

Some Selected Cultures

Indian society reflects certain characteristics of Eastern societies which have led to much lower levels of parent-adolescent differences and conflicts. Indian society has traditionally placed a great deal of emphasis on the quality of interpersonal relationships. Relatively little value is placed on the quality of material existence. Indian society has been relatively static for centuries, and as such the traditional values and norms of the society have remained unchanged and unchallenged. Generational gaps have not developed because life styles have remained constant for centuries.

A characteristic of Indian society which alleviates the strain of potential adolescent-parent conflicts and is consistent with the societal emphasis upon quality of interpersonal relationships is the Indian "extended family." This is the characteristic type of household in India.[12] The extended family allows for a number of interpersonal relationships whereas the nuclear family tends to produce more developed and intense relationships with fewer people. The extended family allows for more distribution of emotions and feelings over a greater number of family members, as compared with the nuclear family with fewer members, a potentially higher concentration of emotions per relationship, and therefore a higher likelihood of potentially explosive intergenerational relationships.[13]

In India the institution of "Asrama" insures the smooth transfer of authority from generation to generation. According to this tradition, sons owe complete obedience to their fathers, while fathers are required to eventually relinquish their authority so as to avoid conflicts between adolescent males and their fathers. The Indian male is supposed to go through four Asramas, or "age grades" in Western terminology. At the age of eight the young enters the first Asrama or the celibate stage. During this period the child is guided by one or more instructors who teach him about the sacred love, the arts, the use of weapons, and the profession which he will eventually take up.[14]

After his educatory period, the adolescent (age of twenty) is admitted into the next age grade—that of "householder." In this stage, he gets married, starts a home, and establishes himself in the profession for which he has been trained. This stage lasts until his own son reaches the "householder" stage or, if he has no children, until he is middle-aged (determined when his hair turns grey). At this point, he enters the third age grade, or "Vanaprasthasrama." During this stage the man is relieved of his household duties and is now free to devote his time to meditation or to worship. The man may live at home if he so desires, although an orthodox Hindu usually lives in a nearby house. The Vanaprasthasrama age grade ensures that by the time a man's son is able to enter the householder stage and accept the responsibilities of manhood, the man of the house is leaving the household, thus allowing a smooth transfer of authority.[15]

Underlying the social structure of the Asrama is the basic Hindu ideal of "dharma" (ideal duty) which consists of the right behavior appropriate to one's particular stage and station in life. According to the Hindu religion a man is born with three debts: (1) a debt to the gods; (2) a debt to the sages; and (3) a debt to his ancestors. The individual pays these debts by worship and ritual, learning and teaching, and by raising children.[16] These activities comprise the various age grades or asramas. Indian life is carefully organized so that each individual is constantly "paying" one of his three debts, while respecting his elders at all times.

Another underlying current in Indian life which reduces intergenerational tension is the emphasis on the ascetic life (as manifested in worship and meditation). Respect for the aged, retired Indian who devotes his last years to religious meditation is encouraged because of the widespread Hindu ideal of respect for the ascetic life. The Hindu belief that self-denial is superior to self-indulgence unites all Hindus in their respect for the aged and further contributes to the lessening of conflicts between adolescents and parents.[17] Respect for one's elders is an absolute requirement of behavior for adolescents in India.

The above description of the nature of the parent-adolescent conflict in the Indian family is perhaps most applicable to the more orthodox Hindus and higher caste families and somewhat less descriptive of lower caste families, Muslim communities in India, and families influenced by Western ideals of family relationships. Even for the orthodox Hindu families and the upper caste families, the pattern has been changing very slowly. With increasing trends toward urbanization and the greater availability of factory jobs, the adolescent has begun to liberate himself from complete dependence on the extended or joint family. For thousands of years the extended family has been the main economic force in Indian society. All members of the joint family operate as a single unit, contributing all their earnings to the entire extended family. One might expect greater resistance to this tradition, once individuals are able to establish themselves independently of the family. Although India is still far from being an industrialized nation, increasing industrialization and concomittant social change would appear to be factors in the breakdown of the traditional structure of adolescent-parent relationships.

Unlike the family structure of the United States or that of India, the traditional Chinese society (before the Communist takeover) placed great emphasis on differences of generation as the basis of role differentiation. In the traditional

Chinese society the father was respected not because he was older, or a "good father," but because he was of a different generation. The younger generation always had the burden of responsibility toward the older generation. Traditional Chinese society was a patriarchal society in which the men dominated the women and the older generation dominated the younger generation by control of the economic roles of the family. This control was often exercised by the older generation to restrain adolescents from leaving the household or otherwise causing problems, since these adolescents often had no alternative employment except that offered by their parents.[18] (This situation began to change in the twentieth century as Chinese cities became industrialized and more job opportunities were available to Chinese adolescents.) The fact that the traditional Chinese family was the primary economic unit in the society placed a great deal of power and authority in the hands of the older generation who controlled the extended family. Respect for one's elders by the Chinese adolescent was not only a traditional value of the society but was also a socioeconomic necessity.

Traditional Chinese society also had a hierarchal chain of power. Older brothers were more important than younger brothers and were responsible to their parents for insuring the proper behavior of their siblings. Through these chores and responsibilities, adolescents usually gained experience in controlling young children which would be of enormous assistance when raising their own children. In the determination of this power hierarchy, traditional Chinese society emphasized "particularistic" factors such as one's generation and the sequence of siblings. Certain Western societies (e.g., United States) place more emphasis on "universalistic" factors such as the ability to run a family.[19]

In comparison with the United States, the dominance of the older generation in the traditional Chinese society holds for both the early and later years of life. In American society the older generation dominates for the early years of growth and not necessarily for the later years.[20] This dominance of the Chinese youth and adolescents was the result of traditional Chinese inheritance customs. The land and property were not supposed to be divided among the children until the father died. No matter how ineffectively an old man carried out the household chores or the work in the fields, he was still recognized as the head of household with authority over all those of a younger age. This remained the case even though younger family members might, in fact, be doing more efficient work than the titular household head. The inheritance customs buttressed the tradition of ancestor worship and respect for one's elders. There was no problem dealing with the rate of recession of parental authority as in the United States, since parental authority was virtually a lifelong fact. Thus, three main factors established the locus of power in traditional Chinese society: (1) sex (males dominate); (2) generation; and (3) relative age (oldest siblings have preference).

From childhood on, the youth and adolescent was virtually immersed in a culture which stressed filial piety. There simply were no exceptions to the rule—that is, no successful exceptions. The child and the adolescent were confronted with examples of filial piety in their daily lives, in novels, textbooks, and in nursery schools. Veneration of the older generation and of one's ancestors was possible in traditional China because it was essentially a static culture in which the life experiences of one generation were almost identical with those of any other generation.[21] Perhaps the respect for age and experience rather than youth and imagination is best expressed by Confucius. When asked how to farm, Confucius replied: "I do not know as much as an old farmer."[22]

Respect for the aged was in part derived from the Confucian ideal that the good life consisted of the proper behavior between individuals. According to Confucius there were five types of interpersonal relationships: (1) the parent-child relationship; (2) the King-minister relationship; (3) the husband-wife relationship; (4) the older brother-younger brother relationship; and finally (5) the friend-friend relationship. Confucius felt that the highest form of respect that could be shown was between father and son. More specifically, that respect which the son showed for his father. Confucius felt that the father-son relationship was the archetypal pattern for the other four basic types of relationships. This Confucian ideal was in fact quite practical, in light of the traditional Chinese inheritance customs.[23]

The traditional precedence of the older generation has broken down in modern China. With the coming of industrialization and its emphasis on youth and adolescents as those best qualified to learn the new technological and factory jobs, the unqualified respect for the older generation began to deteriorate. The role of the traditional Chinese family as the educator of the young for occupational work is being usurped.

The gradual transition from an emphasis on age to an emphasis on youth was one of the primary trends which the Communist movement seized upon to gain support among Chinese youth. When the Communists finally gained control of China after World War II, they encouraged progressive young people to disregard existing kinship ties and to ignore the concept and practice of ancestor worship.[24] Soon progressive adolescents and young people became feared throughout China because any word from them to the Communist leadership of the practice of age prestige rituals could lead to stiff reprisals against the "guilty." Youth in "transitional" China have also rebelled against the traditional parental control of marriages and divorces. Marriage soon became focused upon the husband-wife relationship, the nuclear family (husband and wife rather than the clan), and on the free choice of one's mate. Thus, with the collapse of the family as the main economic unit in Chinese society and with the subsequent decline in age prestige, traditional adolescent respect for parents has begun to wane.

Indian and Chinese societies have traditionally had considerably less adolescent-parent conflicts primarily because the traditional concept of adulthood has been coherent and meaningful to the adolescents of these societies. There has been little difficulty in becoming an adult because the prerequisites for adult status were within the reach of virtually all adolescents. The only requirement was that adolescents become reasonably proficient in such tasks as working in the fields and caring for the young—tasks for which they had been trained all their lives. There were no frustrations involved in the choice of vocation, level of education, or style of life since there were virtually no alternatives to the existing social framework, nor any notion that society could be any different than it always had been. In short, where there were no choices and virtually everyone accepted the existing life styles, there were fewer individual frustrations—that is, fewer frustrations caused by having to choose between two or more alternative styles of life and accepting the consequences of that choice, but probably more frustrations in getting enough food—and fewer frustrations in interpersonal relatices such as the principle of reciprocity (the societal norm of mutual duties and

The traditional simplicity and integrated social organization of China and India have permitted an emphasis on the quality of interpersonal relationships. In Erikson's terms, the attainment of "ego identity" has been so much easier in India and China that much smoother interpersonal relationships have been more likely to occur than in societies where ego identity is more difficult. Where there is greater agreement on the means and ends of life, and where adulthood is both possible and meaningful, adolescent-parent respect is greatly facilitated. Respect for one's elders and one's ancestors is more likely when one can be sure that Confucius' old farmer does, in fact, know more about farming than Confucius.

In some African societies the parent-adolescent relationship is linked primarily to an individual sense of reciprocity with the groups of which one is a member.[25] Principles of mutual rights and mutual duties—especially between the father and his adolescent son—run strongly through interpersonal relationships in many of the African societies such as the Tallensi.[26] Religious beliefs often further reinforce the influence of social groups ("spirits"—especially the spirits of the wrongdoer's ancestors—are frequently seen as forces of retribution.[27] Further, many African societies have clearly delineated systems of courts and methods of hearing disputes between the various clan members and between the various clans or joint families. Many African societies handle the problem of parent-adolescent conflict with clearly spelled out age grades or initiation ceremonies or some combination of the two (e.g., the pastoral Massai tribe). Other societies such as the Tallensi or the Mossi, lacking in highly specific age grades or initiation ceremonies, rely on other societal institutions or practices such as the principle of reciprocity (the societal norm of mutal duties and rights in interpersonal relationships, the practice of parent-adolescent avoidance, or refined methods of parent-adolescent dispute arbitration).[28]

In the social system of the pastoral Massai, the male sex is divided into boys, warrior, elders, and old men.[29] The first age grade ("boys") lasts until circumcision somewhere between the ages of thirteen and seventeen. Those who are circumcised automatically become members of a peer group of circumcised youth called by some distinctive name such as "white swords."[30] The newly circumcised youths do not become adults until they reach a senior age grade and are then allowed to marry. The newly circumcised live in a separate age-set village and are entirely under the authority of the elder age grades. Circumcision is thus only a first step to becoming an adult man. It might be thought of as the prerequisite for passage from childhood to adolescence, whereas marriage is the "rite de passage" from adolescence to adulthood.[31] Once the youth acquires a wife he becomes a "junior elder." Junior elders are mostly concerned with family matters (their interests are mostly of a private nature) and they play a rather small role in tribal politics. Their major concern is to increase their status by having more wives and by having their own children circumcised and then initiated. Once junior elders have had one of their children initiated, they become "senior elders" with a full share in the tribal political life and assume the prime responsibility of initiating new members into Massai manhood.[32]

The major importance of the Massai system of age grades and initiation rites is that they regulate an individual's conduct in relation to those in similar or different age grades. The basic prerequisite for effective initiation ceremonies and age grades—the solidarity of the initiators in their concept of the role of an adult—is fulfilled in Massai society.[33] Unless this solidarity exists, initiation rites do not have a very powerful effect. Solidarity exists not only among the elders but also among the lower age grades, all of whom subscribe to the personality type preferred for each of the various age grades. These strong forces of cohesion among the various age grades direct attention away from inter-age grade struggles and channel it toward fulfillment of the specified goals of each age grade.

Not all societies require such rigid maintenance of age grades and "rites de passage" as do the pastoral Massai. For example, in the Tallensi society the transition from childhood to adulthood is a gradual process. The Tallensi do not have rigidly defined age grades but two rather loose criteria on which they determine the developmental level of an individual: the physiological criterion of pubescence and full physical maturity, and the economic criterion of development of skills requisite to doing a man's work in the fields. The attainment of adulthood in Tallensi society is based on a rather variable and somewhat flexible schedule. There is no social break to mark the transition from childhood to adulthood.[34]

For the Tallensi, the cornerstone of filial piety is the basic notion that the bearing or begetting of a child is a difficult matter commensurate with a great deal of respect. Because childbearing is thought to be so difficult, children are taught to have respect for the fact that their parents brought them into the world, even though they may or may not have respect for their parents as persons. In this respect, the Tallensi believe that the bonds between the youth and the parents can never be obliterated or repudiated. The Tallensi concept of filial piety is a diffused norm rather than a specifically elaborated doctrine as in the case of the Chinese.[35]

The principle of reciprocity, or the mutual rights and duties of fathers and their adolescent sons, operates to prevent conflict between them.[36] In the parent-adolescent relationship, reciprocity operates in two notable instances to prevent conflict: (1) in the case of the use of the son's property by the father; and (2) in the arrangement of a marriage by the father. In the former case, if a man uses one of his son's cattle or goats he must eventually make an equivalent return. In the second instance, while the son is always under the authority of the father, it is the latter who arranges the son's marriage. Thus, the principle of reciprocity helps to reduce parent-adolescent conflicts by ensuring that, in the long run, mutual services and favors must balance.

Other means of reducing parent-adolescent conflict are held in common with the Mossi tribe of the Voltaie Republic just north of Ghana. The basic economic activities of the tribe include the production of livestock and cotton, and caravan trading. If a wife gives birth to a male child as her first born, the father sends the child away to live with the boy's maternal relatives until he reaches puberty. The father then has little contact or conflict with his future successor. This practice of initial removal of the newborn child is a rather drastic means of avoiding parent-adolescent conflict.

Since the Mossi firstborn males eventually inherit the wives of their fathers, it is often those wives who produce friction between fathers and their adolescent sons. In order to avoid conflict, Mossi firstborn sons are usually allowed to visit the family compound for important purposes only. It is an unfortunate circumstance when a father encounters his oldest adolescent son at the doorway of the household compound. In order to avoid this situation, the father or son usually shout something loud upon entering or leaving the household.[37]

Mossi fathers usually do not want to procure wives for their sons because the Mossi father sees the possession of a wife as representing the advantages of majority status. Since the son will gain this status only when the father dies, the antagonism between father and son is increased. It is not surprising that fathers often become extremely jealous of the development of their adolescent sons since this usually signals—at least in their minds—their eventual demise and decline in power. Mossi fathers are careful to avoid situations in which they are or could be directly compared with their adolescent sons. Often a Mossi father will not be seen walking with his eldest son in the village, lest someone accidentally fail to acknowledge their difference in age.[38]

Thus, in Mossi society the major indication of majority status and the major basis for adolescent-parent conflict is the possession of a wife by the adolescent. This state of affairs is further complicated by the fact that the eldest adolescent son usually inherits his father's wife upon the death of the father, and also because the adolescent youth remains completely dependent upon the father for obtaining a wife. (Adolescent sons do not possess the property to trade for a wife and must rely upon their fathers to get them one.) Interpersonal relationships between fathers and their adolescent sons are strained because of the social structure of Mossi society in which a confrontation between father and son is openly acknowledged, although adequate precautionary measures are taken. Perhaps the adherents of the Oedipal theory of generational conflict would point out that the very fact that the Mossi mothers are the subject of conflict between fathers and sons is evidence that the Oedipus complex produces adolescent-parent conflict. The problem with such a position is that there is little or no objective evidence that what appears to be a manifestation of the Oedipus complex in Mossi society does in fact occur in other cultures with any kind of regularity.

Besides the Mossi practice of avoidance, their system of intergenerational conflict resolution helps to alleviate adolescent-parent conflicts. When a serious decision or problem arises between parents and their adolescent youth, the head of the extended family is consulted. No decision regarding such intergenerational problems is ever made without consulting the head of the extended family. This type of decision-making process helps to improve parent-adolescent relations by (1) insuring that adequate reflections and thought precede each decision, and (2) by preventing direct confrontations between parents and adolescents.[39]

Further supplementing the practices of avoidance and dispute arbitration in maintaining smooth relationships between parents and adolescents is the practice of shared adolescent discipline. Disciplining of Mossi adolescents is a shared responsibility between the father—the family head—and the head of the adolescent work group. After Mossi adolescents are circumcised at about the age of thirteen or fourteen, they join a "work group" composed of other youths recently circumcised. In the work group the youths perform work in the fields and are given their food and lodging for the seven-month period that they work there. During this work period the adolescents are responsible to the work group leader—usually an adult—for proper behavior and may be disciplined by the leader for improper actions. The division of disciplinary responsibility in this fashion reduces the possibility that conflict regarding disciplinary practices will develop between father and adolescent.[40]

Should the situation ever arise that the eldest adolescent has been given a wife, the son is required to leave the family household and build his own "soukala" or hut, even though he will eventually return to his father's house to inherit his property when the latter dies. The Mossi do this because of a strong belief that young married adolescents have a strong need for independence and that if the father were to keep his married son at his side, the latter might wish his father's death so that he could achieve independence.[41] (Mossi fathers maintain the right to control the lives of their children until the death of the children or the father. Mossi children rarely achieve independence until the death of the father, at which time the property is divided and the father's wives are inherited.)[42]

Thus, the problem of adolescent-parent relationships is kept within reasonable bounds by means of several rules of behavior which govern the relationship of the adolescent and the parent. Potential intergenerational rivalry is fostered primarily by the social custom of inheritance of wives. In order to prevent such conflict the following rules of behavior have been established: (1) the enforced removal of the eldest male child from the household; (2) the dispute arbitration process; (3) the divided disciplinary responsibility; and (4) the departure of the eldest married son from the household. The parent-adolescent conflict is rooted deeply in the structure and functions of the Mossi culture. The primary economic and social force in Mossi culture is the joint family. In order to maintain the solidarity of the family unit, the act of acquiring wives is reserved for the elders of the extended family—the father, and the clan chief or head. All marriages are arranged to enhance the extended family and to ensure that all new family members and new wives will remain loyal to the extended family. The inheritance of wives indicates not only the relatively low position of women in Mossi society but also serves to maintain the continuity and effective operation of the Mossi extended family by retaining effective women workers in the family unit. The code of behavior governing adolescent-parent relationships is the Mossi method of successfully integrating personal needs of both fathers and their adolescent sons with the functional prerequisites of a strong and efficient extended family.

In the Arab world the general guidelines for parent-adolescent relationships are laid down in the religious mores and teachings which pervade all aspects of Arab life. Muhammed is reported to have said: "Whoever has a son born to him, let him give him a good name, teach him good manners, and when he reaches puberty get him married . . . if he reaches puberty and has not been married and falls into sin, it is the father who is responsible. . . ."[43] The Qur'an (holy book of the Muslim faith) advocates proper respect toward one's parents and states that filial piety is the highest form of good works. In practice these ideals are usually translated into a rather stern father who exercies his absolute authority over his children in arranging their marriages and ensuring that they do not "fall into sin," lest he be responsible. Adolescents are required to show absolute respect and obedience to their father.

Among Arab families there usually exists a close relationship between mother and youth and a rather loose one between father and adolescent boy. The Arab father has complete authority over both sons and daughters. Generally the father is the disciplinarian while the mother helps to keep the family together as an integrated unit and often acts as a buffer between the father and the adolescent. If the adolescent gets into any kind of trouble, help is usually sought from the mother rather than from the father. The parent-child relationship is usually a short one for girls, who generally marry at a rather early age and leave the household, whereas sons have an extended relationship with their parents. The son eventually marries and brings his wife with him to live with his family.[44]

The rights of Arab youth and adolescents are usually fixed by custom and/or law. The child has the right to food, care, and upbringing. The child has three types of guardianship: (1) guardianship of upbringing (tarbiya) which is accomplished by the mother and usually ends when the child is seven or nine years old; (2) guardianship of education (spiritual guardianship) which involves proper training in the values and rules of the society; and (3) guardianship of property which involves the maintenance of the adolescent's property until he reaches majority status.[45]

In the villages of Turkey, the male population is divided into four groups: (1) children; (2) unmarried youths ("delikani"); (3) young men (genc); (4) old people (ihtiyan). Marriage is the only necessary requirement to move from the delikani age grade to the genc stage. Passage from childhood to delikani is accomplished by growing a mustache, while passage from genc to ihtiyan is accomplished by growing a beard to accompany the moustache. Circumcision is a necessary requirement before one can marry.[46]

Villages in Turkey are made up of households containing extended families. Status in this society is designated by differences in age. Old people are held in greater esteem than the young. Respect for age difference is demonstrated dramatically at household dinners and community feasts. Old men and guests always sit in places of honor and are served first.[47] Adolescents must always address their elders with specific titles or kinship terms indicating their age grade, and are rarely permitted to address them with their actual names. Competition between adolescent and parent is further diminished by the social concept of status in Turkish life. One always outranks those in lower age grades. Status is determined by one's position in his age grade (a man is judged by his achievements in comparison with other individuals of his age).[48] In this manner, interage grade comparisons are usually avoided and are relatively insignificant if they occur at all. Such comparisons between adolescents and adults are relatively unimportant since both are virtually agreed on the basic goals of adolescence—to get the adolescent married and to prevent him from "falling into sin." Thus, there is less intergenerational conflict than in the West because life experience in Turkey has been relatively static for hundreds of years. There is no conflict because there are no alternative styles of life.

Transmission of knowledge to the adolescent presents no real problem because this knowledge is virtually the same from generation to generation and specific social institutions have long satisfied this need. The most important semiformal situation for learning the values of the village culture and the role of man in village society are the evening meetings at the Muhtar's home (the Muhtar is the village head—usually an older experienced man).[49] The meetings are usually attended by all the males in the village although they are primarily directed at the adolescent males.

The relatively homogeneous values of the Turkish village, which promoted a smooth transition from adolescence to adulthood and which greatly lessened the incidence of parent-adolescent conflict, have nonetheless been threatened by social change. Problems have begun to arise due to increasing number of youths going on to higher education after grade school and, therefore, not getting married at the customary age of about nineteen.[50] Most of the towns and villages now send their children to a "grade school" from which qualified graduates continue on to a junior high school and to a high school ("lise"). As of 1961, very few of the villagers had finished grade school.[51] However, with the increasing number of village children and adolescents who complete high school and continue on to some form of higher education, youths are delaying marriage. This failure to marry at the traditional age had perpetuated the status of Turkish adolescence such that, in the future, there may no longer be a clear and distinct difference between adolescents and adults. The implications of this social change for the adolescent-parent relationship are that experiential differences between adults and adolescents and the special opportunities afforded to the educated youth may, in fact, lead to a kind of "generation gap."

Summary, Conclusions and Observations

In conclusion, the behavior norms and role expectations provide the basis for the smooth adolescent-parent relationships surveyed in the several cultures presented in this paper. Nontechnological societies which have relatively clearcut and broadly accepted societal goals and values have less adolescent-parent conflict than societies in which individuals are confronted with the choice of many different occupations and life styles. This is not to suggest that the so-called generation gap has reached the critical stage in industrialized nations or that choice of life styles is a bad thing, but that where conditions exist which could create experiential gaps between parents and adolescents, the stage is set for more numerous intergenerational conflicts.

Surely, not all adolescent-parent conflicts are caused by social change or are avoided merely by the absence of social change. Personality variables, undoubtedly play a role. Perhaps a solution to the problem of the generation gap lies in the development of certain cognitive styles which will promote understanding between the generations in the face of ever increasing social change. Perhaps a feasible method of partially alleviating the problem is for the school to assume some leadership responsibility in teaching individuals to accept change and to adapt to a world of ever increasing change and complexity in which the establishment of "ego identity" (a stance toward the world) in adolescence becomes ever more difficult. Perhaps what is needed—and is most difficult to develop in individuals—is a tolerance for uncertainty. As the rate of social change increases, the shape of the future becomes somewhat fuzzy and the schools become less certain of how to prepare people for that future. To the extent that this kind of situation exists, it would seem logical that the more successful individuals will be those who can tolerate and live with this uncertainty. David E. Hunt of the Ontario Institute of Education Studies has developed a model of cognitive development which has as its goal a conceptual style which permits numerous combinations of information processing, as opposed to monolithic or stereotyped categorization of information.[53] Higher conceptual levels are associated with more advanced information processing and with greater tolerance of frustration and uncertainty.

Perhaps also the school could exercise some leadership in reexamining some basic ideas of human development such as Maslow's notion of self-actualization. Maslow postulates two basic kinds of human needs: (1) deficit-needs, or those shared by all members of the human species (e.g., safety, love), and (2) idiosyncratic or self-actualization needs which are peculiar to each person.

> Just as all trees need sun, water, and foods from the environment, so do all people need safety, love and status from their environment. However, in both

cases this is just where real development of individuality can begin, for once satiated with these elementary, species wide necessities, each tree and each person proceeds to develop in his own style, uniquely, using these necessities for his own private purposes. In a very meaningful sense, development then becomes more determined from within rather than from without.[52]

As the distribution of wealth and the standard of living have increased, notions of self-actualization such as the acquisition of material goods have become somewhat less meaningful to a growing minority of so-called "alienated" youth. Although it may have been very meaningful for a man to work to provide food and shelter for his family during the depression of the 1930s, when food and jobs were hard to get, it is perhaps somewhat less meaningful to work to pay off a mortgage on a split-level suburban home. David Riesman has raised the poignant question: "Abundance, for what?"[54] Perhaps adolescent-parent conflict and the attendant generation gap might be eased if there were some reinterpretation of the notion of self-actualization to encompass the broad range of human possibilities.

References and Notes

1. Eisenstadt, S. N. *From generation to generation: Age groups and social structure.* New York: Free Press, 1956.
2. Davis, K., The sociology of parent-youth conflict. *American Sociological Review,* 1940, 5, 523-535.
3. Freud, S. *An outline of psychoanalysis.* New York: Norton, 1940, pp. 25-33.
4. Muss, R. E. *Theories of adolescence.* New York: Random House, 1962, Pp. 34-39.
5. Muss, Pp. 34-39.
6. Muss, Pp. 84-93.
7. Lewin, K. *Resolving social conflicts.* New York: Harper & Row, 1948, p. 4.
8. Aberle, M. et. al. The functional prerequisites of a society. In A. D. Vilman, *Sociocultural foundations of personality.* Boston: Houghton Mifflin, 1965. p. 396.
9. Kaplan, B. (Ed.) *Studying personality cross-culturally.* New York: Harper & Row, 1961, p.30.
10. Mead, M. *Coming of age in Samoa.* New York: Morrow, 1928.
11. Levy, M. J. *The family revolution in modern China.* Cambridge: Harvard University Press, 1949, P. 155.
12. The "extended" or joint family is to be distinguished from the "nuclear" family which is composed of a husband, a wife, and their children. An extended family is a group of nuclear families living together in the same household. The above definitions are taken from S. A. Queen, R. W. Halsenstein, & J. B. Adams. *The family in various cultures.* Philadelphia: Lippincott, 1961. P. 12.

13. Hsu, F. L. K. *Americans and Chinese, two ways of life.* New York: Abelard-Schuman, 1953. P. 28.
14. Karve, I. *Kinship organization in India.* Deccan College Monograph Series: Poona, 1953. Pp. 60-62.
15. Karve. p. 62.
16. Karve. P. 62.
17. O'Malley, L. SS. (Ed.) *Modern India and the West.* London: Oxford University Press, 1941, p. 59. (Human Relations Area Files: India, Source 1.)
18. Levy. P. 155.
19. Levy. P. 161.
20. Levy. P. 161.
21. Yang, C. K. *The Chinese family in the Communist revolution.* Cambridge, Mass. M.I.T. Center for International Studies, 1953. P. 150. (Human Relations Area Files: China, Source 70.)
22. Yang, C. K. P. 150.
23. Osgood, C. *The Koreans and their culture.* New York: Ronald Press, 1951. Pp. 38-39. (Human Relations Area Files: Korea, Source 22.)
24. Yang, C. K. P. 151. Also see the following for an analysis of the results of this policy: Chen, T. H., & Wen-Hui, C. "Changing attitudes towards parents in Communist China. *Sociology and Social Research*, **43**, 175-182.
25. Ottenberg, S. *Cultures and societies of Africa.* New York: Random House, 1960. P. 57.
26. Fortes, M. *The web of kinship among the Tallensi.* London: Oxford University Press, 1949. P. 209.
27. Ottenberg. P. 58.
28. Fortes. P. 209.
29. Hollis, A. C. *The Massai: Their language and folklore.* Oxford: Clarendon Press, 1905. P. 298. (Human Relations Area Files: Mossi, Source 1.)
30. Hollis. P. 298.
31. Hollis. P. 298.
32. Bernardi, B. *The age system of the Massai. Annali Lateranensi,* 18, Citta de Vaticano: Pontificio Museo Missionanio Ethnologico, 1955. Pp. 257-318. (H.R.A.F.: Massai.)
33. Young, F. W. *Initiation ceremonies—A cross-cultural study of status dramatization.* Indianapolis: Bobbs-Merrill, 1965. P. 141.
34. Fortes. P. 198.
35. Fortes. P. 171.
36. Fortes. P. 207.
37. Skinner, E. P. Intergenerational conflict among the Mossi: Father and son. *Journal of Conflict Resolution*, 1961, **5**, 55-60.
38. Skinner. Pp. 55-60.
39. Skinner. Pp. 55-60.
40. Mongin, E. *Essay on the manners and customs of the Mossi people in the*

Western Sudan. Paris: Augustin Challamel, 1921. P. 92. (Human Relations Area Files: Mossi, Source 2.)
41. Tauxier, L. *The black population of the Sudan, Mossi and Gourounsi country, documents and analyses.* Paris: Emile Larose, Librairie-Editeur, 1912. P. 49.
42. Tauxier. P. 49.
43. Patai, R. Relationship patterns among the Arabs. *Middle Eastern Affairs,* **5,** 180-85. New York: Council for Middle Eastern Affairs, Inc., 1951. (Human Relations Area Files: Middle East, Source 41.)
44. Patai. P. 184.
45. Khadduri, M., & Liebesny, H. J. (Eds.) *Law in the Middle East: Origin and development of Islamic Law, 1.* Washington: The Middle East Institute, 1955, P. 155. (Human Relations Area Files: Middle East, Source 56.)
46. Stirling, P. *Turkish Village.* London: Weidenfeld and Nicholson, 1965, P. 223.
47. Pierce, J. E., *Life in a Turkish village.* New York: Holt, Rinehart and Winston, 1964. P. 83.
48. Pierce. P. 84.
49. Pierce. P. 91.
50. Pierce. P. 91.
51. Pierce. P. 91.
52. Maslow, H. H. *Toward a psychology of being.* New York: Van Nostrand, 1962. P. 31. Maslow presents a possible approach to the question of "self-actualization" by defining motivation in terms of two concepts: "self-actualizing" motivation and "deficiency" motivation.
53. Harvey, O. J., Hunt, D. E., & Schroder, H. M. *Conceptual systems and personality organization.* New York: Wiley, 1961.
54. Riesman, D. *Abundance, for what? And other essays.* New York: Doubleday, 1964.

Recommended Additional Readings

Abu-Laban, B. The adolescent peer group in cross-cultural perspective. *Canadian Review of Sociology and Anthropology*, 1970, **7,** 201-211.
Bronfenbrenner, U. *Two worlds of childhood: U.S. and U.S.S.R.* New York: Russell Sage, 1970.
Grinder, R. E., & Englund, D. L. Adolescence in other cultures. *Review of Educational Research*, 1966, **36** 450-462.
McNassor, D. Social structure for identity in adolescence: Western Europe and America. *Adolescence*, 1967, **2,** 311-334.

13   Of Youth and the Time of Generations

Thomas J. Cottle

Orientation

A common theme in psychology or parent-child relationships is that great amounts of turmoil and trouble exist between parents and their offspring. Storm and stress, rebelliousness, the generation gap—all of these are commonly heard terms in the process of describing the adolescent's new relationship with his or her parents. But is this a true state of affairs? Are most families rocked by anguish and misunderstanding at this time? Or is the storm-and-stress notion only a creation of popular literature and the mass media? These are difficult questions to answer; in some respects they all seem true, and in other respects they seem to be distortions of reality.

One could make a strong case (and indeed many have) that most American adolescents pass through this stage of life very placidly and quietly. This argument would insist that violent disagreement and misunderstanding do not characterize typical parent-adolescent relationships. However, it may be that the techniques of observation available to social scientists are not sensitive to the kinds of distress the adolescent may be experiencing. For the great majority of the time the individual gets along quite amicably with his parents, and major disputes are rare. But what is going on inside the individual? What are the anxieties, frustrations, and uncertainties that the adolescent is trying to cope with internally? Perhaps we just do not see these sources of inner turmoil. Thus, perhaps storm-and-stress does indeed characterize adolescence, but for most youth this is mainly an internal state.

Most American families operate according to a consensus approach. Problems among family members generally are resolved in a fashion characterized by mutual agreement and compromise. A main determinant in reaching decisions—such as, How late can Mary Ann stay out on her date?—is what other parents and other adolescents do. If most parents allow their daughters to stay out until midnight, then it is likely that Mary Ann will also be permitted to remain out this late. Her parents thus feel that they

Reprinted by permission of the author and publisher. Copyright ©1971 by Thomas J. Cottle. This article originally appeared in *Saturday Review* under the title "Parent and Child—The Hazards of Equality."

have made a reasonable decision, and Mary Ann can feel reasonably satisfied about the outcome, for she is being treated similarly to her friends.

A further point for consideration is the nature of the generational disagreements. It would seem that most conflicts arise over rather minor concerns—curfews, dress and hair styles, homework, and so on. Seldom do adolescents attack fundamental issues—the legitimacy of parental authority, the values of their parents, or the social system. Conflicts are therefore typically trivial, temporal matters; the passion of the moment quickly diminishes in time, and arguments almost never escalate to levels that seriously strain the parent-adolescent relationship.

But this is not to say that parent-adolescent relationships are all harmonious. Both the youth and the parents hold a conflicting set of attitudes that could be called dual growth ambivalence. For the adolescent, passing out of childhood and moving toward adulthood is an exciting adventure, for it suggests the nearness of new statuses and new privileges, such as driving the automobile, acquiring more independence and greater opportunity to make decisions for oneself. At the same time this transition can be anxiety provoking and perhaps even somewhat saddening to an especially sensitive adolescent. Just what the future will hold for the individual is uncertain, except that it is known for sure that there will be more responsibilities, more duties, more pressures. This transition also spells the end of a significant part of one's life—gone forever is carefree childhood. Adolescents' diaries often contain two recurrent themes: a sense of gain and a sense of loss.

The adolescent's changing nature also causes feelings of ambivalence in parents. On the one hand, they are proud and happy to see their child grow and develop, acquire new skills, move toward the promise of adulthood. On the other hand, their child's changing nature is a constant reminder that they will soon "lose" him or her: The adolescent is going through a process of shedding off his previous identity of being viewed chiefly as the child of Mr. and Mrs. Anderson and forming a new identity in which he is an autonomous, self-reliant individual. Soon he will be leaving home and will no longer be under the control of his parents.

One aspect that contributes to the dual growth ambivalence in parents is that the recognition that *their offspring is rapidly approaching maturity also causes them to evaluate their personal performance as parents: "Did we do a good job of raising our child?" "Will she be a good citizen?" "Will he be a success?" Parents are judged by others (and by themselves) mostly on the basis of how well the behavior of their children conforms to the expectations of society. Their own ego needs are affected considerably by the behavior of their adolescent son or daughter. Thus, if their 16-year-old daughter announces to them one day that she would like to spend the next summer traveling alone around Europe, it is likely that the girl will be met with a resounding "No!" Prevailing attitudes do not comfortably allow for adolescent girls to be traveling around alone—particularly in strange lands. The pressures for conformity are very great.*

To parents having little understanding of adolescenthood, the fact of their adolescent son or daughter engaged in the process of breaking away from the family constellation may cause very intense feelings of anxiety. If the anxiety and confusion passes beyond a certain point, they may frantically attempt to reassert their dominion over their rebellious offspring. The natural course of social development can only be hindered or possibly even permanently retarded in such an event.

A final comment about the generation gap issue. In recent years we seem to have suffered a case of "generation gapitis." The popular press has been filled with accounts and stories about the perniciousness of the supposedly gaping chasm between youth and adults. We are led to believe that this gap has been growing increasingly wider within the past few years. Margaret Mead insists that it is impossible for those who were born before World War II (and the Bomb) to understand the experience of those born post-World War II, and vice versa. All of this may be true; or perhaps only some of it is true. In any case, the prevailing mood seems to be that the generation gap is somehow a very bad and depressing state of affairs. Consider the possibility that the differentness of generations is a good development, one that offers strength and vitality to society. (This is the point of view offered by Cottle in the subsequent article.) We typically are caught up in the current affairs of the day; it is most difficult to take a more detached, historical view of changes in society. That our society is immersed in a great number of serious social and environmental crises cannot be denied, and it is the youth of the society who are continuing to remind us most vocally about these problems. Whether the present generation of youth will be able to solve these dilemmas remains to be seen. In any case, it would seem that the real hope for the human species lies in the ability of future generations to arrive at solutions to pressing problems that are at least slightly better than those attempted by earlier generations.

The next three articles, by Cottle (#13), Woiwode (#14), and Neugarten (#15), focus upon this problem of generational conflict.

Thomas J. Cottle is now associated with the education research center at the Massachusetts Institute of Technology. Dr. Cottle has researched and written extensively on the question of psychological perception of time.

13

Young people's involvement with adult authority—it's an old theme hammered to life almost daily in studies published on parents of adolescents, hippies, dropouts, druggies, militants, and the rest. Recently some writers "on youth" have openly chastised parents for failing to assume assertive roles with their children. Even some psychiatrists now argue for parental toughness, perhaps as a reaction to an oft-blamed emphasis on permissiveness.

Authority implies an inequality or what some prefer to call an asymmetry between the old and the young. There is no even exchange between generations, nor is there ever a possibility for it. Parents are by definition not peers, and their concern does not imply that they become colleagues. Yet the asymmetric structure of authority is not all bad, although parents and children are more than a bit ambivalent about it. Longing for the taste of adolescence, parents in many instances overstep the bounds that the asymmetry purports to guard. In some cases their intrusions are nothing short of disastrous. For some young people, a quiet inner strength vanishes when their parents trespass on the property of time and destroy the very same asymmetry that they themselves once wished to destroy.

The theme of authority is complicated, therefore, because young and old alike wish to tamper with the time of generations but realize the potentially devastating results of such an escapade. The asymmetry implies restraints on behavior, and the young, being today so profoundly aware of all the facts of life, recognize these restraints as well as anyone. Generally, the young seem more open than ever before, just as social reality seems more translucent. Perhaps there are fewer secrets today than yesterday, and perhaps too, our society honors revelation more than confidential trust.

There is little doubt that young people extend, prolong, or simply react to their parents' demands, be they uttered or silently passed on. Erik Erikson, the American psychoanalyst, has said that one generation revives the repressions of the generation before it. But, equally important, adolescents have become brilliant readers of parental intentions, or adults generally—including parents, teachers, ministers, deans, and psychotherapists—have become predictable or transparent in their dealings with the young. High school students now portray the "shrink scene" with ease. They anticipate, with frightening accuracy, the words and moods of churlish school administrators. A fifteen-year-old Negro boy told me that he could not get help from his school guidance counselor: "I wouldn't say this to his face, but he doesn't like Negroes. He may not even know this, but we know it." I spoke to the counselor in question. The student had not only correctly interpreted the man's attitude—his impersonation of the man's behavior, right down to the speech pattern, was perfect.

All of this suggests that the cat of the authority relationship is out of the bag. The young understand and appreciate adult motivations, and, significantly, the sociological rationalizations for their actions in authority contexts. While they may protest against school principals and programs, they confess a sympathy for their elders' plight of being trapped in the policies of some greater bureaucratic establishment, "the system." They recognize a "sell-out" or "game Player" a mile away, and a heady college freshman, if the matter concerns him at all, can differentiate between the authentic liberal and institutional brand from the last row of a lecture hall. Their language simplifications, such as "smarts," "head," "cool," "cop-out," are illustrations of an almost social-scientific terminology, which functions in reducing complex action patterns to succinct and manageable levels. Their language shows, moreover, the swiftness and clarity with which they can first interpret and then act upon personal and institutional demands. (Most students know that their parents' social class is still the best predictor of their own school success, and that the poor, and particularly the poor blacks, cannot hope to compete even with the omnipresent mediocrity found among the advantaged. Hence, their understanding of local school competition and mobility channels is profound, although frequently disillusioning and uninspiring.)

Perhaps the best illustration of language reflecting social sophistication and the apparent translucency of social reality is the expression "psyche out." A college junior assured me: "It's so easy to know what the teacher wants, or what he'll ask on a test. They never change. Give 'em what they want. You make them happy and you win." Even modest Phi Beta Kappa students claim they have "psyched out" their teachers and have emerged superior merely because they are better game players. The fact remains that to "psyche out" something is to stay one slender step ahead even of expectation. It is the ability to perceive the expression on the face of the future.

While it is hard for young people to be duped by authority figures, it is easy to be damaged by them, an act so often occurring when the superordinate—the elder, the parent, the teacher—wants to equalize what must remain as that asymmetric relationship. Again, asymmetry refers to relationships wherein the commodities exchanged are of unequal and, therefore, incomparable content, and the behavior of one person is not a call or demand for identical behavior in the other. In its most fundamental form, asymmetry describes relationships in which one of the members represents unquestioned authority in a particular context; hence, it refers to interactions engaging parents and children, teachers and students, and doctors and patients.

Several years ago, while leading a self-analytic group, I was invited to a party given by the members. As it was early in the group's history, it seemed reasonable that an informal evening together might loosen up and simplify all relationships. I was tempted to go, but a wiser man suggested that I not go. The asymmetry, he urged, ultimately must be preserved by the person holding authority. I may have lost something by declining, but I probably protected a valuable tension in the leader-member relationship. Moreover, the symbolic nature of the refusal reaffirmed the asymmetry, or inequality, which some of us working in groups feel is essential, and which members often confess, in their way, is preferred. The leader (or father) must in some sense forever remain the leader, and while this angers many, particularly those in groups, "humanness" is in no way automatically precluded by such a philosophy.

More recently, members of a self-analytic group observed their leader's participation in a political demonstration. At the following meeting they spoke of him with a newly discovered reverence. How good that he shares the same values; that he shows the courage to speak out against administrations. But they spoke, too, of a disgust for their mothers wearing mini-skirts and parents generally who act like kids. Anna, a mature young woman, told of a feeling of nausea that came over her when her roommate's mother reviewed the college courses she, the mother, was attending. Upon returning to her dormitory, Anna made a long-distance phone call home and luxuriated in the relief that her own mother still was pursuing mother-type activities: luncheons, museum visits, and food budgets.

The ambivalence is evident. Young people want to attack authority, and this is probably the way it must be. But in matters of human dealings, although not in issues of strict ideology, authority is not to "come down" to the child's level, as parents once perceptively felt it necessary to kneel down, if only to attain a spatial equality of the generations. Authority is not to give in; it is to remain firm in its commitment to preserve the essential asymmetry and the indelible generational separation, even if this means being seen as a "square" or "straight arrow."

When a small child orders his parent out of his bedroom, he necessarily fears the enormity of the act. In a tearful rage, he can only pray that the parent will go no farther than the living room. Similarly, when members express the intense desire to kick out the leader of self-analytic groups (in symbolic re-enactment of the primal horde story perhaps), invariably they want to know would he really go and would he return.

There is, then, a primitive core, developing first in interactions with parents, that pleads for the overthrow of authority, yet simultaneously for the inability to do it by nature of the superordinate's strength in resisting. Parents simply cannot break down or retreat. They must prevail, and no one wants this more than the child. In terms of this infantile core that stays with us, parents are perfect, without problems, immortal. Relationships with them preclude both equality and peership. A college student said it this way: "No matter what I do in the face of authority, I end up a child. It happens even when I don't know the authority. Are we forever children to older persons?"

For children to out-achieve their parents, an event not uncommon among college students (let us not forget that women, too, are confronted with career aspirations and the ensuing competitions as much as men), means that they, the younger, must delicately initiate revisions in parental relationships so that the older generation will not interpret the younger's accomplishments as their own dismal and static ineptitude. What an incredible task it is for these young and talented students to return during Christmas and summer vacations to the rooms and persons of their childhood; to return where all of us know we cannot again return, then to battle the very essence of an unjust but immutable temporality.

Why is it that each of us believes in the development, even in the successes, of our surging expectations, but see only aging in our parents? Perhaps the eternal danger of the immediate future is that while it guarantees reports of our most present investments, it brings first our parents, then us, closer to some inexplicable end. But for the handful of "right nows," our youthful preoccupations make only our own movement in the life space visible. All the rest, parents and teachers included, remains unchanged, timeless: "It's like they've stood still. They bring me back to my childhood hang-ups. They know I've grown up; they know I'm at college, but they're used to me as I was when I was last there."

These last phenomena are so clearly not the sensations of regression. Although we all have fought back urges to feel once more, for even a bittersweet interval, the winds of childhood, returning must not be mistaken for regressing. On the contrary, returning is resuming. This is what is meant by bringing one back to childhood "hang-ups." It seems like regression, for only in our direct involvements does family time again move ahead. In our separation, that certain time stops, and the stillness augurs death. But the student returns and time jolts forward again, alive, just as the family itself becomes vitally alive, although now life becomes a bit more cumbersome.

The predicament confronting the child at these times is to help his parents resolve the problems that occur when the young out-achieve their elders. Variations in accomplishment must be reconciled in ways that legitimately reinforce parents' ultimate authority and special superiority. Regardless of their attainments, son and daughter want to remain in the child's role, at least in this one context. The parents know the child's task and, like the vaudeville joke, the child knows the parents know, and the parents know the child knows they know.

It is in interpersonal dilemmas and gestures of this sort, gestures made and carried out in such public yet at the same time secretive ways, that families reaffirm health. The gestures imply the mutual recognition and trust of which Professor Erikson has spoken so poetically and firmly. By these gestures the social and temporal gaps are preserved, sociologic and psychologic genes are somehow passed from one generation to the next and one is, in Erikson's words, able "to see one's own life in continuous perspective both in retrospect and in prospect" (*Young Man Luther*). The division, made first by time, permits the evolution of the adult and sanctifies the appropriateness and truth of the confirmation and the bar mitzvah. For sociological reasons, the gap between generations stays open. But it is all right because distance need not be construed as distrust, nor separateness as desertion.

For two years, I saw Kathy, who is now thirteen, in a hospital therapy setting. Her language and psychological test performance indicated a possible psychotic diagnosis. She had a recurring dream, one that intrigued us both, that she was in a forest being chased by a large bear. Up on its hind legs, it pursued and often caught her. The dream had become so terrifying that to prevent the bear from appearing, Kathy had resorted to magical powers symbolized in ritualized bedtime behavior. How terribly symbolic was this content: the personification of impulses at the same time sexual and aggressive. How literal was the content: her father, an alcoholic for all of Kathy's life, returning home at night, pitifully drunk, staggering toward her, his shirt off, the hair on his chest plain and exposed, his smell, his pants open; pleading for sex at a locked bedroom door, being rejected by his wife, until he promises to "grow up and behave like a man," masturbating as a little girl watches, bewildered and horrified.

Like Kathy, too many children have been freaked out by some form of family drama. Now, although nascent and unconscious, their strategy is to get out of their homes, out of their lives, and out of their minds. What a miracle it is that some stay, conjuring up reasons for the necessity of their remaining close. But the muffled aggression in their loyalty is unmistakable. The children and their parents are like the envied lovers in the old story who never stopped holding hands until just once, whereupon they beat each other to death. Holding on to a mother's skirt after all, may be more than a wish to remain near and in touch. It may be playing the boxer who by staying in a clinch, prevents himself and his opponent from manning battle stations at arm's length.

When a thoughtless and angry Cambridge mayor's purge on young people led him to chastise hippies for having run away from home, I reacted by thinking on the contrary. The parents must have run away first, in some fashion, hence the children merely followed suit. Now, after examining life stories, I wonder whether, like the most domesticated of pets, "pre-hippies" ran because their parents rushed them and frightened them and got too close too soon. I wonder whether it was because they felt emotionally crowded by their parents that they "split." Still, even in unabandoned escape and angered protestation, children may be responding to or fulfilling some communicated need or directive. How curious is the thought, therefore, that protest and escape represent obeisance turned upside down.

It is equally curious that the familiar "need-to-escape-from-it-all" explanation of intoxication is used again when referring to serious drug-taking as a desire to repress. Mickey is a handsome, young high school dropout with an exceptional literary talent. When he was eleven, his parents fought so bitterly that he often found his mother lying in a pool of blood. Mickey would have to call for the ambulance and later on, after nursing his mother back to health, he would turn his attention to reuniting his parents.

During one cryptic account of a drug experience, he practically went into a swoon: "But when you come down, man, you come down hard, and that taking each moment one by one dissolves into that rotten other present, the one where you say, I gotta go back to my job. And you ask yourself, why do I do it, and you know, you gotta feel responsible. But It's OK because you think about the next high." I suggested to Mickey that coming down means having to think about tomorrow. "Wrong, man," he smiled for he had one on the shrink, "it's the past. It's on your back like you know what! . . . You say why did it have to happen to me?"

In speaking with Mickey and boys like him, one senses an ironical and twisted searching for insanity. Where the shocks of childhood were merely flirtations with craziness, by sixteen they have reappeared as an open willingness to consider "steady dating." At first only a couple of times a week; later on, every day and every night. The apparent psychotic quality or "way-outness" of the drugs is at once terrifying and exhilarating. The downs hurt but serve to affirm the lingering presence of sanity, or at least the ability to call upon it. If the user is sure it's still there, he goes back up on top again.

Not ironically, the very same strategy—"blowing the mind"—is used as a way of keeping out the mind-blowing experiences that might have urged persons toward this action in the first place. But just as drinking fails to induce forgetfulness, drugs seem to be failing many persons in their efforts to "repress" the past and keep it off their backs. If Timothy Leary is right, the next state will be electronic brain stimulation; hence, when pharmaceutical repression fails, attempts may be made at total memory ablation. At that time, a metaphysical present will evolve, free of any recollections and expectations, free of all regrets and despair.

Failing to understand so many of these complicated and gifted people, I often forget myself and remind them of their futures as parents. It is not that easy. For one thing, their very sense of future differs from mine. Moreover, the option to "start again" in marriage is highly problematic. Many fear they will repeat the desecrating scenes of their childhood: "I'll ruin my kid a helluva lot more than the drugs I take will"; "Are you kidding, man? Can you see me as a father? You gotta be nuts! And you a shrink!"; "A freak kid's gotta better chance than I did!"

If starting again were possible, most would probably opt for total recommencement. Knowing full well that their parents never wanted them in the first place, some almost cannot go back far enough to reach a time when their own histories might have started off on a good footing. No one admits it, however, for that would be to proclaim absolutely one's non-being. It would be to break the slim and delicate threads that now barely hold the generations pridefully together. Kathy told me that her mother was informed by doctors that she could have no more children after the birth of Kathy's nearest older sister. In fact, two more children were born. The mother admitted she had not wanted either one. Her "not wanting" became the daughter's description of herself as the "unexpected surprise." Kathy and I knew that she understood the conditions of her origin and the facts of her life. Indeed, I felt that her rather protracted inability to comprehend how children are born might have symbolized an even more profound reluctance and self-protection.

Regrettably, the concept of insanity pervades the worlds, however expansive, of many young people. What many want to know is utterly predictable: "Just tell me one thing, man; am I crazy? I mean, you know, am I crazy?" The word "crazy" is ubiquitous. It has lost its primeval jolt, but it holds on to an unmodifiable message. There is, however, plenty of insanity left over in television scripts and movies. Insanity is feared when witnessing the inexplicable behavior of those around us as they do nothing more than fight aggressively for social and private rights too long in coming. It is also feared when witnessing those well meaning men who seek to control those who protest. The young hear the President called mad and the war insane, and they puzzle over insanity's bewildering function in jury trials, and partly because of this they seek it as a way of getting out of the draft.

In my day, not so long ago, a "joking" admonition for guaranteed military deferment was, when the army doctor examines you, kiss him. Now it's insanity. Naturally, the worry exists that they might carry forever the brand of insanity on their sleeves just about where the private stripe might have gone. To be crazy is to avoid military service. Like kissing the doc, it is the avoidance of maleness. An often cruel society rubs this in: A real man fights for his country. Ideologies and spirit react against this, but the doubt stays. American socialization patterns, normally instituting strict sex-role differentiations, take care of that. There will be a lingering doubt, although in much of their questioning and concern, perceptions and anguish, the young are supported. Many of the "knowing class," they come to learn, now prefer to think of "business as usual" as the real insanity course, and jail as an undesirable but honorable way out.

Earlier I spoke of a resistance to bearing children and the feeling that one cannot successfully assume responsibilities of parenthood. In some cases it seems as though the diffidence displayed in "going on" masks a wish to start anew. The present urge to keep the cycle from repeating and the intention to keep fresh life from beginning must be considered from the point of view of sexuality. Although the language remains unchanged, industries of "procuring" and "scoring" today refer to drugs. The prophylactic, its slick package dirtied by months in the seams of an old wallet, has been replaced by the nickel bag: "Always be prepared." A funny reversal regards sex-role functions in a new economic market, as girls now solicit funds to pay for their boyfriends' stuff. I was stopped by one of these girls in the street on a beautiful October day: "Excuse me, Sir," she began her proposal. "How about a quarter for a cup of God knows what?"

One cannot be certain of the sexual habits of the persons of whom I speak. Anyway, it's no one's business until they mention it. The subject, however, is close to the conversational surface. It is as intimate as it ever was, but beginning to be freed of its irrational ties to some mysterious and primordial secrecy. As with much of their behavior, many of the young merely make overt what their elders do covertly. In so doing, they seem much more honest and far less foolish. However, the conspicuous consumption of other youngsters is little more than a mimicry of their parents.

Many young men on drugs confess their apprehensions about homosexuality. It is not simply that they fear their impulses; this seems more common among those actually engaged in heterosexual relationships. Instead, they tell of a lack of sexual impulses and a concern that perhaps drugs have destroyed the sex drive. Because of their sophistication, they comprehend the possibility that their activities generally could be interpreted as homosexual, but they manifest little panic about this. Some admit that they are able to "make it" with girls only when "high." They confess to fright, but it does not compare to the fear that they may be crazy.

This is the supreme danger, as it suggests again the complex reversal of not only competence in drug and sex work, but the associated interchange between the organs of sex and the "organ" of drugs, the mind. One almost wants to assert that a phallic phase of development has been temporarily supplanted or postponed by a "cephalic" phase. All life is fixated in the mind, and Leary spoke for the generation at least once when he advertised that each brain cell is capable of brilliant and repeating orgasms.

This then leaves one issue: the "freak-out," the ultimate reward—the ultimate punishment. It is total destruction, at once implosion and explosion. In their own words, it is brain damage and disintegration. It is, simultaneously, conception, pregnancy, childbirth, castration, and death. Some continue to believe that from the womb of the mind, a new child, a freak child is born.

By their own admission, the freak-out is also a premeditated cop-out. Like living with a woman unmarried, it is anticipated endingness and the preparation for a later recourse. Demanding no commitment, it is an out, permitting the luxury of retiring as undefeated champion. No one can find fault with the last-minute term paper writer or the hospital patient. Both have their excuses. Both wonder, presumably, about what their competence might be like void of recourse. Both wonder, too, about the lack of preparation for the equivocal future and the minimal confidence displayed in present endeavor.

Depicted in most of these notions is the mass communicative society in which we survive. The accomplishments by so many are so great, the knowledge and awareness so swift in arrival and so deep in meaning, that in a way we leave the young no excuse for failure other than severe illness and total collapse. Adlai Stevenson once confessed relief that career decisions were behind him. It's hard to be young today, he observed. So many good people are already so advanced in practically any area that one might choose for himself. Perhaps this is the reason why some drop out.

In sexual relations, the excuse that probably maintained the sanity of frightened generations of men no longer exists. Girls have "the pill," and aggressive action now swings both ways. Students offer apologies for not smoking pot and agonize over an inability to get excited, much less involved, in political enterprises. To be straight is to be square, and like it or not, the straight become defensive and tempted.

Our televised and instant-replay society also allows fewer secrets. We see the war; we see men murdered; and we become frustrated when we cannot discover the exact frame on which is recorded a President's death. Our newspapers pry and our movies reveal, and so too, apparently, do some parents. Where many children fantasize that the secrets they guard preserve some mysterious family integrity, others, in fact, are maintaining this integrity by biting a quivering lip in fear that exposure of their treasured secrets will cause their families to unravel. All the while, performance-demands shriek for attention. One must compete and succeed often enough; make it on his own; and react to the war and the fact that he or a boyfriend will soon be drafted and, not so unlikely, killed! One must be good in school, good at home, good at sports, good at pot, and good in bed. Life becomes unmanageably meaningful. It is enough to make one (want to) go insane.

Most make it, however, even with the knowledge that their culture warns of belligerent Chinese, overkill, and an equivocal future. One cannot know when the next and final war will come, or when past experience with drugs will suddenly re-erupt in the form of a grotesque child or one's own psychotic demise. Unmistakably near, death becomes a real reality. Less fuzzy than ever before, its shape and sound hover about self-analytic groups, bull sessions, and coffee dates. Damn the future and the inevitable! It was better in the Thirties when gravelly throated heroes sang into megaphones. It was better, too, in the last century when men wore frock coats, beards, and long hair. It was better and easier because it was the past, and perception of the completed proves the validity of survival, if not achievement. At the very least, the past means having got this far. It also means the seat of much of the trouble.

Some young people reveal a peculiar attitude about the past. It is not merely a time that was, but the series of events that once were, yet somehow continue to remain as the present's lining. Neither recalled nor retrieved, the past has become the stuff of moment-to-moment encounter and the routine of day work. The past has not yet become past in the sense of being over, because its foundation, like a child's body, remains soft and unfinished. There are no completions, no triumphs, no guaranteed deferrals or subsistence.

No one as yet has studied the notes written by parents to their runaway children in New York's East Village or San Francisco's Haight-Ashbury district. These pitiful missives document so well the lack of generational space and the confession of failure in parenthood and adulthood. They could be the letters of children, who, wishing to come home, promise never again to misbehave. If they did not cause guilt or confusion in the recipients, hippies would have little need to prevent them from reaching the runaway child. (Those people whose self-appointed task is to maintain the separation and lack of communication between parent and child must fear the fruits of love's temptation, the very philosophy they profess. Moreover, they are reminiscent of professional mourners, who periodically remind the congregation or family of the recent loss by crying when others attain momentary composure.)

The "come back home; all is forgiven" notes stand as a testament to what must be seen by the young as a crumbling structure, or a tragic reversal of intentionality and interpersonal competence. They reflect adult pleas for help and forgiveness, and as such they represent a far worse social fact than hippie farm colonies or pot parties. The notes only document what the poets know so well: Of all rewards, youth is a supreme ideal. The old wish to be young, and the young are happy exactly where they are.

Few parents are able to accept the passing of adolescence, especially when their own children dramatize more vibrantly than ever the former gratifications and projected incompleteness of their own lives. It is inconceivable to think that young people have ever been simultaneously idolized and despised, worshiped and envied as they are presently. Without doubt, the problem of age-grading is now of paramount significance in the United States. It is *the* dimension: Good or bad, the old are preoccupied with the young, and the young are preoccupied with themselves.

When the activities of the young were secretive, adults were compelled to deal with their own imaginations. Now, when sexuality, in particular, screams at us from advertisements, fashions, television, movies, and magazines, it becomes increasingly difficult to decline youth's unintended invitation and accept the process and reality of aging. Adults must work hard to avoid the eternal seductions of the young, for these affairs simply do not work out. Time inevitably chaperones such liaisons, and the primordial strain that comes about through the separation of generations never will permit a successful consummation of these two hearts, the young and old.

The seduction does not stop with parents, however, for the succulence of youth is dreamed of each day by teachers, counselors, therapists, ministers, etc. A most dangerous tack for any of these persons is to be uncritically won over by youth's stated demands and ideologies or interpretations of them. An example of this point seems in order. We are emerging from an unfortunate era in which psychotherapy was viewed as either panacea or black magic. Psychotherapists finally have undertaken critical self-scrutinization, and for the most part, attacks on theory and procedure have resulted in clarifying statements for the practitioners. Still, there are some critics who expend a suspiciously great amount of energy communicating to youth the evils of psychotherapy and even more benign adult interventions. By acting this way, they signify their "stand with youth," a stand normally introduced by some phrase which seems an apologia, but which in truth is more a boastful pledge to be young like the young.

Frequently, these critics demonstrate a striking accuracy in their realignment of youth's goals, ambitions, and philosophies. Just as often, their arguments are indecorous and evil. Many young people, in fact, do find illness in themselves and do seek help. They despise the proverbial "shrink scene" and rightly so, but in their quest of a "hip shrink," they wish for a modification—or, better, modernization—of the psychotherapeutic relationship, but not its annihilation. They know it is no panacea, but in anticipation they feel it has worth and are willing to try. The best adults can do, therefore, is to experiment with the helping apparatus and not discourage the trying.

Those who aspire to speak for or understand youth must be aware of the seductive nature of their interests so that they will not reach the point where speaking for youth means no longer needing to listen to it. Genuine representation, after all, does not require reliving; it requires recalling.

One final point regards the heightened sophistication of the young, their eagerness to speak, their access to recesses of an experienced childhood, and their poignant observations of adulthood. While each generation can expect to live longer, much of society, as Erikson points out, demands that individuals be allotted less time for youth. Earnest young proto-professionals especially uphold this ethic. Scattered not so infrequently about, however, are those whose parents have denied them even this minuscule tenure. For most, the awareness is simply a function of a precocious curiosity and creative need to experience. For the ones knowingly in trouble, the most immediate and pressing action resembles an attempt to complete some poorly understood mission started long ago by someone else.

That time repeats itself is but a comforting saying. The concept of a family cycle, moreover, is misleading as it tends to slur over the individual cycles unwinding at various tempi within it. Individual cycles never repeat themselves, for in progressing or carrying on in any guise, healthy or sick, the young, as ingenious as they are, do little more than obey the wishes of others and the demands that time imposes. Typically, the directions given by those who were here before us are to wait patiently and not walk so fast.

Sociologists have written that a major function of social structures is to direct its members to appropriate goal states, means of attaining them, and attitudes that may be taken in evaluating goals and means. The desire to become a doctor or lawyer, indeed the need to achieve, does not come from out of the blue. These are learned. So too is the desire to rebel, have sex, take drugs, escape, and even "freak out." In their way, all of these actions are creative because they develop out of social forms of, as well as private needs for, expression. But they have not "sprung up"; like instincts, they have evolved.

For many today, the evolution is not satisfying, and the internal excursions and elaborations have become (and probably started out as), in David Riesman's terms, "other-directed" movements. Knowing exactly this, many young persons continue nonetheless, in their other-directed patterns, and thereby show themselves most willing to listen outward and upward. Considering much of our adult behavior, this fact is remarkable.

Recommended Additional Readings

Cottle, T. J. *Time's children: Impressions of youth.* Boston: Little, Brown, 1971.

Feuer, L. S. *The conflict of generations: The character and significance of student movements.* New York: Basic Books, 1969.

Lorenz, K. The enmity between generations and its probable ethological causes. *Psychoanalytic Review*, 1970, **57**, 333-377.

14   The Contest

L. Woiwode

Orientation

Reprinted by permission;
©1969 The New Yorker
Magazine, Inc.

The general introduction to this article—one of three in this book on parent-offspring relations—is to be found in the preceding introduction. This article is unique in at least two respects, however, in that (1) it pertains to parent-child relations, as opposed to the more typical parent-adolescent generation gap preoccupation, and (2) it is a short story, not the scientific prose form usually found in books of this type. The article was included in this set of readings on developmental psychology because it suggests that generational problems may occur far sooner than that point in the life cycle when we typically locate them—the adolescent/young adult years, and because it so eloquently illustrates that significant events across the life cycle can be portrayed in ways other than the scientific approach to investigation.

The author of this story, L. Woiwode, presents this little drama through the eyes of the principal character, the young boy Charles. In a powerful yet sensitive manner, the story suggests that the problem of personal identity—which the developmental literature would lead us to believe is a problem solely of the adolescent—occurs also in young children. It would seem that novelists and other nonscientific writers have been most forthright in reminding us of that universal paradox that reoccurs throughout all parts of the life span: Social interactions—even with those persons whom we love most dearly—lead to a sense of loss as well as to a sense of gain. Clearly, the problem of intergenerational relations is not restricted to the adolescent years.

The short stories of L. Woiwode have appeared frequently in the pages of The New Yorker.

14

Held at arm's length, he watches her eyes travel over the front of his body and fill up with pure, unabashed pride. Thanks to her trip to the attic and her work with awl, needle, and thread, he at last feels properly dressed; the chaps ("*sh*aps," she instructed him, is the correct pronunciation) covering his overalls are the same chaps that her father wore when he rode through the buck brush and the buffalo grass of the unbroken Dakota plain. The stiff cowhide legs, turned up at the bottom to fit him, are faced with shaggy goatskin which she has brushed until its finger-length strands of silvery hair glisten. The holster belted below his hip is weighted with a silver pistol. He is also wearing a plaid shirt and a blue bandanna. Their eyes meet and there is a smile of complicity between them.

"Here," she says, and takes a felt hat off a hook on the wall.

"But it's Dad's!"

"His old one. Come here once." She squats so that her face is on a level with him.

"But it's too big!"

"With that head of yours, I wouldn't be surprised if it fit."

Her tone of voice makes him feel as unpleasantly exposed as when she examines him to make sure he has bathed well. She tilts up his chin, sets the hat on the back of his head, rises up and steps back a step, placing her hands on her hips and crossing one leg over the other at the ankle. "Well!" she says. "Not *quite*."

If she were not his mother, the boy would say she was giggling—something he detests. And even though that's precisely what she's doing, it doesn't register that way. What she is blinds him to what she does.

"I'm anxious to go over it now that we've got you in costume," she says. "Aren't you?"

He nods eagerly, the hat falls over his eyes, and there is another outburst of laughter from her.

"See, Mom, I can't wear this big old thing."

"Stop whining! You sound like a girl. And keep it on for now. We'll put some tissue in the crown before tonight. Where's the script? Where did I put the script?"

"On the table. Where you always put it."

"I'm so nervous I can't think," she says, her voice revealing more delight than nervousness, and she goes to the table and sits down. She has been working with him for two weeks, an hour in the morning and an hour in the afternoon, on this reading. Even though the boy's father has been through college (where he dabbled in speech and dramatics) and she has not, it was assumed that she would train the boy. Her husband, who is kind and cheerful, aware of his limitations, and wholly ingenuous, defers to her in fields where he feels she is superior, and in matters of taste and critical judgment. She is brilliant, as her husband, who doesn't have her kind of pride, would be the first to admit, and she has an innate air of refinement about her, as some people are born with enviable features. Whatever she does, from letter writing to housework, she does with a professional attitude, conscientiously, self-critically, always striving for perfection, and she expects others to do the same. The other women of this North Dakota village, plain, rural women, look upon her learning and refinement with mistrust, and none of them will make friends with her. As a result, she assumes an aloofness toward them, though she is warm and sympathetic within the walls of her own home.

She folds back the top page of the script, spreads it flat on the table, and inclines her head over it, chin in her hand. To the watching boy, her profile against the bright square of the window is no more than a silhouette, dark and featureless, and when she turns and looks at him he bends back the brim of the hat and squints to see her better.

"You look like a fool!" she says, and now there is no doubt that she's laughing. He takes off the hat and throws it on the floor. "No, no," she says. "Don't do that. You don't look so bad looking like a fool!"

Perhaps, the boy thinks, his father is right when he says to her, "You know, sometimes you act just like a kid"—a criticism that makes her turn white with indignation and walk out of the room. But his father is just as bad as she is, or worse. He plays games and works jigsaw puzzles with an exuberance that exasperates the boy, who is dead serious about everything he does. Unlike other fathers, he doesn't smoke or drink or swear. Instead, he sits at home and reads books, and sometimes even cries over them, which embarrasses the boy. And his father is as easily affected by radio dramas, which provoke his mother to remarks like "Nonsense, I don't believe you!" Or, "Would you hurry up and get this miserable scene over with!" Or, turning from the radio to her husband, "How did she get a job as a professional? Listen to that shoddy articulation. Listen to that twang!" But in the end she usually succumbs to the drama in the same way that the boy's father does, and when this happens she stalks into the kitchen and busies herself with something there until her emotions are under control.

Two or three nights ago, when she was out shopping, the boy's father took him through a rehearsal. Studying his father's large eyes as they swept back and forth over the page, the boy went through the entire monologue without once being interrupted. That surprised him. He didn't think he was doing well, and his mother always stopped him the moment he made a mistake and had him correct it. He waited, with dread, for his father to criticize his phrasing, his comic timing, his emphasis, his pronunciation—wondering if his father would be as severe a perfectionist as his mother and make him go over and over the speech until he had it right. Instead, his father said, "I don't know what to say. I really don't know what to say to you. You've got this whole thing memorized word for word!"

"Ready?" his mother asks now. "Hat and all?"

"It's not funny."

"I know, I know. I'm sorry. If you're ready, let's begin. Remember, now, try and do it the best you can. Most likely, this will be your last rehearsal before tonight."

The word "tonight" settles below his lungs and sends out long, thrilling appendages that cut off his breath, upset his self-control, and threaten to destroy everything that he's mastered in the last two weeks. Feeling somewhat ill, but knowing he must not show the least sign of weakness, he begins:

"I'm a buckaroo from the Wild West,
I rope and I ride and all the rest. . . ."

The words are no longer words. Repetition has robbed from them every trace of meaning, and they affect only his vocal mechanism, which grapples with them, subdues them, and molds them into familiar shapes—patterns and pictures he can visualize as he speaks. At first the words won't join up with one another, and each one comes out as a separate, clumsy geometric—an oblong, an oval, a triangle—and drops to the floor. Then, as he grows more sure of himself, the shapes begin to link together and flow from him to his listener, making a bond between them. That sentence, done up in garish primary colors, strings together the apples, bushel baskets, marbles, and coins in his first-grade arithmetic book. That pause, coming before the last fat, rhymed word of the line, is a beheading. In this sentence a locomotive takes shape, colored cars link behind it, the train starts down the tracks, and just as its wheels settle into a regular clacking, a metallic sing-song, the train explodes and the scattering fragments become flushed quail roaring toward the cover of trees.

These images and rhythms, these colored phrases, he uses only for one purpose: to affect his mother. Giving back to her his version of what she has taught him, he watches for her slightest response, and when she has to shield one side of her face to hide her pleasure, or when her laughter draws her head toward the script and her loose hair swings over her cheeks, he feels himself growing huge and adult, filling with adult sensations—overpowering, inexplicable sensations, which, for some reason, despite their vastness and desirability, are not quite satisfying.

"No, no, no, no, *no*! You brandish the pistol in the air. You *brandish* it. You do not *click* the trigger." The tone of her voice and her stare—uncompromising, deep blue, a blue he can see and feel even though her features are in shadow—make him shrink back into the constricting limits of his six years. "It's distracting. If you do that, no one's going to pay any attention to that line. You understand? All right, no more of it. Let's take it from there. Don't bother to go back to the beginning. You were doing fine with the speech itself—it was just that pistol." Turning back to her script, she adds, "The next thing I know, you'll be wanting caps for it."

"Can I?"

This time all it takes is the stare. He lowers the pistol into the holster, dwelling on the action as though he hopes it will last forever. He focuses his eyes on the linoleum and goes on with the piece, feeling each word ring thin and colorless in the air because he's disappointing her. In despair, he reaches the end.

"Now, let's have a look at that hat," she says, in a voice that, to him, sounds suspiciously nonchalant. There has been no praise, no encouragement; not even the usual criticism. Because he knows she can't lie, he thinks that in her silence and her concern for the hat (she puts on an imitation of one of her scowls as she studies it) she's withholding the truth of his failure out of kindness, knowing it would upset him so badly he couldn't perform in the evening. But her silence is just as destructive. He wants to run somewhere and hide, to vanish. He doesn't even care about the contest, even though his secret goal for two weeks has been to win first place, which is to be his gift to her.

He is more than surprised then, and wonders more than ever about his mother, when she tosses the hat into the air with a cry of delight and gives him, most rare and consummate of her rewards, a kiss on the lips.

People are appearing on the main street of the village, which is a block long, and moving toward the Town Hall, a great long gray building covered with sheets of pressed tin that simulate blocks of stone. A few feet from the foundation the surface of the tin has become polished a bright silver from children running their hands along the building as they walk past, making a miniature thunder. Tonight, burnished by the orange sun, the building looks portentous. Tonight it will not house the usual friendly social gathering: a card party, a carnival, a basketball game, or the weekly serial movie, which are never announced in the newspaper—because there is no newspaper—or by posters; everybody knows of their occurrence. Tonight is the annual talent contest, with prizes donated by the local businessmen.

With a cloth laundry bag containing his costume slung over his shoulder, the boy steps into the entry. He looks at the ticket booth (knotty pine, the same as the walls) and sees a half-moon cut out of the bottom of the glass, and on another half-moon—the wooden tray for counting out change—a white hand is resting. Because of a reflection in the glass, the hand is all the boy can see, and the anxiety he has been repressing during the walk from home to the Town Hall, now that the disembodied hand is there to serve as a focus for it, rises up in him and makes him weak.

"Don't just stand there," his mother says, nudging him gently. "Go on in. They know we're competing. We don't have to pay."

As they go toward the double doors that lead to the auditorium, the reflection on the glass slides to one side, revealing the heavily made-up face and the bejewelled breast of Mrs. June Koenig, the wife of the village dentist. "What's the matter?" Mrs. Koenig cries, her shrill voice made miniature and flat-sounding by the enclosure. "Is he scared already?"

"No, he's not," the boy's mother says, and guides him through the double doors. Inside the auditorium, he is surrounded by a familiar smell compounded of old varnish, damp crêpe paper, mildew, and human sweat. He sniffs it as carefully as an animal would, testing it. Tonight it will be carrying the sound of his voice. He looks toward the far end of the auditorium. The basketball backboard has been hoisted against the ceiling, and the curtains are drawn shut across the stage. The smell is so much a part of the building, and so overpowering, it seems responsible for the curtains' particular shade of purple, for the way the folds fall in them, the way the shadows are pressed between the folds, and the boy, too, feels affected by this force, locked within it, unable to move. Then a pair of tiny feet patter across the stage, the curtain billows out, and he is conscious of voices beside him: ". . . Children's Division is first, and that's, of course, the one he'll be in."

The speaker, Dr. Koenig, possessor of the biggest hands and broadest fingers the boy has ever seen (a judgment that has nothing to do with seeing them close up when the Doctor works on his teeth), is the master of ceremonies for the evening. Having got through the preliminaries with the boy's mother, he turns to the boy, and says, "Got the jitters?"

"Not half as bad as I have," she answers. "In fact, I don't think he's worried at all. He has no reason to be. He's well prepared."

"Well, you can't blame a kid for a case of nerves. I've got them myself, just knowing I've got that introducing to do."

"He'll be all right."

"I'm glad to hear it!" the Doctor says (in the same tone of voice he would say, holding a tooth in front of the boy's eyes, "I got it!"). "And I hope you end up with that first prize. I'm rootin' for you! Ten silver dollars is a pile of money. Why, I bet it's more than you've had in your hand at one time."

"No," she says. "It's not. Anyway, money isn't the important thing."

"Just a second," the Doctor says. He shakes a quart Mason jar filled with scraps of paper, and his loose bluish jowls also shake. As he holds the jar out to the boy, he explains to her, "He has to draw a number to ascer . . . to certi . . . to find out when he'll come up on the program."

"Don't get your hand stuck in there," she warns.

After unfolding the scrap of paper, the boy looks at his mother, then at the Doctor's round eyes, then at his mother again, and says, "Five."

Dr. Koenig sets the jar on the seat of a folding chair, takes a piece of paper from his shirt pocket, takes a pencil from behind his ear (each movement displaying the same precision and mysterious significance the boy finds so frightening when he's in the dental chair), traps the paper against the wall with one big hand, and says as he writes, "Number Five, Children's Division, Chuckie Neumiller. . . . You'll go on right after Nick Hammes. The boys and their mothers are using the dressing room on the left of the stage, so that's of course the one you'll use. The girls are in the other one. Oh, does your play-act have a name?"

" 'Broncho Boy,' " she says.

The Doctor tilts back his head and laughs his resonant laugh, which, as always, ends as abruptly as it began, with a look of bewilderment and perplexity on his face. " 'Broncho Boy'! Sounds like a good one, all right!"

"I didn't title it," she says, and takes the boy by the hand and, after a jerk to get him started, leads him toward the dressing room at the far end of the auditorium. As yet there are no people seated in the wooden folding chairs set up on the floor of the auditorium, and the stage lights have not been turned on, but the cold blue twilight filtering through the high windows fills the building with a dim theatrical atmosphere, and the chairs themselves seem to the boy to be the audience, cold and implacable, waiting for the slightest mistake, the least betrayal of fear.

He looks away from them and sees, on the walls and the ceiling, dangling from thumbtacks, scraps and short strips of crêpe paper of every imaginable color, some of it faded and stained, some of it new-looking. At the carnival last winter he was allowed, for the first time, to stay up until the children were told they could tear the decorations down. But he didn't really enjoy it. His mother wasn't there. She had sat at a side table most of the evening, toying with some trinket he'd won at the Fish Pond, and finally she said she was tired and went home, leaving him with his father.

Now, as he and his mother draw near the dressing room, he hears a murmur of excited voices, and the sound increases in volume when his mother opens the door. The boy takes in the sight of feminine legs (the dressing room is up a few steps), the bright lights, the smell of makeup. Inquisitive faces turn toward the door, there is an instant of recognition, and the voices drop to a lower pitch. She takes him by the hand, precedes him up the steps, and leads him down the long dressing room, narrow as a corridor, past mothers who are putting rouge on their sons' pale cheeks, wiping away streaks of dirt with handkerchiefs moistened with saliva, going over the *do re mi* of the scales, and, in one instance, gingerly taking the Scotch Tape off a swatch of hair that's been taped in place to a boy's forehead. In spite of all this animation and forced intimacy, no one in the dressing room greets his mother.

Near the back, one woman is telling her son, who is sitting on a bench crying, to grow up and act like a man, because he has to go out there and sing whether he likes it or not. Nick Hammes. Charles's mother stops and releases his hand. "What's wrong?" she asks.

"Oh, don't ask *me!*" Mrs. Hammes says, and in her exasperation blows a stream of air from her lower lip, causing a wisp of gray hair on her forehead to fly up. "I've been after him for weeks now about the contest, and a couple of days ago, he seemed all for it. He said yes, he'd love to be in it, sure enough. Everybody else was going to be in it, so why shouldn't he? And now at the last minute he wants to up and quit on me."

"He decided to be in it only two days ago?"

"That's right."

"When has he had time to rehearse?"

"Well, you know, he's been singing in school all along. And at home he sings, too—so much that it gets on the husband's nerves. So all there was to do was settle on a piece he liked. One I could play on the piano."

Charles's mother squats in front of Nick and takes his hand in hers. "Are you scared?" He nods at her, his lower jaw trembles, and his tears, having wet paths already prepared for them, slide easily down his cheeks.

"What grade do you get in singing at school?"

Nick Hammes takes a deep breath, opens his mouth wide, and bawls, "AAAAAA's!"

"Do you know what Charles gets? C's. I try to comfort him by telling him a C means he can sing like a canary, but I know it's a lie and he does too. Have you ever heard him sing?"

Nick Hammes snuffles, and then stares at her with a mixture of skepticism and fascination.

"What does he sound like instead of a canary? Think now. What does he sound like? Think of an animal that has a name that begins with C."

Nick Hammes wipes the tears off his cheeks with the back of his hand, screws up his eyes, and looks at Charles. Then a slow grin spreads over his face. "A cow! He sounds like a cow!"

"Exactly. Now, knowing there are people who sound like that, and knowing how good you sound, and what a wonderful gift it is to have a voice like yours, don't you think it's a little strange for you to be afraid? Aren't you even ashamed of it some, old as you are?"

Nick Hammes lowers his head and after a minute or two he nods.

"Good," she says. "That means you're sorry for acting the way you've been acting, causing your mother so much worry. Which means you'll sing twice as well—see if I'm not right!—to make up for it."

She pats Nick on the knee, rises, and leads Charles to the back wall of the dressing room, where she sits him down on a wooden bench. As she begins to undress him, she says quietly, "Don't be jealous. It's not his fault. She had no right bringing him here so ill prepared. If you'd like, you can be a little angry with him. I didn't like the way he said that, either—about the cow. But he had to forget about himself and it's best he said something about you, because you know what you're doing. And you're grown-up enough to understand why I talked to him or I wouldn't have. Here, now! Stand up straight."

Then, as she helps him on with his costume, which feels chilly and stiff after his street clothes, the two of them luxuriate in a silence that makes the boy feel shielded from the tumult of the room. He's aware only of her closeness, of her hands moving up the buttons of his shirt, her eyes travelling over his face and clothes, examining them unconsciously yet with care to each detail, her breath passing over the side of his neck, pausing, passing over the side of his neck again—and once more he feels he's an adult, and he thinks, This will last forever. And, as if in confirmation of the thought, a pleasant sensation passes over him.

Through the gentleness of his mother's touch, more than from her words, he understands what she meant about Nick Hammes. Then he realizes that this same gentleness, present in her no matter how straightforward or unpredictable she might be, was also communicated to Nick Hammes when she took him by the hand. And then jealousy strikes him.

When he is finally dressed to her satisfaction, and she has rechecked the tissue in the hat, the bandanna, the belt of the holster ("Whatever you do, don't click that thing!"), and the wide tooled belt of the chaps, she says, "Just do what you've learned," and leaves him.

To preserve the serenity she has given to him, he goes out the side door of the dressing room and stands in the darkened wing of the stage. Still, he feels vulnerable. He walks over to a travelling drape gathered at the back of the cyclorama, takes hold of the edge of the drape with one hand and turns in several slow circles, wrapping himself up in its folds. He remains there until he's heard applause for the first performer, and when he unwinds himself and reappears, there is a detached, trancelike expression in his eyes.

Mary Beth Stahl, in the opposite wing, makes a hurried sign of the cross, and walks into the brilliant light in a white dress (her First Communion dress minus the veil) and sings "Dear Hearts and Gentle People." Applause. Everett Hyerdahl, dressed as Uncle Sam, does a tap dance to "I'm a Yankee Doodle Dandy." Applause. Nick Hammes, with black-face covering whatever remnants of fear his face might betray, stands with his back to the audience, polishing imaginary shoes, and sings "Chattanoogie Shoe Shine Boy." Applause. Then Dr. Koenig's bass voice: ". . . doing a thing called 'Broncho Boy'!"

A nameless feeling he's had only a hint of before—the presence that brushes over him whenever he thinks of performing, making him feel hollow in the stomach—falls over him with its full weight. His tongue is hard and dry; there's a rope around his throat. He feels stifled, yet at the same time detached and weightless.

He walks into the harsh light, blinking against it, and even though the presence exerts so much pressure it seems to him he is growing visibly smaller with each step, his composure keeps it from becoming a part of him. The pressure, however, has an exhilarating effect. Bearing down on all his senses, it compresses them, makes them more keen. He takes in details that he wouldn't have noticed under normal conditions—the deep shade of his thumbnail (why is it so purple?); a particular strand of goat's hair on the chaps, longer than the others, that is slightly discolored (what from?); the intricate grain in the boards of the stage; the specks of dust, brightly lighted, that seem to have been arranged over the grain of the wood with an eye for symmetry; a piece of blue thread stuck to the sole of his boot, and the shadow of the thread, also blue.

Though the houselights are off, he sees row on row of faces and identifies them. He even sees that the contestants who have finished are now sitting on the floor of the auditorium in front of the stage. And then he realizes he has not only made it to the front of the stage and executed his bow but is smiling at the audience, ready to open his mouth and begin, absolutely under control, prepared for anything but what happens.

There is giggling from the audience, and some children in the front row are pointing at him. The pointing (bad manners) and the giggling (he hates it) are directed at him because of the shaggy chaps, the chaps that once belonged to his grandfather (who would thrash them all if he knew they were laughing at anything he wore), the chaps his mother has taken pains to sew up—and from all these impressions, and others hinted at in his mother's attitude, he renders a judgment: The people out there are stupid. He hates them. And once emotion has broken through his composure, the vivid details fall away and he sees only white.

He races through his lines without any thought to the phrasing and coloration he's practiced. His anger gives the lines a fierceness and spontaneity they've never had before, and at the climax of the speech, to shut himself off from those stupid fools, he grabs the hat and pulls it down over his eyes. The large hand of Dr. Koenig settles across the width of his back and propels him in the proper direction, offstage.

Safe in the wings, he lifts the hat free, and the tissue lining its crown falls to the floor. With the return of his sight, his sanity returns and he's aghast at what he's done. What will his mother say? He tries to recall whether or not he clicked the trigger of the pistol and can't. He can't even say for sure whether he finished the speech. And that isn't all. Above the steady noise of the audience, he hears Dr. Koenig's voice, beseeching, joking, saying something about wool over the eyes, and finally, after another swell of laughter, the audience quiets down. Now he's positive he has done something wrong, and since he has no memory of anything, he's sure it must have been something awful, and he starts trembling.

He bends down and picks the tissue off the floor, as though it's essential to salvage everything he *does* have, and goes into the dressing room. His mother is there, alone, waiting for him. Before he can speak, she grabs him by the scruff of the neck and, half dragging him (a button pops off his shirt), half lifting him, she takes him to the rear of the room and sits him so hard on the wooden bench that a warm pain branches from his tailbone.

"Why did you pull that hat over your eyes?" she asks, shaking him by the shoulders. "What kind of second-rate performer are you?"

A wave of emotion breaks over his words before they can form.

"I've never been so mortified in my life! Do you think I took you through two weeks of rehearsal just to see you ruin it at the last minute with a cheap piece of slapstick!"

She begins to undress him, and his shirt rips under the armpit. No one but a child—whose earliest memories of comfort are associated with touch, whose only way of expressing love is through touch—is capable of feeling with such acuteness what is communicated to him through hands, especially when the hands are those of his own mother. As she undresses him, her touch is so rough and unloving that his body, and his whole being, too, feels desecrated. In her harsh handling of him there isn't even the underlying gentleness, the tremor of understanding he can usually feel even when she spanks him. For the first time, her hands give him no quarter. The tears that have been building up beneath his eyelids gather on his lower lashes, and when she starts tugging wrathfully at one of the legs of the chaps without unbuckling the belt, they fall off and start down his cheeks.

"After pulling a stunt like that," she says, "you don't deserve to win first place."

"I don't want to win now!"

"Whether you want to or not, it makes no difference. You've performed, and you'll be judged. Except for that bit of slapstick at the end, it's the best you've done the speech. I thought they'd never stop applauding!"

"I did that with the hat because they laughed at me!"

"That makes no difference. When you're performing, you have a duty to yourself and your audience, and you can't afford to lose your temper. That's something you better learn right now. (Oh, you were so good!) And when you're onstage you might well expect to be laughed at. Isn't that one of the reasons you're out there? You're certainly not there to vent your temper. There's opportunity enough for that in real life. And if you don't start learning to control your temper there, too, in real life, I mean—your temper and that terrible pride of yours—life's going to give you a whole lot of trouble, believe me!"

She squats down in front of him, picks up the piece of crumpled tissue he's dropped to the floor, puts it in his hand, and closes his fingers around it. "Here," she says. "Dry your eyes. Pretty soon they'll be announcing the winner. You."

He is not consoled. He is not freed from his grief. Her confidence in him does not restore the complicity that bound them together before. Something has changed between them, and it stays changed.

A few minutes later, he walks onto the stage to receive, as his mother had predicted, the first prize of ten silver dollars. Though the houselights are now on, he sees the faces of the audience only vaguely, and their applause has no effect on him. They don't have the power over him he believed them to have. He takes the roll of silver dollars from Dr. Koenig and walks across the stage into the wing and places the money in the hands of his mother, his real, his unequivocal, and, as the slow unfolding of years will come to prove (in spite of the sudden, unexpected end, within three years' time, of her physical life), his only competitor.

Recommended Additional Readings

Berger, T. *Vital parts*. New York: New American Library, 1970.

"Take advantage of every opportunity to enjoy yourself before you get into kindergarten, Brian. That's when they start breathing down your neck."

Reprinted by permission of the artist; copyright ©1970 Saturday Review, Inc.

The Old and the Young in Modern Societies

Bernice L. Neugarten

Orientation

This is the final article in the series of three on the subject of generational issues and problems. (The reader is referred to the general introduction to this trio prior to article 13.) Neugarten here explores the area of relationships among age groups in modern industrial societies, but she places special emphasis upon the role of the aged in such societies. This, of course, is in sharp contrast to the accent on young people in typical generation-gap investigations.

The number of elderly persons in modern societies is growing rapidly, due to the ability of these societies to provide adequate nutrition and medical care to the greater proportion (but, alas, not all) of their population. Perhaps the manner in which a society deals with its elderly is an index of its humanitarian commitment to providing the best possible life for all its citizens. Does it ensure older persons a meaningful and dignified existence, or does it shut them away in nursing homes and lonely apartments?

Neugarten makes a number of important points in her article. One of them pertains to the concept of "age-ism," another form of prejudice to be added to the list already containing racism, sexism, and many others. Another issue has to do with the intriguing possibility of the aged population becoming a political force in modern societies. But perhaps the point of greatest import is the plea for the breakdown of the distinctions we create among people because of their chronological age; the need is for a greater tolerance—or rather, acceptance and respect—for individuals and their uniquenesses.

Bernice Neugarten is chairman of the committee on human development at the University of Chicago. Professor Neugarten has researched extensively in the area of personality development across the life span.

Reprinted from *American Behavioral Scientist*, Volume 14, No. 1 (September/October 1970), 13-24, by permission of the authors and the publisher, Sage Publications, Inc.

Social scientists are interested in two broad themes or aspects of aging. The first is how any society functions as an age structure and how changing age distributions over time affect economic, political, and other aspects of social organization. The second is how attitudes and roles change over the life cycle of the individual or in cohorts of individuals. The remarks to follow relate to both these themes, to the position of older people in the society at large and intergenerational relations, and to the social role patterns of older persons as delineated in recent empirical studies. In taking both perspectives regarding the roles of older persons in the modern industrial society we shall venture certain generalizations about the present and certain predictions about the future.

Relations Between Age Groups

Relations between age groups have been given little consideration as an area of sociological inquiry. Only recently has attention been focused upon the so-called "generation gap" or "the revolt of youth," now a highly publicized phenomenon in countries all over the world, although still one which is seldom systematically investigated. The extent to which this is a new or merely the latest version of a very old phenomenon is a question too complex for analysis here. The historian Feuer (1969) draws a basic distinction between, on the one hand, conflict of generations as a universal theme in history, and on the other hand, student movements which have recently become the chief expression of generational conflict.

The intensity of the conflict fluctuates according to given historical, social, and economic factors. Under fortunate circumstances, it is resolved within an equilibrium in which all generations receive an appropriate share of the goods of the society and an appropriate place for their different values and conceptions of the world. Under other circumstances, the conflict becomes bitter, as when the older generation, through some presumable historical failure, becomes de-authorized in the eyes of the young.

Feuer argues—although many take issue with him—that student movements of today have much in common with student movements throughout modern history: while they are idealistic and passionate, and often represent the conscience of the society, they are also in large part dominated by unconscious drives (the will to revolt against the de-authorized father, together with the guilt that is its accompaniment); they become irrational, nihilistic, and self-destructive; and in most, though not all, instances, they end in failure because they do not succeed in attaching themselves to "carrier" movements of more major proportions, such as peasant, labor, nationalist, racial, or anticolonial ones. Thus, in Feuer's view, student movements are not synonymous with the more underlying phenomenon of generational conflict, but have a distinctive psychological base. There are other points of view, of course. John Seeley (1969) views the present youth movements as being much more fundamental: as a rebellion against technology, against the corruption of men's minds by mass propaganda in small matters and large, against the dominance of technological thought patterns in which means come to determine social ends, and against the progressive alienation and constriction of the individual in such a society. If these youth movements are not crushed by the Establishment, they may be the beginning of a transformation of society "more closely analagous to the Reformation or the Renaissance."

Whether present student movements are to be interpreted as extreme but relatively superficial forms superimposed upon underlying generational conflict, or whether they are to be viewed as the true wave of the future, they have taken the limelight in the attention of social scientists and the population at large. With the exception of professional gerontologists, it may be fairly said that the society is preoccupied with the young, not with the old.

In the long historical perspective, as well as in the context of the immediate present, this preoccupation with the young has multiple explanations. Not only does the hope of any society lie in its young, and not only are the young more vociferous than the old in proclaiming their grievances, but it is a relatively recent phenomenon that large numbers of the old are present at all.

From this historical perspective, societies—as opposed to individuals—age as the benefits of industrialization, rising standards of living, and modern medical research lead to longer life expectancy, and as the proportions of old people therefore increase in relation to the proportions of the young. Societal aging is common to all industrialized nations. For example, if we consider an index of societal aging that reflects the proportions of persons aged sixty-five and over in relation to the number of children under fifteen, that index moved from twenty to seventy in Great Britain in the hundred years from 1850 to 1950; from twenty-four to sixty-four in Sweden; and from ten to forty-five in the United States. More recent changes in birth rates have reversed the trend in the United States, where increasing numbers of the young are resulting in a stabilization of the *proportion* of the old in the total population.

As the *numbers* of older people increase, however, it is inevitable that their social visibility increases. In some countries the appearance of large numbers of old people has been more sudden than in others, and dislocations have occurred because those societies are not yet prepared to meet the needs of large groups of older people. In such countries—and the United States is one—a sizable percentage of older people suffer from poverty, illness, and social isolation; a group of needy aged appears which creates acute problems in the fields of social and medical welfare. In other countries, the historical process has been more gradual, and the needy among the aged are no more numerous than the needy among the young. In all industrialized societies, however, all members, young or old, must adapt to new phenomena such as multigenerational families, retirement from work, and trends toward gerontocracy; in this sense, increased numbers of older persons, whether needy or not, pose new social problems for the total society and new alignments between age groups.

Movements of the Aged

While it is the student movements that are now in the limelight, some social scientists have begun to broaden the perspective and ask if generational conflicts are indeed increasing, and if they will also appear in the other direction—that is, between the old and the society at large. Perhaps a more realistic way of putting the question is whether or not conflicts will appear between the young, the middle-aged, and the old. As age groups become more differentiated, is political and social competition between the groups becoming more evident? Do the political movements of the young represent a more general age-divisiveness that is appearing in various parts of the world?

In considering the position of the old in this context it is tempting to speculate upon the analogies posed by student movements and the extent to which they do or do not represent a major phenomenon of social change. (It is too soon to evaluate the long-term significance of the latest student movement that was precipitated in spring, 1970 with the extension of the Vietnamese war into Cambodia.) The political movements which have thus far arisen in the United States in which older people have been in the forefront have been relatively short-lived and narrow in focus (for example, the Townsend movement, or the fight over Medicare). While these political movements of the aged are perhaps free from accusations of irrationality and unconscious guilt, they must be seen as relatively superficial and by no means as setting a wave of the future, either with regard to the position of older people in the society or with regard to the future course of the larger society. In short, political movements among the aged have thus far been like student movements of the past—either ending nowhere, or absorbed into a more general move on the part of the society to right an immediate wrong (as when the adoption of Social Security programs put an end to the Townsend movement). We have seen as yet no major movement on the part of older people that might be regarded as the expression of underlying generational conflict.

But what is true of the past may not be true of the future. A revolt of the elders seems unlikely, although the novelty of the idea may be itself a reflection of our accustomed patterns of thought about older people and our traditional view that as a man ages, he becomes unequipped as well as unwilling to do battle.

Whether a revolt of the elders is or is not an apt way to pose the problem, increased tensions between age groups may well be forming, and in the process, the old may be expected to become a more vocal group.

Certain antagonisms toward both the young and the old are now appearing that were not present earlier. Anger toward the young is on the increase as witnessed by the threats against universities arising in Congress and in state legislatures; hostile attitudes reflected in public opinion polls and in physical attacks of workers against students; and the actions of those middle-aged who feel estranged from their children or those who, presuming their own children to be singularly different from other children, feel free to censor the young as an age group.

Anger toward the old may also be on the increase. Because industrial societies become gerontocratic societies, with an increasing proportion of power positions in the judiciary, the legislative, the economic, and professional areas occupied by older people (a trend which has had some marked exceptions in the past decade), it brings with it resentments on the part of the young and the middle-aged. Among lower-income groups, as the number of old increases, the economic burden is perceived as falling more and more upon the middle-aged worker and taxpayer.

Age-ism

As generational equilibrium becomes an increasing problem in one direction, then, it may also appear in the other direction, and the phenomenon that has been called "age-ism" (that is, negative or hostile attitudes toward an age group different from one's own) may be directed toward both young and old. (I am indebted to Dr. Robert N. Butler who suggested the use of the term "age-ism.")

Covert forms of age-ism toward the old are, of course, nothing new in American society, as witnessed by the pervasive pattern of attitudes that make us slow in providing meaningful work or nonwork roles for older people and slow in raising their status as an age group; by the small percentage of the welfare dollar that is spent for services to the aged; and even by the fact that research on aging is so slow to develop in both the biological and social sciences. (But one example of the latter fact is that while 25 percent of admissions into public mental hospitals is of persons aged 65 and over, less than 1 percent of the federal mental health budget goes to research on aging or the aged.)

It is possible that age-ism toward the old may take more overt forms in the near future and may be countered by action on the part of older people. For one thing, while a larger proportion of the aged than of the young now suffer from poverty, there are other segments of the society in which old people are becoming a visible leisure class. While a large proportion of the present old are disadvantaged in terms of education as well as income, numbering many foreign-born and many who have spent their lives as unskilled workers, future generations of old will be quite different: better educated, healthier, longer-lived, in higher occupational levels. As they become accustomed to the politics of confrontation they see around them, they may also become a more demanding group. There are signs that this is already so, with, for example, appeals to "senior power" (in some ways analagous to the appeal to "black power"), and with more frequent newspaper accounts of groups of older people picketing and protesting over such local issues as reduced bus fares or better housing projects. Whether these incidents will remain isolated and insignificant, or whether there is an activist politics of old age developing in the United States is a debatable question, but it would be a mistake to assume that that which characterized the political position of the old in past decades will be equally characteristic in the future.

Age-ism is not likely ever to take the virulent forms of racism, for there is an adhesive between the generations that results from the biological inevitability of the life cycle, as all young become old; a psychological adhesive as adults remain the socializers of the young and as the middle-aged and old are forever invested in the fate of their offspring; and a social adhesive as the future of the society must always remain in the hands of the young. Nevertheless, future forms of age-ism toward both young and old and the relations between age groups may well be one of the "sleeper" problems of our times, perhaps in the way that race relations was a sleeper problem no more than twenty years ago when the urgency of the issues was only dimly recognized by social scientists. It may be an exaggeration to compare age relations to race relations, but more than one observer has suggested that the Western world may now be entering a century of social change in which, like an earlier century of major struggle for political rights, followed by a century of struggle for economic rights, it is now a struggle for age rights. If so, it may become not only a struggle of the young, but also of the old.

If, however, we are now engaged in a struggle for age rights, it is a struggle which will probably end in greater freedom for both young and old from the constrictions of age norms. Industrial societies, like preindustrial ones, create a system of norms and expectations in which duties and obligations, rights and rewards, are assigned according to age. We have lived with a long tradition that youth is the time for education; adulthood and middle age, the time for work; and old age, the time for leisure. But various economic and technological changes are now occurring in postindustrial societies that are leading to a shake-up of these long-held traditions.

For one thing, the rapid growth of technology is accompanied by rapid obsolescence of work skills, and the necessity for workers to undertake new jobs at several different times in their careers. It is estimated that in the United States, only one man in five will remain in the same occupational category throughout his work life, and that the majority of job changes will not be "orderly" in the sense that they can be arranged in an ascending order of prestige and economic income.

The future will probably see greater experimentation and greater flexibility in the timing of education, work, and leisure, with more middle-aged and older people being retrained for new jobs or engaged in formal education as the means to self-fulfillment; with more leisure spread throughout adulthood; with long work sabbaticals; and with periods in and out of the workplace. Women are already entering and leaving the labor market at various points in their lives—the same may become true for men who may study for a few years, work for a few years, return for more education, and so on. We are also turning to the concept of life-cycle education. All this is likely to lead to more age mix in our educational institutions and in our economic institutions.

Another factor is that the young today are recognized as being more sophisticated and more mature than in earlier decades. (A recent example is the lowering of the voting age to eighteen in Great Britain, and a similar movement now growing in the United States.)

Chronological age may become less important in the future, with young, middle-aged, and old participating in the society according to each person's special abilities and special experience. With greater physical vigor in the middle-aged and old, with mass education and communication influencing the old as well as the young, with the young maturing more quickly, and the old aging less quickly, with political and economic rights being more evenly distributed across age groups, the result is likely to be a society in which age categories and age restrictions will be relaxed.

In the short-run future, then, we may see increased competition between age groups, but in the longer-run future, a blurring of age groups. To the extent that the quality of life can be improved for all members of a society, the aged will gain along with everyone else.

Social Role Patterns

Turning now from the relations between age groups in the society at large to the lives of individual older people, and considering social roles in a narrower sense, what are the social role involvements likely to be in the future for successive groups of older men and women? How are people likely to structure their lives after about age 65? Under what conditions will they achieve life satisfaction?

Here, too, a historical perspective is enlightening. Because gerontology as a whole is a relatively young science, we have operated, understandably enough, with mistaken interpretations that are only now being corrected.

Twenty and thirty years ago, the general view was that as societies became industrialized and urbanized, they would inevitably create situations unfavorable to the old: increased geographical mobility would bring the breakdown of the extended family; rapid rates of technological change would bring obsolescence for older workers; increasing productivity would mean increasing retirement. In short, the result would necessarily be what some gerontologists described as a "roleless role" for the aged, and associated with it would be poverty, isolation, loneliness, and alienation.

This point of view, it turned out, was in error. For one thing, empirical studies had been few and had been based on atypical samples: the hospitalized, the institutionalized, the recipients of social welfare, in other words, the "problem" aged or the needy aged rather than the typical aged. More recently, when large representative samples of older men and women have been studied, the overall interpretation has begun to change.

Nonisolation

Large-scale surveys have shown that older people are not isolated from other family members. To the contrary, while most older people prefer to live in their own households, they live near a child or a relative and see other family members regularly and frequently. This is as true in the United States as in Denmark and in Great Britain—indeed the overall patterns of life for older people are remarkably the same in at least these three countries (Shanas et al., 1968). On the average, men do not decline in health or in morale after retirement; neither do they, after a short period of readjustment, fail to establish meaningful patterns of activity. While attitudes vary according to occupational level, the majority of workers seem increasingly to welcome retirement and to choose leisure over work just as soon as they have sufficient income on which to live (Barfield & Morgan, 1970). Although there are some signs of increased age segregation, as in the retirement communities that have multiplied in the United States, this trend involves only a small proportion of older persons, and that proportion is a self-selected group who appear to be exercising a larger rather than smaller degree of freedom in choosing where to live. Furthermore, such studies as are available indicate that in such communities where the density of older people is relatively high, social interaction of the older person is increased (Rosow, 1967). On the whole it cannot be said that urban industrial societies preclude the social integration of the old.

Variety of Life Style

A second major conclusion is that there is a wide variation of life styles among older people, enormous diversity, and multiple patterns of successful aging. There is no single pattern of disengagement, and no single or modal pattern that produces life satisfaction. These overall findings apply not only to the younger aged—those in their sixties—but also to persons in their seventies.

For example, in one large-scale study of persons aged seventy to seventy-nine in the United States, followed over a seven-year period, a number of patterns were delineated (Neugarten et al., 1968). One group called the "reorganizers" was composed of persons who substituted new activities for lost ones—who, when they had retired from work, gave time to their families, to community affairs, or to church or other associations. They reorganized their patterns of activity, were competent in a wide variety of activities, and had high levels of satisfaction. One such person was a retired schoolteacher, who, at age seventy-five, was busy selling life insurance and making more money than ever before. He was an elected officer in an association of retirees, he attended concerts and the theatre with his wife, and he visited with friends regularly.

Another group which showed high morale had only medium levels of activity. It was called the "focused," because members had become selective in their activities, and now devoted energy to one or two role areas. One such case, for instance, was a retired man who was now preoccupied with the roles of homemaker, parent, and husband. He had withdrawn from work and also from club memberships, but welcomed the opportunity to live a relaxed life with his family, seeing his children and grandchildren, gardening, and helping his wife with homemaking which he had never done before.

A third group was called the "disengaged." This group was composed of persons who had high life satisfaction, but low role activity, persons who had voluntarily moved away from role commitments, not in response to social losses or physical deficits, but because of preference. This pattern might be called the "rocking-chair" approach to old age—a calm, withdrawn, but contented pattern.

There were other patterns, of course: people who were less content, some of them striving and ambitious, and still eager to hold onto the patterns of their middle age; some who were busily defending themselves against the threats of aging by constricting their social interactions and closing themselves off from new experience; some who had strong dependency needs but who maintained themselves fairly well by having one or two other persons whom they depended upon for emotional support.

In another study—this one based on small groups of relatively healthy seventy to seventy-five-year-old men in six different countries (United States, Germany, Austria, Holland, Italy, and Poland) who were either retired schoolteachers or retired steelworkers—wide individual differences were found, as well as differences between the two occupational groups and differences between countries (Havighurst et al., 1969). Overall, however, there was little social isolation. The most common pattern was one in which there was highest activity in family roles (parent, grandparent, spouse, and kin member); intermediate activity in the roles that relate to informal social relations (friend, neighbor, acquaintance); and low activity in the roles that reflect formal social relations (club member, church member, citizen, and political participant). But other patterns were frequent, also, indicating again that there is no single pattern of social roles that characterizes men in their seventies.

Studies like these are based on persons who live in their own homes in the community and who are still relatively healthy. It is true that there are lonely and isolated old people, sick old people, and old people who live in hospitals and homes for the aged, for whom the picture is more dismal. There are studies, however, which show that many of the people who are socially isolated in old age are persons who were relatively isolated also when they were younger.

What we certainly cannot do is to generalize about social role patterns even in 75-year-olds, to say nothing of 65-year-olds. What is striking to the investigator are the wide variations and the different patterns that are associated with contentment or high morale.

If there is no single social role pattern for the aged in 1970, the diversity is likely to become even more true in the future. Differences between older people will probably become greater. With better health, more education, and more financial resources, older men and women will have greater freedom to choose the life style that suits them. Future generations will also have had continuing experience with change—in jobs, in living arrangements, and in the timing of education, work, and leisure—and this experience is likely to result in greater adaptability. Persons who have adapted to change in their youth and middle age are usually those who adapt to change in their old age.

All this is not to deny the fact that at the very end of life, there will continue to be a shorter or longer period of dependency, and that increased numbers of the very old will need care, either in their own homes or in special institutional settings. For persons who are terminally ill or incapacitated, it will be idle, in the future as in the present, to speak of meaningful social roles. While it must be left to the biologist to predict the future with regard to the prolongation of life and the extent to which men will gain greater controls over death, the problems for the society will continue to be those of providing the maximum social supports, the highest possible levels of care and comfort, the assurance of dignified death, and an increasing element of choice for the individual himself or for members of his family regarding how and when his life shall end.

The future for the aged will depend, of course, upon conditions of world stability; upon increasing economic productivity, rising standards of living, and equitable distribution of goods and services; and above all, upon population control. However, within the limits of these larger uncertainties, and to the extent that man will continue to create a viable physical and social environment, the position of the aged should improve in the long-run future.

Summary

From the broad perspective of the relations between age groups, the near future may witness a rise in the phenomenon called age-ism in at least those industrialized countries like the United States, in which there is not yet a tradition of valuing older people. This trend may not occur at all in other countries where the traditions and the value systems are different, or where there is a longer tradition of distributing the goods of the society more equally among age groups. In those places where it will appear, it is likely that such a trend toward increased competition among age groups will be short-lived; in the distant future, the position of the old will improve as they become successively younger in body and mind, and as age differentiations tend to diminish in importance throughout the whole of adulthood.

From the perspective of social role patterns for older individuals, the future will probably bring increased social permissiveness, increasing diversity of life styles, and increased freedom to develop idiosyncratic patterns that provide for higher levels of life satisfaction. In the long run, we may come to diminish the importance of chronological age as a major distinguishing feature between individuals, and instead of speaking of social roles for *the aged*, come to speak of the social roles of *individuals* who happen to be young, middle-aged, or old, but more important, who happen to have different tastes, different goals, and different ways of enhancing the quality of their lives.

References

Barfield, R. A., & Morgan, J. N. Early retirement: The decision and the experience. Ann Arbor: Institute of Social Research, University of Michigan, 1970.

Feuer, L. *The conflict of generations.* New York: Basic Books, 1969.

Havighurst, R. J., Munnichs, J. M. A., Neugarten, B. & Thomas, H. *Adjustment to retirement.* Assen, Netherlands: Van Gorcum, 1969.

Neugarten, B., Havighurst, R. J., & Tobin, S. S. Personality and patterns of aging. In B. Neugarten (ed.) *Middle age and aging.* Chicago: University of Chicago Press, 1968. Pp. 173-177.

Rosow, I. *Social integration of the aged.* New York: Free Press, 1967.

Seeley, J. R. Youth in revolt. In *Britannica book of the year.* Chicago: Encyclopedia Britannica, 1969. Pp. 313-315.

Shanas, E., Townsend, P., Wedderburn, D., Friis, H., Milhj, P., & Stehouwer, J. *Older people in three industrial societies.* New York: Atherton Press, 1968.

Recommended Additional Readings

Maxwell, R. J., & Silverman, P. Information and esteem: Cultural considerations in the treatment of the aged. *Aging and Human Development*, 1970, *1*, 316-392.

Mead, M. *Culture and commitment: A study of the generation gap.* New York: Doubleday, 1970.

Riley, M. W. Social gerontology and the age stratification of society. *Gerontologist*, 1971, *11*, 79-87.

Troll, L. E. Issues in the study of generations. *Aging and Human Development*, 1970, *1*, 199-218.

16   Physics, Femininity, and Creativity

Herbert J. Walberg

Orientation

Reprinted from *Developmental Psychology*, 1969, 1, 47-54, by permission of the author and the American Psychological Association.

Harvard Project Physics supported the present research. The author thanks Mary Fisher for statistical assistance and Rhoda Baruch, Robert Bridgham, Banesh Hoffmann, David Moment, David Riesman, Nancy St. John, Fletcher Watson, Eleanor Webster, and Wayne Welch for comments on a draft of the manuscript.

The author thanks the Consulting Psychologists Press, the Psychological Corporation, and Milton Rokeach for permission to reproduce items from their instruments for research purposes. Since there is evidence to indicate that item responses obtained to selected items isolated from the context of a personality inventory may not be comparable to those obtained from within the context, the results of this research should not be considered applciable to the standardized complete forms of the instruments.

There is a difference between boys and girls. Around this obvious biological fact there have been constructed roles and stereotypes; and psychological notions of masculinity and femininity pervade our thinking about the persons we know. In this society, a real boy plays baseball, climbs trees, hates school, and likes blue as his favorite color. A real girl plays with dolls, jumps rope, is polite to adults, and likes pink dresses. Children who do not fully conform to these stereotypes are sissies or tomboys.

Although we are told that it is distasteful and illiberal to ascribe status to persons, the fact remains that there exists a difference in rank between male and female. Women are allowed to vote only long after the establishment of the nation, and they have never produced a president. (Indeed, most young children do not even have the concept of a female president—"that's impossible" is their usual protest when this idea is proposed to them.)

In short, it is a little "better" to be male than female. A girl who acts like a boy is doing something that we find more understandable than is the boy who acts like a girl. We may explain the girl's behavior as motivated by her desire to improve her status; the "sissy," on the other hand, is to be ostracized because he is imitating a group that ranks below himself. There is nothing in our set of values that will make his behavior acceptable.

Somehow all of these attitudes and stereotypes are conveyed to children. Indeed, it is clear that the concepts of maleness and femaleness are learned quite early in childhood. The following article explores certain dimensions of the meaning and significance of these sex-role stereotypes for the adolescent female. There is no need to provide elaborate justification for this allocation of explicit attention to the developing female: It is clear that the scientific study of human development has been overly preoccupied with the male.

First of all, how does this girl react to the dramatic physical and physiological changes taking place within her at the onset of puberty? It seems clear that the menarche *(the first menstruation) is not well received by most girls. Very few indicate that they are delighted, happy, or proud of this change in status; most say that they are upset, worried, or even ashamed about it. Menstruation would seem to signify much more to the adolescent girl than just a simple physiological readjustment; it is a symbol of her newly acquired sexual maturity, and it is suggestive of her future role. Why, then, should this event not be received with great enthusiasm? The reasons seem closely tied to our usual rather negative view of womanhood. Therefore, one reason for negative attitudes in the girls themselves is that they have been influenced by the negative reactions of others. Adults now express sympathy for the girl's "plight"; she will be told by some that she is "unwell"; others may refer to her "curse." (It is well to note that in a great many cultures the menarche—and the menses in general—is regarded as symptomatic of sickness or evil.) Another important reason for the less-than-enthusiastic welcome for this change is the physical discomfort which accompanies it. Headaches, cramps, backaches, abdominal pains, and assorted other annoyances typically are found, especially during the early years before the menstrual cycle becomes stabilized.*

Inadequate preparation for this event is still another cause for bewilderment and perhaps shame surrounding this physical change. Although it appears that girls receive better advance instruction about the physical changes that will occur within them than do boys, very often this information is deficient or inaccurate. Parents (usually mothers) who adequately handle this instruction often supplement their conversations with their daughters with well-written and well-illustrated pamphlets.

A final reason that has been suggested by some professionals for this negative attitude about menstruation is that the girl simply lacks a satisfactory feminine identification. The menarche is supposedly symbolic of the girl's future role of wife and mother, and therefore, according to these experts, she should eagerly anticipate her change in status. Most likely this orientation is universal, but there are signs that within this society many persons are denouncing this attitude because of its demeaning, degrading assumptions about the female sex.

Whatever the girl's feelings about the menarche

might be, it seems clear that her attitudes toward the external social world change after this event. Postmenarchial girls' thoughts tend to be preoccupied with social activities with the opposite sex. Personal adornment and display increases in importance; the incredible sales of Seventeen (mostly read by girls considerably younger than 17) and similar magazines attest to these concerns.

We are socialized into our sex-appropriate roles very early in life. Even preschoolers indicate a considerable knowledge of which activities are appropriate for girls and which are appropriate for boys. The editor of this book has found in his research activities that over 75 percent of first-grade girls indicate that they want to be either nurses or teachers in adulthood (do you remember the persistent question: "What do you want to be when you grow up?"), while the same-aged boys nominate a wide variety of occupational roles, including doctor, lawyer, scientist, astronaut, football player, and even president of the United States. Six- and seven-year-old girls also often say that they will be mothers when they grow up, but no boy ever says he will be a father.

Our schools continue the process of socializing children into sex roles that was begun by parents. However, in schools this process is complicated somewhat in that contradictory attitudes are implicitly communicated to boys and to girls. This duality is especially clear in the case of the boys, for they are expected to do well in academic matters, but it is also assumed that they will find intellectual activity unmanly and will therefore dislike it. ("Boys shouldn't be sitting inside doing homework; they should be outside and be active.") The contradictory attitudes conveyed to girls consist of, first of all, the expectation that they will be docile and will accept the schooling process happily; but secondly, it is assumed that they probably are not capable of serious intellectual achievement. In any case, whatever their performance might be is quite irrelevant and perhaps even detrimental to their femininity.

Despite the strong pressures against intellectual pursuits, girls in fact do better in school than boys, particularly in elementary and junior high school. This may be due to the fact that the preponderance of teachers at these levels are women. Schools offer an effeminate and prettified curriculum, and teachers reinforce "feminine" behaviors—in boys as well as girls. However, these effects seem to become less potent toward the high school years.

Differences between the sexes in school performance are also to be found in specific subject matter areas. Girls in the United States are supposed to be less good in mathematics than boys, and boys are supposed to be less good in verbal skills than girls. This is indeed the way it is, according to available data (of which there are plenty). Do differences of these kinds arise because the female sex has less natural ability for mathematics, or because girls in this country learn that it is not feminine to be very good at numbers? Certainly they do learn the latter lesson. Probably the fact that almost all elementary teachers are women fosters this learning, for they are children's first teachers of mathematics, and they most likely convey, in subtle ways, their own distaste for the subject.

There is little doubt that socializing children into these stereotypic sex roles is very functional for the society: Males become the dominant and productive elements, and females assume the rearing and care-taking roles for the successive generations. To a great extent this process works smoothly, and the society is continued and maintained quite nicely. Thus, these sex roles are functional, but are they necessary? More and more persons in this society (mostly women, for obvious reasons) are questioning these traditional orientations. Are there any compelling reasons why females cannot take equal part in the productive and professional roles in society, and why males cannot assume equal responsibility for the rearing of children? Those who insist that there are compelling reasons (mostly men) typically refer to biological dissimilarities between the sexes: Males are more physically active (even shortly after birth); males are physically stronger; females tend to be "unstable" because of their menstrual cycle; and so on. On the other hand, a strong case can be made that the female is clearly a biologically superior being. Women live far longer than men (females in this society have an average life span of seven years longer than that of males); females mature physically much more rapidly than males; many more females than males survive prenatal and neonatal life. In fact, it appears that in order to maintain a relatively equal sex ratio, the conception ratio of males to females is *about 160 to 100; the actual birth ratio of males to females is only 105 to 100; after age 65 years, the sex ratio has changed to 100 men for every 135 women.*

But how important is biology? Are these factors sufficiently important to justify the rigid sex-role stereotyping that exists? It would seem that almost all the occupations typically held by men could also be equally well filled by women; likewise, the functions typically fulfilled by women could be done by men (with some obvious exceptions, such as breast-feeding an infant). Freud claimed that "anatomy is destiny." Such a rigid view is unnecessary. Society seems to be a more important determinant of our individual destinies than is biology. We cannot do much about altering our biology, but society certainly can be transformed to allow each individual to attain his or her intellectual and emotional potentialities.

Professor Herbert J. Walberg, formerly associated with Harvard University and the research and development center for cognitive learning at the University of Wisconsin, is now at the University of Illinois at Chicago Circle.

16

It is puzzling why so few women enter scientific fields and why even fewer achieve eminence. The problem has been noted in the past but appears to have become more serious in recent years. From 1950 to 1960 the proportion of women employed in the natural sciences declined from 11 to 9%. During this period the absolute number of men employed rose 30% while the number of women in the natural sciences rose only 10%. (Rossi, 1965) Many women sacrifice their careers for marriage, child rearing, and homemaking; and stereotypes have it that science, like law and medicine, is a masculine occupation. However, the problem is obviously more complex than these observations suggest. The paradoxical relationship between school success, and later achievement in men and women may be a key to deeper understanding. Girls get consistently better grades than boys in school even in subjects in which boys score higher on standardized tests of achievement. However, after graduation from high school, men write more books and articles, produce more works of art, and make more scientific contributions. Moreover, a long-term study of gifted children (Terman & Oden, 1947) showed a substantial correlation between IQ measured during the school years and later occupational levels for males, but no correlation for females. These findings lead one to suspect that by learning more about boys' and girls' reactions to elementary and secondary education, one may find some of the reasons for women's problems in science and the resultant loss of scientific woman power.

The exploratory research reported here focuses on a crucial time in education and career planning—the last 2 years in high school—and examines the differences in abilities, motives, and outlooks between boys and girls studying physics. This subject affords a good opportunity to investigate science and femininity since physics is more "masculine" and presumably more conflict laden for girls. In 1960, 4% of all employed physicists were women. The corresponding percentages in biology and mathematics were 27% and 26%, respectively. Moreover, the number of employed male physicists increased 93% from 1950 to 1960 while the corresponding number of women increased only 20% (Rossi, 1965). Unpublished figures from the American Institute of Physics show that only about a quarter of high school physics students are girls and only 13% of high school physics teachers are women. The last figure is higher than one might expect since women physicists are more likely to enter teaching than research or industrial positions.

Method

Testing Procedure

Test data were collected on a national sample of 705 girls and 1,369 boys for the preliminary evaluation of Harvard Project Physics, a new physics course for high school students. Seventy-six classrooms in 45 schools located in 17 states and two provinces of Canada were represented. But since the participating teachers had been selected from a pool of volunteers to teach the new course, the sample is to an unknown extent unrepresentative of teachers in the nation. On the other hand, because the teachers are select does not mean the students are.

The teachers administered the tests according to standardized instructions in their regular physics classes on 2 days during the first week of school in the fall of 1966 and on 2 days late in November. Random halves of each class took each of two sets of two tests on the first 2 days, and random fourths of each class took each of four tests later in the term. This method of randomized data collection increases the number of tests administered and decreases the total testing time per student while permitting a random sample within a larger sample (Walberg & Welch, 1967). In this study using three class periods, random subsets of students took nine instruments yielding 56 subscores.

Instruments

Among the cognitive instruments administered was the Haenmon-Nelson Intelligence Test-Form B-Twelfth Grade Level. This is a group-administered, multiple-choice test and has an equivalent-form reliability of .91 (Henmon & Nelson, 1957). The Science Process Inventory (Welch & Pella, 1968) and the Test On Understanding Science (Cooley & Klopfer, 1961) purport to measure general understanding of scientific methods and have Kuder-Richardson Formula 20 reliabilities of .76 and .86 respectively. The Physics Achievement Test is a locally constructed, five-option, multiple choice of physics achievement with a KR20 reliability of .77.

The Pupil Activity Inventory (Cooley & Reed, 1961) contains a list of 70 self-initiated science activities such as "Using the microscope at home." Students indicate frequency of participation in these activities on a 5-point scale from "never" to "often." A recent factor analysis of the items (Walberg, 1967) revealed five clusters with the following corrected split-half reliabilities: Academic, .94; Nature Study, .86; Tinkering, .88; Cosmology, .80; and Applied Life, .77. The test is scored by obtaining the mean rating for each cluster.

Value and personality measures were also included. The Study of Values (Allport, Vernon, & Lindzey, 1960) has six scales with the following split-half reliabilities: Theoretical, .84; Economic, .93; Aesthetic, .89; Social, .90; Political, .87; and Religious, .95. With permission from the publishers, three scales used by Rokeach (1960) and four by Edwards (1959) were reproduced in a booklet with 4-point scales from "strongly disagree" to "strongly agree." These scales and their alpha reliabilities are: Dogmatism, .65; Authoritarianism, .69; Rigidity, .68; and the needs for Achievement, .78; Order, .82; Affiliation, .84; and Change, .75.

Semantic differential scales (Osgood, Suci, & Tannenbaum, 1957) were employed to obtain the images of four physical science concepts (See Table 2). The instructions require the subject to rate the concepts on 5-point scales between bi-polar adjectives such as "safe/dangerous." Factor analyses revealed 16 dimensions for the concepts and cluster scores based upon the factors have a median alpha reliability of .73.

The last measure is the Classroom Climate Questionnaire. The items on the questionnaire describe characteristics of the socioemotional climate of learning in school classrooms. The respondent agrees or disagrees on a 5-point scale with such statements as "There is a high degree of participation in the class on the part of all students." Walberg and Anderson (1968) showed that the scales are valid in predicting cognitive, affective, and behavioral learning. The factor analytically derived scales and their corrected split-half reliabilities are: Satisfaction, .53; Goal Diversity, .64; Personal intimacy, .58; Classroom Intimacy, .79; Egalitarianism, .67; Goal Direction, .80; Democratic Policy, .80; Friction, .86; Social Heterogeneity, .79; Speech Constraint, .41; Group Subservience, .57; Organizational Formality, .51; Responsibility Stratification, .55; Alienation, .75; Disorganization, .55; Interest Heterogeneity, .51; Group Status, .68; and Strict Control, .51. (See Walberg & Anderson, 1968, for additional psychometric information and sample items.)

Statistical Analysis

The means and standard deviations were calculated for boys and girls separately and t tests were used to estimate the significance of difference between the two groups. The girls differed significantly (probability of no difference less than .05) from the boys on the majority of variables, 36 out of a total of 56. As one might expect from such a large sample, many of the significant differences are small in absolute terms. Thus a more appropriate metric, w^2, the variance in a dependent variable associated with or accounted for by the independent variable, in this case, sex differences (Hayes, 1963) shown for all comparisons in Tables 1, 2, and 3.

Results

Turning first to the cognitive variables in Table 1, one sees that the girls scored higher on three of the four measures: IQ, Scientific Processes, and Understanding Science. The boys scored high on a test of physics achievement. The first three tests tap mainly verbal factors while the achievement test contains more quantitative and spatial items. These sex differences are consonant with previous research (Maccoby, 1966). However, if the tests measure what they purport to measure, girls in physics have more scholastic aptitude and understanding of science and scientific process. While the differences are statistically significant ($p < .05$), sex accounts for only 1 or 2% of the variance in the cognitive measures. It is important to note that both the boys and girls in these classes scored distinctly higher on the IQ test than did the standardization group for norms (Henmon & Nelson, 1957). The mean IQ is 115 which is about one standard deviation above the norm. Thus, the average physics student in the sample exceeds 84% of high school students in the standardization group.

Table 1 also shows that girls indicated greater participation in activities involving nature study and applications of science to everyday life, for example, "Worked on a collection of insects, bird nest, or other animal specimens" and "Found out about the science of nutrition and how the body uses food." Boys expressed more participation in cosmological activities: "Thought about such questions as: What is time? Gravity? Space? Energy?" and many more tinkering activities: "Built or repaired radio sets or other electronic equipment." Girls' activities predominate in life sciences and their application; boys engage more often in abstract ideation and physical manipulation of objects.

Girls in physics differ from boys more on values than on the other measures, and the differences are consistent with the test norms (Allport, Vernon, & Lindzey, 1960). The girls scored higher than boys on Social, Aesthetic, and

Table 1
Statistical Differences Between Girls and Boys on Psychological Measures

Measures by group	Girls M	Girls SD	Boys M	Boys SD	T	w^2
Cognitive						
IQ	118.1	15.0	113.3	15.1	3.2***	2.0
Science Processes	105.5	10.8	102.9	14.1	3.1**	.8
Understanding Science	34.4	6.9	32.9	8.0	2.9**	.8
Physics Achievement	16.5	4.4	18.6	5.3	− 6.3***	.8
Science Activities						
Applied Life	1.1	.7	.9	.6	4.6***	2.4
Nature Study	1.6	.7	1.3	.7	6.9***	.8
Academic	1.5	.8	1.5	.8	1.0	.0
Tinkering	.8	.5	1.5	.8	−14.0***	18.7
Cosmos	1.9	.7	2.0	.8	− 2.1*	.4
Values						
Social	41.4	8.7	35.0	7.1	8.0***	13.0
Aesthetic	42.8	9.2	36.6	8.4	6.9***	9.8
Religious	41.9	8.9	35.5	9.6	6.6***	9.1
Economic	36.1	7.4	43.6	7.5	− 9.7***	17.9
Political	39.2	6.1	44.6	6.8	− 8.0***	12.9
Theoretical	38.6	8.3	44.6	6.8	− 7.9***	12.7
Personality						
Affiliation	3.4	.5	3.0	.6	8.0***	12.9
Change	3.2	.5	2.9	.5	4.9***	5.1
Order	2.6	.7	2.5	.7	1.5	.3
Rigidity	2.1	.5	2.1	.4	.9	.0
Achievement	3.1	.6	3.1	.5	.2	0
Dogmatism	1.9	.4	2.1	.4	− 2.8**	1.5
Authoritarianism	2.0	.5	2.0	.5	− .1	.0

Note:
Random subgroups of students within classes took the tests. For the Science Processes, Understanding Science, Physics Achievement, and Science Activities instruments the numbers of girls and boys are approximately 675 and 375. The corresponding numbers for the other tests are 310 and 140. The figure w^2 estimates the variance accounted by the sex differences.
*$p < .05$.
**$p < .01$.
***$p < .001$.

Religious implying, respectively, greater valuing of people, form and harmony, and transcendental unity. They scored lower on Economic, Political, and Theoretical implying lower values for businesslike practicality, power, and truth. The greatest differences between the sexes are on Economic and Social. They may reflect students' preoccupations during the last year in high school: boys' preparation to make a living and girls' concern for homemaking and family life. Compared to college norms (Allport, Vernon, & Lindzey, 1960) both groups have more "masculine" values; they both scored higher on Theoretical and Political and lower on Social and Aesthetic than the respective norms for college males and females.

On the personality measures, the girls scored higher on the needs for Affiliation and Change: "I like to do things with my friends rather than by myself" and "I like to do new and different things." They scored lower on Dogmatism, "Obedience and respect for authority are the most important virtues children should learn." It is worth noting that sex differences accounted for the greatest amount of variance in Affiliation, 13%, among the personality variables. There were no significant differences on the needs for Order and Achievement, Rigidity, and Authoritarianism.

Semantic and Classroom Climate Ratings

Table 2 contains the separate statistics for boys and girls on the semantic dif-

Table 2
Statistical Differences Between 285 Girls and 565 Boys on Semantic Differential Ratings

Ratings by concept	Girls		Boys		T	w^2
	M	SD	M	SD		
Universe						
Friendly	3.6	.9	3.0	.8	9.5**	9.4
Beautiful	4.6	.5	4.3	.7	6.3**	4.3
Secure	2.9	.8	2.9	.8	0.1	.0
Physics						
Important	4.5	.6	4.5	.7	.0	.0
Complex	4.2	.6	4.2	.6	.9	.0
Safe	3.2	.6	3.3	.7	−3.2*	1.0
Interesting	3.5	.7	3.6	.7	−1.5	.1
Laboratory experiments						
Important	4.5	.6	4.2	.8	4.3**	2.0
Fun	3.4	.7	3.3	.8	1.7	.2
Ordered	3.8	.9	3.7	.9	1.4	.1
Simple	2.9	.9	3.1	.9	−3.3**	1.2
Myself as a physics student						
Starting	4.7	.5	4.5	.7	4.6**	2.3
Pleasure	3.8	.7	3.8	.8	.1	.0
Facile	2.4	.9	2.6	.8	−3.7**	1.5
Valuable	4.1	.9	4.2	.8	−1.9	.3
Ordered	3.5	.9	3.6	.8	−1.2	.0

*$p < .01$.
**$p < .001$.

ferential ratings. In comparison with boys, girls rated Universe as more friendly and beautiful. Girls perceived Physics as less safe and saw Laboratory experiments as more important but less simple. In rating Myself as a physics student, the girls saw themselves as less facile and more apt to be starting.

The last series of subscores are measures of the students' perception of the social and emotional climate of their physics classroom. Table 3 shows that more often than boys, girls rated their classes as more Satisfying, Egalitarian, and Diverse in its goals. They also indicated greater Intimacy among the class members. Boys felt more Friction between class members and perceived greater social differences among their fellow students. In addition, they felt more constrained in what can be said in class and saw a greater degree of group subservience than did the girls.

Thus, a ragbag group of differences between the sexes in high school physics classes has been seen. Yet childhood characteristics foreshadow these differences and the parallels help to explain some adult differences between men and women in science.

Table 3
Differences Between 145 Girls and 310 Boys on Ratings of Classroom Climate

Aspects	Girls		Boys		T	w^2
	M	SD	M	SD		
Satisfaction	2.5	.8	2.3	.8	2.6**	1.2
Goal diversity	2.3	.7	2.1	.7	2.5*	1.1
Personal intimacy	1.6	.8	1.6	.7	2.3**	.9
Classroom intimacy	2.7	.7	2.6	.7	2.1**	.8
Egalitarianism	3.0	.8	2.8	.8	2.0**	.6
Goal directed	2.3	.8	2.1	.9	1.5	.3
Democratic policy	1.7	1.1	1.5	1.0	1.4	.2
Internal Friction	.9	.8	1.4	.8	−5.9***	6.8
Social heterogeneity	2.0	1.0	2.3	.9	−3.0**	1.7
Speech constraint	1.0	.9	1.2	.9	−2.5**	1.1
Group subservience	1.0	.8	1.2	.9	−2.0*	.7
Organizational formality	1.2	.8	1.4	.8	−1.6	.4
Responsibility stratification	1.3	.7	1.5	.8	−1.5	.3
Alienation	.8	.9	1.0	.9	−1.3	.1
Disorganization	.8	.7	.9	.8	−1.0	.0
Interest heterogenity	3.3	.8	3.3	.7	−0.9	.0
Group status	1.5	.8	1.5	.9	−0.5	.0
Strict control	2.1	.7	2.1	.7	−0.5	.0

*p < .05
**p < .01
***p < .001

Discussion

The sex differences appear to be coherent in at least four frames of reference: verbal aptitude, social values, cautiousness, and aestheticism with the girls being higher on all four. Turning first to verbal aptitude, it is important to note that both the boys and girls in the study scored distinctly high on intelligence; the mean IQ is 115. One's suspicion that girls in physics are even more select is also confirmed, but only on the three verbal measures. In a recent study (Kagan & Freeman, 1963) girls who rejected traditional sex role behavior scored higher on intelligence than girls who conformed to the feminine role. Another study (Bing, 1963) points to maternal child rearing practices as one possible cause of discrepant verbal and nonverbal aptitude. Mothers of children with high verbal aptitude but low scores on number and spatial abilities as contrasted with their counterparts (mothers of children with high number and spatial abilities but correspondingly lower verbal aptitude) gave their children more verbal stimulation, criticized them more for poor academic accomplishment, restricted object experimentation, and aroused anxiety to make the children more cautious and less exploratory. Thus closeness to a verbally demanding, intrusive mother fosters verbal ability; nonverbal aptitude develops best when the child is free to explore and experiment on his own. Mothers are commonly more restrictive in raising daughters, and consistent with this observation, the girls scored lower on the physics achievement test which emphasizes nonverbal skills.

Two studies bear obliquely on the ties between "masculinity" on one hand and quantitative aptitude and lack of caution on the other. All the autobiographies of eminent women mathematicians analyzed by Plank and Plank (1954) showed their strong attachment to, and identification with, their fathers rather than mothers. In another study by Kass (1963), girls, when given the option of playing three slot machines with various payoffs, chose much more often than the boys the machines that paid small amounts steadily rather than those which paid large longshots more infrequently. One shall see that these cognitive differences between the sexes relate to emotional and behavioral characteristics as brought out below.

The girls value people and religious unity more than boys while boys value practicality, power, and theoretical ideas more. The girls are less dogmatic and have higher needs for affiliation and to be outgoing, and they regard the universe as more friendly. Their activities in science—involving people and animate objects as opposed to cosmological ideas and tinkering—reflect their social dispositions. Moreover, these dispositions are further reflected in their general feelings of satisfaction, intimacy, and equality in their perception of classroom climate.

These findings coincide with a number of studies and previous research on sex differences in early childhood reviewed by Mischel (1966). Girls are more dependent and conforming; boys are more aggressive, independent, and autonomous. For example, in one study nursery school girls became more passive after encountering difficulty on an intelligence test; boys became more active

(Crandall & Rabson, 1960). In another study reported by Moriarity (1961), girls approached tasks in a more organized way, they became less integrated when they encountered difficulty and failure, and they wanted to quit or return to early successes. Other studies reviewed by Mischel showed that girls improve their intellectual performance when socially approved; boys improve more under competitive conditions. These personality characteristics may be related to differences in thinking. Dependent children are distracted more easily perhaps because they are more oriented to external interpersonal stimuli (Maccoby, 1966). This may lead to the female intuition but it also may interfere with internal, sequential processing required for abstract, quantitative, and spatial abilities so necessary in the physical sciences. The tension between internal abstraction and external interpersonal values may be all the more acute in bright girls such as those in the present sample. In a study by Crandall, Dewey, Katkovsky, and Preston (1960), examiners asked children how they thought they could do on a new task. Among boys, IQ was positively correlated (r - .62) with expectation of success, but the brighter the girl, the less well she expected to do ($r = -.41$). Also, although the brighter boys believed success was an outcome of their own efforts, there was no correlation between intelligence and self-reliance for girls. Thus bright girls may hold back intellectually for conformity to the feminine role and social approval.

The third pattern of differences suggests greater cautiousness on the part of the girls. Studies showing less risk taking on the part of the girls and the pattern of child rearing which fosters caution have already been discussed. The semantic differential and classroom climate measures bring out further related differences. Compared with boys, the girls find laboratory experiments important but not simple. They saw physics as less safe, and in rating themselves as physics students, the girls saw themselves as beginners on a difficult task. While the girls rate their classrooms as interpersonally satisfying, the boys seem to be bursting the class seams: they feel more constraint, subservience, friction, and saw greater social differences between class members. The boys also scored higher on dogmatism. These differences are consistent with the common observation that girls are more passive and docile and with studies showing more caution and dependence and less aggressiveness on the part of girls.

The fourth pattern is the girl's greater aesthetic concerns. The girls scored higher on aesthetic values and also rated the universe as more beautiful. These findings are significant for the present study only in that they appear to reflect the external and perhaps more superficial orientations of girls.

Two important though tentative explanations can be derived from these findings. The first is that girls get higher grades and are generally more content in high school science perhaps because of their greater verbal aptitude, their docility, dependency, and willingness to work for social approval. These characteristics are probably very attractive to teachers, and they may reward girls more for their verbal knowledge and conformity to the norms of the school. The student role is more consistent with the sex role for girls; they adjust to it more easily and find more satisfaction in it.

But the second consequence is that the very factors that correspond in the feminine and student roles may penalize the girl's chances for later eminence in science. A summary article of the research on creativity in science (Taylor & Barron, 1963) concluded with the findings common to the seven major studies of the distinguishing traits of creative scientists. Among the 13 traits were the following which also appear to distinguish the boys from girls in the present select sample: (a) a high degree of autonomy, self-sufficiency, self-direction; (b) a preference for mental manipulations involving things rather than people: a somewhat distant or detached attitude in interpersonal relations, and a preference for intellectually challenging ones; (c) a high degree of personal dominance (dogmatism) but a dislike of personally toned controversy; (d) a high degree of control of impulse, amounting almost to overcontrol: relatively little talkativeness, gregariousness, impulsiveness; (e) a liking for abstract thinking with considerable tolerance of cognitive ambiguity; (f) marked independence of judgment, rejection of group pressures toward conformity in thinking; and (g) a special interest in the kind of "wagering" which involves pitting oneself against uncertain circumstances in which one's own effort can be the deciding factor (adapted from Taylor & Barron, 1963, pp. 372-373). On one characteristic the girls did score higher, superior general intelligence, but this is attributable to select and relatively small number of girls who elect to take physics. No comparable measure of the 5 other traits were available for the present sample. The 13 distinguishing traits summarized by Taylor and Barron are not exactly the same conceptually or operationally across the seven studies; nor are the 7 distinguishing traits of girls found here exactly the same as the above group taken from the Taylor-Barron list. Nevertheless, the apparent trait discontinuities in feminine and scientific roles help to account for the relatively poor showing of women in science.

References

Allport, G. W., Vernon, P. E., & Lindzey, G. *Study of values manual.* Boston: Houghton Mifflin, 1960.

Bing, E. Effect of childrearing practices on the development of differential cognitive abilities. *Child Development*, 1963, **34**, 631-648.

Cooley, W. W., & Klopfer, L. E., *Test on understanding science.* Princeton, N.J.: Educational Testing Service, 1961.

Cooley, W. W., & Reed, H. B. The measurement of science interests: An operational and multi-dimensional approach. *Science Education*, 1961, **45**, 320-326.

Crandall, V. J., & Rabson, A. Children's repetition choices in an intellectual achievement situation following success and failure, *Journal of Genetic Psychology*, 1960, **97**, 161-168.

Crandall, V. J., Dewey, R., Katkovsky, W., & Preston, A. Parent's attitudes and behaviors and grade school children's academic achievements. *Journal of Genetic Psychology*, 1960, **104**, 53-66.

Edwards, A. E. *Personal preference school manual*. New York: Psychological Corporation, 1959.

Hayes, W. *Statistics for psychologists*. New York: Holt, Rinehart and Winston, 1963.

Henmon, V. A. C., & Nelson, M. J. *The Henmon-Nelson Tests of Mental Ability Manual*. Boston: Houghton Mifflin, 1957.

Kagan, J., & Freeman, M. Relation of childhood intelligence, maternal behaviors and social class to behavior during adolescence. *Child Development*. 1963, **34**, 899-911.

Kass, N. Risk in decision making as a function of age, sex and probability preference. *Child Development*, 1963, **35**, 577-582.

Maccoby, E. E. Sex differences in intellectual functioning. In E. E. Maccoby (Ed.), *The development of sex differences*. Stanford, Calif.: Stanford University Press, 1966.

Mischel, W. A social-learning view of sex differences in behavior. In E. E. Maccoby (Ed.), *The development of sex differences*. Stanford, Calif.: Stanford University Press, 1966.

Moriarity, A. E. Coping patterns of preschool children in response to intelligence test demands. *Genetic Psychology Monographs*, 1961, **64**, 3-127.

Osgood, C. E., Suci, G. J., & Tannenbaum, P. H. *Measurement of meaning*. Urbana, Ill.: University of Illinois Press, 1957.

Plank, E. H., & Plank, R. Emotional components in arithmetical learning as seen through autobios. In R. S. Eissler (Ed.), *Psychoanalytic study of the child: Volume 9*. New York: International Universities Press, 1954.

Rokeach, M. *The open and closed mind*. New York: Basic Books, 1960.

Rossi, A. S. Women in science: Why so few? *Science*, 1965, **148**, 1196-1199.

Taylor, C. W., & Barron, F. A look ahead. In C. W. Taylor & F. Barron (Eds.), *Scientific creativity: Its recognition and development*. New York: Wiley, 1963.

Terman, L. M., & Oden, M. H. *The gifted child grows up*. Stanford, Calif.: Stanford University Press, 1947.

Walberg, H. J. Dimensions of interests in boys and girls studying physics. *Science Education*, 1967, **45**, 320-326.

Walberg, H. J. & Anderson, G. J. Classroom climate and individual learning. *Journal of Educational Psychology*, 1968, **59**, 415-420.

Walberg, H. J., & Welch, W. W. A new dimension of randomization in experimental curriculum evaluation. *School Review*, 1967, **75**, 369-377.

Welch, W. W., & Pella, M. O. The development of an instrument for inventorying knowledge of the process of science. *Journal of Research in Science Teaching*, 1968, **5**, 64-68.

Recommended Additional Readings

Garai, J. E., & Scheinfeld, A. Sex differences in mental and behavioral traits. *Genetic Psychology Monographs*, 1968, **77**, 169-299.

Kohlberg, L. A cognitive developmental analysis of children's sex-role concepts and attitudes. In E. Maccoby (Ed.), *The development of sex differences*. Stanford, Calif.: Stanford University Press, 1966.

Mischel, W. Sex-typing and socialization. In P. H. Mussen (Ed.), *Carmichael's manual of child psychology*. (3rd ed.) Vol. 2. New York: Wiley, 1970. Pp. 3-72.

Mussen, P. H. Early sex-role development. In D. A. Goslin (Ed.), *Handbook of socialization theory and research*. Chicago: Rand McNally, 1969. Pp. 707-731.

Stein, A. H. The effects of sex-role standards for achievement sex-role preference on three determinants of achievement motivation. *Developmental Psychology*, 1971, **4**, 219-231.

17 The Black Youth's Self Concept

Alan Ziajka

Orientation

A naive, uninformed person could easily develop the impression, as he worked his way through most textbooks in the behavioral sciences, that behavior and development are essentially identical for everyone. It is true, of course, that most of these texts devote some amount of space to the special circumstances of those persons who are not white and not middle class; nevertheless, the reader is seldom left with an understanding of what it is like to grow up outside the dominant culture. It is entirely likely that it is impossible to present an accurate description in words of just what it means to possess a skin color or ethnicity that does not belong to those of the majority within a society, but recent autobiographical accounts by black Americans (e.g., The Auto-

Printed by the permission of the author.

biography of Malcolm X, *Claude Brown's* Manchild in the Promised Land, *Anne Moody's* Coming of Age in Mississippi*) perhaps have come closest to achieving this goal.*

The following paper by Alan Ziajka surveys a wide assortment of literature—including theoretical treatises, psychological and sociological research, and personal accounts—in an attempt to construct a picture of the nature and development of the self-concept in young black Americans. Ziajka offers many penetrating observations on what this experience is all about. Of course, a written account cannot present the exact feelings of despair and anguish that are a part of being black in America, but this article does offer some insights into the nature of human development out of the mainstream of American culture.

The following article is a revised version of a paper written by Alan Ziajka when he was a student in the editor's course in child development at the University of Wisconsin. Mr. Ziajka, a former graduate student in political science, now holds a teaching position in California.

17

The black psychologist, Kenneth B. Clark, and the black psychiatrists Alvin Poussaint, William Grier, and Price Cobbs, have studied extensively the development of the black youth's self concept. Their studies, and the work of other social scientists, can be usefully viewed from the perspective of theoretical models provided by George Herbert Mead, Charles Cooley, and Erik H. Erikson. In these models one can find the analytical framework to look at the following topics: The studies carried out by Kenneth B. Clark, Gordon Allport, and Martin Deutsch on the perception which black children have of themselves; studies done by psychiatrists on the perception which a developing black youth has of his physical characteristics; the values, attitudes, and norms, which the dominant white society has projected to the black throughout American history; and the way in which the existing school system reinforces those values and norms in the developing black youth.

Three Developmental Models

G. H. Mead hypothesized that an individual's self concept develops as a result of experiences with other human beings. What Mead defined as the *self* develops through interaction with other members of society by taking the role of the "other." The self is formed by internalizing the attitudes, values, and norms of one's peers, parents, teachers, and all other representatives of the society. What a person does, thinks, and feels depends upon the projection of oneself into the minds of others. Further, the individual experiences himself as such, not directly, but only indirectly, from the particular standpoint of other members of the same social group to which he belongs. This social group, or community, that gives the individual his unity of self is the *generalized other* (Mead, 1934).

Several writers have drawn upon and clarified Mead's theoretical model. Personality, in terms of Mead's framework, can be viewed as a function of social learning. The myriad of experiences which constitute an individual's personal history serve as the learning framework from which the self is derived (Hall & Lindzey, 1957; Wylie, 1961). This learning process provides the individual with many definitions of the self: "the conception of the self as he really believes he is, the self to which he realistically aspires, the self which he believes others see him in, the self as he hopes he is now, and the self he fears now he is" (Phillips, 1969, p. 32).

Mead's theory of self has been related to the definition of culture as the largest social unit acting upon the individual. In childhood, the individual begins to build meanings for his world. Impressions are formulated and extended as his interactions with the social environment proceed. His progressive socialization is in compliance with the expectations and demands of superordinate members of his culture (Vontress, 1966).

Charles H. Cooley has provided, through his *looking glass* model, a theory that is similar to Mead's construct. Cooley perceived the self as having three principal elements: the imagination of our appearance to the other person, the imagination of his judgment of that appearance, and some sort of self feeling. The element that moves one to pride or shame is not the mere mechanical reflection of oneself, but an imputed sentiment, the imagined effect of this reflection upon another person's mind. This is evident from the fact that the character and weight of that other, in whose mind we see ourselves, makes all the difference with our feelings (Cooley, 1956).

Closely related to the frameworks provided by Mead and Cooley is the concept of *identity* as expounded by Erik H. Erikson. Personality, according to Erikson, can be said to develop according to steps predetermined in the human organism's readiness to be driven toward, to be aware of, and to interact with a widening social radius. Each step in a person's psychosocial growth has a special developmental crisis associated with it. This crisis takes the form of a struggle between the negatives and positives at each stage of development (Erikson,

1968). In a process similar to the one outlined by Mead, the individual internalizes a set of attitudes, values, or "prototypes." "Every person's psychosocial identity contains a hierarchy of positive and negative elements, the latter resulting from the fact that the growing human being, throughout his childhood is presented with evil as well as ideal prototypes by reward and punishment, by parental example, and the community's typology as revealed in wit and gossip, in tale and story" (Erikson, 1966, p. 237).

The Development of Racial Awareness in Black Youth

The development of the black youth in American society can be viewed from the perspective of the three models outlined above. With reference to Mead's theory, one can look at the ways in which a black youth internalizes the values projected by the dominant white society. One can use Cooley's model to analyze the image reflected back to the black youth as he gazes into the mirror held up by the dominant white society. Finally, one can see how Erikson's concept of identity can be applied as a frame of reference to explain the way in which a black youth perceives his world as a result of being presented with certain prototypes.

The black youth's self concept, his identity, and the social mirror in which he sees himself are all structured around one myth expounded by the dominant white culture: He is inferior because he is black. In describing his youth, Malcolm X provided a concrete example of this point. He recalled his early experiences with whites:

> What I am trying to say is that it just never dawned upon them that I could understand, that I wasn't a pet, but a human being. They didn't give me credit for having the same sensitivity, intellect, and understanding that they would have been ready and willing to recognize in a white boy in my position. But it has historically been the case with white people, in their regard for black people, that even though we might be with them, we weren't considered of them. Even though they appeared to have opened the door, it was still closed. Thus they never really did see me (Malcolm X & Haley, 1964, p. 27).

The process of internalizing the values projected by the white society begins in early childhood. The self developed as a result of incorporating these values has been the focus of several studies. In a study done by Martin Deutsch (1967), the findings indicated that in all comparisons Negro children had significantly more negative self images than did white children. Some examples illustrate this point. In completion of the sentence "If someone makes fun of me—," 47 percent of all white children responded with the suggestion of some kind of counter action, while only six percent of the Negro children responded in this way. In completion of the sentence "When I look in the mirror I—," the most frequent response in both groups was "I see myself." But 20 percent of the Negro boys gave such responses as "I cry," "am sad," "look ugly," and the like, while such responses occurred in only 9 percent of the white boys.

Another study by Kenneth Clark (1952) produced similar results. Black children ranging in age from three to seven years were presented with four dolls, identical in every respect except for their skin color. Two of the dolls were white and two were black. Clark then asked these black children questions such as: "give me the doll that looks like you," "give me the doll that looks like a colored child," "hand me the doll that you like the best and is nicest." In response to these questions, approximately two-thirds of the subjects indicated they preferred the white doll best, or that they would like to play with the white doll in preference to the colored doll, and that the white doll was a nice doll. Further, 59 percent of the children indicated that the black doll looked bad. The rejection of the black doll by the black children was explained in rather simple, concrete terms: For white doll preferences, "cause he's pretty" or "cause he's white"; for rejection of the black doll, "cause he's ugly," or "cause it don't look pretty," or "cause him black." Moreover, Clark reported that some of the children who were free and relaxed in the beginning of the experiment broke down and cried or became somewhat negativistic during the latter part when they were required to make self identification.*

The studies undertaken by Clark and Deutsch are supported by a number of studies done by Gordon Allport (1954) in nursery schools. Allport found that black children were racially aware earlier than white children and that they tend to be confused, disturbed, and excited by the problem. This interest and disturbance, Allport found, can take many forms. Black children asked more questions about racial differences, often fondled the blonde hair of a white child, and were often rejective toward black dolls. Like Clark's study, Allport found that when black children were given a white and a black doll to play with, they almost uniformly preferred the white doll; many slapped the black doll and called it dirty or ugly. Although they were too young to understand the nature of the trouble, some of them were already in various ways defensive, overreactive, and tense as a result of their vague feelings of disadvantage.

Other studies dealing with the self concept of black children found results analogous to the findings of Deutsch, Clark, and Allport. One study (Stevenson & Stewart, 1958) conducted among black and white children ranging in age from

---

Editor's note: Three replications of Clark's classic study have been reported since Mr. Ziajka wrote this paper. Asher and Allen (1969), using both black and white interviewers, obtained results similar to the original study. With a sample of four- and five-year-old Canadian children and using only white interviewers, Crooks (1970) also confirmed Clark's findings. However, in another study using both black and white interviewers, Hraba and Grant (1970) found that the majority of black children preferred black dolls. Perhaps this last finding is an indication that self-perceptions historically held by black American children are changing within the most recent generation.

three to seven used a series of tests related to attitudes toward self and race. The results showed a greater frequency of own race rejection on the part of blacks as compared to whites. The tests revealed a lower proportion of black children making own race choices in selecting playmates, in choosing companions to go home with, in selecting guests for a party, and in selecting dolls. Further, black children placed blacks in negative positions in stories more frequently than white children placed whites in such positions. Another study of racial attitudes concluded with this observation: "Many of the Negro children who replied that they were colored did so reluctantly and with emotional strain. Indicative of such reluctance and strain was the reply of a Negro boy who hung his head and said quietly, 'I guess I'se kinda colored' " (Morland, 1958).

These studies can be viewed from the perspective of the models proposed by Mead, Cooley, and Erikson. The black child, as he develops, gradually internalizes values projected by the "generalized other," which makes him aware of the societal meaning of his black skin. This analysis of the development of a self concept in black children is, obviously, a generalization which undoubtedly has many exceptions. Nevertheless, the studies in the field seem to indicate considerable support for this generalization.

The looking glass held up by American society reflects back at the black youth one central image: the derogation of blackness. This derogation can be related to Erikson's concept of evil and ideal prototypes. When a black person looks into the mirror held up by white society, he sees black skin and kinky hair being represented as evil prototypes and white skin and straight hair being valued as ideal prototypes. The values, attitudes, and norms espoused by the dominant white culture say to the black man that he should strive to approximate the ideal prototype of light skin and straight hair. In reflecting upon his youth, Malcolm X provided a penetrating insight into the desire by the black man to approximate whiteness through hair straightening. In describing his own "conk," he reflected,

> ... and on top of my head was this thick smooth sheen of shining red hair, real red, red and as straight as any white man's. How ridiculous I was. Stupid enough to stand there simply lost in admiration of my hair now looking "white." I vowed that I'd never again be without a conk, and I never was for many years. This was my first really big step toward self-degradation: when I endured all that pain, literally burning my flesh to have it look a white man's hair. I had joined that multitude of Negro men and women in America who are brainwashed into believing that black people are inferior, and white people superior. They will even violate and mutilate their God created bodies to try to look pretty by white standards (Malcolm X & Haley, 1964, p. 54).

The studies that have been cited with respect to the self concept of the developing black youth were undertaken in the 1950s and earlier. Further, the experiences described by Malcolm X took place in the 1940s. Perhaps a significant question at this point concerns the extent to which racist values and attitudes projected by the dominant white society still shape the self concept of the black youth. In the face of the Black Muslim, Black Nationalist, and Black Power movements of the 1960s, the concept which a developing black youth has of himself may have changed considerably. Nevertheless, a study done by Kenneth Clark in 1965 suggested the continuing preoccupation of many black youths with hair straighteners and skin bleaches. Clark recorded the following conversation between a group of teenage boys:

"You know, if he go in there with his hair slick up like white, they might go for him better, you know."
"They might use him for a broom or a mop."
"Well, why do you wear brushes."
"Why do I wear brushes? It's a blind, a front. Are you saying that I'm ignorant?"
"He's a playboy. He like to do his hair like that. He's ashamed of his own hair, you know. He feels that he's black and now he wants to be half and half. He wants to be a half-breed" (Clark, 1965, p. 65).

William Grier and Price Cobbs (1968), black psychiatrists, explored the effects on the black woman of having internalized values that derogate her physical characteristics. After a series of interviews with a black woman the psychiatrists wrote, "She thought it a fundamental truth that black women's thick lips and short kinky hair were ugly. Intellectually, she could discuss varying standards of beauty, but they had no relevance for her—she was ugly" (p. 9). Based upon their interviews and studies, these psychiatrists have developed several generalizations with respect to the way black women view their physical characteristics. In this society, the ideal type is the blonde, blue-eyed, white-skinned girl. Women exert great efforts to approximate this ideal. The developing black girl finds that her blackness is the opposite of the ideal of creamy white skin, her lips are thick, her hair is kinky and short. The United States, in Grier and Cobbs' view, presents to all its citizens, but most clearly to the black woman, a negative as well as a positive ideal. The positive ideal is in many respects unobtainable for her, because it really involves trying to become less black and more white. The negative ideal is the black.

What are the psychological effects for a black person in striving to be white? Two psychiatrists, Kardiner and Ovesey (1962), in attempting to come to grips with this problem, presented this view:

Accepting the white ideal is a recipe for perpetual self hatred, frustration, and for tying one's life to unattainable goals. It is a formula for living life on the delusioned basis of "as if." The acceptance of the white ideal has acted on the

Negro as a slow but cumulative and fatal poison. Its disastrous effects were due to the fact that the more completely he accepted the white ideal, the greater his intrapsychic discomfort had to become. For he could never become white. He had, therefore, to settle for the delusion of whiteness through an affection of this attributes that most closely resemble whiteness. This also means the destruction of such native traits as are susceptible of change. In its most regressive form, this idea becomes the frantic wish to be reborn white (p. 310).

Dreams and Racial Identity

The effects of the values and attitudes projected by the dominant white culture eventually come to be reflected in the dreams of the black person, as recorded in clinical interviews. These dreams reveal the extent to which the black man in American society has internalized whiteness as an ideal prototype and rejected black as an evil prototype. A psychiatrist described an interview with a black woman in the following terms:

> One day, during her pregnancy, she was irritated by a neighbor, who boasted of a light grandson. She promptly had a dream. "All I remember is seeing my new baby. I saw him as he looked when he was born. He was very light. He looked white." She rejected the obvious wishful quality of the dreams. She insisted she had no preferences. However, in the first interview after the delivery, she triumphantly proclaimed, with evident satisfaction. "Remember I told you I was going to have a white baby? Well he came out pure white." (Kardiner & Oversey, 1962, p. 136).

One can never conlcude that two cultural situations are exactly the same. Nevertheless, certain similarities can be found between the dreams of the black in America, and the dreams recounted by Charlotte Beradt in *The Third Reich of Dreams* (1968). In describing Hitler's Germany, she worte, "The theory of the superiority of the blonde race comes at night to seek out victims among darkhaired individuals or persons who possess some physical traits that deviate from the state-approved norm" (p. 168). Beradt described a "nineteen year old girl whose hair, eyes, and complexion, deviated sharply from the blonde ideal. All her dreams deal with the so called 'inferiority of the dark race!' 'At a gathering of exclusively blonde, blue eyed people, a two-year-old child who couldn't talk yet opened its mouth and told me, the likes of you don't belong here at all' " (p. 84).

Beradt's interpretation of the dreams of those in Nazi Germany who did not exemplify the Aryan ideal is a generalization that can be usefully applied in certain respects to the developing black youth in America. Kardiner and Oversey (1962), in their interviews with blacks living in the United States, described dreams that reflect anxieties and feelings of inferiority similar to those recounted by Beradt. They described a 20-year-old black girl:

> The deflation of her self esteem is now on a completely permanent basis and there is not a minute of the day that she is free from such awareness. Her opening dream focuses on the underlying conflict. "I was in a hospital. Someone forcibly reached into my vagina and brought out a white baby. I denied it was mine." Her associations all dealt with her shame over a sexual affair with a white man. The dream is a rebirth fantasy in which she sheds her blackness and is reborn white (p. 287).

Freud called dreams the "royal road to the unconscious." Bruno Bettelheim (1968) expressed similar thoughts when he described how dreams portray what our real feelings are, whether we like them or not. They reflect those fears and desires we wish consciously not to know about, or dare not express. In comparing the dreams recounted by Beradt with the experiences of black youths in America, Bettelheim concluded, "The dreams reflecting inner anxiety about being dark haired or dark skinned echo the American Negro's attitudes toward his own skin. It is the feeling that there is something undesirable within us that opens us to defeat by those who see things the same way" (p. 168).

The dream analysis described above, the doll and ghetto youth studies by Clark, the studies undertaken by Grier and Cobbs, Deutsch, Allport, and others, all attempt to analyze the self concept of the black youth from the perspective of the attitudes held by the wider society. The "generalized other" projects negative values and attitudes to the developing black youth, and the black youth may easily develop a negative self concept as a result of having internalized these values.

A Historical Perspective

One must look to the history of the United States to find the origins of the negative self concept which the black has of himself. The decades before the Civil War witnessed the emergence of a set of justifications for slavery that became the foundation for much of the racist doctrine still in existence today. The theory that the black man was an inferior being developed as a legitimating myth to support the social conditions of white superiority achieved through slavery. The idea of the inferiority of the black man enjoyed wide acceptance in the South and the North. It was the central component in the theory of society expounded by Southern thinkers and leaders. It was developed into a well defined body of thought by the biological and social scientists and ministers of the South, out of which evolved a theory of racial inferiority justifying white domination over the black man and control of the slave (Franklin, 1966; Grimes, 1964; Meir & Rudwick, 1966).

The black is the victim of this theory of black inferiority, because he has internalized its components. More than a century ago Alexis de Tocqueville described the effects this racist doctrine had upon the self concept of the nineteenth-century black:

> The Negro makes a thousand fruitless efforts to insinuate himself among men who repulse him; he conforms to the taste of his oppressors, adopts their opinions, and hopes by imitating them to form a part of their community. Having been told from infancy that his race is naturally inferior to that of the white, he assents to the proposition, and is ashamed of his own nature. In each of his features he discovers a trace of slavery, and if it were in his power, he would willingly rid everything that makes him what he is (de Tocqueville, 1851, p. 338).

In the view of Grier and Cobbs, the passage of more than a century has not fundamentally changed the situation described by de Tocqueville. In 1968 these black psychiatrists addressed themselves to the question of the type of attitude developed against blacks during slavery, and the extent to which it dominates current society. They concluded, "Persisting to this day is an attitude, shared by black and white alike, that blacks are inferior. This belief permeates every facet of this country and it is the etiological agent from which has developed the national sickness" (Grier & Cobbs, 1968, p. 25).

A second major effect of slavery was the fragmentation and destruction of black culture, history, family ties, and tribal loyalties. Beginning in 1619, with the first introduction of slavery into America, the black person in this country has been a man without social, linguistic, or historical foundations. The institution of slavery obliterated his African past. His family and tribal ties were severed, and he was prohibited from initiating new family ties within the slave system. Slave owners neither encouraged marriage among their slaves nor hesitated to separate slave couples on the auction block. The slave household tended to develop in a fatherless, matrifocal pattern. For the black slave, all conditions for maintaining his old culture, or for developing a new one, were lacking. The only participation in any culture that the black man had was in the culture of his master. In this culture, his only participation was as an inferior being. Unlike the European immigrant, the black slave coming to this country had no loyalties, history, or culture to draw upon or identify with. The black man, after slavery, had no conscious awareness of the past, because the white man had destroyed that past (cf. Pettigrew, 1964; Thompson, 1968).

A third major consequence of slavery is related to the major tenets of the Protestant Ethic. The Protestant Ethic contains as some of its major tenets the following: the belief that hard work and effort brings success, that self-assertion and expression designed to remake the environment is desired, and that from hard work and self-assertion one's environment can be controlled. In short, the Protestant Ethic, as it related to secular affairs, furnished the moral commitment to achieving control over one's world (McClelland, 1961; Tawney, 1926; Weber, 1958). The institution of slavery cut the black man off from the dominant cultural ideology insofar as it called for the internalization of the major assumptions of the Protestant Ethic. The slave was trained to be nonassertive, unaggressive, and docile. The structure of the slave system prevented the black slave from trying to achieve any sense of control over his environment. Blacks were placed in a completely dependent role. All of their rewards came, not from assertion or achievement, but from absolute obedience and dependent compliance. This kind of relationship, between master and slave, all but eliminated the desire to achieve.

The above remarks obviously constitute a set of generalizations for which there are many exceptions. Slave revolts, such as those led by Denmark Vesey and Nat Turner, escaped slaves, such as Frederick Douglass and Harriet Tubman, the underground railroad, and the need to enact slave codes, all bear witness to the fact that not all slaves were docile and unassertive. Nevertheless, the brutality of the slave system beat the majority of black slaves into submission and prevented most from trying to assert personal control over their world.

Many of the values, norms, and attitudes, projected to the black man during slavery continued to bombard him as a free black. The dominant white culture, after slavery, still called for the black to be nonassertive and unaggressive. Some of the characteristics of this socialization process were described by Poussaint (1968b) in the following terms:

> The Negro most rewarded by whites has been the "Uncle Tom," the exemplar of the black man who was docile and nonassertive, who bowed and scraped for the white boss and denied his aggressive feelings for his oppressor. In order to retain the most menial of jobs and keep from starving, black people quickly learned such servile responses as "yassuh, massa": passivity was a necessary survival technique. To be an "uppity nigger" was considered by racists one of the gravest violations of racial etiquette. Vestiges of this attitude remain to the present day, certainly in the South but even in the North: blacks who are too outspoken about racial injustices often lose their jobs or are not promoted to higher positions because they are considered unreasonable or too sensitive (p. 243).

Competence and Control of the Environment

Related to the apparent inability of the black man to feel that he has a sense of control over his environment is the black's perception of the environment as being a hostile and threatening place. One study looked for possible differences between black and white teenage boys in terms of the way they perceive their environment (Mussen, 1953). The results showed that blacks saw the general environment as more hostile and threatening than whites. A series of pictures were used about which the boys were to tell a brief story. It was found that a greater number of blacks than whites told stories in which the hero was hated, scolded, reprimanded, or was the victim of physical assault. Further, tests showed that blacks, less often than whites, saw themselves establishing friendly relations, respecting others, or being kind and considerate of others. Blacks infrequently saw themselves as being respected, followed, or obeyed by others. The general picture emerging from this study was one of feelings of inferiority, helplessness, and indifference among black youth.

The degree to which a person will accept personal responsibility for what happens to him is a function of the extent to which he feels he can extract material and social benefits from the environment through his own efforts. A child's feelings about whether his own efforts will produce rewards affects his perception of a given goal, and his expectancy of attaining that goal. In this light, a few studies have attempted to relate a child's sense of control over his environment to his ambition to achieve. The typical finding is that black children are often unlikely to develop ambitions and needs to achieve comparable to those found among white children. These differences in ambition and needs to achieve are a function of the differences in the sense of control which black and white children hold over their respective environments. In an unresponsive and hostile world, hard work and extended effort toward achievement appear to the black youth to be unrewarding (Coleman, 1966; Mingione, 1965).

Black Youth and Schools

American culture and society during slavery and since emancipation has provided the developing black youth with three interrelated negatives: a negative self concept, a negation of his original culture, and a negative environment in the sense of it being hostile and beyond his control. In terms of the conceptual model of Mead, an institution to which the black youth brings these negatives, and an institution that largely reinforces these negatives in the developing black youth, is the school.

Several studies have shown that one of the strongest indexes of academic achievement is the extent to which a youth has a feeling of being in control of his environment and his destiny (e.g., Coleman, 1966; Katz, 1967; Slaughter, 1969). Yet the black youth, as we have seen, living in an environment which is hostile and threatening, does not feel that his world is either responsive or controllable. He brings to the classroom this concept of being unable to evoke a positive response from his world. In the classroom, the black youth finds that the social environment further restricts his ability to control his world. The existing school system insists upon the maintenance of order and the imposition of strict control over individual assertion and initiative. In describing the attitude toward control which one school administration gave to teachers in a junior high school in a black ghetto, James Herndon (1969), a former teacher, observed that:

... you were supposed to figure out the real attitude of the administration toward the behavior of students in your classrooms, with an eye to your own evaluation. That is, what degree of control you were being ordered to maintain or what degree of disturbance and chaos would be acceptable. On that afternoon, it was really made easy by Miss Bentley, the vice-principal. Miss Bentley offered us the example of the Army. The Army, she submitted, was an organization of people given certain tasks to perform. So was the school (p. 8).

Attempts to maintain order and discipline sometimes take a brutal form in ghetto schools. Jonathan Kozol (1967), a former teacher in a predominantly black grammar school in Boston, described the frequency with which corporal punishment in the form of beatings and whippings were administered to pupils. Reliance upon physical punishment is the ultimate means to maintain order and discipline. In the view of Kozol, it also grows out of deeply seated racial hate.

It has already been hypothesized that the black youth, as he develops, internalizes the racist values and attitudes of the wider society. Through this process the black youth acquires a negative self concept. In several ways, the existing school system contributes to and reinforces the black's negative self concept. Kenneth Clark (1963, 1965) has illustrated how segregation and racial isolation perpetuates a black's feelings of inferiority and contributes to a negative self concept. Whether segregation is supported by law or custom, by public apathy, or by lack of understanding, its effects are the same: it debases and distorts the black man's personality. In 1954, the Supreme Court used a brief submitted by Clark and other psychologists and social scientists in declaring *de jure* segregation to be unconstitutional. In *Brown* v. *Board of Education* the court stated, "to separate them from others of similar age and qualifications solely because of their race generates a feeling of inferiority as to their status in the community that may effect their hearts and minds in a way unlikely ever to be undone."

Since the Brown decision little desegregation has taken place in the South. In the mid 1960s most students continued to attend schools that were 100 percent white or black (Coleman, 1966). Moreover, racial isolation continues to increase in the urban centers of the North, Midwest, and West. The 1967 Civil Rights Commission found that in 75 major central cities 75 percent of all black students in elementary grades attended schools which had a majority of black students. By 1975 it is estimated that, if current trends continue, 80 percent of all black students in the 20 largest cities, comprising nearly one-half of the nation's black population, will be attending 90 to 100 percent black schools (Report of the National Advisory Commission on Civil Disorders, 1968).

"Cultural Deprivation"

Kenneth Clark has proposed that "residual psychological prejudices" underlie the continued inability of society to eliminate *de facto* racial isolation and *de jure* segregation. For many years the existence of segregated schools was supported by law or justified on the basis of the existence of segregated neighborhoods. Today, more subtle explanations are used. These include doctrines of "cultural deprivation" and myths that the culturally determined educational inferiority of black children will impair the ability of white children to learn if they are taught in the same classroom. This cultural deprivation theory assumes that, because of their background, black children will not only be unable to compete with white children but will also retard the white children. This theory supports the continued rejection of black children (Clark, 1968).

In Cooley's terms, the image that the dominant white culture projects through the looking glass at the black man continues to say he is inferior. "Cultural deprivation" has merely been added as another component in this concept of inferiority. The black man is expected to be inferior because he emerges from an inferior cultural environment. This expectation on the part of the dominant white culture can be related to the entire question of self concept, academic achievement, and the *self-fulfilling prophecy*.

Robert K. Merton has provided the fundamental statement on the self-fulfilling prophecy. Merton feels that people respond not only to the objective features of a situation, but also, and at times primarily, to the meaning this situation has for them. The self-fulfilling prophecy is, in the beginning, a false definition of the situation evoking a new behavior which makes the originally false conception come true. In short, if men expect something to be true, they will govern their behavior accordingly, and may in the end make what was once false become true (Merton, 1948).

A study conducted by Rosenthal and Jacobson (1968a, 1968b) provided a concrete example of the self-fulfilling prophecy operating in a school situation. In their study a group of teachers were told that on the basis of test scores, certain of their students should be expected to show rapid increase in academic achievement through the next year. Actually, no tests had been given, and the children designated as potential spurters had been selected at random. Nevertheless, achievement tests at the end of the year revealed that the designated children did show large gains in academic achievement in comparison to the other children. The authors proposed that the explanation for the results of their test could be found in the subtle features of teacher-student interaction. A teacher's tone of voice, facial expression, touch, and posture, may be the means by which expectations are communicated to the pupil. Such communications change a child's conception of himself, his anticipation of his own behavior, his motivation, or his cognitive skill. Children who are expected to gain intellectually, and have these expectations communicated to them, indeed do gain.

Several other studies have looked at the expectations which the school has of lower-class black youth. One study of teachers in a Northern city found that the majority of teachers and their supervisors rejected lower-class youth and looked upon them as inherently inferior. The children were seen as not being capable of profiting from a normal curriculum and not capable of learning (Trow, 1968). Another study showed that teachers place the burden of academic change on the minority student. The inadequacies of the minority child were deemed the most important factors in his low achievement and general poor school performance (Carter & Hickerson, 1968). Another study, conducted by Howard Becker (1951-1952) in the Chicago school system, found that teachers generally felt that the lowest group—slum children—were difficult to teach and morally unacceptable. Another investigation found that urban teachers tended to describe black pupils as talkative, lazy, fun loving, high strung, and rebellious (Banks, 1969).

Jonathan Kozol recounted a member of the Boston School Committee's statement on this subject: "We have no inferior education in our schools. What we have been getting is an inferior type of student" (p. 60).

These studies and accounts indicate that teachers often have negative attitudes and expectations toward poor and lower-class black youth. Just as the study by Rosenthal and Jacobson showed that positive attitudes and expectations of the teacher can raise student achievement, several other studies have shown that the converse is also true. If teachers have negative attitudes toward their students and expect little from them, students will likely fulfill these expectations. Negative teacher attitudes, as described above, act as self-fulfilling prophecies. In the words of one social scientist: "Our hypothesis is that the chief cause of the low achievement of the children of alienated groups is the fact that too many teachers and principals honestly believe that these children are educable only to an extremely limited extent, and when teachers have a low expectation level for their children's learning, the children seldom exceed that expectation" (Kvaraceus, 1965, p. 22).

The School Curriculum

For generations, black culture and history has been denied or distorted by the schools, its teachers, and textbooks. The school and its curriculum have projected the myth of white superiority and black inferiority. Black youths in the schools have been led to believe that they never had a culture or civilization of their own. No mention has been made of the black African civilizations of Ghana, Mandingo, and Songhai, nor of the cultural, commercial, and intellectual centers at Jenne and Timbukto. The civilization, culture, and history taught to the black youth, has been white, Western European, and Christian.

History textbooks have largely distorted the position of the black man during slavery and reduced the harshness and brutality of the slave system. From the end of the Civil War until the 1950s, judging by the way textbooks treat this period, the black man was virtually nonexistent. A 1968 survey in Michigan revealed that current textbooks still presented false impressions of the black person, mostly through errors of omission. Textbooks were characterized by oversimplification, inadequacy, and incorrectness in their portrayal of the black man (Trezise, 1969). In response to a question asking if Negro history had been taught in her school, one black student in New York replied, "no more than George Washington Carver and his peanuts. I am sick of George Washington Carver and his damn peanuts" (Hill & Burke, 1968, p. 139).

The result of all this is that the black youth's self concept is negated, his academic motivation is destroyed, and his school experience becomes meaningless. As the Kerner Report stated:

> The quality of education offered by ghetto schools is diminished further by use of curricula and materials poorly adapted to the life experiences of their students. Designated to serve a middle class culture, much educational material appears irrelevant to the youth of the racial and economic ghetto. Until recently, few texts featured any Negro personalities. Few books used or courses offered reflected the harsh realities of life in the ghetto, or the contributions of Negroes to the country's culture and history. This failure to include materials relevant to their own environment has made students skeptical about the utility of what they are being taught (National Advisory Commission on Civil Disorders, 1968, p. 434).

Values and School Desegregation

Several black social scientists have set forth a position with respect to the existing school system which challenges several formerly held assumptions. One assumption that has been attacked concerns the legitimacy of the values inherent in the schools themselves. The values intrinsic in the existing school system are of the dominant white middle class, according to this argument. From the "liberal" perspective of the 1950s and early 1960s, the assumption has always been that black children need to integrate into the existing school structure in order to improve the situation. Derived from this assumption, the major goal has been to destroy the *de facto* and *de jure* barriers to integration. Hence, the rationale for desegregation with "all deliberate speed" in the 1954 Supreme Court decision, and "at once" in the 1969 decision. Hence, the rationale for school busing on the belief that integration into the dominant white culture was a positive good. Efforts at integration have been based in large part upon the assumption of white cultural superiority.

Integration into the existing school system is not necessarily a positive good for several reasons. In integrated as well as segregated schools, white supremacy dominates the pattern of teaching and the entire educational process. The existing school system holds up the model of the white middle class to be emulated and aspired to achieve. To attain high scores on tests geared for the white middle-class student, black students in integrated schools are thus induced to emulate the culture of another ethnic group. Further, integration has always taken place on the white man's terms. It is the black who must seek out and travel to the white man's institutions. This situation implies the inferiority of the black man (cf. Hamilton, 1968; Poussaint, 1968a).

The above remarks can easily be placed within the Mead-Cooley-Erikson framework. The "generalized other" still holds up the white prototype as the one to be attained. Integration, as defined by whites and to the extent it projects a white ideal as the one to be emulated, still espouses a looking-glass reflection that says the black man and his culture are inferior.

As a means of overcoming the racism in the existing school system, the concept of control by the local black community has been proposed. One aspect of community control would be the creation of local governing boards having the power to hire and fire superintendents, teachers, principals, and to control funds and fiscal policy. Another aspect of community control is to turn the local school into a center of community life. The local school should be family and community oriented; a center for recreation, education, and social events; and should be open as a year round, day and night, center for the entire family. Public welfare and law enforcement could also be centered at the local school.

As another way of altering the existing racism in the school system, courses and curricula emphasizing black culture and history in grammar and secondary schools have been demanded. This demand grows, as Hare (1969) put it, out of a quest for black consciousness, pastness, and collective destiny. Having thus acquired a new self image as a result of pride in his race and its pastness, the black student will seek to convey this image to others. Black history and culture should therefore involve or integrate the student into his community, augmenting his functioning in the community. The fostering of identity with the black community and the development of community consciousness would commit the black student more to the task of helping to build the black community, when once his studies are over, in contrast to the desire to escape the black community.

Implications

In the past, the relationship between blacks and whites has always been defined by the white society. The white person has always detemined when, if, and in what form the black person could take part in the larger society. Many blacks seem to be saying that the black person should define his own relationship with this larger society. Even integration, if continued to be defined by the white man, will perpetuate racism. To build a developing black youth's self concept, and to give him a sense of control over his environment, many blacks now argue that what is needed is separate control over their own institutions. Only by controlling their own institutions, and thereby gaining an internal sense of control and a positive self concept, will black people of America be able to provide the opportunity for their youth to rise from the position the white society has created and perpetuates.

References

Allport, G. *The nature of prejudice.* Cambridge, Mass.: Addison Wesley, 1954.
Asher, S. R., & Allen, V. L. Racial preference and social comparison processes. *Journal of Social Issues, 1969,* **25,** 157-165.
Banks, J. Relevant social studies for black pupils. *Social Education,* 1969, **33,** 69-79.
Becker, H. S. Career of the Chicago public schoolteacher. *American Journal of Sociology,* 1951-1952, **57,** 472-489.
Beradt, C. *The Third Reich of dreams.* Chicago: Quadrangle, 1968.
Bettelheim, B. Essay on dreams. In C. Beradt, *The Third Reich of dreams.* Chicago: Quadrangle, 1968.
Carter, T. P., & Hickerson, N. A California citizen's committee studies its schools and de facto segregation. *Journal of Negro Education,* 1968, **37,** 105-109.
Clark, K. B. Racial identification and preference in Negro children. In G. Swanson (Ed.), *Readings in social psychology.* New York: Holt, Rinehart and Winston, 1952.
Clark, K. B. *Prejudice and your child.* Boston: Beacon, 1963.
Clark, K. B. *Dark ghetto—dilemmas of social power.* New York: Harper & Row, 1965.
Clark, K. B. Alternative public school systems. *Harvard Educational Review,* 1968, **38,** 105-109.
Coleman, J. S. *Equality of educational opportunity.* Washington, D.C.: U.S. Government Printing Office, 1966.
Cooley, C. H. *Human nature and the social order.* New York: Free Press, 1956.
Crooks, R. C. The effects of an interracial preschool program upon racial preference, knowledge of racial differences, and racial identification. *Journal of Social Issues,* 1970, **26** (4), 137-144.
de Tocqueville, A. *American institutions and their influence.* New York: Barnes, 1851.
Deutsch, M. *The disadvantaged child.* New York: Basic Books, 1967.
Erikson, E. H. Concept of identity in race relations. In T. Parsons & K. B. Clark (Eds.), *The Negro American.* Boston: Houghton Mifflin, 1966.
Erikson, E. H. *Identity, youth, and crisis.* New York: Norton, 1968.
Franklin, J. H. The two worlds of race: A historical view. In T. Parsons & K. B. Clark (Eds.), *The Negro American.* Boston: Houghton Mifflin, 1966.
Grier, W. H., & Cobbs, P. M. *Black rage.* New York: Basic Books, 1968.
Grimes, A. P. *Equality in America.* New York: Oxford, 1964.
Hall, C. S., & Lindzey, G. *Theories of personality.* New York: Wiley, 1957.
Hamilton, C. V. Education in the black community. *Freedomways,* 1968, **8,** 319-324.
Hare, N. The teaching of black history and culture in the secondary schools. *Social Education,* 1969, **33,** 385-389.
Herndon, J. *The way it spozed to be.* New York: Bantam Books, 1969.

Hill, B., & Burke, N. Some disadvantaged youths look at their schools. *Journal of Negro Education*, 1968, **37**, 135-139.
Hraba, J., & Grant, G. Black is beautiful: A reexamination of racial preference and identification. *Journal of Personality and Social Psychology*, 1970, **16**, 398-402.
Kardiner, A., & Oversey, L. *The mark of oppression: Explorations in the personality of the American Negro.* New York: Harcourt, 1962.
Katz, I. Some motivational determinants of racial differences in intellectual achievement. *International Journal of Psychology*, 1967, **2**, 10-21.
Kozol, J. *Death at an early age.* Boston: Houghton Mifflin, 1967.
Kvaraceus, W. C. *Negro self concept: Implications for school and citizenship.* New York: McGraw-Hill, 1965.
Malcolm X., & Haley, A. *The autobiography of Malcolm X.* New York: Grove, 1964.
McClelland, D. C. *The achieving society.* Princeton, N.J.: Van Nostrand, 1961.
Mead, G. H. *Mind, self, and society.* Chicago: University of Chicago Press, 1934.
Meier, A., & Rudiwick, E. M. *From plantation to ghetto.* New York: Hill & Wang, 1966.
Merton, R. K. The self fulfilling prophecy. *Antioch Review*, 1948, **8**, 193-210.
Mingione, A. Need for achievement in Negro and white children. *Journal of Consulting Psychology*, 1965, **29**, 108-111.
Morland, K. J. Racial recognition by nursery school children in Lynchburg, Virginia. *Social Forces*, 1958, **37**, 132-137.
Mussen, P. H. Differences between the TAT responses of Negro and white boys. *Journal of Consulting Psychology*, 1953, **17**, 375.
National Advisory Commission on Civil Disorders. *Report.* Washington, D.C.: U.S. Government Printing Office, 1968.
Pettigrew, T. F. *A profile of the Negro American.* Princeton, N.J.: Van Nostrand, 1964.
Phillips, R. Student activities and self concept. *Journal of Negro Education*, 1969, **38**, 32.
Poussaint, A. F. Education and black self concept. *Freedomways*, 1968, **8**, 334-339. (a)
Poussaint, A. F. Negro youth and psychological motivation. *Journal of Negro Education*, 1968, **37**, 243-269. (b)
Rosenthal, R., & Jacobson, L. *Pygmalion in the classroom.* New York: Holt, Rinehart and Winston, 1968. (a)
Rosenthal, R., & Jacobson, L. Teacher expectations for the disadvantaged. *Scientific American*, 1968, **218**, 19-23. (b)
Slaughter, C. H. Cognitive style: Some implications for curriculum and instructional practices. *Journal of Negro Education*, 1969, **38**, 105-111.
Stevenson, H. W., & Stewart, E. C. A developmental study of racial awareness in young children. *Child Development*, 1958, **29**, 399-409.
Tawney, R. *Religion and the rise of capitalism.* New York: Harcourt, 1926.

Thompson, C. H. Race and equality of educational opportunity. *Journal of Negro Education*, 1968, **37**, 192-202.
Trezise, R. L. The black American in American history textbooks. *Social Studies*, 1969, **60**, 164-169.
Trow, M. Two problems in American public education. In R. Bell & H. R. Stub (Eds.), *The sociology of education*. Homewood, Ill.: Dorsey, 1968.
Vontress, C. E. The Negro personality reconsidered. *Journal of Negro Education*, 1966, **35**, 210-222.
Weber, M. *The Protestant ethic and the spirit of capitalism*. New York: Scribner, 1958.
Wylie, R. C. *The self concept*. Lincoln: University of Nebraska Press, 1961.

Recommended Additional Readings

Brown, C. *Manchild in the promised land*. New York: Crowell-Collier-Macmillan, 1965.
Grier, W. H., & Cobbs, P. M. *Black rage*. New York: Bantam Books, 1968.
Malcolm X, & Haley, A. *The autobiography of Malcolm X*. New York: Grove, 1964.
Moody, A. *Coming of age in Mississippi*. New York: Dell, 1968.

Part Three The Course of Human Development

The Infant Separates Himself from His Mother

Harriet L. Rheingold
Carol O. Eckerman

Orientation

The initial article in the third section of this book, "The Course of Human Development," concerns the issue of the separation process between infant and mother—that inevitable point in time when the individual first sets out on his own into the exciting world before him. As the authors, Drs. Rheingold and Eckerman, emphasize, during the first few months of life the infant develops a strong attachment with his caretaker (and, of course, the parent develops a similarly strong attachment with the infant). Later in infancy the individual begins the detachment process, which is marked by those first faltering ventures away from the close proximity of mother. This latter process is one of the landmarks of ontogeny.

For purposes of comparison, also presented here is a short review of mother-infant interaction in nonhuman primates. This aspect of the article is significant, for it indirectly points to the fact that, in actuality, developmental psychology can be considered as a subset of the broader field of comparative psychology, which typically is viewed as pertaining to the comparison of behaviors between various species of the animal world. Developmental psychology, of course, usually is concerned with comparing behaviors at different ages or points in time, either between or within individuals. However, the comparison approach can be used in a broader sense, for not only can we compare behaviors between ages, but we can also compare behaviors between animal species (as Rheingold and Eckerman have done here) and between cultures (cross-cultural psychology).

The reader is invited to consider carefully the experimental setup used by these researchers to study the detachment process. While this methodology is novel and provides a rich setting for examining an important kind of behavior, the question remains as to its relationship to a natural (i.e., the home) setting. Do infants and mothers behave in the same way at home as they do in psychologists' laboratories? As in all psychological laboratory studies, the meaning-

Reprinted from *Science*, 1970, **168** (3927), 78-83, by permission of the authors and the publisher; Copyright © 1970 by the American Association for the Advancement of Science.

The preparation of this article and the laboratory research were supported by a Public Health Service research career program award (HD-23620) and research grant (HD-01107) to one of us (H. L. R.). We thank Judy M. Cathey for assistance in collecting the data and Don W. Hayne for statistical advice.

fulness of data obtained in artificial settings can only be speculated.

One final consideration: This article concerns the separation process between infant and mother. Keep in mind that separation between loved ones occurs at all points in the life span: The child leaves home for his first day in school; the adolescent/young adult leaves home for work, marriage, or college; and the aged parent leaves his offspring once and for all through death.

Dr. Harriet L. Rheingold is a research professor and Dr. Carol O. Eckerman is a research associate in the department of psychology at the University of North Carolina.

18

At some point in time an infant leaves its mother. It is our purpose to call attention to this behavior in man and animal, to examine its biological and psychological consequences, and to relate the behavior to current principles of behavior theory. The argument is first presented very generally. The general presentation is followed by a review of the literature on nonhuman primates, and then by an outline of some procedures for experimental analysis of this behavior, based upon recent work in our laboratory.

The infant's leaving the mother is well-nigh universal behavior throughout the animal kingdom. Here, however, we limit the presentation to species of the class Mammalia. And within that class the focus is the human infant, although the discussion moves freely between him and infants of other species. Recent studies of nonhuman primates 1-5 offer some data on the infant's separating himself from his mother, but the behavior has not often been precisely documented in other mammals, even in the human infant.[6] With only a few exceptions, the behavior has seldom been the primary subject of study.

Let us consider this behavior. At some point in his life the mammalian infant leaves his mother's side. The first excursion is typically short in extent, and brief. In many species the mother promptly retrieves the infant making his first excursions, but the excursions are not thereby suppressed; they occur again and again. With time, and experience, the distance traveled increases, and so does the time spent away from the mother.

The infant's separating himself from his mother depends, of course, on his ability to move his body by his own efforts. As soon as the human infant is able to move thus—it takes him all of 7 months—he does so, even if he can progress only by inching along on his belly. Later he creeps, and then walks away from his mother. He goes out the door and enters another room. In time he walks out of the house, plays in the yard all morning, goes to school, goes still farther away to high school, then to college and to work. He crosses the country, and now he may go even to the moon. Eventually he sets up his own home and produces infants who, in turn, repeat the process.

Biological and Psychological Significance

The infant's separating himself from his mother is of biologic importance. It is of consequence for the preservation of both the individual and the species—of the individual, since it confers the advantage of greater familiarity with the environment and thus increases the likelihood of adaptation to the environment; of the species, since it allows the mother to care for the next offspring and leads eventually to the formation of breeding pairs.

The infant's separating himself from his mother is also of psychological importance for it enormously increases his opportunities to interact with the environment and thus to learn its nature. For, while he is in physical contact with his mother, his universe is confined to her person and the environment near her There are limits to what the most attentive mother can bring to him. Even when he is carried about, his contacts with the universe are necessarily circumscribed. When, however, he leaves her side by himself, many new kinds of learning can occur.

The infant comes in contact with an increasing number and variety of objects. Through touching them he learns their shapes, dimensions, slopes, edges, and textures. He also fingers, grasps, pushes, and pulls, and thus learns the material variables of heaviness, mass, and rigidity, as well as the changes in visual and auditory stimuli that some objects provide. He moves from place to place within a room, and from one room to another. From the consequent changes in visual experience, coupled with his own kinesthetic sensations, he learns the position of objects relative to other objects.[7] He also learns the invariant nature of many sources of stimulation. In a word, he learns the properties of the physical world, including the principles of object constancy and the conservation of matter.

Although in considering what can be learned by the infant as he moves away from his mother we have been speaking of the human infant, parallels can be drawn for the infant of other species. Similarly, although we have been considering what can be learned about the *physical* environment, parallels can be drawn for the *social* environment.

Relation to Attachment

Up to this point we have presented a class of behavior rather generally; now its relations with some other classes of behavior can be considered.

The first of these other classes of behavior is the infant's attachment to his mother, and of course for some species this also includes attachment to the nest and the littermates. It is clear that mammals of necessity stay with their mothers for some time; that at an early age they distinguish their mothers from other individuals; that they often respond more positively to their mothers and to other familiar individuals than to less familiar individuals; and that they are upset by the departure of these familiar social objects. Attachment, furthermore, persists throughout the life of some species, although the form of the behavior changes.

We use the term *detachment* for the behavior of interest here, for balance with *attachment*, and for contrast.[8] Detachment occurs later in the life of some infant mammals than the first evidence of attachment. This is the case with the rodents, carnivores, and primates. But detachment does not signal the end of attachment, nor is it simply the opposite of attachment. Attachment and detachment should be viewed as an interplay of classes of behavior, developing side by side and coexisting for the life of the individual.

Finally, the kind of separation we are talking about is not to be confused with the separation of "separation anxiety."[9] We know from observation and from our laboratory studies that the infant who *separates himself* does so without anxiety.

Relation to Exploratory Behavior

When an infant leaves his mother, moves toward objects, and touches and fingers them, his behavior may be characterized as exploratory. Exploratory behavior has proved a troublesome class of behavior to handle conceptually[10] (Is the organism exploring or just active?). But exploratory behavior is so obvious in the young animal that the concept cannot be ignored just because it does not yet fit easily into traditional behavior theory.

The psychological advantages proposed as resulting from the child's leaving his mother's side are those very products assumed to result from exploratory behavior—an increase in a store of perceptions; new opportunities to learn what can be done with an object and what results from manipulating it; and an increase in new techniques for controlling external events. Furthermore, in our early attempts to study the infant's leaving his mother, we see that some of his behavior appears to be under the control of those same factors that control exploratory behavior—among them, novelty, complexity, and change.

Mother's Role in the First Separations

What is the human mother's role in the infant's leaving her side? In our culture, even from the beginning, the mother often physically separates herself from the infant, in contrast to some other mammals who seldom leave when the infants are young, or to the primates, whose young maintain physical contact with them. The mother's leaving cannot be responsible for the infant's leaving her, because even in cultures where the mother separates herself less often, the infant also leaves at some point in time. The human mother *permits* the infant to leave. Although she is watchful, she nevertheless appears to retrieve the infant less often than many other mammals do. Later, as the human infant progresses farther and faster, the mother does restrain and retrieve him more often, but she is generally ingenious in constructing an environment where restraining and retrieving are less necessary, since such foresight reduces her caretaking duties.

Primate Studies

In the last decade or so, many investigators of nonhuman primate behavior have followed the lead of Harlow and Zimmermann[11] in studying the interaction between mother and infant. We have drawn data from the reports of field and laboratory studies of several primate species to support the thesis that the infant does separate himself from his mother. These studies supply information on when the infant leaves, how far he goes from his mother, how long he stays away, how often he leaves, and how the mother responds to his first departures.

  Measures of separation. Several investigators have reported that, as the primate infant matures, he goes farther and farther from his mother. Kaufmann[12] provides explicit data concerning the rhesus monkey (Macaca mulatta) on Cayo Santiago, Puerto Rico. The distance infants *walked away* from their mothers increased rapidly over the first 2 months of age (Fig. 1). In Figure 1 the data points for the curve labeled "first seen" are the lower limits of the age range Kaufmann reported at each distance, and those for the curve labeled "commonly seen" are the midpoints of the age range, for a sample of 30 infants observed from birth.

  Vessey[13] provides additional data for rhesus monkeys; he found that the *average* distance between mother and infant increased linearly after the first 8 weeks to 30 feet (9 meters) at the end of the first year. The measures were based on minute-by-minute observations, averaging 1 hour per week, of nine rhesus infants born into the colony at La Parguera, Puerto Rico. Six of the infants were male and three were female; no difference by sex was found.

  Data for some other primates are shown in Figure 2. The curves are based on the distances between mother and infant reported by Jay[14] for langur monkeys (Presbytis entellus), by De Vore[15] for baboon (Papio, olive and yellow), and by Schaller[16] for the gorilla (Gorilla gorilla beringei). We have used the exact ages

Figure 1.
Distance traveled from mothers by infant rhesus monkeys of different ages (see text).

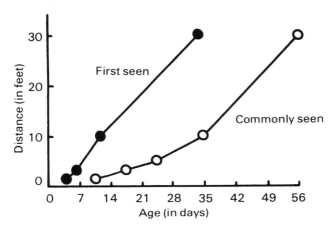

Figure 2.
Distances between infants and mothers of some primate species. The data points represent the earliest ages and the maximum distances reported. (The curve for the baboon reaches 90 feet at the end of 10 months.)

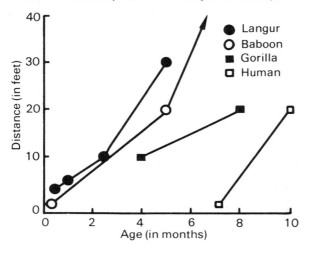

and distances reported by the investigators, but it should be pointed out that these observations were not the subject of their main interest. In general, the text of the reports implies that the distances resulted from the infant's movement away from his mother, rather than from the mother's movements. The first point of the curve for human infants in Figure 2 is based on Nancy Bayley's [17] age placement for forward progression; the second point is based on the distance that 10-month-old infants traveled from their mothers in our laboratory. This point differs from the other distances in that, given the dimensions and arrangement of the particular rooms used, it is the farthest the infant could travel. Informal observation of infants in their own homes suggests that the distance can be much greater.

The *time* a primate infant spends away from his mother at one time also increases with age. Altmann [18] recorded in the field that the howling monkey (alouatta palliata) at the age of 15 days stayed away from his mother for 10 seconds; at 26 days, for 1 minute; and at 1 month (only 5 days later), for as long as 4 minutes. Infant langurs in the field, according to Jay, [14] also separated themselves from their mothers for 4 minutes at the age of 1 month, but at 1 year they stayed away, playing for more than 20 minutes at a time. In the laboratory, Kaufman and Rosenblum[4] found that, in the case of both bonnet (Macaca radiata) and pig-tailed (M. nemestrina) infants, the mean duration of their vertical separations from the mother (separations involving progress to a different level of the cage) increased up to the fourth and fifth month; that the duration was stable thereafter may be attributed to the space limitations of their living quarters.

As he grows older, not only does the nonhuman primate infant go farther from his mother and stay away longer, he also leaves more frequently. Kaufman and Rosenblum [4] reported that the *frequency* of the bonnet infants' moving away from their mothers, while remaining on the same level of the cage (horizontal departures), increased from 4.5 departures at the age of 1 month to 9 departures at 4 months, per 1000 seconds of continuous observation. Concurrently, the frequency of going to another level of the cage (vertical departures) increased from 0 at the age of 1 month to 16 at 15 months. Similar changes in the pig-tailed infants' behavior also occurred; horizontal departures increased over the first 11 months. Furthermore, throughout the 15 months of observation, both bonnet and pig-tailed infants left their mothers more frequently than their mothers left them. [19]

The data summarized so far show that the nonhuman primate infant leaves his mother, and that, as he grows older (and more experienced), he leaves more frequently, goes farther, and stays away longer.

The mother's response. As the primate infant matures, the mother is not passive. Her behavior undoubtedly contributes to the changes seen in her infant's departures. It has been shown by various investigators that she restrains him from departing and retrieves him once he has left. Her thwarting of his attempts to suckle and her punishing him when he approaches occur much later; although they affect his subsequent behavior, they are not relevant to the central issue of this article.

Data on the mother's restraining and retrieving are shown in Figure 3. Curve A, derived from the data of Doyle, Andersson, and Bearder,[20] shows the frequency with which the galago mother (Galago senegalensis moholi) retrieved an infant that had left the nest. Curves B and C are based on data for the rhesus monkey. (Because of the differences in the measures and time-sampling procedures of the investigators, the scores for the various sets of data are expressed

Figure 3
Frequency with which primate mothers restrained and retrieved their infants (see text).

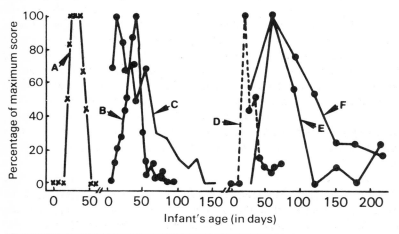

as percentages of the maximum score reported, a procedure that permits comparisons of events over time but not, of course, comparisons of actual frequencies.) Curve B is based on the data of Harlow, Harlow, and Hansen[1] for the rhesus monkey and combines frequencies for the mother's restraint and retrieval of the infant. Curve C, from the work of Hinde and Spencer-Booth,[2] represents the percentage of rhesus infants restrained by their mothers. Curves D, E, and F are derived from the work of Rosenblum and Kaufman[21] with bonnet and pig-tailed macaques. Curve D shows the percentages of time the pig-tailed mother prevented her infant from leaving; curves E and F give the frequency of retrieval by bonnet and pig-tailed mothers.

The data of Jensen, Bobbitt, and Gordon[22] on this subject are not presented in Figure 3 because they are given not as actual frequencies but as frequencies *relative* to other classes of behavior. Still, the data of these workers also show that the mother's "retaining" of the infant relative to all her manipulations increased over the first few weeks, and then decreased.

Despite the differences in procedures of observation and in classes of behavior reported, the data show (i) that these nonhuman primate mothers restrain their infants from leaving and retrieve them once they have left, and (ii) that the frequency of the behavior increases over the early days of the infant's life and then decreases. It is clear that the mother's behavior reflects the increasing frequency of the infant's attempts to leave—the topic of central concern here— and that, over time, he gets his way.

Studies of Human Children

In the past 2 years we have studied children in the process of separating themselves from their mothers. Two studies are reviewed here to show that the behavior can be subjected to experimental analysis. The first was carried out in a seminaturalistic setting with children between 1 and 5 years of age, the other in the laboratory with 10-month-old infants.

Relationship between age and distance. To measure the relationship between a child's age and the distance he will travel from his mother, 48 children were studied, three boys and three girls at each half-year age between 12 and 60 months. We placed a mother and child at one end of an unfenced lawn, with the mother sitting in a chair and the child starting at the mother's knee but left free to roam for the 15 minutes of the study. The mother was instructed to remain in her chair but was encouraged to respond otherwise to the child in her usual manner. Neither mother nor child had been in the yard before. The yard contained a couple of trees, two birdbaths, a set of planted terraces to one side, and a small paved patio and the house to the other. Aside from these usual objects, no lures were provided. The yard was L-shaped, running along the back of the house and around its corner to the road. The first leg of the L, in front of the child, was 27.4 meters long and 12.2 meters wide; the second leg began 15.2 meters from the child's starting position and extended 39.6 meters to the road; it too was 12.2 meters wide. Thus the child could get out of range of the mother's vision; the trial was ended, however, for any child who went 15.2 meters past the corner, a precaution taken to guard against his getting into the road.

Observers stationed at windows in the house traced the child's path on a map of the yard. Small rods were inserted at the borders of the lawn, inconspicuous in the tall grasses, to mark off 10-foot (3.05-meter) squares for the use of the observers in plotting the child's course. Distance was calculated as the midpoint of each square the child entered. In 25 of a sample of 26 records, the two observers, working independently, agreed exactly on the farthest square the child entered.

The mean farthest distance traveled from the mother by 1-year-olds was 9.9

meters; by 2-year-olds, 15.1 meters; by 3-year-olds, 17.3 meters; and by 4-year-olds, 20.6 meters. Firgure 4 shows that variability after the second year of life was considerable; for example, one 2½-year-old boy went 31.5 meters but another went only 7.5 meters. Nevertheless a linear regression of distance relative to age was significant at P less than .01. The equation for the estimated regression line was

$$Y = 2.43 + 0.35X$$

This suggests that, for each added month of age, the children went about a third of a meter farther.

Figure 4
Distance traveled from mothers by children of different ages.
(Solid circles) males: (open circles) females.

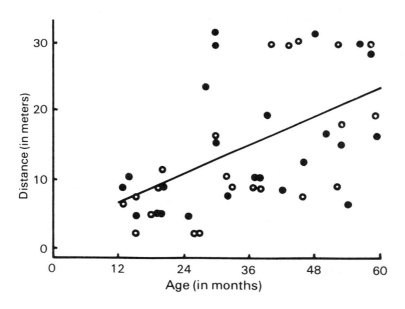

Clearly this relationship cannot be linear for ages much below 1 year; the predicted average of 4.9 meters at 7 months (the average age of first forward progression) would be too great. Furthermore, the boundaries imposed not only by the topography of the lawn but also by the experimenter's stopping a child at an arbitrary point appear to have restricted the distance the older children might have gone.

No evidence of a difference by sex appeared in these small samples. The regression for each sex was significant, and the two regressions did not differ reliably ($F = 0.37$, 2 and 44 d. f.). Of the ten children who went out of sight of their mothers, it is interesting to note, two were males and eight were females. Unlike rhesus infants, as reported by Kaufmann,[12] the human child, according to the regression equation, would not attain a distance of 30 feet until the age of 19 months, a distance reached by rhesus monkeys at 2 months. But the comparison can be only suggestive because the children were placed in an environment with which they were unfamiliar; the mothers reported that at home children went much farther.[23] One could of course use a longer session or repeated sessions to increase familiarity. Interesting objects could be placed at varying distances from the mother to lure the child farther, or another child could be present, as a means of measuring the effect of social facilitation.

Environmental stimuli and the infant's leaving his mother. In the laboratory we have investigated some properties of the environment that lead the 10-month-old human infant away from his mother.[24] A simple situation composed of mother, infant, two adjacent rooms, and a few toys provided the laboratory setting. The mother placed the infant beside her in one of the rooms, and the door to the second room was left open. The properties of the environment were altered by varying the number and location of toys within the second room, and by having the toys sometimes present from the start and sometimes added later.

Two observers behind windows fitted with one-way glass independently recorded how long an infant took to enter the second room, how long he stayed there, how far he went, what he touched and manipulated, how often he returned to his mother, and how much time he spent in contact with her. Vocal behavior was tape-recorded and subsequently analyzed as either distress or nondistress sounds. The agreement between observers was substantial, product-moment correlations being in the neighborhood of .95 for the duration of actual events.

The subjects were normal, home-reared infants, 10 months of age, an age that insured that most of them could locomote by some means. In fact, about two-thirds of the infants could creep proficiently on their hands and knees. Of the other third, half were still crawling on their bellies and the other half were already toddling. The infants were selected by age alone from the register of births at the University Hospital at Chapel Hill, North Carolina, and thus reflected the socioeconomic characteristics of a small university town.

The experimental area consisted of two rooms—a small room in which the test was started, called the "starting room," and a larger room, called the "open field"; both were unfurnished (Fig. 5). The small room measured 2.7 by 2.7 meters; the large room, 2.7 by 5.5 meters. The floor of the large room was divided, by narrow masking tape, into cells approximately 0.9 meter (1 yard) square. Neither the mother nor the infant had seen the rooms prior to the test.

To start the test, the mother sat on the floor of the starting room and placed the infant facing herself. She had been told that she could look at and smile at

Figure 5
The Experimental situation for studying the human infant's departure from his mother.

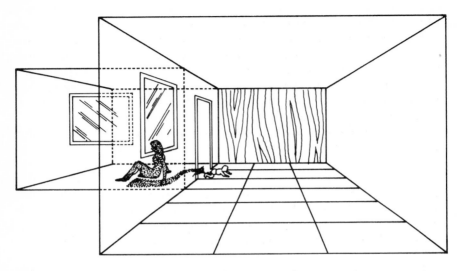

her infant, and that she could talk softly to him in short phrases when he was near her, but that she should allow him to leave or to stay.

Twenty-four infants were studied. In experiment 1 the open field was empty for 12 of the subjects (group 0); for the other 12 (group 1) it contained a toy—a plastic pull toy—in the cell just outside the starting room (the cell that the infant has reached in Fig. 5). The experiment lasted for 10 minutes.

All 24 infants left their mothers and, without fussing or crying, crept out of the room in which the mother sat and entered the larger room, whether it was empty or contained a toy.

The two groups, group 0 and group 1, spent similar amounts of time in the larger room (Fig. 6A). (This statement, and all succeeding statements of difference are supported by P values of less than .05, obtained in Wilcoxon two-tailed matched-pairs signed-ranks tests.) The presence of the toy did not keep group 1 infants in the open field longer than group 0 infants. This result may be attributed to an unanticipated response of group 1 infants: they brought the toy *into* the starting room.

The major difference between the two groups was a difference in the amount of time spent with the mother. When group 0 infants were in the starting room they spent more time with the mother than group 1 infants did; group 1 infants spent at least half the time playing with the toy.

Figure 6.
(A) Experiment 1: effect of a toy. (B) Experiment 2: effect of previous experience. (C) Experiment 2: effect of number of toys. SR, *Starting room;* OF, *open field;* M, *mother;* T, *toy or toys. The sizes of* SR *and of* OF *denote time spent in each; the sizes of* M *and* T *denote time spent in contact with that object in each environment.*

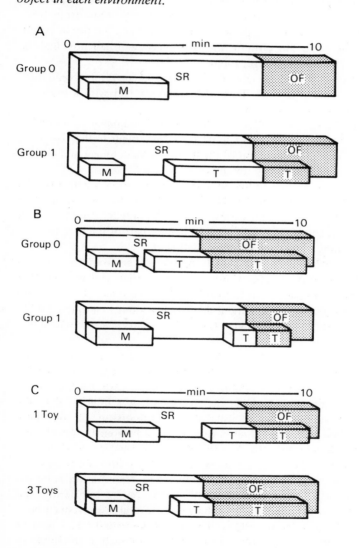

Of interest, also was the observation that infants did not go out, come back, and stay in the starting room; later, they went out again and returned again, some infants alternating many times whether or not the environment contained a toy. One infant went out of the starting room 13 times.

Experiment 2 followed directly after experiment 1. The same infants were now exposed to essentially the same environment but with certain changes in the toys. Half the infants who had no toy in experiment 1 now had one toy, and the other half had three toys, arranged diagonally across the larger room from upper left to lower right. Similarly, half of those who previously had the toy in experiment 1 now had the same single toy in the same place, and half now had the three toys, also spaced across the larger room.

Group 0 infants now entered the open field sooner and spent more time there than group 1 infants did (Fig. 6B). They also made contact with a toy more quickly and spent more time playing with it.

An independent analysis of the effect of three toys as compared with one toy for both groups of infants showed that three toys drew them farther from the mother. Three toys also kept them out of the mother's room longer and elicited more play in the open field than one toy did (Fig. 6C).

The results, in summary, showed that infants left their mothers and entered a new environment. How quickly they entered it, how far they went, and how long they stayed away were responses controlled by the number and location of the stimulating objects and by whether the objects were part of the new environment from the beginning of the test or were added later.

That infants leave their mothers and, with no distress, go from one room to another is a matter of everyday observation. But the finding that infants crept into the experimental environment of this study and moved freely about with no distress contrasts sharply with the marked distress and almost complete inhibition of locomotion shown, in an earlier study,[25] by infants *placed* alone in that same environment.

An infant's entering a room that contains a toy would seem to demand no explanation. But an infant's entering a room that does not contain a toy, or any other prominent object, raises the question of what evokes entry. Devoid though the room was of any prominent object, it nevertheless was brighter than the starting room, and it contained many visual stimuli—a doorstop, curtains, lines and angles. If an infant can creep at all, creeping into a new environment must often have been reinforced previously by such changes in visual stimulation.

The infant's return to the starting room and his reentry into the large room may be considered illustrative of Ainsworth's (p. 78)[6] "exploration from the mother as a secure base," and is reminiscent of the report by Harlow and Zimmermann[11] of infant rhesus monkeys that, after several sessions in an open field, used the cloth cylinder on which they had been raised "as a base of operation," moving away and returning to it between contacts with stimuli in the environment. But if the mother is considered a base, the data of our study show that the infants did not always *touch* base; on one-third of the returns, to *see* seemed sufficient. Furthermore, Ainsworth's term *secure* implies the affording of safety, and would be more appropriate here if the return to the mother were a flight from the larger environment. Of this there was no evidence. Quite the contrary; the return was often accompanied by facial and vocal expressions of pleasure and not by signs of fear or of relief from fear.

Problems for Further Study

Many questions about the process by which the infant separates himself from his mother still await investigation. Questions about the role of environmental stimuli in effecting his departure lend themselves most easily to experimental analysis. We propose that the visual properties of objects, both social and nonsocial, control the infant's leaving the mother and making contact with the objects. As we have shown, such properties also determine how long it takes him to leave, how far he goes, and whether he gets out of sight of his mother. The feedback properties of objects determine the duration of contact, the nature and extent of manipulation, and hence the time spent away from the mother; they may also control the child's subsequent return to the objects. The properties of the ambient environment may similarly affect the infant's leaving his mother. For example, we have noticed in the laboratory that the infant is more likely to enter a brightly illuminated room than a dimly lighted one. The mother's behavior in laboratory studies will certainly be a determinant of the infant's behavior. Here talking, caressing, or playing games may keep him at her side. If she moves to a new position or leaves the room, will he follow?

So far we have been reporting laboratory studies. Naturalistic studies in the child's own home can supply complementary information. As in the primate studies, the frequency of contacts, approaches, and departures by both infant and mother can be charted at different ages, and charted in a given child over a period of time. Which behaviors of the infant evoke maternal restraining and retrieving behavior? Alternatively, how does the mother foster or encourage his departures? These questions specify variables of maternal behavior that would be viewed as a set of environmental stimuli modifying the infant's leaving her side. Although the first movement away from her side is his, her response may alter its subsequent occurrence. Similarly, the behavior of other members of the family deserves attention.[26]

Once again, the details of study have been outlined for the human infant. It is clear, however, that the same questions and the same procedures, with minor variations, apply to other mammals. Although many accounts of the development of behavior in young mammals contain some information on the infant's leaving the mother and nest,[27] we found no substantial body of data on this topic for mammals other than the primates.

The lines of inquiry proposed up to this point seem straightforward and clear.

One sees how the stimulating conditions can be varied and the behaviors measured. Less clear are the procedures for demonstrating the psychological advantages that moving away from the mother may confer on the infant. Reasonable and likely as these advantages seemed when proposed earlier in this article, they nevertheless await confirmation in tests on the young, developing organism.

If we look beyond the period of infancy, the full significance of the child's separating himself from his mother comes into view. Leaving her side is but the first step in the continuous process of achieving psychological independence.

Summary

In this article we have defined a class of behavior which has not often been the subject of formal study. Its universality among infants of all species is not basis enough for its engaging scientific attention. Far more important are its biological and psychological consequences; we propose that among these consequences are increased opportunities for learning on the part of the infant. Primarily by his own physical contacts with objects, near and distant, he learns the structural arrangements of objects in space and the tactual and other feedback properties of objects, both social and nonsocial.

The human infant, unlike many other mammals, separates himself from his mother at the first moment any mode of locomotion is possible. He does not wait until he can creep or walk efficiently. The separation, once effected, increases in distance and duration over the life of the individual.

The behavior is patent, it can be measured, it need not be inferred, and, as we have demonstrated, it can be experimentally manipulated. Further, it lends itself nicely to comparisons among species.

We do not propose that the infant's detachment from his mother is a negation of attachment to his mother, but current preoccupation with the attachment of the young to the mother should not obscure the importance of detachment. Its study can present the same challenge that the infant seems to find in going forth on his own.

References and Notes

1. Harlow, H. F. Harlow, M. K., and Hansen, E. W. In H. L. Rheingold (Ed.), *Maternal behavior in mammals.* New York: Wiley, 1963. pp. 254-281.
2. Hinde, R. A., and Spencer-Booth, Y. *Animal Behavior,* 1967, **15**, 169.
3. Jensen, G. D., Bobbitt, R. A., and Gordon, B. N. In S. A. Altmann (Ed.), *Social communication among primates.* Chicago: University of Chicago Press, 1967.
4. Kaufman, I. C., and Rosenblum, L. A. In B. M. Foss (Ed.), *Determinants of infant behavior IV.* London: Methuen, 1969.
5. Van Lawick-Goodall, J. *Animal Behavior Monograph,* 1968, **1** 161.
6. Ainsworth, M. In B. M. Foss (Ed.), *Determinants of infant behavior II.*

London: Methuen, 1963. In this report Ainsworth describes the Ganda infant's separating himself from his mother.
7. Held, R., and Hein, A. (*Journal of Comparative and Physiological Psychology*, 1963, **56**, 872) showed the importance of active, rather than passive, movement for the development of visually guided behavior in kittens.
8. In the psychoanalytic literature, the term *detachment* means the suppressing of emotions and has been used by J. Bowlby (*Attachment and Loss:* Vol. 1, *Attachment.* London; Hogarth, 1969) to describe a phase in the child's response to his mother after extended separation.
9. Bowlby, J. *International Journal of Psycho-Analysis,* 1960, **41**, 1; *Journal of Child Psychology and Psychiatry,* 1961, **1**, 251.
10. Cofer, C. N., and Appley, M. H. *Motivation: Theory and research.* New York: Wiley, 1964. Fowler, H. *Curiosity and exploratory behavior.* New York: Crowell-Collier-Macmillan, 1965.
11. Harlow, H. F., and Zimmermann, R. R. *Science,* 1959, **130**, 421.
12. Kaufmann, J. H. *Zoologica (New York),* 1966, **51**, 17.
13. Vessey, S. *American Zoology,* 1968, **8**, 740; personal communication.
14. Jay, P. In H. L. Rheingold (Ed.), *Maternal behavior in mammals.* New York: Wiley, 1963. Pp. 282-304.
15. DeVore, I. In H. L. Rheingold (Ed.), *Maternal behavior in mammals.* New York: Wiley, 1963. Pp. 305-335.
16. Schaller, G. B. *The mountain gorilla, ecology and behavior.* Chicago: University of Chicago Press, 1963. Pp. 265-266.
17. Bayley, N. *Bayley scales of infant development.* New York: Psychological Corp., 1969.
18. Altmann, S. A. *Journal of Mammalogy,* 1959, **40**, 317.
19. Jensen, et al. (ref. 3), for the pig-tailed macaque. Hinde, R. A., and Spencer-Booth, Y. [In D. Morris (Ed.), *Primate ethology.* Chicago: Aldine, 1967. Pp. 267-286.] for the rhesus macaque, also reported the movements of mother and infant toward and away from each other; the data are given in terms of relative and not absolute frequencies. Both sets of data do, however, show that the frequency of the infant's departures relative to the sum of the infant's and the mother's departures increased as the infant grew older. That the data also show similar increases in the relative frequency of the infant's approaches and that his approaches outnumber his departures are findings that offer additional information about the behavior of young animals but in no way detract from the main thesis of this article.
20. Doyle, G. A., Andersson, A., and Bearder, S. K. *Folia Primatologica,* 1969, **11**, 215.
21. Rosenblum, L. A., and Kaufman, I. C. In S. A. Altmann (Ed.), *Social communication among primates.* Chicago: University of Chicago Press, 1967. Pp. 38-39.
22. Jensen, G. D., Bobbitt, R. A., and Godon, B. N. *Behaviour,* 1968, **30**, 6.
23. J. W. Anderson, according to J. Bowlby (*Attachment and loss:* Vol. 1,

Attachment London: Hogarth, 1969. P. 253) observed the behavior of children with their mothers in a park, but the findings as reported do not lend themselves to direct comparison with ours.
24. Rheingold, H. L., and Eckerman, C. O. *Journal of Experimental Child Psychology,* 1969, **8,** 271.
25. Rheingold, H. L. In B. M. Foss (Ed.), *Determinants of infant behavior IV.* London: Methuen, 1969. Pp. 137-166.
26. Jensen, *et al.* (ref. 3) Hinde and Spencer-Booth [In D. Morris (Ed.), *Primate ethology.* Chicago: Aldine, 1967.] and Kaufman and Rosenblum (ref. 4) describe the effect of the rearing environment upon the infant macaque's departure from the mother.
27. Bolles, R. C., and Woods, P. J. *Animal Behavior,* 1964, **12,** 427.

Recommended Additional Readings

Ainsworth, M. D. S. Object relations, dependency and attachment: A theoretical review of the infant-mother relationship. *Child Development,* 1969, **40,** 969-1025.

Ferguson, L. R. Origins of social development in infancy. *Merrill-Palmer Quarterly,* 1971, **17,** 119-138.

Rheingold, H. L. The social and socializing infant. In D. A. Goslin (Ed.), *Handbook of socialization theory and research.* Chicago: Rand McNally, 1969. Pp. 779-790.

19

A Study of Social Stereotype of Body Image in Children

J. Robert Staffieri

Orientation

One of the most dramatic aspects of all developmental events is the host of interrelated physiological and morphological changes that occur through childhood and adolescence. These physical changes do not occur in a vacuum; it is clear that in the course of human relations, regardless of the desirability of such practices, it is typical that many of the initial impressions made about a person are based upon corporal assessments. "Short and fat," "tall and skinny," "lean and muscular"—all of these and many other body configurations carry certain connotations regarding temperament and other aspects of personality. Staffieri, in the following article, provides convincing evidence that stereotypes related to body image are learned quite early in life. Moreover, he demonstrates that the young child himself, as well as those with whom he interacts, incorporates and comes to believe in these perceptions. Very soon in one's life a vicious circle is established: The individual learns that certain body builds are deemed superior to others; evaluational attitudes form around these body perceptions; the individual discovers which body build he possesses and the expectations attached to this body image; and he begins to behave in accordance with these expectations.

Thus, the following report of an empirical study provides insights into the early development of the complex set of generalizations that relate to one's body image, personality, and social expectations. There can be little doubt that these early learnings carry over into adolescent and adult life; for some persons, particularly those with body builds deemed more or less desirable by the standards of their culture, these stereotypes can have profound implications.

Dr. J. Robert Staffieri is presently the director of Saint Elizabeth's Child Development Center, Portland, Maine.

Reprinted from the *Journal of Personality and Social Psychology*, 1967, 7, 101-104, by permission of the author and the American Psychological Association. This article is based on a dissertation submitted in partial fulfillment of the requirements for the degree of Doctor of Education at Indiana University. The author wishes to express his appreciation to Boyd R. McCandless, under whose direction the investigation was conducted.

19

The role of an individual's body configuration in social interactions and the effects of these interactions on self-concept is an important part of the total process of personality development. It is not uncommon to be the initiator or recipient of such statements as "He certainly is a crafty, wiry little thing," or "He is all skin and bones—must worry a lot." It is reasonable to assume that the individual who is the recipient of statements which are based on another person's perception of his body is likely to incorporate these perceptions into his own body concept. As a direct result of an individual's body configuration, he typically receives rather consistent reactions from others. These reactions thus provide a framework for his body concept, which becomes a significant part of the total self-concept.

One implication from this process is central to the concerns of this investigation. Since body configuration is objectively definable (e.g., tallness versus shortness, fatness versus thinness), it is reasonable to hypothesize a definable range of consistent and stable reactions to a particular body configuration.

Evidence linking body build and personality is at best tenuous. Perhaps the etiology of such a relationship (if it exists) lies in the realm of social learning rather than physical constitution per se. McCandless (1960) stated that ". . . there are suggestions that if substantial and consistent personality-body type relations should be demonstrated, patterned types of social response may constitute the responsible factor as reasonably as genetics (p. 47)."

Evidence suggests that individuals will behave to some degree in a manner consistent with the expectations of others. If these expectations are consistent over people and time, it is reasonable to expect emitted behavior consistent with the expectations. Once the individual accepts an expectation as being true of himself, the literature suggests that he will act in a manner to fulfill the belief, thus providing added continuity to behavior which was originally emitted because of expectation (e.g., Payne & Farquhar, 1962).

Most investigations relating body build to social-personal behaviors suggest that, although the magnitude of the correlations is not high and the proportion of correlations is not much higher than would be expected by chance, some support exists for acceptance of the concept of body-built-personality interrelations (e.g., Hanley, 1951; Walker, 1962, 1963).

Previous research has provided some evidence to suggest that a body type is capable of eliciting rather common reactions from adults in the form of descriptions of personality/behavior traits. Is the role of body type as a stimulus in social situations limited to the adult population? What is the relation of age to this process? Is there a relationship between body image stereotype and individual process of perceiving? What part does body type play in social interactions? The purpose of this study was to investigate the role of body-image stereotypes in children with general reference to three areas: development, interpersonal functions, and social functions.

Method

Subjects

The subjects for this study were 90 male children from 6 to 10 years of age in the elementary division of the University School, Indiana University. The sample population consisted of 18 subjects at each age level (6 to 10 years), who were grouped (according to their relative fatness, muscularity, and thinness) on the basis of the ponderal index

$$\frac{\text{height in inches}}{\sqrt[3]{\text{weight in pounds}}}$$

and teacher rating of body types. Although these groups (six subjects in each group at each age level) are referred to in the study as endomorph, mesomorph, and ectomorph, these are meant as descriptive characteristics of typically fat, muscular, and thin boys, respectively, and do not stringently represent a particular method of body typing. In addition, there was a sample of 12 4-year-old and 12 5-year-old subjects included in a part of the study, but the results utilized from this group are considered as a tentative pilot investigation.

Measures

Three measures were obtained for each subject who participated in the study: (a) assignment of adjectives to silhouettes representing three variations of body type, (b) determination of body type preference of each subject, and (c) a sociometric measure of each class.

Adjectives. A list of 39 descriptions (primarily one-word adjectives) was developed, each of which could be applied to any of the silhouettes. The descriptions, hereinafter referred to as adjectives, were selected on the basis of (a) a prior pilot study, (b) apparent relatedness to behavior/personality variables, and (c) the likelihood of their meaning being known. The adjectives were not designed to survey all possible descriptions which could be applied to individuals, but were intended to be a reasonable sample of a more extensive universe.

Silhouettes. The stimuli to which the above mentioned adjectives were applied were three full body silhouettes (black on white). Each silhouette profile was approximately the same height (13 inches), head shape, and facial outline. The distinguishing features were those which are commonly associated with the three gross bodily characteristics of extreme endomorph, mesomorph, and ectomorph. Essentially, the body curvature represented fat, muscular, and thin body types. In this study, there were two sets of three silhouettes each. One set was designed to represent child figures. The silhouettes were counterbalanced for both order in body-types presentation and child-adult form. This was also done for assignment of adjectives and body-type preference. In each age level, half of the subjects (nine) responded to child forms and half to adult forms. With three silhouettes, there were six possible orders of presentation, so that each subject of a particular body type in each age group responded to a different order of silhouettes. All three silhouettes were hung on a white background card (30 X 20 inches). They were clearly discriminable.

Body-type preference. In order to determine the body type preferred, each subject was presented with five variations of body-type silhouettes. In addition to the three used in the assignment of adjectives, two silhouettes were cut representing endo-mesomorph and ecto-mesomorph. These five silhouettes represented a range of body type and were also counterbalanced for order in presentation and child-adult form.

Sociometry. The sociometric measure used in this study was responded to by most boys and girls in all classes from Grades 1 through 5. It was administered to all classes separately. The children were asked to indicate (in writing) who, in that class, were their five best friends and who were the three children that they "did not like so well." The instructions were constant for all classes and the list was ordered from their "very best friend" (first choice) to their fifth best friend (fifth choice), and from the person whom they disliked most (first choice) to the person that they disliked, but did not dislike as much as the first or second choice (third choice). The sociometric data for 6, 7, and 8-year old children were collected (using the same procedure) about 5 months prior to this study, and were available to the investigator. This measure provided data from each child in the class which indicated the number of times each subject was chosen as first best friend, second best friend, and so on, and most disliked, second most disliked, and third most disliked.

Results

The results of the study indicate that:
   1. Boys from 6 to 10 years old demonstrated a common concept of be-

havior/personality traits which were associated with various body types. These stereotypes began to appear in children 4 and 5 years old, but the evidence is tentative (Table 1).

2. All the significant adjectives assigned to the mesomorph image were favorable (16); the adjectives assigned the endomorph were unfavorable (socially) and primarily socially aggressive; the adjectives assigned to the ectomorph were primarily unfavorable (personally) and of a generally socially submissive type.

3. The assignment of adjectives to the images was not related to the body type of the subject who assigned them.

4. Subjects showed a clear preference to look like the mesomorph image. This preference became apparent between 6 and 7 years. (Table 2).

5. A selected group of subjects demonstrated reasonable accuracy in perception of their own body types (Table 3).

6. Mesomorph subjects received a consistently high number of acceptance choices and endomorph subjects received a consistently low number of acceptance choices (Table 4).

Discussion

The results of this study indicate a rather clear stereotype pattern for the three body images within age levels and across age levels. The mesomorph image is perceived as entirely favorable. The ectomorph image is basically unfavorable, but different from the unfavorable concept for endomorph.

It is clear that the mesomorph body type for both children and adults is representative of an ideal male physique. The favorable stereotype of the mesomorph is evident at 6 years of age. However, the preference to look like the mesomorph does not appear until 7 years and is not clearly established until 8 years. Data suggest reasonable accuracy of self-perception of body type from 40 subjects, mostly 7- and 10-year olds. The 10-year-old boys were more accurate in their self-perceptions than the 7-year-old boys. If the younger boys do not perceive themselves accurately, there may not be sufficient motivation for them to report a preference to look differently. Subjects 8 years and older appear to report self-perceptions quite accurately, and clearly prefer to look like the mesomorph image. The point at which accuracy of self-perception becomes apparent (probably 8 to 9 years of age) may also be the beginning of dissatisfaction with one's body, and the degree of dissatisfaction may well be proportional to the extent that one's body differs from the mesomorph image.

While the results of this investigation do not provide an answer to the problem of whether body type constitutionally determines personality, or that body type is a determinant of personality through the process of social learning, there are some implications which support the latter point of view. Most authors would agree that, at least minimally, an individual's physical constitution determines certain behaviors which can be emitted. Beyond this point, however, the

Table 1
Frequency of Assignment of Adjectives to Silhouettes by Subjects 6 to 10 Years

Adjectives	Silhouettes			x^2
	En	M	Ec	
Strong	15	74	1	100.067***
Best friend	9	67	14	68.867***
Quiet	21	10	59	44.067***
Fights	40	45	5	31.667***
Kind	24	35	31	2.067
Cheats	63	9	18	55.800***
Clean	3	54	33	43.800***
Worries	30	11	49	24.067***
Lots of friends	8	69	13	74.467***
Nervous	42	5	43	31.267***
Happy	19	54	17	28.867***
Helps others	17	55	18	31.267***
Polite	13	50	27	23.267***
Argues	58	17	15	39.267***
Remembers	10	26	26	26.400***
Gets teased	60	4	26	53.067***
Lonely	33	9	48	25.800***
Sick	41	7	42	26.467***
Forgets	51	14	25	24.067***
Lazy	67	5	18	71.267***
Healthy	4	77	9	110.867***
Lies	56	10	24	37.067***
Sneaky	25	24	41	6.067*
Honest	13	61	16	48.200***
Sloppy	72	9	9	88.200***
Brave	8	68	14	72.800***
Teases	35	30	25	1.667
Naughty	46	18	26	13.867***
Good looking	1	74	15	100.067***
Mean	54	19	17	28.867***
Afraid	27	7	56	40.667***
Ugly	77	4	9	110.867***
Sad	31	11	48	22.867***
Smart	5	67	18	71.267***
Dirty	60	16	14	45.067***
Tired	39	12	39	16.200***
Stupid	58	8	24	43.467***
Weak	21	0	69	83.400***
Neat	2	73	15	95.267***

Note:
En = Endomorph; M = Mesomorph; EC = Ectomorph.
 *$p < .05$.
 ***$p < .001$.

Table 2
Body Type Preference of Subjects by Age

Age	Silhouettes			x^2
	Endomorph	Mesomorph	Ectomorph	
4-5	8	6	10	1.000
6	5	7	6	.333
7	6	12	0	12.000**
8	1	16	1	26.833**
9	4	11	3	6.333*
10	0	16	2	23.333**

*$p < .05$
**$p < .01$.

Table 3
Accuracy of Self Perception of a Selected Group of Subjects

Self ratings	Subjects			
	1-2	3	4-5	ϵ
1-2	7	4	3	14
3	3	10	8	21
4-5	0	1	4	5
ϵ	10	15	15	40

Note:
$x^2 = 10.367$ (<.05), $C = .454$.

Table 4
Number of Acceptance Choices Received by Subjects

Order	Subjects			x^2
	Ectomorph	Mesomorph	Endomorph	
1st	59	76	41	10.443**
2nd	76	101	58	11.906**

**$p < .01$.

results of this study suggest that personality correlates of body type may be reasonably explained on the basis of expected behavior. Thus social expectation could explain the low but rather consistently found correlations between body type and behavior/personality traits.

If children behave in a manner consistent with expectations, even to a minimal degree, some reinforcement of the expectations occurs and gives continued support to the stereotype.

References

Hanley, C. Physique and reputation of junior high school boys. *Child Development,* 1951, **22,** 247-260.

McCandless, B. R. Rate of development, body build, and personality. *Psychiatric Research Reports,* 1960, **13,** 42-57.

Payne, D. A., and Farquhar, W. W. The dimensions of an objective measure of academic self-concept. *Journal of Educational Psychology,* 1962, **53,** 187-192.

Walker, R. N. Body build and behavior in young children: I. Body build and nursery school teachers' ratings. *Monograph of Social Research in Child Development,* 1962, **27,** No. 84.

Walker, R. N. Body build and behavior in young children: II. Body build and parents' ratings. *Child Development,* 1963, **34,** 1-23.

Recommended Additional Readings

Dwyer, J., and Mayer, J. Psychological effects of variations in physical appearance during adolescence. *Adolescence,* 1969, **3,** 353-380.

Garn, S. M. Body size and its implications. In M. L. Hoffman and L. W. Hoffman (Eds.), *Review of child development research.* Vol. 2. New York: Russell Sage, 1966. Pp. 529-561.

Lerner, R. M. and Gellert, E. Body build identification, preference, and aversion in children. *Developmental Psychology,* 1969, **1,** 456-462.

Tanner, J. M. Relation of body size, intelligence test scores, and social circumstances. In P. H. Mussen, J. Langer, and M. Covington (Eds.), *Trends and issues in developmental psychology.* New York: Holt, Rinehart and Winston, 1969. Pp. 182-201.

Jean Piaget's Developmental Model: Equilibration-Through-Adaptation

Guy R. Lefrancois

Orientation

The following article concerns the lifelong work of a single person—Jean Piaget. The significance and the prodigiousness of Piaget's work boggles the mind of the ordinary reader and scientist; it is likely that several more years will pass before we come to appreciate fully this man's contribution to our understanding of intellectual growth during ontogeny. In addition to his work in developmental psychology, Piaget has written extensively on the topic of the philosophy of science. It seems that few psychologists yet recognize his contributions in this latter realm.

Piaget's influence upon the study of human development has waxed and waned over the past five decades. Through the late 1920s and early 1930s he received considerable attention by other persons in the field of child psychology, but then during the next several years his works fell into relative obscurity, especially in the United States and England. The reasons for this lack of interest are many: the lack of translations (Piaget writes in French), the difficulty level of his treatises, the unpopularity of his views of cognitive development. However, since the late 1950s his writings have been widely translated and have been enormously influential in guiding research and educational practice. Many psychologists have come to realize that the prevailing psychological theories simply are inadequate to capture the richness and complexity of human behavior and thought. Piaget's theorizations and experiments have provided the needed substance to help fill this gap.

Perhaps Piaget's most important message—to educators and scientists alike—is that we must attend to the individual's spontaneous activities, for these are indicative of whatever we mean by "intelligence." The organism does not merely react to environmental stimuli; in Piaget's words, the act, or the "response," comes first. Thus, the growth of intelligence occurs as the individual progressively acts upon his environmental surroundings. (Perhaps during the next few years we will find that Piaget's position is criticized *for not paying sufficient attention to the reciprocal nature of organism-environment interactions.)*

It is hazardous, and perhaps even foolhardy, to describe Piaget's theory in a short article. Nevertheless, in the following article Guy R. Lefrancois made a noble attempt to summarize Piaget's position with respect to human development. His article represents an excellent introduction to this complex area; the next step for the reader is to start to read the voluminous writings by Piaget himself.

Guy R. Lefrancois is a member of the faculty of the department of educational psychology at the University of Alberta.

20

The Man

No descriptive and exploratory account of the work of Jean Piaget would be complete without some consideration of the man himself—of a man who over the past four decades has poured such a wealth of information and theorizing into the coffers of psychological knowledge that, as yet, only a small part of it has been assimilated by his audience.

Piaget was born in Neuchatel, Switzerland on August 9, 1896. The often related instances of his precocity include his publication of a scientific paper at the age of ten—a one page note relating to a partly albino sparrow which he had observed. Soon thereafter he volunteered as a laboratory assistant in the local museum of natural history. By 1930 he had published around twenty-five papers on zoological topics—particularly on molluscs. He studied at the University of Neuchatel where, at the age of 22, he received a doctorate in the natural sciences (with a dissertation on molluscs). During the next two years he worked at various centres in Europe: at Bleuler's clinic, at the Sorbonne, at the Laboratories of Wreschner and Lipps and eventually under the direction of Simon in Binet's laboratory at a Paris grade school. It is probably this latter position that marks the beginning of his interest in the development of children. His first series of articles in this area appeared in 1921 and 1922. These were followed in 1923 by his book—*Language and Thought of the*

Child. Since then, he and his collaborators have published more than 20 books and over 180 major studies, the bulk of which are found, written in French and untranslated, in *Archives de Psychologie,* of which Piaget is currently co-editor. In addition he is co-editor of the *Revue Suisse de Psychologie,* director of the Insititut des Sciences de l'Education, founder of the Centre d'Epistemologie Genetique, and director of the Bureau Internationale de l'Education, an affiliation of UNESCO.

The Theory

Orientation

Consistent with Piaget's early training and interests, the position he advances stems very directly from a biological orientation. The central questions in biology have been concerned with the classification of various species in phylogenetic order, and with the development of adaptive mechanisms and processes in these species. Piaget's attempt to relate these same questions to human ontogeny—that is, to development from birth to maturity—has resulted in a theoretical position which purports to explain the central features of development. Indeed, the really essential aspects of Piaget's theoretical position are reducible to the logical implications of these questions. Accordingly, this paper examines the Piagetian position as it derives from these biological considerations. This is followed by a brief summary of Piaget's description of human ontogeny.

Adaptation

Any theoretical account of human development must be concerned with the progress made by individuals in understanding their environment and in their ability to cope with it. Whether or not a theory uses the term adaptation, if it deals with human development, then it is concerned with at least some aspects of adaptation insofar as the term denotes changes in the way an organism deals with the environment, changes in the organism *per se,* or both. Piaget has evolved a theoretical system which explains development directly in terms of the adaptative process and in terms of progressively changing intellectual makeup. The inferences he makes regarding human development are based on observed behavioural events (content). From the fact that there is behaviour, and that it is *this* particular behaviour, he infers that there must be some organized or organizational properties of the intellect which determine behaviour (structure). The crucial question then becomes one of discovering how structure arises, how it functions, and how and why it changes. Initially he assumes that an individual is born with the capability to perform certain simple acts, ordinarily referred to as reflex acts. Thus a child can, if given the opportunity, suck from a nipple or grasp an object pressed into his palm. Initially, therefore, mental structure is defined in terms of these simple acts. Whenever an object is presented to a young child and he acts toward it in terms of some activity which he has already mastered (e.g., a nipple is placed in a child's mouth and he sucks on it) he is said to *assimilate* the object (or at least some aspects of the object) to the sucking activity. This overt activity together with whatever structural connotations it has is termed a *schema*. It is readily obvious that when Piaget is describing the exercising of a pre-existing schema, he is talking about only one aspect of adaptation, albeit, an essential aspect. The other aspect of adaptation is the one which allows for the modification and growth of the organism. Again, it is readily apparent that before long, an infant will suck not only short "bulb" nipples but long thin ones, or even objects that are not nipples. The significant thing here is not so much that new or different objects are being assimilated by the sucking schema, but that the activity of sucking is no longer the same. The mouth and tongue are held differently for different objects, and the movements performed are different if the object sucked provides food or if it does not. There is, in other words, an *accommodation* of the organism (or more specifically, of an activity of the organism) to the environment—and consequently there is a change in the structure (or intellectual make-up) of that individual. It is, then, these two processes, assimilation and accommodation, which define adaptation and which will continue, throughout life, to serve as the *modus operandi* of intellectual growth.

Intelligence

To facilitate interpretation of the preceding section a brief look at Piaget's views on intelligence is necessary. Unlike many of his contemporaries who have viewed intelligence as a relatively fixed but nebulous quantity or quality, Piaget sees it as being inherent in the activity of individuals.

Intelligence is not only observable through overt behaviour, but it *is* this behaviour together with the structures which direct and determine the behaviour and utilization of the structures in the face of environmental demands. The whole concept of intelligence-in-action can be diagrammed as follows:

CONTENT
STRUCTURE
FUNCTION
ASSIMILATION   ACCOMMODATION
ENVIRONMENTAL DATA

The model presented above can be arrived at logically in two ways. One can begin with content, which is behavioural data, and infer from this data that certain structures are present in the child's mind which lead to the appearance of this behaviour. Alternatively, one can look at the environment and at the behaviours which result from an attempt to adapt to it. As has already been explained, the processes of adaptation are defined by the functional invariants assimilation and accommodation—"functional" because they are the mechanisms of human intellectual functioning; "invariants" because as functions they do not change throughout development.

Structure

The account of Piaget's theoretical position presented thus far serves to outline his attempt to address himself to the question of how human functioning tends toward adaptation. His second chief concern stemming from biology has to do with the classification and organization of behaviour. Piaget sees human intellectual development as marked by a series of ordered qualitative changes—changes which are brought about through adaptive interaction with the environment, but which are defined primarily in terms of changing structure. It is the nature of this mental structure, as he has described it, which leads him to define a series of developmental stages. The bulk of that which has been written about Piaget's theory has to do with the characteristics of these stages rather than with those aspects of the system upon which the stage concept is predicated. These stages are outlined later.

Equilibrium

Central to Piaget's developmental position is his concept of *equilibrium*. In order to account not only for the development of increasingly complex intellectual structures, but also in order to explain the organizational properties of these structures, Piaget utilizes what is yet a rather nebulous concept, that of equilibrium. At the simplest level, equilibrium denotes a state of balance between assimilation and accommodation. It is evident that the preponderance of one or the other of these functions with respect to a given object in the environment will result in less than optimum adaptive behaviour. For example, a preponderance of assimilation (which, incidentally, characterizes the early "play" behaviour of children) will lead to an unrealistic interpretation of the environment. That is, when a child uses a stick, a broom and a spoon as guns he is assimilating these objects to the "gun" schema (or structure) without regard for their more important non "gunlike" attributes. When, on the other hand, there is a preponderance of accommodation (as is illustrated by a child's imitative behaviour) interaction with the environment is less likely to be meaningful. It seems obvious that if a child does not assimilate at least some aspects of the environment to established structures, this environment will be relatively meaningless to him.

When Piaget says that equilibrium is the state towards which all cognitive functions tend, he is, in fact, saying that development is to a considerable degree governed by a need to arrive at meaningful and realistic understanding of, and ability to cope with, the environment. And this relationship *vis-a-vis* the environment will prevail, according to Piaget, to the extent that there exists a state of equilibrium between assimilation and accommodation.

At a more complex level, equilibrium signifies more than a simple balance—it is a state of development with respect to mental structures where external disturbances can be compensated for within the system of structures without either destroying the essential logical properties of that system or distorting the reality which is being reacted to. Equilibrium can therefore be seen as a motivational concept of first importance in the Piagetian system. Insofar as there is a need to maintain some sort of equilibrium—in other words, a need to understand the environment and to reduce external disturbances—the individual will continue to function as long as he is in a state of disequilibrium. And just as important from a motivational point of view is the thesis that this functioning will be aimed toward the elaboration and stabilization of structures in equilibrium.

The Stage Theory

Piaget's concern with human adaptive processes and with the structure that both influences and results from adaptation is evident in his account of human development. He describes human ontogeny in terms of the logical progression of intellectual structures through development. In other words, his apparent concern with describing what a child can and cannot do at different ages stems not so much from a simple observational and reporting emphasis as it does from the theoretical considerations referred to earlier in this paper, which are not only suggested by developmental data but which also provide a logical framework for the understanding and explanation of this data.

The following sections of this paper deal with Piaget's description of the four major stages which comprise the human developmental sequence: the sensori-motor state (birth to 2 or 3 years); preoperational thinking (2 or 3 to 7 or 8 years); concrete operations (7 or 8 to 11 or 12 years), and formal operation (11 or 12 to 14 or 15 years). This account is admittedly and necessarily incomplete. Hopefully, however, the essential features of each stage have been included. It is important to note that the ages given for each stage are merely approximations. It is not the age at which structures appear but the order of appearance which is of importance.

The Sensori-Motor State (0 to 2 or 3 years)

The newborn child's intellectual structure is defined in terms of reflexes. The child's understanding of his world at this stage will be limited to the actions that these reflexes allow him to perform on his environment. In other words, his world is a world of the here and now which is defined almost entirely in terms of actions. The young child does not separate an object from the actions which he performs on this object. This type of representation of reality is evident in that he does not define objects in terms of the attributes which he will use later to identify them. A ball is not simply a round object—it is "to play with".

The simple reflexes (reaching, grasping, sucking) that the child is born with are called schemata. Piaget says that there is an intrinsic for schemata, once established, to perpetuate themselves through functioning. For this reason all objects that can be assimilated by a schemata will be reacted to. Initially schemata are discrete and unrelated. Eventually, as a result of interaction with objects, schemata will become coordinated so that more than one activity can be performed sequentially in relation to the same object. In other words, the child will not be able to reach for, grasp, bring to his mouth and suck an object. There is still, at this early age, no distinction between the activity and the object. Indeed one of the most important achievements of the sensori-motor period of development will be the establishment of what Piaget calls the "object concept". This involves the simple realization that objects have a permanence and an identity of their own and that they continue to exist even when they are not being perceived by the child.

Early in the sensory motor period there is no language or use of symbols. According to Piaget language will develop first through the internalization of activity. It is not until the child has internalized an activity related to an object that he can think of that object when he is not acting on it. It is this internalization of activity which sets the stage for the development of the intellectual structures which will characterize the later periods of development. Once an activity has become internalized the child has some sort of image which he can use in a symbolic sense. It is then a relatively simple matter to substitute a verbal symbol for this image, and this will be the beginning of language.

By the end of the sensori-motor period the child has acquired language, achieved the object concept, and can represent to himself the outcomes of actions before they occur. His thinking is still characterized by extreme egocentricism. His world is still the world of the here and now and objects continue to exist only in relation to himself. Through continued sensori-motor assimilation of and accommodation to the environment, however, the stage has been set for later intellectual development.

Pre-operational Thinking (2 or 3 to 7 or 8 years)

The term operation is a relatively important one in Piaget's developmental system. It refers to an action which is not only internalized but reversible as well. An action is reversible when the child realizes that an inverse action necessarily and logically nullifies it. As is obvious from the term pre-operational, the child from ages two to seven has not yet reached the level of operational thinking.

Piaget divides the pre-operational period into two sub-periods; the first he calls pre-conceptual thinking. This lasts from ages two to four. From ages four to seven the child is in the period of intuitive thinking. The first of these periods, the pre-conceptual, is characterized by the fact that the child at this stage does not have a true concept of class. He does have some sort of a pre-concept, which enables him to identify objects which belong to classes but which does not allow him to group objects into the classes to which they belong. Piaget illustrates this by refering to an experience he had with his young son while on a walk in the woods one day. Shortly after they began the walk the boy saw a snail and pointed out to his father that here was a snail. When, a short while later, they saw another snail the boy said, "Here is that snail again." For him all snails were one and the same. Among the many instances of this type of reasoning in young children is the fact that a three year old child can see four different Santa

Clauses in four different stores on the same day and not for a moment suspect that there might be more than one Santa Claus.

Another characteristic of this stage of development is that the child's thinking is neither inductive nor deductive, but rather transductive. That is, the child does not reason from the general to the particular (deductive) nor from the particular to the general (inductive) but reasons instead from particular to particular. An extreme illustration of this type of reasoning might be the following:

> Cows eat grass.
> Goats eat grass.
> Therefore goats are cows.

During the next substage of pre-operational thought, the intuitive stage, the child's thinking is still bound to perception. He now has some idea of class membership but cannot apply logical rules to classification. Probably the best illustration of the role of perception in the child's thinking at this stage, and the lack of logic in his thinking, comes from Piaget's experiments on conservation. While the child has now developed an object concept and has achieved some understanding of the quantitative attributes of objects he does not realize that these quantities will not change unless something is added to or taken away from the objects. For example, if a five year old child is presented with two balls that are made up of equal amounts of plasticine he will admit that they do in fact have the same amounts. If one of the balls is then lengthened into a sausage-like shape and the child is asked whether the "sausage" and the ball still have the same amount of plasticine he will probably say that the "sausage" has more plasticine in it because it is longer or that it has less plasticine in it because it is thinner. Not only is the child being misled by the perceptual features of the objects before him, but in addition, he is not applying the logical operations which will characterize his thinking during the period of concrete operations.

Concrete Operations (7 or 8 to 11 or 12 years)

The key feature of the period of concrete operation is action on the environment which becomes internalized and results in the ability to classify, to order and to work with numbers. These abilities are the direct result of ordering, dissociating, combining, and setting up correspondences with real objects. The child's intellectual structure at this stage will be defined in terms of operations (internalized reversible actions) which are coordinated into systems to which certain logical laws apply. In other words the child's thinking can no longer be as ego-centric or as perception-dominated as it was in his earlier periods of development because it must now conform to the following rules of logic.

1. Closure—the child realizes that the sequential application of two operations forms a third operation. In a mathematical sense adding two and three yields a third number, five.
2. Reversibility—for any operation there exists an inverse cancelling operation. For example, adding three to two yields five, but subtracting three from five cancels the first operation and the result is again two.

3. Associativity—if any three or more operations are combined it makes no difference what the order of the combination is. For example, two plus three plus four is equal to four plus three plus two.
4. Identity—this is simply the awareness that there are operations that leave objects unchanged. For example, two plus zero is equal to two.
5. (a) Tautology—the repetition of a relation or a class does not change it. For example, the class of all men plus the class of all men, etc., is equal to the class of all men.
    (b) Reiteration—repetition of a number produces a new number. Three plus three plus three is equal to nine.

While the examples given for these five rules which govern concrete thinking have been related to numbers, illustrations can also be drawn from the realm of non-mathematical thinking. Referring back to the conservation of mass problem described earlier, it can be shown that when a child has achieved reversibility or identity he will have acquired conservation. Using reversibility when faced with the conservation problem, if the child reasons that there now appears to be more plasticine in the "sausage" but this "sausage" could be transformed back into the plasticine ball and there would still be the same amount of plasticine in it, or alternatively that the other plasticine ball could be transformed into a sausage in which case it would be perceptually and actually identical to the given sausage, he would then be able to demonstrate conservation of mass. Or, using identity, the child might simply reason that nothing has been added to or taken away from either of the plasticine in both. Indeed the child at the end of the period of concrete operation has acquired all of the important conservation—mass, number, weight, volume, length, distance, and area.

While the child at the concrete level of operations can deal with classes, numbers and relations, and while his thinking is governed by some of the rules of logic it is still restricted when compared to the level that will be reached in the period of formal operations. In the first place his thinking does not possess complete generality. The various conservations are tied to concrete separate systems. The child cannot, once he has acquired these five rules of logic, apply them to all types of conservation problems. He will acquire conservation of mass around the ages of 7 or 8, that of weight around the ages of 9 or 10, and that of volume only at the end of the period of concrete operations. In addition, his thinking deals only with real and visible objects or with objects that are at least capable of evocation. He does not deal with the hypothetical: he cannot reason from the possible to the actual, or from the actual to the possible. Also, his thinking deals with classes but does not take into account the product of these classes. In other words while he can combine two separate classes into all possible combinations, he will not then systematically combine the products of these combinations.

Formal Operations (11 or 12 to 14 or 15 years)

The major acquisition of the period of Formal Operations is the child's ability to deal with the hypothetical as opposed to the real and concrete. Piaget describes the structures-in-equilibrium characteristic of the adolescent in terms of a logical model which illustrates all possible transformations and combinations of any two (or even three) propositions (statements that can be true or false). What this means, in effect, is that the adolescent can now think hypothetically in a logical manner. He has become freed from the necessity of considering only the real and concrete. His thinking has become potentially completely logical and is characterized by rules which, unlike those of the period of concrete operations, apply to the system as a while. This period is the culmination of an adaptive process which began at birth, and which, through the processes of assimilation and accommodation, has resulted in the development of intellectual structures which are theoretically as sufficient for understanding and coping with the environment as they will ever be.

Logic

A discussion of Piaget and his work would be incomplete without some mention of his use of models derived from Logic.

Piaget's use of logic in psychology is concerned with the explanation of psychological "facts" by reference to the algebra of logic. He has not attempted to explain the formation of structures in the child's mind through the use of logic, but rather, has endeavoured to show that a more complete and clearer understanding of thinking could be obtained by drawing a parallel between pure logical systems and the child's mind where the characteristics of these logical systems become evident.

Piaget's use of logic is not restricted to the level of formal operations but applies as well to earlier periods of development. There is obviously an analogy to be drawn between the thinking characteristic of the concrete level of operations and a logical system defined by the five laws that govern thought at this stage. At the same time, however, it is partly by drawing this analogy that the limitations of concrete thinking are brought to light—and it is possible, at least *a posteriori,* to extend the logical system which provided the analogy for concrete thinking and thereby strive at a prediction of what the state of the mental structures that define formal thought will be. Indeed, Piaget's developmental system can be interpreted as an attempt to describe the evolution of logical structures in the child—a description which could well have been blueprinted *a priori* by his awareness of the possibility of developing a coherent model of development through the use of a framework derived from the realm of logic.

The logical models utilized by Piaget are based on the language of Boolean algebra—an algebra which is sufficient to represent both classes (entities that have similar properties) and propositions (statements that can be true or false), together with the relationships that can exist between classes or propositions and the transformations that can be effected on them. It is not necessary to understand the mathematico-logical models in order to appreciate the contribution these can make to a system which is concerned largely with the development of the ability to deal logically with an environment made up of classes and propositions. It follows that the intellectual structures which define concrete thinking can be described in terms the logic of classes, while those pertaining to formal operations can be described in terms of propositional logic.

More specific illustrations of Piaget's application of logic are provided by the number of his experiments. In one of these, subjects are presented colourless liquids will yield a yellow colour. Solving this problem systematically involves testing the 16 possible combinations. The child at the formal level of operations will do this while a younger subject will likely try a smaller number of combinations. The 16 combinations are analogous to the logical model which Piaget describes in connection with formal operation.

Summary

Piaget's developmental position, as outlined in this paper, involves essentially the application of an equilibrium-through-adaptation model to the description of the qualitatively different intellectual make-up of the child at different stages of development. Accordingly, while the descriptive data pertinent to the stages are relatively unique, extremely interesting and very pertinent to an educational application of the position, it is the adaptation model which is most crucial to the theory, and which has been emphasized in this paper. It must be kept in mind, however, that much of Piaget's four decades of work in this area (and many of his books and articles) has been devoted to the experimental illustration and verification of the central model. Among the many specific areas of research not mentioned in this paper are time, number, movement, space, causality, geometry, chance and perception. Only a book-length review could hope to do justice to all of these topics—and only such a review could adequately transmit the depth and comprehensiveness of the work of Jean Piaget.

Recommended Additional Readings

Flavell, J. H. *The developmental psychology of Jean Piaget.* Princeton, N. J.: Van Nostrand, 1963.

Furth, H. G. *Piaget and knowledge: Theoretical foundations.* Englewood Cliffs, N. J.: Prentice-Hall, 1969.

Ginsburg, H., and Opper, S. *Piaget's theory of intellectual development: An introduction.* Englewood Cliffs, N. J.: Prentice-Hall, 1969.

Piaget, J. *The psychology of intelligence.* New York: Harcourt, 1950.

Piaget, J. *Genetic epistemology.* New York: Columbia University Press, 1970.

Piaget, J., and Inhelder, B. *The psychology of the child.* New York: Basic Books, 1969.

Drawing by Ed Arno; ©1970 The New Yorker Magazine, Inc.

21   Animism Revived

William R. Looft
Wayne H. Bartz

Orientation

The previous article by Lefrancois presented an overview of Jean Piaget's theory of the development of intelligence. The present article expands upon one aspect of the very early work by Piaget, that of animism, and it reviews the subsequent research and theory on this topic by other developmental psychologists.

Very important in our lives are the concepts of life and death. Animism, which refers to the attribution of life or lifelike qualities to nonliving things, is extremely prevalent in the thinking of young children. Indeed, as the following article illustrates, animistic thinking is not uncommon among adults. (Have you ever cursed your automobile for being so "stubborn" or "uncooperative" when it does not start on the coldest days of winter?) Where and when do notions about life and death originate within the individual? It is unlikely that a child engages in deep, systematic thinking about the concept of life, but nevertheless he definitely does have some ideas about it. This is easily verified: Just ask him. His statements may suggest that his thinking about life is naive and illogical, at least from an adult point of view, but it must be recognized that the child is struggling to make sense out of the complex world he lives in as best as he can. And almost all of this learning takes place without the "aid" of formal adult instruction (just as the very young child acquires language proficiency without structured tutoring or lesson plans).

The concept of life serves as an excellent example that a child's thought about the world and about reality is likely to be very much unlike the thought of the adult on these matters. We too often commit the error of **adultomorphizing** the child; we see him and what he does from **our** point of view, not from his. It is all too easy to see the child as only a smaller, weaker, more stupid version of the adult creature. The message contained in the following article (and in all of Piaget's work) is that the nature of thought is qualitatively *(not just quantitatively) different at different points in life.*

William R. Looft, the coauthor of this article and the editor of this book, is an associate professor in the college of human development at The Pennsylvania State University. Wayne H. Bartz is a professor of psychology at Iowa State University and has conducted extensive research in memory and other cognitive processes.

21

In the *Zeitgeist* of current psychological research, considerable attention is conferred to questions initially investigated by Jean Piaget in his developmental studies of cognition in children. Among the more popular Piaget-inspired topics are conservation of mass and number, formation of the object concept, and moral development, to name a few. However, animism, a concept accorded much consideration in Piaget's early work, is an issue which has received meager attention in the recent literature. In the two decades following the first published reports on his studies of animistic thinking in children (Piaget, 1929, 1933), a flurry of investigations by other workers appeared. Since that time, with few exceptions, very little research has been concerned with this question. Indeed, Bronfenbrenner (1963) noted that the content of a recent handbook of child psychology (Stevenson, 1963), in comparison with a similar handbook 30 years back (Murchison, 1931), excludes "the familiar trinity of 'realism, animism, and artificialism' [p. 528]." A perusal of the earlier studies, however, rather clearly indicates that the issues involved were far from settled. In view of the findings of a recent large-scale investigation (Laurendeau & Pinard, 1962), which revealed that animistic thinking is still very much prevalent among children and youth, the question appears to be of sufficient importance to merit revival.

The Clinical Studies of Jean Piaget

Though the existence of animistic thinking has long

been acknowledged (Dennis, 1938), it is Jean Piaget who is credited with the first systematic investigations (Piaget, 1929, 1933). Piaget referred to animism as "the tendency among children to consider things as living and conscious [1933, p. 537]." Actually, his primary concern was with what might be called the development of the concept of life. As a child grows older, Piaget noted that he tends to attribute life to inanimate objects with decreasing frequency, until eventually he reaches a mature cognitive state in which life is attributed only to what educated adults consider alive, that is, animals and plants.

Piaget conducted what he called the "clinical method" of studying the content of child thought, which consisted of the investigator informally questioning the child and attempting to pursue his reasoning. The following is a typical question and answer session:

Vel (8:6): "Is the sun alive?—*Yes.*—Why?—*It gives light, it is alive when it is giving light, but it isn't alive when it is not giving light.*—Is a gun alive?—*Yes, it shoots.*" Vel even goes so far as to say that poison is alive—"*. . . because it can kill us*" [1929, p. 196].

At this stage of development, Piaget observed that a child attributes life to *usefulness* or *activity in general*. In the next stage, life is attributed solely to *things which move*.

Zimm (8:1): "Is a stone alive?—*Yes.*—Why?—*It moves.*—How does it move?—*By rolling.*—Is the table alive?—*No, it can't move . . .*" [1929, p. 199].

According to Piaget, a third stage in this development is characterized by the restriction of life to those *things which move spontaneously.*

Sart (12:6): "Is a fly alive?"—*Yes.*—Why?—*Because if it wasn't alive it couldn't fly.*—Is a bicycle alive?—*No.*—Why not?—*Because it is we who make it go . . .*—Are streams alive?—*Yes, because the water is flowing all the time . . .*" [1929, p. 202].

Piaget indicated that the fourth and final stage is that in which the concept of life is confined to *plants and animals*. In general, the responses given by a child at this level are consistent and approximate to those concepts acceptable to educated adults. Piaget found that Stage One corresponded roughly to ages 4—6 years; Stage Two: 6—7; Stage Three: 8—10; Stage Four: 11 and over.

Variations and Extensions of Piaget's Animism Research

As indicated previously, Piaget's studies on this and other concepts gave the impetus to substantial further research, some of which was highly critical in nature, but most of which was seen as improvements upon Piaget's method. One of the most salient criticisms pertained to the lack of quantitative analysis of his data. An allied criticism suggested that Piaget, by the very nature of his questioning, may have often suggested answers to the children. For example, instead of asking "Who put the snow there?", which suggests some personal agency, he should have asked, "How did the snow get there?" (Munn, 1965). A further criticism was that the stages of animistic thought advanced by Piaget may have been imposed upon the data by Piaget himself. The pertinent question was whether other observers would discern the same stages, given the same data. The remainder of the present paper concerns primarily the subsequent research efforts which attempted to deal with these questions; in addition, attention is paid to the peculiar methodological and theoretical issues intrinsic in them.

Early Attempts at Replication or Refutation

Independent studies. In what was probably the earliest attempt to replicate Piaget's animism findings, Johnson and Josey (1931-1932) used Piaget's technique without change but did not confirm his results. No animism or other forms of precausal thought were found among 6-year-old children. These workers suggested as one possible explanation for their contrary findings that the English language may be superior to the French (as used in Piaget's research) as an instrument for logical thinking.

Oakes (1947) also concluded that her data, collected from a group of 4- to 13-year-olds, refuted the notion that animism is characteristic of child thought. Physical explanations were the predominant mode of response to her verbal questions and simple experiments. However, an examination of her published data reveals that from 4 percent to 19 percent of the various age groups gave nonphysical explanations, many of which were clearly animistic in nature.

Two other studies found evidence in support of Piaget. Bruce (1941) constructed a questionnaire and found that Piaget's four stages describing the evolution of the life concept were easily recognized among a group of white and Negro children in a rural area of southern United States. Klingensmith (1953) also devised a set of questions and demonstrated the presence of animism among his sample of 150 school children.

The Russell-Dennis studies. The criticisms leveled at Piaget's investigations made apparent the need for carefully standardizing his procedures with respect to (*a*) the statement of the questions, (*b*) the specific objects upon which the questions are to be focused, and (*c*) the interpretation of responses with regard to classification into developmental stages. In response to these needs, a rigorous standardization was developed by Russell and Dennis (1939) and was employed in several subsequent studies. This series of studies using the standardized questionnaire approach, in addition ot the one-shot efforts described above, comprise the bulk of research on animistic thought in the United States.

The questions in the Russell and Dennis procedure were focused upon 20 objects (examples: stone, watch, broken dish, comb). Some objects, such as those just enumerated, were present in the interview room; others were merely referred to verbally (e.g., tree, clouds). The child was told the following: "We are going to play a game. I am going to ask you some questions and we will see how many you can answer. You know what living means? A cat is living but if an automobile runs over it, it is dead." A comparison was made of the responses from two comparable groups of children, one group with and the other without this orientation. This statement was not found to be suggestive to the subjects; on the contrary, it appeared to serve the desired function of acquainting the subjects with the nature of the questions to follow (Russell & Dennis, 1941).

The subsequent procedure was as follows: The child was shown an item (e.g., a chair) and was asked, "Is the chair living or dead?" If he said the object was dead and then added that this was so because it did not move, the investigator moved the object and asked, "Is the chair living or dead when it is moving?" If the child attributed life to the item under these conditions, a further question was asked in order to determine whether the child meant imparted or spontaneous movement: "Can the chair move by itself or does something move it?" The same procedure was followed for each item in turn. The answers were classified according to standardized directions. Upon application of this procedure to the responses of 385 American school children (ages 3–15½ years), Russell and Dennis obtained results in general agreement with those of Piaget. At the lowest level no concept was in evidence at all; next, life was attributed to anything of utility or in good condition; then to anything that moved; then to anything exhibiting spontaneous movement; and at the highest level only to animals and plants. Three other independent judges showed 87 percent agreement with Russell and Dennis' assignment of responses to the specific levels of animistic thought, and test-retest reliability was .81 for the test itself. One point of disagreement with Piaget was that any one of the stages of maturity was found to contain at least a few cases from the entire age range. Generally, however, maturity in thought increased with age level.

In further research utilizing this standardized procedure, Russell (1940a) found that 98.5 percent of American school children ($N = 774$) were readily classifiable into the stages of animism suggested by Piaget. The development of this concept was found to be constant over geographic locations and socioeconomic status (urban, suburban, rural). No differences were evident between sexes, and the same stages and sequential development were found in children at every level of intelligence. Generally, however, curves for the percentage of cases in each stage revealed increasing maturity with increasing mental age. This study also gave further evidence of the impossibility of limiting the age range of the stages as Piaget attempted to do. In another investigation (Russell, 1942), 611 older children between the ages of 8 and 20 years were also found to be readily classifiable into one of the stages of animistic thought (98 percent). A definite increase in the percentage of individuals at the adult level (Stage Four) accompanied progressive increases in mental and chronological age.

Animistic Thought in the Mentally Retarded

In addition to these investigations with normal children, the early literature reveals three studies with mentally retarded subjects (Granich, 1940; Russell, Dennis, & Ash, 1940; Werner & Carrison, 1944). The three investigations concur in their conclusions that a high incidence of animism exists in the deficient population. Granich (1940) and Russell et al. (1940) specifically investigated the effects of experience on the development of the life concept. In general, the degree of animism exhibited in any response was negatively related to the retardate's amount of experience with the phenomenon in question. Russell et al additionally found that a comparison between retarded adults and normal children of corresponding mental age showed the retarded to be more advanced with regard to their conception of life, but the subnormal adults were nearer to the children in this respect than to normal adults.

Werner and Carrison (1944) found that a group of retarded, brain-injured children gave considerably more animistic and anthropomorphic responses than did a group of retarded, non-brain-injured children matched for mental age and IQ. They suggested that although animistic thinking may be characteristic of normal personality development, the preponderance of animism among the brain injured is better interpreted in terms of response patterns peculiar to abnormal development.

Animistic Thought in Other Cultures

A number of studies have looked for the phenomenon of animism among children in a wide range of cultures, and thus a comparison is possible with the research on American and western European children.

One of the more notable and more controversial studies is that of Margaret Mead (1932) with the Manus tribes of the Admiralty Islands. She found that these children manifested no spontaneous tendency to animism but responded negatively to explanations couched in animistic terms as opposed to those in terms of practical cause and effect. Mead concluded thusly: "This type of thought was proved to be culturally determined, a potentiality of the human mentality under special cultural conditions, but not the inevitable concomitant of any stage of mental development [1933, p. 915]." It must be noted, however, that Mead's work was done before Piaget published his *Child's Conception of the World,* upon which most subsequent research on animism was based. Here procedure involved eliciting free crayon drawings, interpretation of ink blots, and attempts to attribute will to inanimate objects in test situations. None of her subjects was even specifically asked the direct question, "Is this living or dead?" As Dennis (1943) pointed out, a child takes "life" for granted; questioning is therefore necessary to elicit animistic ideas. Given these considerations, there is no reason to believe that the methods of either Piaget or Russell and Dennis would not reveal the presence of animism among the Manus children.

Further evidence contrary to the Piaget, Russell et al. findings was obtained by Huang and Lee (1945). In their sample of Chinese children, only in a small proportion of cases were inanimate objects said to be alive, and in these few cases the animistic responses appeared to be explained by the apparent characteristics of the specific object rather than by any general tendency. Again, however, for comparison purposes it must be noted that Huang and Lee did not use procedures similar to those of Piaget or Russell and Dennis. Klingberg (1957) used Huang's method with Swedish children and obtained substantial evidence of the animistic mode of thought. In addition he found that the question, "Is it living?", elicited higher percentages of animistic responses than the question, "Has it life?" He concluded that the word "living" is associated with other concepts (e.g., moving, sounding) more so than is the word "life," which appears to be more abstract.

The life concept has also been investigated in Zuni (Dennis & Russell, 1940) and Hopi (Dennis, 1943) Indian children. Though the samples were small, the evidence indicated that these Indian children were retarded in animistic thinking, for a preponderance of "No Concept" and elementary response sets were obtained from children below age 13. The concepts of the Indian children were generally of the same types as those found among white children, however. One question for serious attention in these two studies is the language factor. Both the Zuni and the Hopi subjects were bilingual, with English holding essentially an auxiliary position. The questions in these studies were posed in English, and the investigators conjectured that subjects carried out a considerable amount of translation into their native languages to understand and to answer the question. It is possible that much distortion of meaning could have occurred in the course of his translation. Upon examining the Zuni and Hopi terms referring to the "life" and "alive" concepts, however, Russell and Dennis formed the conviction that in regard to the animistic usage of the language, the Indian practices approximate those of the English language. Therefore, it was concluded that if the examination had been conducted in the natural languages of these people, the results would have been the same.

In a more recent study of American Indians, Havighurst and Neugarten (1955) also obtained a great prevalence of animistic responses among the children of 10 different tribes. In 9 of the 10 tribal groups the percentage decreased in older age groups. However, in view of their inability to find 100% animism at the early age groups, these investigators felt the need to question Piaget's hypothesis.

Jahoda (1958b) conducted an investigation of South African school children. Two approaches were employed: One used an adapted version of a story previously presented to American Indians, and the other was a more flexible technique requiring children to explain the workings of a grammophone. The incidence of animism was found to be low in comparison with results reported from other semiliterate societies, but a progressive decline in animism was found with increasing age, in conformity with Piaget's theory.

In reviewing the cross-cultural literature, Jahoda (1958a) called attention to the apparent absence of consistent trends. The incidence of animism reported ranged from extremely high figures down to zero. The probable causes of these divergences, as outlined by Jahoda, were these: (*a*) The samples were too small, and often no information other than age was reported; (*b*) the likelihood that linguistic factors enter into animism; (*c*) the types of problems presented have varied in many studies; (*d*) the types of administration appeared to influence the results (oral procedures produced higher indices than written procedures); (*e*) personal biases of the investigator may have influenced the results (e.g., those with a highly negative attitude toward Piaget's work). He concluded that cross-cultural standardization of tests of animism is likely to be no more successful than that found for intelligence tests.

Recent Empirical Efforts

As indicated previously, there has been little research interest in animistic thought in recent years. However, since 1960 two related studies and one major research effort have been reported. In a series of studies on children's scientific concepts and interests (King, 1961), a considerable number of animistic responses were noted in a sample of 1811 school children. It was concluded, nonetheless, that there was insufficient evidence for Piaget's evolution of stages. In another study not concerned with the concept of animism per se (Safier, 1964), the characteristic decrease in animism with increase in age was found in three age groupings of boys. It was demonstrated here that a parallel development occurs in the formation of the separate concepts of life and death.

The major recent undertaking, under the direction of Laurendeau and Pinard (1962) of the University of Montreal, had as its primary objective the systematic replication of Piaget's early experiments under stricter methodological requirements. After constructing and administering their own standardized questionnaires to 500 school children, these researchers identified several inadequacies in Piaget's classification scheme. A major criticism had to do with the question of anthropomorphic responses, which are not taken into account by Piaget. An allied criticism dealt with the limited number of stages contained in Piaget's mode. They found that most children, at any age, gave a combination of responses, including generated and spontaneous movement, activity, and anthropomorphism. Consequently classification was problematic. Other problems concerned children who could not explain their responses or made several contradictions, which also led to classification difficulties. Laurendeau and Pinard proposed a new scheme to remove these inadequacies. Stage 0 is labeled "Incomprehension or Refusal," in which the meaning of the questions is not understood; Stage 1 is "Animistic Thinking based on Usefulness, Anthropomorphism, Movement"—these subjects possess imperfect criteria; Stage 2 is "Autonomous Movement with some Residual Animistic Thinking"—this is essentially a transition stage; Stage 3 is "Total Disappearance of Animistic Thinking"—this corresponds to Piaget's Stage Four. This scheme was found by its creators to better classify the responses obtained in their testing.

Methodological Issues

As an introduction to this section, it is to be noted that both Jahoda (1958a) and Laurendeau and Pinard (1962) have very adequately summarized and analyzed most aspects of the criticism which have been leveled at the techniques employed in child cognition studies. Consequently, only a few additional points are commented upon and expanded here.

Grouped versus Individual Analysis

One especially salient feature which becomes evident upon reviewing the literature is the lack of uniformity in quantitative results, that is, the reported percentages of each subject sample which gave evidence of animistic thought. Laurendeau and Pinard (1962) advanced as one explanation, which might account for at least part of this disparity, the differing techniques of data analysis employed. Two general techniques can be identified. One method has been to make a global evaluation of all of each child's responses to a group of problems, all of which were aimed at the determination of the presence or absence of certain types of primitive thinking. Exemplars of users of this technique are Piaget, the Russell-Dennis studies, and Laurendeau and Pinard (1962). The other general technique has been to determine the frequency of answers belonging to the precausal level for each question. This method appears to be more analytical, perhaps resembling an item-analysis technique. Examples of studies using this approach are those of Huang and Lee (1945), Klingberg (1957), Klingensmith (1953), and Oakes (1947).

Thus this problem is one of grouped data versus individual analysis. It is to be noted that the employers of the former technique, which analyzes each child according to the internal coherence of his answers, have been favorable to the hypothesis of precausality. The latter method, in which the data are quantified, object by object, without regard to the subject's explanations, has been employed by a group generally opposed to this hypothesis. It is impossible, of course, to ascertain from the research reports alone a cause-effect relationship; that is, whether a prior attitude led to the use of a particular technique, or whether the use of a particular technique led to a particular attitude.

As Laurendeau and Pinard (1962) have observed, the individual technique seems to have an attenuating, or at least masking, effect. This approach involves counting only "yes" or "no" responses, and therefore is not concerned with the child's justifications for his answers. In attempting to study problems independently, this method loses sight of a child's attitude toward the questionnaire as a whole. As an example, Laurendeau and Pinard illustrated that a child at an intermediate level will respond in quite different ways according to the nature of the problem offered. For a simple question, involving an object close to his daily experience, he might be capable of causal reasoning. Given a difficult problem, however, he is likely to return to a more primitive, animistic mode of thinking. It seems imperative that if one is to estimate correctly any single child's true level of thinking, a global analysis of the entire pattern of his responses is necessary. Even a correct assignment of an object into an animate or inanimate category is often followed by an explanation steeped in a context of very primitive thinking. An analysis based only on frequencies obtained for each item of a questionnaire appears to mask the actual phenomenon and to foster the illusion that there is little animism present within the pool of subjects.

In light of the above discussion it is interesting to note that Strauss (1951) reexamined the original data of Huang and Lee (1945) and decided it was open to interpretations much different from those originally drawn. He concluded from his analysis that the data could also be construed as supporting Piaget's hypothesis.

Concluding his review, Jahoda (1958a) commented that it is "idle to hope for a great deal of quantitative uniformity [p. 211]." The significant question is whether the variations in method are sufficient to explain the divergence in results and conclusions. The answer is in the negative, apparently, for even the same techniques have not been consistent in yielding similar results. While certain regularities can be found in results obtained by a particular type of data analysis, no decisive determinants for the trend of results can be ascertained according to which method of questioning is used (e.g., rigidly standardized questionnaire vs. loose and free interview).

The Semantic Problem

When a child says that something is "living," just what does he mean? To an educated adult, the use of this term most likely implies a number of concepts, including birth, growth, metabolism, reproduction, a progression toward death, and in human beings, perhaps also moral conscience and reasoning. As Huang (1943) aptly observed, "Obviously the child does not mean all that . . . [p. 101]."

Russell (1940b) attacked the semantic problem by posing this question: Is the development of animism by characteristic stages merely a function of a child's usage of the terms "living" and "dead," or is this development actually due to more inclusive ideas concerning the nature of "life" in general? The standardized Russell-Dennis questionnaire was employed with respect to "knowing" and "feeling" rather than "living" or "dead." All but 3 out of 335 subjects followed the general stage progression. The correlations between the stages of animism and the stages of these allied concepts were much higher than the correlations of either series with mental age or chronological age. Since most subjects attributed "knowing" and "feeling" to the "living" objects, and since the development of the two series of concepts followed the same progression, Russell felt justified in the use of the term "animism" as descriptive of the subjects' ideas of life. However, on a conceptual level, it appears that Russell's approach to the problem does not get at the real issue. "Living" and "dead" as two polar categories are considerably different than those of "living" and "*not* living." "Dead" implies that an object at one time had life, but this is not necessarily the case with a "not alive" category. There is evidence that unless specific instructions are given, some children are uncertain as to which of these two non-"alive" categories they are to use (Looft & Charles, in press). Russell and Dennis (1939) themselves found that some children would classify a chipped button or a chipped dish as "dead" because they were broken and therefore of no use.

Contrary to Russell's (1940b) findings, Klingensmith (1953) found that a majority of his subjects who stated that an inanimate object was alive did not attribute "knowing" and "feeling" to that object. Also, his group much more frequently stated that inanimate objects were alive than they attributed sensory and functional attributes to these objects. He suggested one possible reason for this discrepancy is that Russell used the term "Why?" to elicit explanations for the classification of an object as animate or inanimate, which may be suggestive to a child. This may in fact be the case, for it has been shown that subjects asked *why* an object is alive produce more animistic responses than do subjects asked *how* they know an object is alive (Nass, 1956). A possible explanation for this finding might be that "Why?" and "How?" pose two quite different problems for the child. "Why?" may be interpreted by the child that he is to *justify* the categorization he just made (cf. the Russell-Dennis procedure: "Why do you say it is alive?"). This request for immediate justification may very likely have a higher probability of eliciting a superficial, immature (i.e., animistic) reply from the child. On the other hand, "How?" may indicate to the child that he is to support his categorization by more analytical means; therefore he may tend to take more time before giving his reply, and consequently he may produce a more mature explanation.

It appears that a semantic difference may even exist between the words "living" and "life." Klingberg (1957) obtained more animistic responses to the question "Is it living?" than to "Has it life?" He suggested that "living" may have other concepts associated with it (e.g., moving, sounding) than the word "life," which may be more abstract. The difference found here, however, might be better accounted for by an examination of the linguistic properties of these two words. "Living" and "life" represent two different syntactic categories and therefore may contain somewhat different meanings for the child. The -ing ending may indicate an equivalence of the object with the particular condition, whereas "life" may represent a property that the object can possess.

Response bias. It may well be that response bias in younger children is the major methodological problem in this questioning type of research. In this situation, the possibility exists, at least, that a child can be induced to say almost anything. Deutsche's (1937) position was much stronger, for she felt that the individual method of questioning forces a young child to answer at any cost. In such a situation, the child takes recourse to imagination and pure chance, and therefore, Deutsche concluded, nothing proves that the child is thinking of natural phenomena in precausal terms.

Piaget has warned of several pitfalls inherent in any study of children's ideas and beliefs, including the necessity to tread a narrow line between missing or ignoring what is actually there (due to excessive interpretative timidity, methodological inflexibility, etc.) and overinterpreting what the child may have said out of experimenter suggestion or momentary fancy.

> It is as always, open to question exactly how far the children believe what they are saying and at what point they start romancing. But the important thing is to realize that they have nothing with which to replace this artificialism. Whether they make up the details or not they can only explain things by having recourse to human activity and not to things themselves [Piaget, 1929, p. 313].

The child is seen to make associations and connections rather than causal relations between successive elements in a chain of reasoning. As Piaget calls it, he tends to *juxtapose* elements rather than associate them because of logical necessity; the two terms just "go together." Similarly, the child's reasoning is *syncretic* in that diverse terms are imperfectly but intimately correlated within an all-encompassing schema. Because of this juxtapositional and syncretic reasoning, the child tends to find a reason for anything when pressed to do so (Flavell, 1963).

As a guide line for interpreting the data, Piaget (1929) suggested that the specific beliefs which the child expresses, whether elicited spontaneously or through skillful questioning, should be taken as only symptomatic of his turn of mind, a general intellectual orientation toward the world. They should not be accepted as evidence for highly organized and coherent systems of thought.

Admittedly, children have seldom, if ever, reflected on these phenomena about which they are questioned. The ultimate concern, however, is not an examination of ideas which children have already thought out, but rather in determining *how* their ideas are formed and what direction their thinking leads them. What the answer reports is subjective in some proper sense, to be sure, but the fact that *that particular* answer is given is as objective a datum as any which scientific inquiry can yield. It is only necessary to corroborate these kinds of data with other kinds of data.

Cross-Sectional versus Longitudinal Methods

It is well to note at this point that the studies described thus far have been of the cross-sectional type in which different samples of children at various age levels have been compared. Longitudinal research would seemingly provide more meaningful data with regard to the genetic aspect of animistic thought. To achieve its aim completely, such research would deal with the same group of children tested regularly at successive and short intervals. A study of this sort is not easily practicable, especially with reference to the duration of the complete examination, the great number of subjects required, and the necessary limitations in time and personnel under which such an investigation must occur. Given the burdens of this type of research, therefore, it is not at all surprising to discover a dearth of longitudinal evidence to date. One interesting though rather meager bit of evidence comes for Dennis' (1942) observations of his own daughter. Her development was entirely in agreement with the sequence described by Piaget, but the fact that adult answers were given to Piaget's questions by age 6 years 2 months illustrates the great variation possible in the disappearance of childish conceptions.

Theoretical Issues

The literature on animism displays two features which are particularly salient: One is the presence of animistic thought among populations of all age ranges and of great cultural differences. The percentages reported and the interpretations offered for these figures vary widely, but the research clearly indicates, almost without exception, that in each group at least *some* people give *some* responses which appear to be animistic. The other prominent feature is the controversy revolving around the explanation of this phenomenon. There can be little question that animistic-like responses exist; the question which remains unanswered is, what do they mean?

In response to this question, Piaget, once again, is a reasonable starting point, for it was he who provoked the initial empirical interest in the topic. It should be noted that the development of infantile concepts of reality and causality does not occupy a very large place in Piaget's total writings. The two books (1929, 1930) and one article (1933) which deal most extensively with this question date back to the very first part of his psychological work, and Piaget considered these early publications as preliminary endeavors to his total program of research. Commenting upon these works, he recently described them as "un peu adolescents [Piaget, 1959, p. 10]." Nevertheless, these early investigations on the various forms of representation of the world still remain of interest and are still stimulating research. This continued interest is probably due to the fact that these writings were translated early, are readily accessible, and are not filled with the complex logical schemata which characterize his later publications.

For purposes of this discussion, it should be sufficient to provide a most cursory outline of this portion of Piaget's theory; a definitive account is provided by Flavell (1963). In his study of a child's representation of reality, Piaget (1930) sees the child at first as having no image of itself or of the external world; these are two universes compounded into a single reality. These two worlds separate and take shape very gradually through a process of differentiation. Indeed, this period of total egocentrism is a difficult concept for observers to grasp, for it is characteristic of an age at which it is virtually impossible to question a child in any ordered manner. Not until about age 4 can the child be verbally questioned systematically, and by this time the process of dissociation or differentiation is already well underway. In acknowledging the child's inability to communicate verbally his conception of reality at this stage, Piaget

(1951) later emphasized the fact that he formed his notion about the existence of this phenomenon form an accumulation of all types of evidence. At any rate, the child is seen to cling to fewer and fewer subjective notions. "Precausality," the term used by Piaget to describe these adherences to egocentric notions about the world, is an expression which was introduced to represent the explanations intervening between those based on pure psychological and those relating to pure physical causality. Gradually these precausal beliefs are replaced by more objective conceptions of the world.

The Reality of the Animistic Mode of Thought

Among the first criticisms leveled at Piaget's theory were those which disputed the very existence of animism. Upon reviewing the early investigations, Huang (1943) concluded that Piaget and other child investigators were trying to establish the existence of elaborate mental systems in a child's head which are not actually there. Huang proposed that the supposed animism phenomenon could be better explained by a differentiation process quite unlike that proposed by Piaget. His analysis went like this: At first a child is in a neutral or undetermined state, which is not animistic. Gradually the animate-inanimate dichotomy develops, due to repeated definitions and contrasts. It is granted that the child will certainly make errors before reaching perfect differentiation. The more an object shares the typical properties of living things, the greater the tendency for the child to confuse it with these.

Huang made much of the experimenter's inability to investigate adequately very young children. He pointed out that when a young child is closely questioned, he is bewildered, answers at random, and seems entirely at the mercy of suggestion. He felt that it is impossible to pin down the child to a definite statement which could be accepted as representing his belief. "The picture presented is not one of mystic precausality, but the *absence of any definite idea* [Huang, 1943, p.112]." Even under the apparent animistic forms, Huang went on, precausal thinking is neither typical nor universal among children. He referred to his own investigations in which the instances of precausal thinking were rare and often ambiguous (cf. Huang & Lee, 1945). He suggested that animism is readily noted because it is so infrequent and striking that it could not escape attention; physical and naturalistic explanations are so commonplace that they tend not to be noticed. Thus, Huang concluded, the observer is left with an illusion of animism.

Safier (1964) suggested that Huang's notions might be called a "computer theory." She noted that Huang denied any special type of childish logic that would result in different types of animism, and that he considered the child to be different from an adult only to the extent that he lacks information about the "real" world. The child is thus like a computer in that he can give correct answers only to the extent that he has the correct amount of information fed into him.

Klingberg (1957) also challenged the existence of a global animistic trend in very young children. He proposed that pananimism can never be verified empirically because of its presence at such an early age that children cannot be successfully questioned. His position is similar to that of Huang in that he stated that although children may appear to be animistic, nothing proves it as Piaget understood it. Instead of progressing from universal animism toward greater objectivity, mental development is better seen as a transition from a state of total ignorance to a more accurate knowledge of reality.

An examination of the theses of Huang and Klingberg, therefore, seems to reveal the following line of thought: In lieu of predominantly precausal (animistic) thinking in a child, it is preferable to assume the existence of a progression of concept differentiations stemming from a continual accumulation of knowledge. Thus the differences between children and adults, according to this view, are purely quantitative. However, as Jahoda (1958a) has observed, this thesis offers no sound alternative to that of Piaget, for in their postulation of the existence of an undetermined or neutral state (or initial state of ignorance), they have avoided giving any explicit definitions. Just what a "neutral" or "undetermined" state might represent is not at all clear. One quite valid interpretation would be that a child in such a state would respond in the same way whether confronted by either an animate *or* inanimate object. But, as very aptly pointed out by Laurendeau and Pinard (1962), this is precisely one of the salient features of animistic thought!

Perhaps a return to Piaget's (1929) original work can lend some amount of clarity to the issue. Until the ages of 4 to 5, Piaget postulates a state of *diffuse* animism, in which a child's thought begins with a lack of differentiation between living and inert bodies since it possesses no criteria by which to make the distinction. As long as this egocentric indissociation holds sway, the child fails to distinguish between action and purpose, fusing them into a primitive continuum. Animism is held to be a special form of this indissociation, "which consists in attributing to things, characteristics similar to those which the mind attributes to itself—such as consciousness, will, etc. [Piaget, 1929, p. 237]." After age 5 animism is seen to be more transitional and no longer all pervasive. It thereafter appears spontaneously only when the child encounters phenomena which are unexpected and incomprehensible. Obviously, this is quite different from Huang's "neutral" state.

The Notion of Stages of Thought

Numerous objections have been raised to Piaget's proposition that the animistic mode of thought progresses steadily and invariably through clearly delineated levels or stages. Deutsche (1937), Bruce (1941), and Huang (1943) all have observed that at no age is it possible to classify all the responses of those subjects into the same category; each age level always includes children whose responses belong to stages of thought more descriptive of other age levels. Indeed, Russell and Dennis (1939) and Russell (1940a) demonstrated quite conclusively that the and Dennis (1939) and Russell (1940a) demonstrated quite conclusively that the age ranges assigned to the stages of animism by Piaget were not precise. It is important to note, however, that they also demonstrated quite conclusively that virtually every child could be classified readily into one of Piaget's stages, disregarding age. Using a somewhat different classification scheme, Laurendeau and Pinard (1962) also found that each child could be categorized readily, according to his response pattern.

Nevertheless, a number of persons have still found reason to criticize the hypothesis of stages of animism as opposed to the notion of a gradual transition in precausal thought. One critic recently made the following observation: "There is reason to question any breaking up of the data into rigid stages ... one who was not biased by Piaget's original analysis into stages might discern a gradual transition [Munn, 1965, p. 361]." Regardless of whether the actual progression is saltatory or gradual, however, it does appear that such statements, of which the preceding is typical, reflect a superficial reading or a misunderstanding of Piaget's intentions. It is therefore important to know how Piaget arrived at the notion of the stage concept.

His thesis is that thought passes through distinctive stages before adult ideas are attained. It is not that a child grasps adult ideas with varying degrees of adequacy, but rather that a child forms ideas which differ *qualitatively* from those of the educated adult. Furthermore, this transition from precausal toward natural explanations occurs by a succession of substitutions; that is, the final belief is not derived from an initial belief, but is *substituted* for it. Therefore these cognitive stages are essentially discontinuous and qualitatively distinct; each one represents an equilibrium state derived from an equilibration process which is continuous throughout ontogenesis (Flavell, 1963). Piaget agrees that much overlapping among stages is involved, but he emphasizes that the total *pattern* of a child's responses must be considered. Two stages are defined largely by their differences. Thus the positing of stages is basically a process of abstracting highlights, within some frame of reference, from a panorama of gradual change.

Kessen (1962) has made the admonition that one does not fully describe growth merely by narrating the various states through which it passes; the total process must also be contended with. In defense of the use of stages, however, Bruner (Bruner, Olver, & Greenfield et al, 1966) has made an equally cogent point: That discontinuities exist is indisputable, and it is entirely legitimate and necessary to note such discontinuities, however ill equipped we may be to explain them. It should be noted that Piaget himself does not subscribe to the view that child and adult are completely different entities with no commonalities between them. What he does emphasize is that there are significant differences as well as obvious similarities, and that a developmental theory above all must deal with these.

It appears that those critics who reject the concept of stages have not approached the critical question, and that question is essentially one of verifying the *constancy* in the sequential order of a child's modes of thought. This constancy is the important element, and the most satisfactory procedure to confirm or refute its presence is the longitudinal research program.

The Influence of Specific Experience

That the cognitive functioning of a child is susceptive to his specific environmental experiences seems obvious. A considerable amount of effort has been exercised by some to support this point. On an empirical level, correlative evidence has been obtained to illustrate relationships among assorted environmental variables and degree of animistic thought. On a more academic level, a number of the theoretical arguments which have been issued in support of, or contrary to, the notion of animism appear to be interpretable in an influence-of-experience or degree-of-familiarity framework.

Empirical support. Ezer (1962) conducted a study to determine the effects of religion upon children's explanations of cause-effect relationships. He reasoned that since the three major religions in the United States (Protestantism, Catholicism, Judaism) have both animistic and anthropomorphic concepts within their tenets, the children who have had a greater amount of formal religious training or who come from intensely religious homes would be more likely to give precausal responses than children with a less intensive religious background. His results generally supported this hypothesis, which suggests that causal explanations are a function of training as well as maturation.

Another hypothesis is that animism may be transmitted to offspring by parents. From observing his own daughter over a number of years, however, Dennis (1942) rejected this hypothesis. He remained convinced that the animistic tendencies are autogenous and are developed from the child's own experiences and reaction tendencies. Another bit of evidence comes from a study in which parents' differential responses for animistic and magical explanation, as opposed to physical and natural explanations, were compared with the extent of animistic thinking observed in their children (Tynni, 1958). The correlations were of zero order.

A child's degree of familiarity with the items used in traditional interview studies has often been noted to be an influential factor in the degree of animism evinced by that child (Granich, 1940; Huang & Lee, 1945). More specifically, Nass (1956) observed that nonnaturalistic responses occur by more than a 2:1 ratio to remote or unfamiliar stimuli as compared to familiar phenomena.

Speculative support. As another argument for denying the existence of the precausal mode of thought, Deutsche (1937) and Huang (1943) have emphasized the inconsistency in a child's responding. If he had a general mode of thinking, they argued, he would attack problems in the same manner and thus produce consistent responses. The degree-of-familiarity notion appears to serve as an effective counterargument to this inconsistency objection. If a child is confronted with an easy problem, it is likely he will give adult-like answers. For a difficult problem, however, he will revert back to old and familiar arguments. It is quite natural that the first objects to lose their animistic character are those which the child has seen and handled frequently. To some extent, therefore, animistic responses will persist for a more or less prolonged period depending on the nature of the objects used.

Piaget did not specifically deal with experience in his early writings on causal concepts, but in a later writing he made the following observation: "It is true that the more active experience is, the more the reality on which it bears becomes independent of the self and consequently 'objective' [Piaget, 1952, p. 367]." Piaget's later derived processes of assimilation and accommodation appear to be amenable to this discussion also. One might ask the question of what prevents a child from mastering, in one fell swoop, all that is cognizable about a given concept. According to the theory, one can assimilate only those things which past assimilations have prepared him to assimilate. There must be in existence a system of meanings sufficiently advanced that it can be modified to admit items for assimilation which accommodation places before it. Assimilation is a conservative process; phenomena whose interpretations require a complete reorganization or expansion of the existing mental structure simply are unable to be accommodated and thereby assimilated (Piaget, 1954).

An early writing by Nathan Isaacs (in S. Isaacs, 1930) appears to anticipate this composite process of adaptation proposed by Piaget. Isaacs' thesis was that a child builds up from his experiences certain assumptions or "anticipatory schemas" regarding reality. He acts upon these unreflectingly until he encounters a new experience which is not congruent with them. He is then stimulated to revise and extend his knowledge. Isaacs considered a child's "Why?" question as a way to resolve this conflict by appealing to adults for help. Piaget (1931) recognized this thesis as complementary to his theory and saw no real difference between Isaacs' anticipatory schemas and his own structures. Isaacs studied them in the process of formation, at the moment when on contact they either disintegrate or are strengthened, whereas Piaget studies them in their origins as crystallizations of the past activities of the whole subject.

To summarize, a statement by Tiedemann in 1787 still seems appropriate to the present discussion on experience and animism: According to him, the animistic tendency

> is certainly due to the fact that one always envisages an unknown thing through one that is known; ... now there is nothing nearer and more familiar to us than ourselves, wherefore images of our own reactions, our own way of doing things, are constantly mingled with our ideas of external objects; therefore, we conceive all things as being like us, alive as ourselves, and acting by the same powers and motives as we do [in Murchison & Langer, 1927, p. 229].

Animism Beyond Childhood?

One perplexing problem which remains is the nature and meaning of the animism that has been reported in adult populations. Although this discussion has considered animism as basically a characteristic of child thought, percentages as high as 50 percent—sometimes up to 75 percent—have been reported for adults. This finding appears to deny the supposition of a progression toward consecutively more sophisticated modes of thought, and therefore to deny the Piaget theory. It should be noted, however, that Piaget never questioned any adults.

Dennis and Mallinger (1949) administered a concept-of-life questionnaire to a group of people 70 years and older. Surprisingly, 75 percent of this group gave animistic responses, many of which were similar to those given by children. The investigators suggested that a *regression* to infantile modes of thinking may have occurred in this group due to neurological deterioration, senility, and the like. This reasoning may be valid in this instance, but it does not reasonably apply to adults in general. The incidences of animistic responses from adults in subsequent research are much too high to be attributable to mental deterioration or other pathological conditions. Other factors need to be called upon to explain adult animism.

Following the Dennis and Mallinger study, several investigations have revealed the presence of a persistent cast of animism among adults. These data raised considerable doubt as to the veracity of the notion of the infantile and primitive character of this mode of thought. Dennis (1953) found that about one-third of the students sampled in various American colleges and universities attributed life to one or more inanimate objects. A perusal of detailed responses convinced Dennis that the students were not being poetic, philosophic, or whimsical. Typical examples were these: lighted match—*"Living because it has flames which indicate life"*; sun—*"Living because it gives off heat* [pp. 248-249]." Dennis (1957) found a much higher percentage (79 percent) of a sample of college students in Beirut, Lebanon, gave one or more animistic answers. Further evidence came from Crannell (1954), who found results verifying those of Dennis. Bell (1954), Lowrie (1954), Voeks (1954), Crowell and Dole (1957), and Simmons and Goss (1957) all conducted investigations with college populations and obtained analogous results: Considerable percentages of subjects gave at least a few animistic responses.

Dennis (1953) found that college students with courses in biology gave fewer animistic responses than those without this background. He similarly concluded that the high incidence of animism among college students in the Near East could be accounted for in terms of weak scientific instruction (Dennis, 1957). A number of other studies failed to confirm these results, however (Bell, 1954; Crannell, 1954; Crowell & Dole, 1957; Simmons & Goss, 1957). All of these investigators reported relatively high frequencies of students with courses in biology giving animistic answers.

An analysis of these experiments leaves the impression that adult animism does not have the same meaning as that in children. It appears that in many cases the animism observed was artificially induced by the experimental setup. Some of the factors may have been these: The questions posed were greatly different from those given to children; some items appear to be excessively difficult and perhaps even give rise to philosophical issues (e.g., atom, seed); some questions were posed in a suggestive manner (e.g., "When an automobile tire blows out, does the tire feel it?"); subjects were usually not informed that they were to make a distinction between scientific and metaphoric language. Questionnaires in multiple-choice form obtained smaller numbers of animistic answers than those offering a simple alternative (Bell, 1954). The frequency of precausal answers could also be modified according to the instructions given (Simmons & Goss, 1957). Again, the semantic problem is undoubtedly influential. Voeks (1954) reported that many of her subjects admitted that "living" simply meant to them "real" or "existing."

Obviously, "living" and similar words are used abundantly in metaphorical expressions: a living language, a live battery, a live wire, etc. The prevalence of such lay descriptions led Simmons and Goss (1957) to conclude that animistic responses could be attributed to response-mediated generalization. The use of these characteristics as reasons for animistic response, according to them, indicates that the same or similar responses were also aroused by the test words. As a consequence, to the degree that verbalization of these reasons elicits responses such as "living" and "alive," generalization of these responses to the test items would be anticipated.

Lowrie (1954) and Voeks (1954) refuted the notion that these answers are genuine manifestations of animistic thinking. In their studies, the ambiguous character of this hypothetical animism was often revealed through some spontaneous remarks from their subjects. For example, half of Voek's subjects who attributed life to "ocean" indicated that they included all organisms living therein. Such a comment is not likely to be given by a child. If asked about "ocean," it is likely that a child would refer to its movement or usefulness if he were to classify it as animate.

Lowrie (1954) called attention to the almost universal use of anthropomorphic terms, even though this may not reflect on actual belief in anthropomorphism. There is a tendency of people, educated or not, to attribute human qualities to almost all kinds of animals and plants and also to a goodly number of inanimate objects. Probably most people, in using such terms, do not actually mean precisely what they say, but nevertheless they use these analogies.

Ignorance as to what constitute the criteria of life is widespread. Apparently, even college science courses do not present this information in such a way that it can be transferred to the world outside textbooks and the classroom (or at least to the questionnaires devised by psychologists). As Simmons and Goss (1957) have remarked, the emphasis in biology is more likely to be upon characteristics of living things rather than upon the absence of such characteristics in nonliving things. Therefore, the amount of background in science may be largely irrelevant to the extent of experience in distinguishing between living and nonliving things. Furthermore, possibly, and quite probably, many people have just never thought about the matter.

New Directions for the Study of Animism

The number of studies on animism has been decreasing over the past few years; animism is dying. Fresh conceptualizations and, more important, improved methods of investigation are needed to give impetus and direction to the revival of research on animism. One general observation of the present review is that the methods employed in the study of animism have been limited to either interview or questionnaire methods. While the "dynamic-affective" framework of Honkavaara (1958) is a novel approach toward generating hypotheses in that the affective state of the perceiver commands central attention in the comprehension of concepts, it still suffers in that the methods employed (mostly questionnaire) perpetuate the ambiguities in findings and theory noted here. A more fruitful course might be to consider animism and other forms of precausal thought in the contexts of other theories of cognition. Even more important, however, is the necessity of improved methods of investigation; it appears that experimental methods will begin to disambiguate the methodological difficulties noted in this review.

While there are many alternative theories and methodological approaches, a few suggestions are given here for future research on animism.

Animism as Attributive Behavior

Heider's (1958) theory analyzes interpersonal behavior as a function of responsibility attribution, the process by which an individual holds another responsible for events in the life space. This attributive process is conceived as an inferential one, a judgmental process in which responsibility is assigned to persons or to environmental events. Heider's formulations are based on Piaget's theory but represent an extension and improvement in methodology. We suggest that the process now be reversed: The methods employed in the study of responsibility attribution can be employed in the study of animism and precausal thought to great advantage.

Heider outlined five levels in the development of responsibility attribution. Each level represents levels of maturity of thought and reflects the principle that environmental events, rather than individuals, are held responsible as this growth continues. At the first level, the basis for attributing responsibility is association; if a person is in any way associated with an event, he is held responsible. The remaining levels add to this principle. Accordingly, at the second level, a person is held responsible if associated with that event and it is produced by his actions; at the third level, foresight is also necessary; and at the fourth level, the person responsible must be associated with the event, produce it by his actions, and must have foreseen the outcome and intended to produce the effect. The fifth and final level of responsibility attribution is the fully mature stage. A high degree of intentionality is involved in the attribution of responsibility at this level and requires sophisticated reasoning on the part of the subject, thereby precluding the attribution of consciousness or will to an inanimate object.

Shaw and Sulzer (1964) have empirically investigated Heider's theory of the development of responsibility attribution and their results support it. Their experimental method required subjects to view cartoonlike drawings as a story was told (or read). The stories were constructed to reflect and contrast the various features of the levels in Heider's theory. The stories all involved the degree of association of a central character with the outcome. In addition, this character's involvement in the production of events and whether he foresaw or intended it was contrasted. Subjects of different ages were to judge this character's responsibility for the outcome of each story.

The Shaw and Sulzer method has the advantages of greater objectivity and control over that in questionnaire and clinical interview methods. The stimulus materials were designed to be appropriate, as well as similar, for all age levels. In addition, the response requirements were the same at all age levels. These features are not always present in questionnaire and interview methods. Subjects were not required to justify or explain their judgments, although verbal reports were obtained as a heuristic for the production of other stories.

It is suggested that similar methods could be employed in the study of animistic thought. Stimulus materials with the characteristics of those of Shaw and Sulzer would be an improvement over those traditionally employed. The pattern of judgments at different ages might well provide greater insights into animistic and precausal patterns of thought. In addition, such a method would provide an excellent basis for cross-cultural studies, one that would be less likely to be influenced by linguistic factors.

Another method employed in the study of responsibility attributions has been used by Wright (1968) in assessing the influence of personality characteristics on the process. In this method, movies depicting interpersonal events are shown to the subject, who is required to judge the person responsible for the event depicted. This method suggests that many of the characteristics of animistic thought found by Piaget can be investigated experimentally. By means of moving pictures or closed-circuit television, both living and nonliving objects can be made to show autonomous movement, a characteristic of living things. The differences in animistic judgments made by subjects at different ages might then be assessed to evaluate the influence of perceptual features of the stimulus object upon the attribution of life to objects.

The study of attribution of responsibility may be considered to be representative of other attributive behavior. The superior level of control over stimulus materials and response requirements, in comparison to those employed in the animism studies, would be of obvious benefit to a more thorough understanding of animistic thought.

Animism as Conceptual Rules and Strategies

Contemporary approaches to concept learning emphasizes two aspects of concepts: rules and attributes (Bruner, Goodnow, & Austin, 1956; Haygood & Bourne, 1965; Hunt, 1962). A concept is defined as a rule specifying the attributes (discriminable characteristics) and their combination (rule) for categorizing stimulus events as equivalent. Concept learning is viewed as a process of categorization, one of isolating and using these two structural components. The sequence of decisions used to isolate or otherwise deal with a concept is known as a strategy and is considered to be a hierarchial process (Bruner et al., 1956; Hunt, 1962).

The experimental methods used in the study of concept learning generally require that subjects sort or categorize stimulus objects (instances) in some manner. In most studies, the instances are geometric shapes generated by factorially combining a number of stimulus attributes (shape, color, etc.), thus forming a matrix. The subject is required to categorize the individual instances into positive and negative categories where the investigator has designated a subset of instances as the concept. This general procedure is very flexible and the effects of a number of variables upon categorization have been investigated: response requirements, nature of feedback, number of attributes, size of the concept relative to the number of instances in the matrix, the nature of the conceptual rule, strategies, means of presenting the instances, and others (Bourne, 1966).

The theory, data, and methods used in concept learning are relevant to the study of animism. It is of interest that a number of people (e.g., Haygood & Bourne, 1965; Hunt, 1962), like Piaget, emphasize the role of symbolic logic as the rule component in conceptualization. Bourne (1966) has pointed out that Piaget's investigations of conservation have been directed toward isolating the *rule* component of the concept. The basic procedures, too, are similar; subjects in animism studies have been required to categorize objects as living and nonliving (or dead), or to justify such a categorization. Nonetheless, no experimental studies of animism have been conducted. While there are many possibilities, a few can be suggested.

Greater control could be achieved in animism studies if subjects were to categorize a number of stimulus objects which are instances into a matrix. For example, a stimulus matrix might be constructed with familiar objects, using the attributes of life (living and nonliving objects) and movement (autonomous and imparted movement and inert objects), with each cell of the matrix having a number of instances. Under conditions of free classification, the relevant attributes and the rule for categorizing living and nonliving objects could be inferred from subjects' placements. The agreement of these placements with verbal reports and other data would offer a broader base for understanding the thought process underlying animism. In addition, it would be expected that the data resulting from such a study would be consonant with other studies of children's categorizations (Bruner & Olver, 1963; Vygotsky, 1962).

Bruner (1964, 1965) and associates (Bruner et al, 1966) have recently undertaken the study of cognitive growth. Their theory is based upon Piaget's but differs in two important ways. First, the Bruner group emphasizes the role of language in cognitive development. While both theories stress that growth is a process of attaining symbolic representation of events, Bruner argues that symbolic representation comes about through the use of language. Once a child learns to use language to represent events, he attains remoteness from the moment and relies less upon the appearance of things. A second major difference between the theories is that Bruner's is an information-processing theory. Where Piaget emphasizes the processes of accommodation and assimilation, Bruner stresses techniques or strategies for dealing with information, for solving problems. There are, of course, other differences and similarities, but that discussion is beyond the scope of this paper.

Of particular interest here, however, are the methods used by the Bruner group. While varied, the methods provide cues which can be organized in terms of their appearance (perceptually) or their characteristics (symbolically). The task requirements are such that the child's strategy for solving the problem can be abstracted from his choices, judgments, and categorizations. These methods are based upon the principles outlined above in concept learning studies.

For example, Kenney (in Bruner et al., 1966) modified the Piaget water-puring conservation-of-volume procedure. After completing the standard procedure, the children estimated water levels before pouring. The containers were partially hidden from view by a screen; the volumes of water were not visible to the child. Under this condition, more children of ages 5 through 8 demonstrated conservation. However, when the screen was removed, the younger children changed their minds. This study demonstrates that children can deal with the problem in a symbolic manner. However, the visual appearance is a stronger cue for the younger children, who have not yet come to use language as a means of representing their world. In addition, this study shows that a child's thought pattern is not locked in a stage but is a process dependent upon the nature of the task, its appearance, and its requirements.

The Bruner point of view, too, offers a fresh approach to the study of animistic and precausal thought with methodological techniques which allow inferences to be made from choices and categorizations, as well as verbal reports. The results of the studies conducted by the Bruner group suggest that what the young child says and what his other behaviors are can be markedly dissonant. This is a consequence of the nature of the young child's means of knowing, which is primarily through the action and image of things, not abstractions. And abstraction, or symbolic thought, is prerequisite to overcoming animistic notions.

The present paper has been but a brief sampling of experimental methods and other theories relevant to animism. The reader can, no doubt, supply more. If so, then the implicit recommendation of the paper has been made. The child's concept of life is central to his development. The study of this concept, which has centered about Piaget's early formulations of animism, has been investigated by only a few methods and with disparate results. This paper has shown that more powerful methods are available and amenable to the study of animism. It is our intent to apply them.

References

Bell, C. R. Additional data on animistic thinking. *Scientific Monthly,* 1954, **79,** 67-69.
Bourne, L. E. *Human conceptual behavior.* Boston: Allyn and Bacon, 1966.
Bronfenbrenner, U. Developmental theory in transition. In H. W. Stevenson (Ed.), *Child psychology, the 62nd yearbook of the National Society for the Study of Education.* Chicago: National Society for the Study of Education, 1963.
Brunce, M. Animism vs. evolution of the concept "alive." *Journal of Psychology,* 1941, **12,** 81-90.
Bruner, J. S. The course of cognitive growth. *American Psychologist,* 1964, **19,** 1-15.
Bruner, J. S. The growth of mind. *American Psychologist,* 1965, **20,** 1007-1017.
Bruner, J. S., Goodnow, J., & Austin, G. A. *A study of thinking.* New York: Wiley, 1956.
Bruner, J. S., & Olver, R. Development of equivalence transformations in children. *Monograph of the Society for Research in Child development,* 1963, **28,** (Whole No. 86), 125-141.
Bruner, J. S., Olver, R., & Greenfield, P. M., et al. *Studies in cognitive growth.* New York: Wiley, 1966.
Crannell, C. N. Responses of college students to a questionnaire on animistic thinking. *Scientific Monthly,* 1954, **78,** 54-56.
Crowell, D. H., & Dole, A. A. Animism and college students. *Journal of Educational Research,* 1957, **50,** 391-395.
Dennis, W. Historical notes on child animism. *Psychological Review,* 1938, **45,** 257-266.
Dennis, W. Piaget's questions applied to a child of known environment. *Journal of Genetic Psychology,* 1942, **60,** 307-320.
Dennis, W. Animistic thinking among college and high school students in the *normal and Social Psychology,* 1943, **38,** 21-36.
Dennis, W. Animistic thinking among college and university students. *Scientific Monthly,* 1953, **76,** 247-249.
Dennis, W. Animistic thinking among college and high school students in the Near East. *Journal of Educational Psychology,* 1957, **48,** 193-198.
Dennis, W., & Mallinger, B. Animism and related tendencies in senescence. *Journal of Gerontology,* 1949, **4,** 218-221.
Dennis, W., & Russell, R. W. Piaget's questions applied to Zuni children. *Child Development,* 1940, **11,** 181-187.
Deutsche, J. M. *The development of children's concepts of causal relations.* Minneapolis: University of Minnesota Press, 1937.
Ezer, M. Effect of religion upon children's responses to questions involving physical causality. In J. Rosenblith & W. Allinsmith (Eds.), *The causes of*

behavior: Readings in child development and educational psychology. Boston: Allyn and Bacon, 1962.

Flavell, J. H. *The developmental psychology of Jean Piaget.* Princeton, N. J.: Van Nostrand, 1963.

Granich, L. A qualitative analysis of concepts in mentally deficient boys. *Archives of Psychology,* 1940, **35**, No. 251.

Havighurst, R. J., & Neugarten, B. L. Belief in immanent justice and animism. In R. J. Havighurst (Ed.), *American Indian and white children: A sociopsychological investigation.* Chicago: University of Chicago Press, 1955.

Haygood, R. C., & Bourne, L. E., Jr. Attribute- and rule-learning aspects of conceptual behavior. *Psychological Review,* 1965, **72**, 175-195.

Heider, F. *The psychology of interpersonal relations.* New York: Wiley, 1958.

Honkavaara, S. The "dynamic affective" phase in the development of concepts. *Journal of Psychology,* 1958, **45**, 11-23.

Huang, I. Children's conception of physical causality: A critical summary. *Journal of Genetic Psychology,* 1943, **63**, 71-121.

Huang, I., & Lee, W. H. Experimental analysis of child animism. *Journal of Genetic Psychology,* 1945, **66**, 69-74.

Hunt, E. B. *Concept learning.* New York: Wiley, 1962.

Isaacs, S. *Intellectual growth in young children.* New York: Harcourt, 1930.

Jahoda, G. Child animism: I. A critical survey of cross-cultural research. *Journal of Social Psychology,* 1958, **47**, 197-212. (a)

Jahoda, G. Child animism: II. A study in West Africa. *Journal of Social Psychology, 1958,* **47**, 213-222. (b)

Johnson, E. C., & Josey, C. C. A note on the development forms of children as described by Piaget. *Journal of Abnormal and Social Psychology,* 1931-32, **26**, 338-339.

Kessen, W. "Stage" and "structure" in the study of children. In W. Kessen & C. Kuhlman (Eds.), Thought in the young child. *Monograph of the society for Research in Child Development,* 1962, **27**, 65-82.

King, W. H. Symposium: Studies of children's scientific concepts and interests: I. The development of scientific concepts in children. *British Journal of Educational Psychology,* 1961, **31**, 1-20.

Klingberg, G. The distinction between living and non-living among 7-10-year-old children, with some remarks concerning the animism controversy. *Journal of Genetic Psychology,* 1957, **90**, 227-238.

Klingensmith, S. W. Child animism. *Child Development,* 1953, **24**, 51-61.

Laurendeau, M., & Pinard, A. *Causal thinking in the child.* New York: International University Press, 1962.

Looft, W. R., & Charles, D. C. Modification of the life concept in children. *Developmental Psychology,* 1969, **1**, 445.

Lowrie, D. E. Additional data on animistic thinking. *Scientific Monthly,* 1954, **79**, 69-70.

Mead, M. An investigation of the thought of primitive children with special reference to animism. *Journal of the Royal Anthropological Institute of Great Britain and Ireland,* 1932, **62**, 173-190.

Mead, M. The primitive child. In C. Murchison (Ed.), *Handbook of child psychology.* (2nd ed.) Worcester, Mass.: Clark University Press, 1933.

Munn, N. L. *The evolution and growth of human behavior.* (2nd ed.) Boston: Houghton Mifflin, 1965.

Murchison, C. (Ed.) *A handbook of child psychology.* Worcester, Mass.: Clark University Press, 1931.

Murchison, C., & Langer, S. Tiedemann's observations on the development of the mental facilities of children. *Journal of Genetic Psychology,* 1927, **34**, 205-230.

Nass, M. L. The effects of three variables on children's concept of physical causality. *Journal of Abnormal and Social Psychology,* 1956, **53**, 191-196.

Oakes, M. E. Children's explanations of natural phenomena. *Teachers College contributions to Education,* 1947, No. 926.

Piaget, J. *The child's conception of the world.* New York: Harcourt, 1929.

Piaget, J. *The child's conception of physical causality.* London: Kegan Paul, 1930.

Piaget, J. Retrospective and prospective analysis in child psychology. *British Journal of Educational Psychology,* 1931, **1**, 130-139.

Piaget, J. Children's philosophies. In C. Murchison (Ed.), *A handbook of child psychology.* (2nd ed.) Worcester, Mass.: Clark University Press, 1933.

Piaget, J. *Play, dreams, and imitation in childhood.* New York: Norton, 1951.

Piaget, J. *The origins of intelligence in children.* New York: International Universities Press, 1952.

Piaget, J. *The construction of reality in the child.* New York: Basic Books, 1954.

Piaget, J. Esquisse d' autobiographie intellectuelle. *Bulletin de Psychologie,* 1959, **13**, 7-13.

Russell, R. W. Studies in animism: II. The development of animism. *Journal of Genetic Psychology,* 1940, **56**, 353-366. (a)

Russell, R. W. Studies in animism: V. Animism in older children. *Journal of Journal of Genetic Psychology,* 1940, **57**, 83-91. (b)

Russell, R. W. Studies in animism: V. Animism in older children. *Journal of Genetic Psychology,* 1942, **60**, 329-335.

Russell, R. W. & Dennis, W. Studies in animism: I. A standardized procedure for the investigation of animism. *Journal of Genetic Psychology,* 1939, **55**, 389-400.

Russell, R. W. & Dennis, W. Note concerning the procedure employed in investigating child animism. *Journal of Genetic Psychology,* 1941, **58**, 423-424.

Russell, R. W., Dennis, W., & Ash, F. E. Studies in animism: III. Animism in feeble-minded subjects. *Journal of Genetic Psychology,* 1940, **57**, 57-63.

Safier, G. A study in relationships between the life and death concepts in children. *Journal of Genetic Psychology,* 1964, **105**, 283-294.

Shaw, M. E. & Sulzer, J. L. An empirical test of Heider's levels in attribution of responsibility. *Journal of Abnormal and Social Psychology,* 1964, **69**, 39-46.

Simmons, A. J., & Goss, A. E. Animistic responses as a function of sentence contexts and instructions. *Journal of Genetic Psychology,* 1957, **91**, 181-189.

Stevenson, H. W. (Ed.) *Child psychology, the 62nd yearbook of the National Society for the study of Education.* Chicago: National Society for the Study of Education, 1963.

Strauss, A. L. The animism controversy: Re-examination of Huang-Lee data. *Journal of Genetic Psychology,* 1951, **78**, 105-113.

Tynni, H. Is animistic thinking of children transmitted by adults? *Report of the Department of the Psychological Institute of Pedagogy, Jyväskylä,* 1958, No. 14, 21-28. (*Psychological Abstracts,* 1959, 33, No. 8028)

Voeks, V. Sources of apparent animism in students. *Scientific Monthly,* 1954, **79**, 406-407.

Vygotsky, L. S. *Thought and language.* Cambridge, Mass.: M.I.T. Press, 1962.

Werner, H., & Carrison, D. Animistic thinking in brain-injured, mentally retarded children. *Journal of Abnormal and Social Psychology,* 1944, **39**, 43-62.

Wright, J. M. The relationship of abstractness-concreteness and of Michotte's categorization of causal impressions to the attribution of responsibility in human interaction. National Science Foundation Final Project Report, May 1968.

Recommended Additional Readings

Berlyne, D. E. Children's reasoning and thinking. In P. H. Mussen (Ed.), *Carmichael's manual of child psychology.* (3rd ed.) Vol. 1. New York: Wiley, 1970. Pp. 939-981.

Flavell, J. H. Concept development. In P. H. Mussen (Ed.), *Carmichael's manual of child psychology.* (3rd ed.) Vol. 1 New York: Wiley, 1970. Pp. 983-1059.

Maurer, A. Maturation of the concept of life. *Journal of Genetic Psychology,* 1970, **116**, 101-111.

Sigel, I. E. & Hooper, F. H. (Eds.) *Logical thinking in children: Research based on Piaget's theory.* New York: Holt, Rinehart, and Winston, 1968.

22 Identification of Creativity: The Individual

Marie Dellas
Eugene L. Gaier

Orientation

Reprinted from *Psychological Bulletin*, 1970, 73, 55-73, by permission of the authors and the American Psychological Association.

Creativity *is a troublesome concept. Nearly all psychologists are concerned about it, and nearly all educators hope to foster it within their students, yet it seems that very few persons really have it. Perhaps we may soon discover that we have set out to investigate the notion of creativity in a totally erroneous manner; like "intelligence," perhaps we have assumed that "creativity" is quantitatively distributed among the human species: Some people have a lot of it, some people have a moderate amount of it, and some people have very little of it.*

Speculation about and interest in creativity is nothing new; a great deal of research and theorization has taken place over the past several decades. What we have learned conclusively from all this out-pouring of energy, however, is limited. The typical procedure has been to construct some kind of test of "creativity" (similar to tests of "intelligence") and then to garner test scores from persons of varying ages and statuses in order to obtain normative data (similar to research on the growth of intelligence). Perhaps all this work has been guided by the wrong set of assumptions; perhaps all—or nearly all—persons are capable of generating an original and novel set of responses to the questions we pose to them, but they are inhibited from doing so because of prior socialization, response styles, or personality orientations. In other words, many persons may not be "creative" (at least on tests devised by psychologists) because they have been punished or nonrewarded for such behavior in earlier experiences. Indeed, research conducted by the editor of this book and by others (e.g., Hudson, 1968; Price, 1970) has provided evidence that if persons are asked to role-play *(i.e., take the "role" or part of some* other *person), they are capable of attaining very high scores on traditional tests of creativity. Thus, tests of creativity may suffer from the same problem as tests of intelligence: The subject's response may be determined at least as much by his response style as by his native sense of "creativity" (or "intelligence").*

The following article by Dellas and Gaier offers a comprehensive review of research on the creativity construct. The reader is invited to consider these findings in light of the qualifying factors suggested in these introductory remarks.

Marie Dellas is now associated with Eastern Michigan University, and Eugene L. Gaier is a professor of educational psychology at the State University of New York at Buffalo.

22

The thesis of the present paper is a conceptualization of the creative person. The purpose of this critical review of the research and theory is to probe the psychological makeup of the creative individual within the context of the question: Does an uncommon or unusual, but appropriate, production occur fortuitously, or does it occur as a result of a particular constellation of personal traits? That consistent, creative production is manifested in a particular psychological condition is the position of the authors.

Most economically, the literature on creativity can be classified into four major orientations: (*a*) the nature and quality of the *product* created, (*b*) the actual expression of creative acts and the continuing *process* during the "creation," (*c*) the *nature* of the individual, and (*d*) environmental factors and *press* that tend to initiate and foster creativity. Since the appearance of Golann's summary statement on creativity in 1963, studies have focused on two tangential but highly relevant topics dealing with the creative. First, is creativity independent of intelligence? Second, is personality, per se, a vital aspect of creativity?

In an effort to order the empirical evidence concerning the creative individual, the present paper is organized within two general trends. One may be designated the *cognitive* orientation in which analyses are made within the framework of intellectual dimensions—singular, intellectual traits and cognitive functioning and styles. The other position may be identified as the *personological* approach in which personality and motivational traits are the foci.

Specifically, consideration is given to characteristics consistent with creative performance and issues involved in these domains. These include (*a*) intellectual factors and cognitive styles associated with creativity, (*b*) creativity as related/ unrelated to intelligence, (*c*) personality aspects of creativity, (*d*) the potential creative, and (*e*) motivational characteristics associated with creativity.

Cognitive Characteristics

Attempts to identify the creative individual by means of cognitive variables appear directed toward the investigation of either singular intellectual factors or cognitive styles.

Intellectual Abilities

By means of multivariate methods of factor analysis, Guilford and his associates supported 16 of 24 hypothesized intellectual abilities postulated to be related to creative production. Among these are fluency of thinking, comprised of word, ideational, expressional and associational fluency; flexibility of thinking, composed of spontaneous and adaptive; originality; sensitivity to problems, redefinition; and figural and semantic elaboration. Beginning with the first major analysis of Wilson, Guilford, Christensen, and Lewis (1954), in which air cadets and student officers were tested, a series of investigation (Guilford & Merrifield, 1960) isolated 15 of these separate factors in young adult populations. More recently, studies have recognized them at student levels. Guilford and Hoepfner (1966) identified all 16 at the ninth-grade level; Merrifield, Guilford, and Gershon (1963) found 7 at the sixth-grade level, and 6 were demonstrated by Lauritzen (1963) at the fifth-grade level.

Using students rated highly creative in the visual arts, Lowenfeld and Beittel (1959) also identified essentially the same factors. Five attributes—fluency, flexibility, redefinition, sensitivity to problems, and originality—were identical to those reported by Guilford.

Guilford (1967b) has collectively defined these factors as "divergent" thinking, a mode of productive thinking, typical of the creator, which tends toward the novel or unknown. It is this novel output which he considered the essence of creative performance. As compared to convergent thinking which is oriented toward the known or "right" solution, divergent thinking occurs where a problem has yet to be defined or discovered, and where no set way of solving it exists. Guilford has developed tests purported to measure these creative abilities and which have been used to discriminate between the creative and noncreative. Although these measuring devices have some face validity, the issue that arises with their use pertains in part to construct validity—whether or not these divergent production abilities are, in fact, responsible for creative potential. In order to demonstrate that these are, indeed, creative abilities, scores from instruments assessing divergent thinking have been related to other indices of creative performance. The results, thus far, have been contradictory and far from conclusive

An analysis of the data in Drevdahl's (1956) study of arts and science students revealed that those rated as highly creative by independent judges on personal and objective creativity rating scales demonstrated superior performance on Guilford originality tests, the scores for originality correlating .33 with the ratings. When divergent production scores of high school students obtained on Guilford-like tests were correlated with teacher nominations for creativity (Merrifield, Gardner, & Cox, 1964; Piers, Daniels, & Quackenbush, 1960; Torrance, 1962), the correlations were generally low, on the order of .2. Yammamoto (1964a) noted similar low correlations between Torrance creativity measures and peer nominations as criteria.

Using instruments measuring three aspects of divergent thinking—redefinition, semantic spontaneous flexibility, associational fluency—Skager, Klein, and Schultz (1967) concluded that they yielded low and inconsistent relationships with artistic achievement at a school of design. This appears to substantiate Beittel's (1964) findings which indicated a lack of relationship between divergent thinking measures and performance in art.

When employed in the study of eminent creative adults, these divergent production tests have also correlated low or negligibly with criterion ratings of creativity. With respect to architects judged highly creative by experts in their own field, MacKinnon (1961) established that whether scored for quality or quantity of responses, the Guilford tests neither correlated highly nor predicted efficiently the degree of creativity demonstrated in their creative production. Gough (1961), substantiating this claim, presented evidence indicating that for research scientists, rated creativity correlated low or negligibly with various Guilford tests: Unusual Uses correlated $-.05$; Consequences, $-.27$, and Gestalt Transformations .27. With Air Force officers, Barron (1963a) found a positive multiple correlation of .55 between rated originality and a composite divergent production score including Guilford tests. Using semantic divergent production tests with public relations personnel nominated by their superiors as high or low creative, Elliott (1964) also noted that the groups were significantly discriminated on the basis of each of five of the eight Guilford-type tests.

The lack of success of these instruments in predictive efficiency and in correlating with demonstrated creativity and other indices of creative performance may be attributed to several factors. First, there is no consistency among the creativity ratings which appears inseparable from the major problem—the absence of an ultimate criterion for creativity. Gough (1961) rated his scientists on general science and research competence, mathematical-theoretical orientation, efficiency and productiveness, original potential, and general sophistication. MacKinnon (1964) used the indices of publicity and prominence, plus a 5-point scale including such items as originality of thinking, constructive ingenuity, and ability to set aside established conventions and procedures. The differences in these two examples are obvious, to say nothing of the ambiguity of such terms as "prominence" and "competence."

The ratings themselves are only as good as the criteria used for selection and are subject to bias, halo effects, and the judgment abilities of the raters. With respect to teacher ratings, the basis for making judgments is suspect since they typically see students only in the classroom situation. They may also lack an understanding of the dimensions to be rated. Sociometric peer nominations are subject to similar limitations. Thus, valid differential ratings may then be precluded.

Divergent production tests may also not be suited to the group investigated. Initially, it seems reasonable to assume that the particular content of divergent thinking measures—semantic, figural, and symbolic—would be significant in yielding differential results for persons in different fields. Appropriately, Elliott (1964) used semantic divergent production tests with advertising and public relations personnel, obtaining a significant positive relationship with the selected criteria. A significant relationship, however, did not obtain when Beittel (1964) correlated semantic test scores with criteria of creative performance of college art students.

The majority of Guilford's tests also leave unilluminated the personological context in which the creative process functions. As a great number of investigators now agree, personality factors are of great significance despite the fact that, for the most part, little is known regarding their contribution to creative production. Guilford (1966) appeared aware of this when he explained the low reliability of tests of divergent production. "This probably reflects the general instability of functioning of individuals in creative ways . . . and therefore, high levels of predictive validity should not be expected [p. 189]."

It is not surprising, therefore, in light of the above, that assessment instruments may not correlate with criterion ratings. Yet, despite the fact that the validity of Guilford-like creativity measures is still incomplete and unclear, the available evidence does suggest a relationship between creative performance and the divergent thinking attributed to the creative. Furthermore, the diversity of these intellectual aspects of creativity and the fact that they are relatively distinct components indicate that perhaps different types of creative talent exist, the scientific creative probably excelling in different abilities from the aesthetic creative. Consequently, the versatile, creative person may be the exception rather than the rule.

Creativity and Intelligence

Since the recognition of the divergent production abilities, the many studies conducted to probe the relationship of creativity and intelligence concern the empirical distinction between them—whether or not they are separate domains.

Several prominent investigators in the field, Getzels and Jackson (1962), Guilford (1967b), and Torrance (1962), maintain that a valid distinction exists between the cognitive function designated "creativity" and the traditional concept of general intelligence. To buttress their position, they cite the relatively low correlations between IQ and creativity measures. Guilford and Hoepfner

(1966) reported a mean correlation of .32 between 45 divergent production test scores of the ninth graders cited previously and the California Test of Mental Maturity. Studying highly gifted adolescents, Getzels and Jackson (1962) found the correlation between IQ and Guilford-derived creativity measures to be on the order of .3. Using his own tests in a replication of the Getzels and Jackson work at the elementary school level, Torrance (1962) obtained essentially similar results, his correlations with various intelligence tests ranging from .16 to .32.

Other investigators have reported consonant findings. A correlation of .3 was found by Yamamoto (1964b) between Torrance creativity test scores of creative adolescents and the Lorge-Thorndike measure. Investigating scientifically oriented, bright students, Herr, Moore, and Hansen (1965) did not find any relationship between the Guilford tests of creativity and scores on the Lorge-Thorndike, Watson-Glaser Critical Thinking Appraisal, and Terman Concept Mastery measures. Using seven creativity instruments and taking the Getzels and Jackson study as his starting point, Flescher (1963) reported the average correlation with California Test of Mental Maturity scores was .04 for his sample of 110 sixth graders, suggesting definite independence of creativity indices from intelligence.

Within the framework of the Campbell and Fiske (1959) concepts of convergent validity and discriminant validity, Wallach and Kogan (1965) reviewed many of the cited studies and a number of others which have attempted an operational distinction between creativity and intelligence. In a penetrating analysis, they concluded that few of these studies demonstrated both convergent validity, in which intercorrelations of many different measures of creativity were high, and discriminant validity, in which intercorrelations of the creativity measures with intelligence measures were low. These conclusions appear to be in essential agreement with the analyses by Thorndike (1966) and Marsh (1964) of the Getzels and Jackson (1962) investigation, with several Guilford studies, and with the more recent Wallach (1968) analysis of the Torrance tests.

Wallach and Kogan (1965) attributed his failure to obtain convergent and discriminant validation for creativity to the presence of either the time pressure, or evaluation pressure, or both. Therefore, the major innovation in their very comprehensive study of fifth-grade students involved a change in the assessment situation—the establishment of a relaxed, nonevaluative atmosphere with no time limit. An analysis of the data based on a Guilford-derived creativity battery, designed to test Mednick's (1962) associative theory of creativity, revealed that creative thinking was strikingly independent of the conventional realm of intelligence, and was a unitary and pervasive dimension in its own right. The correlation among the creativity factors was on the order of .5, among the intelligence measures on the order of .5, but the average correlation between the two sets of measures was .1. This distinction obtained, despite the fact that the measuring instruments required the use of verbal skills which are accepted as playing an important role in the measurement of intelligence. Cropley (1968), who also used this creativity battery in a group-test situation, came up with similar results. The Wallach and Kogan tests manifested a high degree of internal consistency and were relatively independent of the intelligence tests. However, he also found a substantial general factor loaded highly on both kinds of tests.

The assumption that pressures of time and evaluation may, indeed, be significant influences in the creativity-intelligence differentiation gains some support from Dentler and Mackler (1964). Although the rigid 5-minute limitation of the Torrance Tin Can Uses Test for originality was maintained, they found a greater mean originality produced by subjects in their "safe" group as compared to those in "routine," "indifferent," and "unsafe" groups.

Investigators at the Institute of Personality Assessment and Research (IPAR) provide additional support for the creativity-intelligence dichotomy where the creativity dimension was based on ratings. Among artists, correlations between creativity (rated quality of work) and measured intelligence (Terman Concept Mastery Test) were zero or slightly negative (Barron, 1963a). MacKinnon (1961) provided additional substantiating evidence gathered from creative architects. Within this group, the correlation of intelligence, measured by the Terman Concept Mastery Test, and creativity in architecture, as rated by experts, was −.08, which is not significantly different from zero. Even in fields such as mathematics and science where intelligence is a prerequisite for the mastery of the subject matter, the correlation, although positive, was a low .07 (Gough, 1961).

Barron (1963a) has observed that commitment to creative endeavors is already selective for intelligence, so that the average IQ is already a superior one. Consequently, he has suggested that over the total range of intelligence and creativity, a low, positive correlation of about .4 exists. Beyond the IQ of 120, however, measured intelligence appears to be a negligible factor in creativity. McNemar (1964) appeared to be in agreement with this theory when he stated that "at high IQ levels there will be a very wide range of creativity whereas as we go down to average IQ, and on down to lower levels, the scatter for creativity will be less and less [p. 879]." It is hardly surprising that this "fan shaped" hypothesis, more recently reiterated by Guilford (1967b), was not confirmed by Mednick and Andrews (1967) and by Ginsberg and Whittemore (1968), who found a positive linear relationship throughout the range. The creativity measure was the Remote Associates Test which has been severely criticized as being a measure of convergent thinking rather than the divergent thinking attributed to creatives.

The IPAR data, along with findings previously cited, appear to provide some evidence for describing creativity as a concept at the same level of abstraction as intelligence. Even Thorndike (1966), who so severely criticized the Getzels and Jackson and Guilford studies, conceded that their data suggested that "there is some reality to a broad domain, distinct from the domain of the conventional intelligence test to which the designation of 'divergent thinking' or 'creative thinking' might be applied [p.52]." He considered this distinguishable domain rather "nebulous and loosely formed," any different creativity measures being necessarily less interchangeable and equivalent than different intelligence measures.

In seeking, however, to identify the creative individual, immediate cognizance must be given to the fact that a test of creativity is not "creativity," just as a test of intelligence is not "intelligence." Furthermore, "tests of creativity" not concerned with affect or personality attributes may be an oversimplification and may, perhaps, be tapping a very low level of creativity—reception or differentiation—which may be far afield from the product or performance, and the life activities domain. The inconclusiveness of the data, therefore, seems to suggest that perhaps, until the personological context in which the cognitive variables are embedded is determined, real measures of the dimension of creativity remain elusive.

Cognitive Functioning

Preference for cognitive complexity. One of the stylistic variables emerging as particularly distinctive of the creative is a cognitive preference for complexity—the rich, dynamic, and asymmetrical—as opposed to simplicity. That this perceptual style or aesthetic preference is not limited simply to the creative artist, but is also typical of the creator in other fields, is revealed by the data of the IPAR investigations. Using the Revised Art (RA) Scale of the Welsh Figure Preference Test, that successfully discriminated between artists and nonartists in standardization and cross-validation samples (Barron, 1953; Barron & Welsh, 1952), Gough (1961) demonstrated this cognitive predisposition as characteristic of research scientists, MacKinnon (1961) of creative architects, and Barron (1963a) of creative writers.

Raychaudhuri (1966b), using this same instrument in India, furnished supporting cross-cultural evidence. Investigating musicians and painters, he reported that the mean RA scores of these artists were significantly higher than those of nonartists; these obtained differences were reinforced by the findings of another, comparable study (Raychaudhuri, 1966a).

Eisenman and Robinson (1967) used polygons of varying degrees of complexity to measure this variable. With high school students as subjects, they found that a preference for complexity in polygons was significantly related to a high score on a paper-and-pencil personality measure of creativity. The reported lack of a significant correlation value (.14) between the Stanford-Binet and the polygon test or creativity test scores also suggested that this dimension was independent of the IQ domain. Evidence pointing toward the early development of this characteristic emerged in Rosen's (1955) investigation of art students. Scores on the Revised Art Scale did not increase as a function of training in art—the scores of the first-year art students not differing significantly from those of the advanced.

Thus, a cognitive predisposition for complexity appears to be a distinguishing feature of the creative person, regardless of the field of creative endeavor, at various developmental levels. Apparently, the creative individual has the capacity to integrate this richness of experience into a higher order synthesis that makes for the unusual in creative production or performance.

Empirical evidence also points to the Revised Art Scale as a promising, nonverbal measure of creative potential. It has had relatively consistent success in correlating with independent criterion measures of creativity—.41 with criterion ratings of creative scientists (Gough, 1961), and .48 with those of creative architects (MacKinnon, 1961). It also appears to be relatively independent of intelligence. Harris (1961) reported a correlation of −.07 between the Otis Quick-Scoring Mental Ability Tests and the RA for 390 ninth- and tenth-grade pupils, Welsh (1966), studying 368 gifted high school students in a residential summer school, found that though the correlation between the two intelligence measures—Terman Concept Mastery Test (CMT) and the nonverbal D-48—was significant, the correlation of both tests with the RA Scale was essentially zero, −.07 with the D-48 and −.03 with the CMT. In view of the controvertibility of other creativity tests, the use of this cognitive-perceptual instrument to identify the creative may result in more reliable and fruitful investigations into the domain of man's imagination.

Cognitive flexibility. Accumulating evidence suggests that cognitive flexibility is another stylistic feature distinguishing the creative. Investigations in this area have used an intrapsychic approach in which interactions and relationships among hypothetical mental structures have been examined. Hypothesizing that creative individuals, as compared with noncreative normals, have a greater availability of both the relatively mature and the relatively primitive processes Hersch (1962) studied the cognitive functioning of a group of recognized creators (artists), noncreative normals (firemen, salesmen, entrepreneurs), and schizophrenics. His findings, on the basis of responses to six categories of the Rorschach, confirmed this hypothesis and revealed that the primitive responses involving the association of affective attitudes with sensory stimulation were given most frequently by the creators, and best distinguished them from the other two groups. He attributed the creative individual's extensive use of primitive functioning to optimal, flexible controls which were readily accessible and facilitated quick return to secondary processes. By contrast, the noncreative normals, lacking the ability to function primitively, had rigid, self-limiting controls. An apparent weakness of this study lies in the fact that the control groups, although matched for age and IQ, differed from the artists in education and social class.

An earlier study, however, by Myden (1959), in which he equated his controls on these variables, yielded essentially similar results. Using a battery of tests designed to offer clues to the whole pattern of mental functioning—Rorschach, Thematic Apperception Test (TAT), Vigotsky Concept Formation—he reported that the creative's significantly greater use of the primitive process was not an isolated phenomenon, but rather appeared to be a component of his thinking throughout the records, particularly the Rorschach, with the primary process well integrated with the secondary process, indicating its control by the ego.

Barron (1963a) also noted this biphasic regression-progression capacity in his study of 15 creative Air Force officers. Considering the correlates of originality when intelligence was covaried, he concluded that the impulse or ego-control dimensions emerged most clearly as a determinant of originality. The frequent regression to primary process thinking was attributed to the strong ego of the creator who allowed himself to regress momentarily far into primitive fantasies because he was secure in the control and flexibility of his ego. The importance of the ego-control variable in this process was demonstrated by Pine and Holt (1960). An assessment of expression and control of primary process by means of the Rorschach in 13 male and 14 female undergraduates revealed that these were statistically independent variables. The quality of a person's created production was unrelated to the amount of primary process expression, but was related to the effectiveness with which the expression was controlled.

Studying male undergraduate science majors, designated high creative on the basis of Guilford tests, Garwood (1964) discovered their responses to the Rorschach manifested a greater integration of nonconscious with conscious material than did the responses of the low creatives. The functioning of this style emerged at the adolescent level in the Getzels and Jackson (1962) study. The imaginative productions of the highly creative as manifested in TAT-type stories and drawings pointed to the movement of preconscious material into conscious expressions. That the preconscious was under ego control was evidenced in that their responses were both unique and adaptive to reality. The results of Clark, Veldman, and Thorpe (1965) were entirely consistent with these findings; the fanciful productions of their creative adolescents, as observed on the Holtzman Inkblot Technique, met the demands of adequate appropriateness.

The results of these last three studies, however, must be reviewed with reservation, inasmuch as no external or natural behavior criterion was designated. Creativity was restricted to a Guilford-type test battery whose validity was questionable. The findings of Wild (1965), who also presented evidence that creativity involved the ability to use psychological processes belonging to different developmental levels, must be considered in the same light since there was no interrater reliability on her creativity criterion for art students. Using the Rorschach to probe this dimension in third-grade children whose creativity was assessed by professional artists, Rogalsky (1968) emerged with inconclusive results. This suggests that the Rorschach may be an inappropriate instrument to use for this purpose at such an early age. Not having progressed much beyond operational thought (Baldwin, 1967), nor having developed full ego controls, the combinatory process required in the response may not, as yet, be mastered by children.

One general criticism of these investigations is their use of the Rorschach to measure this cognitive style. Despite a plethora of studies reporting the effectiveness of this instrument in diagnosis, its validity, as well as its reliability, is still recognized as problematic. Furthermore, different scoring systems used in its interpretation (Hersch, 1962, employed the Rorschach Genetic Scoring System and Pine and Holt, 1960, applied a system devised by Holt and Havel, 1960) raise the question of comparability of results.

Although these studies point up certain limitations, the consistent findings, indicating that creative adults exhibit a strong degree of consensus regarding cognitive flexibility as manifested in adaptive regression, suggest that regression is an essential factor in the creator's makeup exploited toward productive, creative expression. Regression is not symptomatic of loss of ego control, but, rather, appears to be a part of the creative individual's thinking development, since primary process seems to be well integrated with secondary process. Barron (1963a) suggested that this seemingly necessary attribute is probably the reason the creator may appear simultaneously crazier yet saner, more naive and more knowledgeable, more destructive and more constructive. Though this stylistic cognitive feature appears at the lower developmental levels, the findings are less conclusive due possibly to the methodological and theoretical reasons discussed.

Perceptual openness. Perceptual openness, conceptualized as a greater awareness of and receptiveness to not only the outer world but also to the inner self, is another distinctive cognitive mode attributed to the creative. Some empirical evidence has accrued supporting its existence in his constitution, difficulties of operationalization notwithstanding.

Applying the Myers-Briggs Type Indicator, MacKinnon (1961) determined that a preference for this perceptual mode differentiated more creative from less creative architects. And Gough (1961) reported this same preference as discriminating between the high- and low-creative research scientists. Mendelsohn and Griswold (1964, 1966) also provided substantiating evidence. Dichotomizing psychology undergraduates as high or low creative on the basis of the Remote Associates Test, they observed that only the high creatives used peripheral cues effectively in solving anagram problems. The creatives, they suggested, deployed their attention more widely, were more aware and receptive, and retained more prior stimulus experiences in usable form, tending not to screen out the irrelevant. A serious limitation of these investigations, however, is their use of the Remote Association Test to identify the creatives. As Jackson and Messick (1965) pointed out, the items in this test have one recognized response and hence have limited value as indicants of originality. Furthermore, the theoretical basis on which this test has been constructed—the concept of the habit hierarchy and the interpretation of creativity in terms of traditional associative and mediation processes—was not supported in a recent study by Jacobson, Elenewski, Lordahl, and Lieroff (1968).

Within the psychoanalytic frame of reference, Barron (1963a) reported a study which may be construed as evidence of the existence of this cognitive style. Original persons, designated by a high composite score based on a variety of measures of originality, rejected suppression as a mechanism for the control of impulse. Since suppression would limit awareness and openness to both the internal and external stimuli, its rejection would make one more "open."

Propst (1962) attempted to demonstrate this mode of perception within the Rogerian framework employing an instrument developed to measure openness to internal experience through an introspection task. She found a positive relationship between this measure of openness and a combined score of originality for a sample of 60 male undergraduates. Some evidence is provided regarding the validity of the new instrument, yet it is apparent that, as any introspective method, it is subject to all the misperceptions and self-deceptions of self-report.

Although these data are far from conclusive, the cited research indicates that certain cognitive characteristics, idiosyncratic of the creative person's mental functioning, contribute to the originality or difference of the end products of his thinking, characterized by a certain intellectual freedom that is not goal-bound, controlled or channeled, but rather seeks the "unknown" and "confusion." It also points up a greater capacity and tolerance for flexibility, complexity, and openness.

Personality Characteristics

Although the foregoing suggests that certain cognitive characteristics are essential to creativity, it is apparent, as Vinacke (1952) asserted, that they function not in isolation, but rather in relation to a total personality system of needs, attitudes, goals, and emotions. The importance of these personality factors is attested to even by those originally committed to a cognitive orientation, who have now modified their opinions. With respect to conditions affecting creative thinking, Guilford (1967a) proposed that consideration be given to "what motivates individuals . . . and needs, interests and attitudes that help the individuals to be productive creatively [p. 12]." Bloom (1963), while engaged in an extensive research of the cognitive dimensions associated with scientific creativity, concluded "rather reluctantly that personality and motivational factors are at least as important as aptitude in determining performance [p. 252]."

The most provocative concepts regarding personality characteristics of the creative individual have been derived from studies of the eminent and well established. Some of the most useful findings have emerged from the Institute for Personality Assessment and Research in California where recognized creative research scientists, writers, mathematicians, and architects were intensively studied. MacKinnon (1961) reported that on the California Psychological Inventory, highly creative architects emerged as self-confident, aggressive, flexible, self-accepting, little concerned with social restraints or other's opinions, and strongly motivated to achieve primarily in those situations where independent thought and action, rather than conformity, were required. They rated high on aesthetic and theoretical scales and their assertiveness and independence were again graphically displayed on IPAR scales measuring these variables. Other instruments revealed their perceptiveness, intuitiveness, and introversion, and though they indicated little desire to be included in group activities, which attested again to their introverted nature, they demonstrated marked social poise, dominance, and a desire to control others when they did interact. That architectural work requires both aesthetic and scientific creativity suggests that the notable personality traits of the highly creative architect would tend to be most generally characteristic of the creative person. Results from studies in other fields tend to support this hypothesis.

Except for the fact that creative research scientists were predominantly judgmental rather than perceptive, Gough's (1961) findings were in essential agreement with those of MacKinnon. The impulsivity, independence, introversion, intuitiveness, self-acceptance syndrome appeared again in the psychometric data of creative writers (Barron, 1963a), although this group differed in their significantly greater originality and emphasis on fantasy consistent with their profession. Creative male mathematicians (Helson, 1961) displayed comparable traits but differed from their female counterparts (Helson, 1967) in that they were more ambitious, sociable, professionally participative, self-accepting, and less eccentric. Raychaudhuri (1966c) established that on the basis of the Rorschach, TAT, and Szondi Test, professional musicians in India were more distinctly marked by emotional and temperamental, than by cognitive, characteristics. As compared to noncreatives, they appeared to be more egocentric, exhibitionistic, more stimulated by frustration, and preferred activities that permitted a greater range of individualism and self-expression.

While the above studies were conducted primarily within a psychoanalytic context, Cattell and his associates, using a factor-analytic approach with the Sixteen Personality Factor Test, which yields more easily quantifiable results, have come up with essentially similar data. Although more bohemian than creative scientists (Cattell & Drevdahl, 1955), creative artists and writers (Cross, Cattell & Butcher, 1967; Drevdahl & Cattell, 1958) demonstrated the same salient traits of ego strength, dominance, self-sufficiency, sensitivity, introversion, desurgency, and radicalism. Creative psychologists (Drevdahl, 1964) displayed the characteristic independence and nonconcern with social environment, but differed in lacking the dominance and aggressiveness manifested by other groups.

Findings by other earlier investigators were consonant with the above. On the basis of the Rorschach, TAT, and biographical data, Roe (1952), in her studies of the development of creative research scientists, described as salient among their characteristics strong curiosity, persistence, high energy level, and a strong need for independence. Stein (1956) found that various psychological tests revealed that more creative industrial research scientists, as compared with their less creative colleagues, were more autonomous and assertive, and had more integrative attitudes and their own value hierarchy. Comparing artists and nonartists on the Rorschach and TAT, Eiduson (1958) reported that the former had interests that were broader, theoretical, and abstract.

While there have been speculations on a possible relationship between creativity and neurosis, the empirical evidence demonstrates no basis for this assumption. Psychopathology in the creative's nature emerges in all the cited IPAR and factor-analytic studies, but the creative's distinctly superior scores on egostrength scales (Barron, 1961, 1963a; Cross et al., 1967; Drevdahl & Cattell,

1958; MacKinnon, 1961) indicate that they presumably have adequate mechanisms to handle these deviations.

Exploring the relationship between job adaptation and creative performance, Heinze (1962) offered additional evidence of the creative's high ego strength. Highly creative scientists, compared to the less creative, exhibited qualities associated with ego strength in that they were relatively unimpeded by conflicting forces in the environment, demonstrating sufficiently flexible controls which enabled them to "transcend" these conflicts. Ego strength also appeared in the cited Wild (1965) and Hersch (1962) studies as the variable that differentiated the creatives for the schizophrenics. This appearance of pathology in the creative, therefore, which may be a necessary concomitant of openness, seems to be more suggestive of high intellect, greater sensitivity, richness, and complexity of personality rather than psychopathology, for the creative individual appears to have adequate resources and control measures to cope with these tendencies.

Sex ambivalence, designating a femininity of interests pattern in the male, and the evidence of a masculine pattern in the female, also emerges with some consistency to distinguish the creative. This high association between creativity and femininity appears on the scales of the MMPI and the California Psychological Inventory in all the IPAR investigations cited, regardless of profession, as well as in other studies using different instruments. Highly original military officers revealed this femininity of interests pattern on the Personal Preference Survey Scales (Barron, 1963a; Myden, 1959); and Hammer (1964) observed a higher incidence of feminine characteristics in artists on the basis of the Rorschach. Although these creative men did not deny expression of this more feminine aspect of their personality, which led them to recognize impulses and interests regarded, in our culture, as more characteristic of women, the data indicated that they were not characterized by an effeminate manner, nor were they homosexuals. Indeed a study by Ellis (1959), in which he confirmed his hypothesis that degree of homosexuality was directly related to loss of creative potential, attested to the contrary, as did MacKinnon (1961), Barron (1963a), and Hammer (1964), who stated that creative men demonstrated a high degree of masculine-associated traits such as assertiveness, confidence, determination, ambition, and power. One might conclude from these findings, that though feminine components appear to a greater degree in the creative male personality, it is the blending of the feminine and masculine, the integration of the necessary sensitivity and intuition with purposive action and determination, that is conducive to creativeness.

Probing this aspect of the female creative, Helson (1967) noted that creative women mathematicians differed from the less creative in that they retained their femininity despite admission of masculine characteristics. The less creative, afraid of their impulsivity, seemed to have donned masculine armor as protection. An earlier study (Helson, 1966), assessing female undergraduates, presented essentially similar conclusions. Although creative nominees were more intelligent and had stronger need for accomplishment, they were not significantly more assertive, skillful in analytical thinking, or generally more "masculine." Rees and Goldman (1961) also reported that though highly creative females displayed more masculine characteristics, they did not score as high on these masculine traits as the moderately creative. These data appear to buttress Barron's (1963a) suggestion that a certain amount of cross-sex identification is prerequisite to creativity in both men and women, with more creative women having salient masculine characteristics, just as creative men have salient feminine characteristics. Yet from these data, one might conclude that certain masculine traits in the female may inhibit rather than facilitate creativity; and may, in reality, be mere indicants of opposition to the feminine cultural role.

In comparing the cited evidence on personality, some limitations of the studies are immediately apparent. There were marked differences in criteria of creativity and nominees were only as good as the criteria upon which their selection was based and the judgment abilities of those making the selections. Sample sizes were often small; and, though males were studied predominantly (a further limitation on generalizability of results) when females were included, sex differences were not often examined (e.g., Barron, 1961). Some subjects were tested by mail (e.g., MacKinnon, 1961), the effects of which procedure are not known, and in this same study, a restriction of the range of indices existed within samples being contrasted. This lack of complete data for all subjects is a notable shortcoming.

Different approaches were used, the measuring instruments were heterogeneous, and the personality inventories utilized have been criticized for ambiguity, susceptibility to faking or malingering, lack of sufficient empirical validation, and liability to contamination of response sets and intelligence. MacKinnon (1964), however, has asserted that probably one of the most convincing checks in science is for different investigators, using a variety of approaches and tests, to come up with consonant findings—a situation that appears to exist with respect to this personality research. Despite the various approaches and heterogeneity of instruments, many similarities in the results can be seen across samples differing in cultural background, eminence, and profession. Independence, manifested not only in attitudes but also in social behavior, consistently emerged as being relevant to creativity, as did dominance, introversion, openness to stimuli, and wide interests. Self-acceptance, intuitiveness, and flexibility also appeared to characterize the creatives, and though they had social presence and poise, they exhibited an asocial attitude and an unconcern for social norms. This may reflect antipathy toward anything encroaching on individuality or compelling conformance. Some of these characteristics appear more pronounced in aesthetic creativity—radicalism, rejection of external restraints—as opposed to scientific creativity; but the data reflect that the majority of these qualities appear to differentiate the creative from his noncreative counterparts. This evidence points up a common pattern of personality traits among creative persons and also that these personality factors may have some bearing on creativity in the abstract, regardless of field.

Personality Traits—Young Creatives

One of the questions with which creativity research is concerned regards the similarity of personality traits of young creatives to those of recognized creative adults. Investigations of undergraduates (Drevdahl, 1956; Garwood, 1964; Rees & Goldman, 1961), high school adolescents (Cashdan & Welsh, 1966; Getzels & Jackson, 1962; Holland, 1961; Littlejohn, 1966; Parloff & Datta, 1966) and elementary school children (Torrance, 1962; Weisberg & Springer, 1967) disclose that highly creative students have personality structures that are congruent to—but possibly less sharply delineated than—those of the mature recognized creatives.

Identifying the highly creative by means of self-report questionnaires (Holland, 1961; Rees & Goldman, 1961), ratings (Drevdahl, 1956; Parloff & Datta, 1965), and psychometrically, these investigators submitted essentially similar findings: High creatives, as compared to the low or noncreatives, manifested greater independence, dominance, autonomy, unconventionality, broad interests, and openness to feelings. Consistent with the evidence gathered from adults, emotional instability also appears controlled in the young creatives.

No significant differences on anxious insecurity versus self-confidence, will control, and instability or nervous tension were discovered by Drevdahl (1956) between creative and noncreative undergraduates. Rees and Goldman (1961) concluded that while the highly creative had high scores on the MMPI and Emotional Stability Scale of the Guilford-Zimmerman Temperament Survey, no indication of maladjustment was demonstrated. They suggested that tendencies toward maladjustment may occur in connection with inhibition of creative production. In comparing high and moderate creatives, Parloff and Datta (1965) found that, though they were both below the norms on the sense of well-being and self-control scales of the CPI, it was the latter who displayed greater defensiveness, self-doubts, and irritability.

As measured by the McCandless Anxiety Scale, Reid, King, and Wickwire (1959) found creative children less anxious than the noncreative. Wallach and Kogan (1965), assessing manifest anxiety and test anxiety, reported anxiety at an intermediate level for their two groups high in creativity, regardless of intelligence level.

With respect to interpersonal relations, creative students also resemble eminent adults in that they display similar social poise and adequacy (Cashdan & Welsh, 1966; Garwood, 1964; Parloff & Datta, 1965). Conflicting evidence, however, regarding their sociability appears. Reporting on this dimension, Rivlin (1959) stated that high school students nominated as highly creative by teacher ratings, emerged as sociable, and more popular with their peers. Reid et al. (1959) lend support to this evidence with similar data derived from an investigation of creative seventh graders selected by peer nominations. Additional substantiation was produced by Cashdan and Welsh (1966). Their creative subjects, identified by the Revised Art Scale, appeared to welcome social contact and had more interest in social activities.

By contrast, Lindgren and Lindgren (1965) submitted evidence that creative undergraduates, selected by the Asymmetrical Preference Test, perceived themselves as asocial and "ornery." Holland (1961) described his creative adolescent subjects as asocial. For Getzels and Jackson (1962), the highly creative adolescent portrayed the outsider, the rejected and rejecting spectator, rather than the welcome and committed participant. Torrance (1962, 1964) also observed that creative elementary school students were isolated from their peers in the classroom situation, and seemed less accessible psychologically. This contradictory evidence may be attributed to the different procedures used to define the creative individual. With respect to the studies of Rivlin (1959) and Reid et al. (1959), the ratings by teachers and peers may have been mere indices of sociability rather than creativity ratings.

Generally, the data indicate that the personality characteristics of young creatives bear similarities to those of creative adults, and, therefore, the conclusion seems tenable that these traits develop fairly early. Their manifestation at this level suggests that these characteristics may be determinants of creative performance rather than traits developed in response to recognition of creative behavior. It may be argued that any overall findings in these groups are ambiguous, since where the creative product is often intangible, idiosyncratically determined judgments of creativity may lack validity. However, since these investigations regarding young creatives appear to corroborate one another, the results and conclusions seem much more decisive. Ideally, of course, data regarding these groups should be obtained from longitudinal investigations. Except for Terman's (1954) pioneering study, these are customarily lacking.

Motivational Characteristics

In identifying the creatives, there is also the "why" aspect which focuses on the nature and degree of motivations—conscious and unconscious—assumed to energize these individuals. That this would be a strong component of their nature is apparent, for with no propelling agent, actual production would be minimal or totally lacking. From one point of view, motivation for creativity is seen as a desire to maximize the experiencing of one's own expressive potentials. For Rogers (1959), the creative individual is attempting to realize and complete himself, to "become his potentialities." This view gains some empirical support from Golann's (1962) study of highly creative males, identified on the basis of the RA Scale, who indicated a preference for activities allowing for self-expression, independence, and the use of creative capacity, while low creatives tended to prefer activities essentially opposite.

Torrance and Dauw (1965) probing this motive in high school seniors through expressed aspirations, suggested that their subjects' greater striving and need for excellence, and greater attraction to unusual and unconventional types of achievement, indicated a strong desire to discover and use potentialities. These findings, consistent with Maddi's (1965) contention that the true creative is motivated by an intense need for quality and novelty, may be viewed as relatively concrete expressions of the general tendency toward self-actualization. He has presented empirical findings to buttress the posited relationship between creativity and need for novelty by citing significant correlations between novelty or imaginative productions and a tendency to prefer novel endings in a newly constructed Similes Preference Test.

Houston and Mednick's (1963) findings also appear germane, further supporting Maddi's speculation. Subjects were asked to choose between pairs of words—nouns that were followed by novel associations and non-nouns that were followed by common associations. The high creative group, identified by the Remote Associates Test, significantly increased, and the low creativity group significantly decreased, in the frequency of noun choices over the series of pairs. However, in the light of the questionable validity of the Remote Associates Test, and the lack of validation on the Similes Preference Test, these data must be interpreted with caution.

Risk-taking tendencies based on the need to achieve and to test limits have also been hypothesized as a form of motivation. McClelland (1963) and Kaplan (1963) suggested that creative research scientists have a greater interest in and are more willing to take long range, calculated risks where their own abilities make a difference in the odds. Although these speculations concern adults, similar contentions have been made concerning younger creatives. Getzels and Jackson (1962) and Torrance (1962) cited the asking of unusual questions, the joining together of dissimilar elements, and the choice of unconventional occupations as manifestations of risk taking in their subjects.

A systematic exploration of the relationship between risk taking and creativity is provided in a recent study by Pankove (1967). Defining creativity operationally in terms of two Wallach and Kogan (1965) tests, and risk taking in terms of scores obtained on three decision-making tasks, where subjects were free to choose the degree of risk under which they were willing to operate, she found a positive relationship between risk taking and creativity in fifth-grade boys. Low defensiveness enhanced the relationship while high defensiveness attenuated it.

Within this sphere Anderson and Cropley (1966) investigated the relationship of various nonintellective variables to originality. Identifying high and low creative seventh graders by means of Guilford-Torrance tests, they confirmed their hypothesis that originality was equated with the inability or unwillingness to internalize "stop rules." Rather than the posited impulsivity factor, it was risk taking that contributed most to the total variance of performance on originality. They suggested, therefore, that there was one superordinate "stop rule"—"Don't take risks" which differentiated the creative from the noncreative. The former apparently ignored or were unable to internalize this caveat.

Barron (1963b) stated that the creative individual's "exceptionally strong need to find order where none appears [p. 160]" may be considered a motivating factor. He proposed that the creative willingly admits into his perceptions the complex and disordered, and is challenged to make new order out of the apparent chaos through the use of his own abilities and experiences. The pleasure derived from the solution motivates him to search for situations defying rational construction and, therefore, to service the need to achieve order. Barron (1963a) also posited "the moral attitude" as a motivational characteristic of the creative. By this, he refers to the creative person's profound commitment to aesthetic and philosophical meanings expressed in work. He has maintained that the creator is constantly involved in the creation of his own private universe of meaning, and will "stake his life" on the meaning of his work. Barron believes that "without this cosmological commitment, no amount of mental ability will produce a genuinely creative act [p. 243]."

Various other motivational factors have been hypothesized as characteristic of the creative individual, yet here again, a paucity of empirical supporting data exists for this provocative aspect of creativity. This may be due to the lack of appropriate measuring instruments in this area (suggested by Taylor and Holland, 1964). Or, it may also be attributed to what Maddi (1965) referred to as the "child oriented" view of creativity. In focusing on the predisposing characteristics of the creative person, this viewpoint stresses the degree to which the creative style of functioning demonstrates playfulness, whimsical childishness, humor, relaxation, and uncritical freedom, rather than any sustained motivational qualities. It neglects to consider the long period of purposeful, relentless, organized thought, and motivated persistence generally preceding the creative insight. Creativity, after all, can only be recognized through some form of production or performance, and therefore, as Golann (1963) suggested, would be highly dependent on motivational characteristics. Though other attributes are relevant, they are not sufficient. The positive force of motivation which serves as an impelling and integrating factor would be one of the prime personality requisites for actual productivity.

Summary Statement

With respect to the primary thesis of the present paper, the data suggest that the creative response, performance, or production is most likely to occur in a specific human condition. Despite differences in age, cultural background, area of operation or eminence, a particular constellation of psychological traits emerges consistently in the creative individual, and forms a recognizable schema of the creative personality. This schema indicates that creative persons are distinguished more by interests, attitudes, and drives, than by intellectual abilities. Whether these characteristics are consequences or determinants of creativity, or whether some are peripheral and of no value is moot. These questions remain insufficiently approached and elucidated.

The cognitive capacities that appear to be most frequently associated with the creative are an above-average intelligence and the effective use of this intelligence, the ability to produce unusual and appropriate ideas, an exceptional retention and more ready availability of life experiences, ideational fluency and the ability to synthesize observation, and a general cognitive flexibility.

In the realm of personality, a clearly differentiating factor that characterizes the creative is the relative absence of impulse and imagery control by means of repression. This relative lack of self-defensiveness seems to accord to the creative fuller access to his conscious and unconscious experiences, and therefore, a greater opportunity to combine dissociated items. An openness to internal and external stimuli is also indicated as a salient characteristic, and this is manifested in various forms. It appears in a predisposition to allow into the perceptual system complexity, disorder, and imbalance for the satisfaction and challenge of achieving an idiosyncratic order. It is also demonstrated in the creative male's lack of denial of the feminine side of his nature, which leads to wider interests, a greater openness to emotions and feelings, a greater aesthetic sensitivity, and self-awareness.

Although the creative appears to be subject to considerable psychic turbulence, empirical evidence has shown no basis for a significant and demonstrated relationship between psychopathology and creativity. Rather, it has demonstrated that the creative individual is possessed of superior ego strength and a positive constructive way of reacting to problems. Intuitiveness also comes through as a hallmark of the creative person. He appears to have an immediate apprehension of what is and what can be; and rather than accepting what is presented to his senses at face value, he searches for deeper meanings and eventualities.

Independence in attitudes and social behavior emerges with striking consistency as relevant to creativity. Possessed of an individualistic rather than a sociocentric orientation, the creative is not concerned with social activities, nor preoccupied with the opinion others have of him. Since he is little concerned with the impression he makes on others, he appears to be freer to be himself, to realize his own potentialities, and hence, may actually be a more fully functioning person than his noncreative counterparts. He appears to have a strong sense of identity and self-acceptance, knowing who he is, and what he wants to achieve. Strong aesthetic and theoretical interests also appear consonant with the creative personality, suggesting that the creative person has a high regard for the values of both truth and beauty.

Though the evidence can hardly be called overwhelming, it does seem to indicate that the creator is endowed with strong, intrinsic motivation, involving a degree of resoluteness and egotism that sustains him in his work. Indifferent to the fear of making mistakes, to social disapproval, and the the "anxiety of separateness." he seems endowed with a sense of destiny, an unshakeable belief in the worth and validity of his efforts that could help him override frustrations and obstacles.

In brief, the roots of creativity do not seem to lie on convergent or divergent thinking, but rather, as Hudson (1966) suggested, in the personality and motivational aspects of character.

One of the great hopes of research on the creative person is the possibility that a finite number of personality characteristics is significant for creativity, as distinguished from those having significance for individual diagnosis, theory, or even academic performance. If some small number of parameters can be isolated, and defined in behavioral terms, great use of this might be mobilized for identifying creative potential.

General Critique

Perhaps the most glaring deficit in the research on creativity has been the absence both of replicative studies, as well as follow-up investigations. The one-shot research study is typical. While little status or manifestation of originality may be accorded replication studies, they are a means of assessing reliability of results contributing both to their validity and generalizability. Follow-up investigations also serve this purpose and, certainly, could be of major import in providing support for the validity of psychometric instruments currently employed to "measure" creativity. Paradoxically, the paucity of replication studies simultaneously exists with the plethora of literature concerning one particular investigation that is merely rewritten with essentially the same information, but with a slightly different emphasis. This is at once disappointing, time-consuming, and exasperating.

Conceptual and semantic ambiguities continue to make communication a problem in reporting. The reader, seeking a clear picture of populations, or wishing to relate findings from one investigation with another, experiences particular difficulty when encountering such terms as "creative thinker," "divergent thinker," and "original thinker," insufficiently or inadequately defined. These same difficulties obtain in the use of such terms as "high," "medium," and "low" creativity. Unfortunately, the phrase "moderately creative," generally used, does not conjure in the mind of the reader the relatively clear picture projected by the words "100 IQ."

A host of studies deal with relatively small samples, thereby increasing the possibility of error and biased results. More important, however, in spite of relatively small, atypical samples and limitations in design, such as the arbitrary group classifications used by Getzels and Jackson (1962), Torrance (1962), and Yamamoto (1964b)—which result in a serious loss of information because of the exclusion of subgroups of considerable size and interest (DeMille & Merrifield, 1962)—the inferences read as though applicable far beyond the studied groups. Age, sex, socioeconomic status, test setting, past experience, and other relevant variables have not always been controlled. Adequate rationale for the subjects selected is also often missing. It is not clear, for instance, whether children or adolescents are chosen in the interest of contributing toward a developmental theory of creativity or merely because of convenience and availability.

Precision is also sometimes lacking in reporting results. Some researchers have a tendency to be anecdotal (e.g., Torrance, 1962) which, while making for pleasant reading, tends to obscure the findings. This obscuration also obtains in those cases where results that do not support the hypothesis or may even contradict it are lost in the profusion of qualifications, interpretations, and explanations. Although data in these predominantly correlational studies merely indicate statistical prediction, use of the phrase "significant correlation" often carries the intentional or accidental implication of a causal condition, which, of course, is misleading.

A spate of researchers continue to rely on factor analysis (Vernon, 1964), multiple correlations, and other powerful statistical tools to isolate and identify the creative person, rather than on well-thought-out hypotheses grounded in adequate theoretical rationale. While broad group factors can be subdivided into smaller ones via factor-analytic techniques, this does not verify their validity, nor demonstrate that they are indeed parameters of creativity. With respect to multiple correlations, as Gough (1961) pointed out, some very dramatic relationships can be generated with sufficient data and sample numbers. When obtained with recognized creatives, however, the validity of these correlations cannot be easily ascertained especially in the absence of comparable sample groups.

One might also view a little critically the intentional or accidental overevaluation of creative abilities, or perhaps, more accurately, the devaluation of other abilities. The implication seems to be that creativity is the cardinal seed of existence and, if someone does not achieve creatively, he is of no particular value. In fact, the accusation of creative inferiority seems to have become a sharp form of criticism and everyone, therefore, wants to avoid this pejorative label. As data are released on creativity, and often couched in Olympian terms, one may find a counterpart to the "organization man"—a new pattern upon which everyone, having been "taught the tests" will model himself. Furthermore, the extolling of creativity to the detriment of mental discipline and the mastery of subject matter is inimical to creativity, and may defeat the ultimate purpose of this research (Kneller, 1965).

It is apparent that the criterion problem—fundamental to other aspects of research, yet far from solution—is of primary importance. The characteristics of such a criterion must come to terms with four dimensions: (*a*) the relevance or pertinence to the actual product or culture, (*b*) comprehensiveness, (*c*) the reliability and discrimination, and (*d*) freedom from bias. It is patent that the quality of research will depend upon the adequacy of the criteria utilized, since these are the standards by which other measures can be evaluated, or can serve to describe the performance of individuals on some success continuum. In spite of this importance, research directed specifically toward criteria is infrequent.

Little time or attention has been devoted to the development of the criteria employed (Brogden & Sprecher, 1964).

Studies of creativity in adults have been primarily limited to men, and the manifestation of this dimension in women remains insufficiently elucidated. Helson's (1961, 1966, 1967, 1968) work, one of the few but major contributions in this area, indicates that different factors may be involved in its operation and emergence in women. Powell (1964) noted differences with respect to female creativity and the social desirability variable. Mendelsohn and Griswold (1964, 1966) concluded that the responses of women varied in creativity tests depending on differences in the experimenter. In the interests of a comprehensive portrait of the creative individual, research designed with a view to the deliberate and adequate investigation of male-female differences is warranted.

The necessity of longitudinal studies is obvious for determining qualities that contribute to creative performance, personality changes within the developing creator, and the interaction of personality, cognitive and environmental variables in the creative versus the noncreative. These longitudinal studies would also be basic to any developmental theory of creativity, and would be dependent for their initiation on the ability to identify early—perhaps preschool—those characteristics that may be related to creativity. Instruments that can be used for early identification are needed and these should be, as suggested by Starkweather (1964), not only of inherent interest to children but also, perhaps, independent of ability. Findings from the investigations concerning the beginnings of creativity in children thus far (Lieberman, 1965; Starkweather, 1964; Ward, 1966) remain inconclusive with little of definite substance reported.

The demonstrated lack of convergent validity, and discriminant validity with respect to intelligence, of various measures of creativity, and Yamamoto's (1966) factor-analytic studies of Minnesota test components, in which he found that scores with common semantic labels did not represent the same thing in each test, boldly underscore the need for more straightforward validation studies of measuring instruments to improve currently used predictors. It also appears that there is a necessity to develop creativity measures based on personality study rather than on task performance. The Preconscious Activity Scale of Holland and Baird (1968) is a step in this direction. That biographical items and past achievement have been rated as the most efficient predictors (Taylor & Holland, 1964) does not mean investigators have no further work in this direction. Rather, it does imply a necessity for a conceptualization and systematization of this evidence.

The implication of greatest significance regarding the psychological constitution of the creative individual is that a change in the dynamics of creativity research seems warranted. If the results of future investigations are to become more meaningful contributions to the cumulative literature on creativity, the data suggest that the assessment of creative potential cannot merely rely on singular intellectual traits, factor-analytically derived, but must also include cognitive styles and personality variables rooted in theoretical concepts. Creativity

ativity research, pursued within this framework of compound criteria from disparate psychological domains, holds promise for more valid findings which may, in addition, contribute toward the resolution of present dilemmas.

References

Anderson, C. C., & Cropley, A. J. Some correlates of originality. *Australian Journal of Psychology,* 1966, **18,** 218-229.

Baldwin, A. L. *Theories of child development.* New York: Wiley, 1967.

Barron, F. Complexity-simplicity as a personality dimension. *Journal of Abnormal and Social Psychology,* 1953, **48,** 163-172.

Barron, F. Creative vision and expression in writing and painting. In, *Conference on the creative person.* Berkeley: University of California, Institute of Personality Assessment & Research, 1961.

Barron, F. *Creativity and psychological health: Origins of personality and creative freedom.* Princeton, N. J.: Van Nostrand, 1963. (a)

Barron, F. The needs for order and for disorder as motivation in creative activity. In C. W. Taylor & F. Barron (Eds.), *Scientific creativity: Its recognition and development.* New York: Wiley, 1963. (b)

Barron, F., & Welsh, G. Artistic perception as a possible factor in personality style: Its measurement by a figure preference test. *Journal of Psychology,* 1952, **33,** 199-203.

Beittel, K. R. Creativity in the visual arts in higher education. In C. W. Taylor (Ed.), *Widening horizons in creativity.* New York: Wiley, 1964.

Bloom, B. S. Report on creativity research by the examiner's office of the University of Chicago. In C. W. Taylor & F. Barron (Eds.), *Scientific creativity: Its recognition and development.* New York: Wiley, 1963.

Brogden, H. E., & Sprecher, T. B. Criteria of creativity. In C. W. Taylor (Ed.), *Creativity: Progress and potential.* New York: McGraw-Hill, 1964.

Campbell, D. T., & Fiske, D. W. Convergent and discriminant validation by the multitrait-multimethod matrix. *Psychological Bulletin,* 1959, **56,** 81-105.

Cashdan, S., & Welsh, G. S. Personality correlates of creative potential in talented high school students. *Journal of Personality,* 1966, **34,** 445-455.

Cattell, R. B., & Drevdahl, J. E. A comparison of the personality profile (16 PF) of eminent researchers with that of eminent teachers and administrators, and of the general population. *British Journal of Psychology,* 1955, **56,** 248-261.

Clark, C. M., Veldman, D. J., & Thorpe, J. S. Convergent and divergent thinking abilities of talented adolescents. *Journal of Educational Psychology,* 1965, **56,** 157-163.

Cropley, A. J. A note on the Wallach-Kogan test of creativity. *British Journal of Educational Psychology,* 1965, **38,** 197-201.

Cross, P. G., Cattell, R. B., & Butcher, H. J. The personality patterns of creative artists. *British Journal of Educational Psychology,* 1967, **37,** 292-299.

DeMille, R., & Merrifield, P. R. Review of J. W. Getzels & P. W. Jackson, *Creativity and intelligence: Explorations with gifted students. Educational and Psychological Measurement*, 1962, **22**, 803-808.

Dentler, R. A., & Mackler, B. Originality: Some social and personal determinants. *Behavioral Science*, 1964, **9**, 1-7.

Drevdahl, J. E. Factors of importance for creativity. *Journal of Clinical Psychology*, 1956, **12**, 21-26.

Drevdahl, J. E. Some developmental and environmental factors in creativity. In C. W. Taylor (Ed.), *Widening horizons in creativity*. New York: Wiley, 1964.

Drevdahl, J. E., & Cattell, R. B. Personality and creativity in artists and writers. *Journal of Clinical Psychology*, 1958, **14**, 107-111.

Eiduson, B. T. Artist and non-artist: A comparative study, *Journal of Personality*, 1958, **26**, 13-28.

Eisenman, R., & Robinson, N. Complexity-simplicity, creativity, intelligence and other correlates. *Journal of Psychology*, 1967, **67**, 331-334.

Elliott, J. M. Measuring creative abilities in public relations and in advertising work. In C. W. Taylor (Ed.), *Widening horizons in creativity*. New York: Wiley, 1964.

Ellis, A. Homosexuality and creativity. *Journal of Clinical Psychology*, 1959, **15**, 576-579.

Flescher, I. Anxiety and achievement of intellectually gifted and creatively gifted children. *Journal of Psychology*, 1963, **56**, 251-268.

Garwood, D. S. Personality factors related to creativity in young scientists. *Journal of Abnormal and Social Psychology*, 1964, **68**, 413-419.

Getzels, J. W., & Jackson, P. W. *Creativity and intelligence: Explorations with gifted students.* New York: Wiley, 1962.

Ginsberg, G. P., & Whittemore, R. G. Creativity and verbal ability: A direct examination of their relationship. *British Journal of Educational Psychology*, 1968, **38**, 133-139.

Golann, S. E. The creativity motive. *Journal of Personality*, 1962, **30**, 588-600.

Golann, S. E. Psychological study of creativity. *Psychological Bulletin*, 1963, **60**, 548-565.

Gough, H. G. Techniques for identifying the creative research scientist. In, *Conference on the creative person*. Berkeley: University of California, Institute of Personality Asessment and Research, 1961.

Guilford, J. P. Measurement and creativity. *Theory into Practice*, 1966, **5**, 186-189.

Guilford, J. P. Creativity: Yesterday, today, and tomorrow. *Journal of Creative Behavior*, 1967, **1**, 3-14. (a)

Guilford, J. P. *The nature of human intelligence.* New York: McGraw-Hill, 1967. (b)

Guilford, J. P., & Hoepfner, R. Sixteen divergent-production abilities at the ninth-grade level. *Multivariate Behavioral Research*, 1966, **1**, 43-64.

Guilford, J. P., & Merrifield, P. R. The structure of intellect model: Its uses and implications. *Report of Psychological Laboratory,* No. 24. Los Angeles: University of Southern California, 1960.

Hammer, E. F. Creativity and feminine ingredients in young male artists. *Perceptual and Motor Skills,* 1964, **19**, 414.

Harris, T. L. An analysis of the responses made by adolescents to the Welsh Figure Preference Test and its implications for guidance purposes. Unpublished doctoral dissertation, University of North Carolina, 1961.

Heinze, S. J. Job adaptation and creativity in industrial research scientists. Unpublished doctoral dissertation. University of Chicago, 1962.

Helson, R. Creativity, sex, and mathematics. In, *Conference on the creative person.* Berkeley: University of California, Institute of Personality and Assessment Research. 1961.

Helson, R. Personality of women with imaginative and artistic interests: The role of masculinity, originality and other characteristics in their creativity. *Journal of Personality,* 1966, **34**, 1-25.

Helson, R. Sex differences in creative style. *Journal of Personality,* 1967, **35**, 214-233.

Helson, R. Generality of sex differences in creative style. *Journal of Personality,* 1968, **36**, 33-48.

Herr, E. L., Moore, G. D., & Hansen, J. C. Creativity, intelligence and values: A study of relationships. *Exceptional Children,* 1965, **32**, 114-115.

Hersch, C. The cognitive functioning of the creative person: A developmental analysis. *Journal of Projective Techniques,* 1962, **26**, 193-200.

Holland, J. L., & Baird, L. L. The preconscious activity scale: The development *Journal of Educational Psychology,* 1961, **52**, 136-147.

Holland, J.sL., & Baird, L. L. The preconscious acitivity scale: The development and validation of an originality measure. *The Journal of Creative Behavior,* 1968, **2**, 217-225.

Holt, R. R., & Havel, J. A method for assessing primary and secondary process in the Rorschach. In M. A. Rickers-Ovsiankina (Ed.), *Rorschach psychology.* New York: Wiley, 1960.

Houston, J. P., & Mednick, S. A. Creativity and the need for novelty. *Journal of Abnormal and Social Psychology,* 1963, **66**, 137-141.

Hudson, L. *Contrary imaginations: A psychological study of the young student.* New York: Schocken Books, 1966.

Jackson, P. W., & Messick, S. The person, the product and the response: Conceptual problems in the assessment of creativity. *Journal of Personality,* 1965, **33**, 309-329.

Jacobson, L. I., Elenewski, J. J., Lordahl, D. S. & Lieroff, J. H. Role of creativity and intelligence in conceptualization. *Journal of Personality and Social Psychology,* 1968, **10**, 431-436.

Kaplan, N. The relation of creativity to sociological variables in research organizations. In C. W. Taylor & F. Barron (Eds.), *Scientific creativity: Its recognition and development.* New York: Wiley, 1963.

Kneller, G. F. *The art and science of creativity.* New York: Holt, Rinehart and Winston, 1965.

Lauritzen, E. S. Semantic divergent thinking factors among elementary school children. *Dissertation Abstracts,* 1963, **24,** 629.

Lieberman, J. N. Playfulness and divergent thinking: An investigation of their relationship at the kindergarten level. *Journal of Genetic Psychology,* 1965, **107,** 219-224.

Lindgren, H. C., & Lindgren, F. Brainstorming and orneriness as facilitators of creativity. *Psychological Reports,* 1965, **16,** 577-583.

Littlejohn, M. A. A comparison of responses of ninth-graders to measures of creativity and masculinity-femininity. Unpublished doctoral dissertation, University of North Carolina, 1966.

Lowenfeld, V., & Beittel, K. Interdisciplinary criteria of creativity in the arts and sciences: A progress report. *Research Yearbook, National Art Education Association,* 1959, 35-44.

MacKinnon, D. W. The study of creativity and creativity in architects. In, *Conference on the creative person.* Berkeley: University of California, Institute of Personality Assessment and Research, 1961.

MacKinnon, D. W. The creativity of architects. In C. W. Taylor (Ed.), *Widening horizons in creativity.* New York: Wiley, 1964.

Maddi, S. R. Motivational aspects of creativity. *Journal of Personality,* 1965, **33,** 330-347.

Marsh, R. W. A statistical reanalysis of Getzels and Jackson's data. *British Journal of Educational Psychology,* 1964, **34,** 91-93.

McClelland, D. C. The calculated risk: An aspect of scientific performance. In C. W. Taylor & F. Barron (Eds.), *Scientific creativity: Its recognition and development.* New York: Wiley, 1963.

McNemar, Q. Lost: Our intelligence. Why? *American Psychologist,* 1964, **19,** 871-882.

Mednick, M. T., & Andrews, F. M. Creative thinking and level of intelligence. *The Journal of Creative Behavior,* 1967, **1,** 428-431.

Mednick, S. A. The associative basis of the creative process. *Psychological Review,* 1962, **69,** 220-232.

Mendelsohn, G. A., & Griswold, B. B. Differential use of incidental stimuli in problem solving as a function of creativity. *Journal of Abnormal and Social Psychology,* 1964. **68,** 431-436.

Mendelsohn, G. A., & Griswold, B. B. Assessed creative potential, vocabulary level and sex as predictors in the use of incidental cues in verbal problem solving. *Journal of Personality and Social Psychology,* 1966, **4,** 423-433.

Merrifield, P. R., Gardner, S. F., & Cox, A. B. Aptitudes and personality measures related to creativity in seventh-grade children. *Report of the*

Psychological Laboratory, University of Southern California, 1964, No. 28. (Los Angeles)

Merrifield, P. R., Guilford, J. P., & Gershon, A. The differentiation of divergent-production abilities at the sixth-grade level. *Report of the Psychological Laboratory,* University of Southern California, 1963, No. 27. (Los Angeles) Cited by S. W. Brown, Semantic memory and creative (divergent-production) abilities of senior high school students. Unpublished doctoral dissertation, University of Southern California, 1966.

Myden, W. Interpretation and evaluation of certain personality characteristics involved in creative production. *Perceptual and Motor Skills,* 1959, 9, 139-158.

Pankove, E. The relationship between creativity and risk taking in fifth-grade children. Unpublished doctoral dissertation, Rutgers State University, 1967.

Parloff, M. D., & Datta, L. E. Personality characteristics of the potentially creative scientist. *Science and Psychoanalysis,* 1965, 8, 91-106.

Piers, E. V., Daniels, J. M., & Quackenbush, J. F. The identification of creativity in adolescents. *Journal of Educational Psychology,* 1960, 51, 346-351.

Pine, F., & Holt, R. R. Creativity and primary process: A study of adaptive regression. *Journal of Abnormal and Social Psychology,* 1960, 61, 370-379.

Powell, M. B. The social desirability aspect of self concept in relation to achievement and creativity. Unpublished doctoral dissertation, New York University, 1964.

Propst, B. S. Openness to experience and originality of production. Unpublished master's thesis, University of Chicago, 1962.

Raychaudhuri, M. Creativity and personality: a review of psychological researches. *Indian Psychological Review,* 1966, 2, 91-102. (a)

Raychaudhuri, M. Perceptual preference pattern and creativity. *Indian Journal of Applied Psychology,* 1966, 3, 67-70. (b)

Raychaudhuri, M. *Studies in artistic creativity: Personality structure of the musician.* Calcutta: Express Printer Private, Ltd., 1966. (c)

Rees, M. E., & Goldman, M. Some relationships between creativity and personality. *Journal of General Psychology,* 1961, 65, 145-161.

Reid, J. B., King, F. J., & Wickwire, P. Cognitive and other personality characteristics of creative children. *Psychological Reports,* 1959, 5, 729-737.

Rivlin, L. Creativity and self-attitudes and sociability of high school students. *Journal of Educational Psychology,* 1959, 50, 147-152.

Roe, A. *The making of a scientist.* New York: Dodd, Mead, 1952.

Rogalsky, M. M. Artistic creativity and adaptive regression in third-grade children. *Journal of Projective Techniques and Personality Assessment,* 1968, 32, 53-62.

Rogers, C. R. Toward a theory of creativity. In H. H. Anderson (Ed.), *Creativity and its cultivation.* New York: Harper & Row, 1959.

Rosen, J. C. The Barron-Welsh art scale as a predictor of originality and level of ability among artists. *Journal of Applied Psychology,* 1955, 39, 366-367.

Skager, R. W., Klein, S. P., & Schultz, C. B. The prediction of academic and artistic achievement at a school of design. *Journal of Educational Measurements,* 1967, **4**, 105-117.

Starkweather, E. K. *Conformity and nonconformity as indicators of creativity in preschool children.* (Tech. Rep. No. 1967) Stillwater: Oklahoma State University, 1964.

Stein, M. I. A transactional approach to creativity. In C. W. Taylor (Ed.), *The 1955 University of Utah research conference on the identification of creative scientific talent.* Salt Lake City: University of Utah Press, 1956.

Taylor, C. W., & Holland, J. Predictors of creative performance. In C. W. Taylor (Ed.), *Creativity: Progress and potential.* New York: McGraw-Hill, 1964.

Terman, L. M. The discovery and encouragement of exceptional talent. *American Psychologist,* 1954, **9**, 221-230.

Thorndike, R. L. Some methodological issues in the study of creativity. In A. Anastasi (Ed.), *Testing problems in perspective.* Washington, D.C.: American Council on Education, 1966.

Torrance, E. P. *Guiding creative talent.* Englewood Cliffs, N. J.: Prentice-Hall, 1962.

Torrance, E. P. *Role of evaluation in creative thinking.* Minneapolis: University of Minnesota, Bureau of Educational Research, 1964.

Torrance, E. P., & Dauw, D. C. Aspirations and dreams of three groups of creatively gifted high school seniors and comparable unselected group. *Gifted Child Quarterly,* 1965, **9**, 177-182.

Vernon, P. E. Creativity and intelligence. *Educational Research,* 1964, **6**, 163-169.

Vinacke, W. E. *The psychology of thinking.* New York: McGraw-Hill, 1952.

Wallach, M. A. Review of E. P. Torrance, *Torrance tests of creative thinking. American Educational Research Journal,* 1968, **5**, 272-281.

Wallach, M. A., & Kogan, N. *Modes of thinking in young children: A study of the creativity-intelligence distinction.* New York: Holt, Rinehart and Winston, 1965.

Ward, C. W. Creativity and impulsivity in kindergarten children. *Dissertation Abstracts,* 1966, **27**, 2127B.

Welsh, G. S. Comparison of D-48, Terman CMT, and Art Scale scores of gifted adolescents. *Journal of Consulting Psychology,* 1966, **30**, 88.

Wild, C. Creativity and adaptive regression. *Journal of Personality and Social Psychology,* 1965, **2**, 161-169.

Wilson, R. C., Guilford, J. P., Christensen, P. R., & Lewis, D. J. A factor-analytic study of creative-thinking abilities. *Psychometrika,* 1954, **19**, 297-311.

Yamamoto, K. Evaluation of some creativity measures in a high school with peer nominations as criteria. *Journal of Psychology,* 1964, **58**, 285-293. (a)

Yamamoto, K. Role of creative thinking and intelligence in high school achievement. *Psychological Reports,* 1964, **14**, 783-789. (b)

Yamamoto, K. An exploratory component analysis of the Minnesota tests of creative thinking. *California Journal of Educational Research*, 1966, **17**, 220-229.

Recommended Additional Readings

Albert, R. S. Genius: Present-day status of the concept and its implications for the study of creativity and giftedness. *American Psychologist*, 1969, **24**, 743-753.

Arasteh, J. D. Creativity and related processes in the young child: A review of the literature. *Journal of Genetic Psychology*, 1968, **112**, 77-108.

Golann, S. E. The psychological study of creativity. *Psychological Bulletin*, 1963, **60**, 548-565.

Hudson, L. *Frames of mind.* New York: Norton, 1968.

Price, A. D. Effect of role-inducing instructions on performance on a new test of creative thinking. *Psychological Reports,* 1970, **27**, 919-924.

Wallach, M. A. Creativity. In P. H. Mussen (Ed.), *Carmichael's manual of child psychology.* (3rd ed.) Vol. 1. New York: Wiley, 1970. Pp. 1211-1272.

School Failures: Now and Tomorrow

Stephen J. Fitzsimmons
Julia Cheever
Emily Leonard
Diane Macunovich

Orientation

Reprinted from *Developmental Psychology*, 1969, 1, 134-146, by permission of the authors and the American Psychological Association. This study was conducted as part of a contract between Abt Associates, Inc. and the United States Office of Education (OEC-1-6-001681-1681). The United States Office of Education provides, under Title One of the Elementary and Secondary Education Act of 1965, "financial assistance to local educational agencies for the education of children of low-income families," But the United States Office of Education does not have unlimited funds for all programs which might improve the quality of this type of education and so needed a measure of the relative cost-effectiveness of alternate Title One programs placed in specific communities. The authors wish to express their sincere appreciation to Clark C. Abt for his many thoughtful suggestions.

In studying the vast bulk of literature on children and youth in school it is possible to acquire a distorted view of the "typical" student. This is so because most of this research has concerned only the academically talented, college-bound student. Perhaps this preponderance of concern with this advantaged group reflects our society's infatuation with success. In the minds of most Americans, high academic achievement and aspirations for college are prerequisites to success. But what about the vast numbers of individuals who do not want (or do not have the opportunity) to go on to the university, who do not do well in traditional academic activities in school, and who drop out of school altogether at the earliest possible moment as defined by law? What do we know about this group? Are their lives reflected in the existing literature on human development?

What do we do for these people to help them develop satisfying lives? It would seem that schools and society abandon concern for these people, apparently for no more reason than that they have chosen (or were forced) not to play the academic/success game any longer. Probably most school districts provide certain "vocational" or "technical" offerings in their curricula that are designed to benefit non-college-bound adolescents. However, upon close examination it is clear that these programs are of secondary importance; this is reflected in budgetary and space allocations, the typical manner by which students are advised to take (or stay away from) these courses, and the prevailing derogatory attitudes in schools about these classes.

Another unfortunate source of stress upon these nonacademically-oriented students is their own parents. In all likelihood parents desire very strongly for their children to succeed in school and to go on to higher study. All levels of society are socialized into the success orientation; however, many are not allowed to achieve it. Thus, the individual is caught between his parents' desires for him and the reality of his own desires or abilities. If and when he does take the ultimate step and leaves school, he immediately is labeled a societal failure. His failure status is continually reinforced in many ways—inability to get a job, reminders from the mass media of how many fewer thousands of dollars he will earn in his lifetime compared to the high-school graduate (not to mention the college graduate), and perhaps even subtle social scorn.

Again, it is important to remind the student about the generality of the research findings he reads. Do these pertain to all persons in our society? What about those out of the mainstream?

The following article by Fitzgerald, Cheever, Leonard, and Macunovich is a report on a large-scale study of students who have already or are likely to drop out of school. Their findings provide insight into the prediction of which students will become "school failures."

Stephen J. Fitzsimmons, Julia Cheever, Emily Leonard, and Diane Macunovich are associated with Abt Associates Incorporated of Cambridge, Massachusetts. Fitzsimmons is vice president of social science research and development of that organization.

23

The problems posed to contemporary American society by both high school dropouts and "low quality" high school graduates have been a subject of major national concern. Books,* magazines and journals, mass media coverage, and recent congressional action,† all indicate growing public awareness and determination to cope effectively with the dropout phenomenon. Cervantes (1965) estimates the dropout rate among United States high school students to be between 30% and 40%. This estimate, if correct, means that more than 7½ million students will fail to complete high school in the 1960s, a figure which is both actually and proportionately greater than in the depression years of the 1930s. It does not even consider the problem of students whose performance records are poor but who still manage to graduate. It is also known that the greatest dropout rate exists among minority groups which can least afford to be educationally deprived.

These dropout statistics are important because of the social and personal problems associated with being a dropout in our society. It is known that crime, drug addiction, high unemployment, illegitimacy, welfare dependency, and alienation from the community exist in a disproportionate segment of this population. Lack of self-esteem and failure to participate in the social and political development of the community are further corollaries. When the direct cost of these problems to the public (police, health, welfare, etc.) is combined with the cost-savings opportunities lost through failure to realize human potential (economic, social, and personal), the impact on society is enormous. While it cannot be argued that dropouts and poor performance *cause* these future problems, failure to complete high school successfully inevitably interacts with the critical factors which caused the dropout in the first place. Thus, it seems reasonable to assume that reversal of this trend could have many desirable consequences.

It has long been recognized that one characteristic of the high school dropout is a history of poor performance, probably beginning early in elementary school. While common sense suggests that there would be numerous published studies of individual student differences over the years, such has not been the case. In part, this may be due to the diversity of student populations in this country, and to different school policies, community characteristics, and different course curricula. It is also possible that a number of studies have been conducted which did not yield significant results and as a consequence were not published. To date, the best analysis of longitudinal research known to the authors has been developed by Bloom (1965).

This study* sought to determine the nature of the dropout's scholastic behavior, as opposed to that of the poorly performing graduate, and to determine quantitatively the relationship of variables in *two* areas:

―――――――
*Abt Associates, Inc. of Cambridge, Massachusetts, developed a computer planning model for estimating the relative cost-effectiveness of alternate Title One programs placed in specific communities. It was assumed that different communities (possessing different socioeconomic characteristics), while frequently having similar problems or symptoms, were in need of different types of programs; since the causes of their problems were different, the solutions might be different. A computer model, serving as a planning aid, had to be able to integrate critical descriptors of a community, its school system, and the instructional process on the one hand, with various programs (e.g., remedial reading counseling, vocational training, new facilities) on the other. Such bringing together of data, if well programmed, would suggest probable outcomes, both qualitatively (better students, better community, etc.) and quantitatively (more graduates, more jobs, etc.). The present study made the necessary predictions about both the quality of graduates and the number of graduates (as opposed to dropouts). Actual records of student performance from which initial computer weightings could be determined were analyzed. Later experience with the consequences of specific Title One programs will provide refinement and validation of these weightings.

1. Did certain patterns exist with respect to poor performance in elementary and secondary school? A basic concern was the subject-grade interdependence. This question focused on the quality of education, regardless of whether a particular student graduated or failed to complete his education. The question also implied that certain patterns of early failure had "downstream" effects which could be analyzed and from which probability indexes could be derived. Such effects might have been subject specific, or have fallen into stages (e.g., seventh grade, ninth grade, etc.).

2. Were there certain patterns of performance which distinguished the dropout from his peers who, despite poor performance, managed to graduate? Here, it was recognized that while external conditions might be an important factor in dropping out of school, that dropouts might also show different academic patterns corresponding to their ultimate failure to complete school. Cervantes (1965) lists 20 characteristics which correspond to dropping out of school. These are broken down into areas of school, family, peer, and psychological test indexes. Cervantes notes that by the seventh grade, there is a depressed performance in math and reading (by 2 years); the majority of letter grades are poor; many have failed at least one grade in school; attendance is poor; students are frequently "underachievers"; there are frequent changes in school and many behavior problems. These were all acknowledged as being important, but the authors felt that it would be interesting to know more about both patterns of failure, and factors which distinguished the dropout from his peers who performed poorly but still managed to graduate.

Quantitative analysis of data on students regarding the above two question areas (covering their elementary and secondary school careers) provided the basis for initial predictions of the consequences of new educational programs.

Keeping in mind the above two considerations (patterns of poor performance and significant distinguishing factors which might identify dropouts) the following questions were of primary importance, with respect to the onset of initial failures: (*a*) How early and in what subjects did initial failures occur among students who did poorly later in their academic careers? (*b*) What were the dominant patterns of failure in terms of grade of onset and by subject of origin? With respect to all failures: (*c*) What were the patterns of failure among graduates and dropouts by year? (*d*) What were the patterns of failure among graduates and dropouts by subject? (*e*) What were the patterns of failure among graduates and dropouts by subject and year? (*f*) In what grade did most dropouts occur? With respect to subject-grade failure interdependencies: (*g*) What were the interrelationships between early performance and midpoint performance? (*h*) What were the interrelationships between single- and multiple-course failures at the seventh-grade level and later graduation? What were the educational policy implications of the above findings?

Method

The data collected for this study were obtained in a suburban New England community of approximately 90,000 people. The median family income was $9,000. Students who were in attendance in the public school system came predominately from middle- to upper-middle-class homes, with a limited segment (perhaps 5%) coming from relatively disadvantaged families (as defined by ethnicity, family education level, and income). Elementary schools served the first six grades; junior high schools served the seventh through the ninth grades; and the high school, the tenth through twelfth grades. The majority of students attending the public high schools went on to further education.

Data were collected on a total of 270 students, drawn from a parent population of approximately 2,500 students in attendance in the high school who graduated (or were scheduled to graduate) during the years 1966, 1965, 1964, and in a few cases, 1963. All students in this sample had attended high school at least long enough to have a record (entered upon the student's initiation of study at the school). The basic criterion of inclusion in the sample was serious performance difficulty during high school attendance or dropping out of school. *Poor performance* was defined as earning at least three Ds or Fs during high school career. Where students had only one or two failures, the grades were considered as "incidental" failures and were not included. Furthermore, data were gathered for only those students with histories which could be traced back through early elementary school. The 270 subjects in the sample included all students who classified as poor performers and for whom records were available back through early elementary school.

A common problem in analyzing data of this type is the "self-selective" nature of the sample; that is, the very students of greatest interest to the study are those most likely to have incomplete or sketchy records, due to moving around from one school to another, dropping back a grade, dropping out of school, etc. In contrast, students who may serve as comparison groups tend to be more stable in these respects. In essence the selection factor makes generalization somewhat more difficult. Certain shortcomings in the data were noted and served to place constraints on the generalizations drawn following analysis of data.

Limitations in generalization of results were imposed by the following considerations:

1. Students who performed poorly in early years of school, but who managed to improve their performance by high school, were not included in this study. Thus, there was no way to analyze the differences in performance patterns of students who remained as poor performers, in contrast to those who improved. Data collected in another survey by Abt Associates on achievement tests in Iowa suggest that students who initially

perform poorly and then improve markedly in later years are in the minority.

2. Students who performed poorly in high school, but whose records had incomplete data for earlier years were not included. We have no reason to believe that they would have records "worse" than those included in the actual sample.

3. In some cases, students classified as "dropouts" may have completed their course work later and elsewhere without informing school officials. This is considered to be unlikely in most cases, however, since requests for records are necessary in cases of transfer and this is noted in the school records.

4. There is a tendency for high school students doing poorly to take less course work in the academic areas. Therefore, it was not assumed that the proportion of students failing English could be taken as comparable data to the proportion of students failing science (since the actual number of enrollees was different).

5. Finally, the data of elementary school records presented some problems. Many student records for elementary school had only limited data in the first through third grades (and in some cases, data were totally absent). In addition, grade-school teachers seemed to be less likely to differentiate performance abilities than were teachers in later grades. Grading was also done on a more objective basis in the later school years. Furthermore, while English and math had an "objective" scoring system for Grades 1 through 3, the same was not true of science and social studies. Because of these problems, considerable care has been taken in drawing conclusions about the exact nature of failure problems in the first and second grades.

Results

How early and in what subjects did initial failures occur among students who did poorly later in their academic careers? The uncorrected distribution of incidents of initial failure (a conservative measure) showed clearly that among high school students who did poorly the vast majority could have been identified early in their academic careers (see Table 1). Of these students, 3 out of 4 had already demonstrated poor performance in the fourth grade, and by the seventh grade, 9 out of 10 had performed poorly. The data indicate that by the time students passed through junior high school, 97 out of 100 students who received low marks had a previous record of failure. Clearly, then, among the high school students who did poorly, the vast majority could have been identified early in their academic careers (in elementary or junior high school).

The limitation of data from the first and second grades means that if anything, these estimates are conservative. In order to correct this distortion, the sample population was divided into subsamples by incidence of first year of data

Table 1
Point of Failure(s), by Subject and Year in School for 270 Poorly Performing High School Students

Subject	1 S	1 M[a]	2 S	2 M	3 S	3 M	4 S	4 M	5 S	5 M	6 S	6 M	7 S	7 M	8 S	8 M	9 S	9 M	10 S	10 M	11 S	11 M
English																						
Reading	7	10	7	42	5	22	2	17	1	5		3		6			1	2				
Oral/written	3	7	5	22	2	26	2	19		4		1					1					
Spelling		5	4	34	3	20	4	15	1	4		3										
Mathematics	2	8	6	26	3	11	7	15	1	4		3	1	2	2	4	2	4		3		
Science		3	1			6		6		2		4	2	9		5		1		4		
Social studies	1			2		10		11		2		5	1	5	1	4		1		2		1
Miscellaneous				2			3				1	4	2	6	2	5	1	1		4		1
Total single: 98	13		23		13		15		6		4		12	3	6	1	5		1			
Cumulative percentage	13%		36%		50%		65%		71%		75%		88%		94%		99%		100%			
Total multiple: 457		33		28		95		83		21		20		28		19		10		18		2
Cumulative percentage[b]		6%		35%		56%		74%		78%		83%		89%		94%		95%		99%		100%
Cumulative failures M & S			53%		61%		75%															
Corrected for data 14[c]	33%																					

[a] S = single initial failure; M = multiple initial failures.
[b] It is assumed that the number of initial multiple failures, by student, is roughly comparable across the years. Thus, the cumulative percentages represent the incidents of failures rather than the percentage of subjects.
[c] See text discussion of "adjusted" failure incidence.

from Grades 1 through 5. For each subsample, the "adjusted" first failure data were assumed to begin the year after the data actually started. From the adjusted data cells, mean first failure occurrences and their standard deviation were determined. These figures were used to establish a probable range of first failure occurrences for the students for whom data were not available. In Grades 1 and 2 the number of occurrences was approximated using the Grade 1 to Grade 2 proportion from the more limited sample for which there were data. Not surprisingly, the adjustment did *not* change the cumulative percentage for the first four grades; that is, a total of 75 % of the sample showed initial failures by the end of their fourth year. However, the adjustment did indicate that Grades 3 and 4 were too heavily biased in the unadjusted data to give the true year-by-year picture. The adjusted data (see Table 1) indicate that in fact slightly over 50% of the total initial failures had appeared by the end of the second grade (while the unadjusted data indicate that this point was not attained until midway through the third grade). This adjustment strengthens the case that performance difficulties appear quite early in the school career.

It is important to note the subject areas which provided the greatest difficulties. The critical performance difficulties fell in the English areas (reading, oral and written usage, and spelling, in that order) and in mathematics. These findings were somewhat biased, for teachers did not give letter grades to the sciences and social studies until the fourth grade. However, these latter two subjects failed to show up as important areas of initial failure at any point (even beyond the third grade, where objective grade scoring was available). Although these subjects were not the source of initial difficulty, this should not be confused with an absence of failures in these subjects. Many students did poorly in social studies and sciences, but they were not their areas of initial failure. This distinction will be discussed later in the paper in sections dealing with downstream effects of early failures.

The raw data were analyzed to determine if the percentage of onset failures differed for graduates and dropouts. The data revealed a clear but limited difference between these two groups. In 21 out of 24 cells representing the first four grades for six subject areas graduates exceeded dropouts in the proportion of initial failures recorded. Dropouts as a group tended to begin failing subjects somewhat later in their academic careers than did poorly performing graduates. (It should be remembered that "graduate" refers to students who completed school with very poor performance records, not to the entire student body.) This finding is important and will be dealt with further in the discussion of results.

What were the dominant patterns of failure in terms of grade of onset, and by subject of origin? It is reasonable to assume that not all children "fail" in the same manner. Some start to fail early in their careers, others do not fail until later on. Some begin having difficulty only in one subject, and still others in a number of areas simultaneously. Furthermore, some begin with many failures and improve themselves. Because of such individual differences, it was hypothesized that over a large sample of subjects, various "patterns" of failure would emerge. These included spread patterns—early poor performance in one or two areas, which later expanded to include other subjects (i.e., the long shadow of achievement gaps); parallel patterns—a pattern of failure which began in three or more areas and continued in those areas; hourglass patterns—early failure in three or more areas followed by improvement and then return to failures in many areas; late failing patterns—onset of failure which did not occur until the sixth grade; and random patterns—no apparent pattern.

No "converging" pattern of reduced failures is included here because such students, by virtue of the selection procedure, were excluded from the sample.

Data separated into these five categories (which contrasted dropouts to graduates) are shown in Table 2.

Table 2
Analysis of Poor Performance According to Pattern of Failure

Pattern	Total		Dropouts		Graduates	
	No.	Percentage	No.	Percentage	No.	Percentage
Spread	108	40	29	40	79	40
Parallel	48	18	10	14	38	19
Hourglass	20	7	5	7	15	8
Late failing	57	21	18	25	39	20
Random or incomplete	37	13	10	14	27	14

Note:
$N = 270$.

Of the 270 records analyzed, 108 were found to have a spread pattern of poor performance. Collectively, this represented 40 % of the total sample, the largest single grouping. This spread pattern was thought to be particularly significant, since even limited failures can be used as "early warning signs."*

With this in mind, the data were then examined to determine the areas which most typically were the origin points of the spread pattern. Analysis of the origins of both single and double area spread patterns revealed that English, and particularly, reading, was the most frequent origin of failures. This was consistent with the data regarding the onset of failure for all students. Beyond English, only mathematics showed any sign of being an important indicator of future spread, and this was usually when in combination with English. The data

*There may be a number of children who do poorly in one or two areas, but then pull
*There may be a number of children who do poorly in one or two areas, but then pull themselves up (these would be absent from this sample).

demonstrated that by the third grade, two-thirds of the students who would have developed spread patterns in future years had their first failure, and these had their origins in English.*

Examination of the distribution of initial failures among dropouts as opposed to poorly performing graduates revealed no interesting differences.

What were the patterns of failure among graduates and dropouts by year? The total number of failures, when corrected for intermittent data in the first 2 years, and for the dropout occurrences in the last 3 years, did not suggest any significant "high points of failure," as for example, junior high school or high school. The occurrence of failure started early, rose rapidly, and then increased

Table 3
Distribution of Failures by Year for Graduates and Dropouts

Percentage of poor performance

Grade	Graduates		Dropouts		Total
1	.7	2.8[a]	.1	4.5	.8
2	3.6	5.7	3.3	5.5	3.5
3	6.2	7.5	5.7	6.4	6.0
4	9.2		9.7		9.3
5	10.2		10.7		9.8
6	7.4		9.7		10.1
7	9.3		10.2		8.2
8	9.3		11.1		9.8
9	12.4		13.7		12.8
10	10.5		13.4		11.3
11	11.8		8.2		10.8
12	9.3		3.2		7.6
Total	99.9		99.0		100%

Note:
$N = 270$.
[a] Approximations based on a full sample size population hypothesized for the first three grades.

slowly over the years (the increase for dropouts being more pronounced than that of the graduates). However, there seemed to be a greater proportion of poor performance among dropouts in the ninth and tenth grades, suggesting that as students came closer to dropping out of school, they dropped further in their performance. (see Table 3).

A comparison of dropouts to graduates suggested that year by year, the

*Specifically, English accounted for almost 70 percent of the single origin spreads (with math making up the remainder), and regarding double area initial failures, English was involved in every case. Reading in combination with other subjects was the most significant origin point of spread.

graduates did relatively better than the dropouts, and that dropouts, starting at about the fourth grade, did progressively worse each year.

What were the patterns of failure among graduates and dropouts by subject? Analysis of the data showed that English was by far the most significant area with almost 30% of the total failures. Mathematics and social studies each accounted for 18% of the failures. Languages (15%) and science (13%) made up the other important areas. Significantly, shop and business, even though started in the seventh grade (rather than in the first), still did not show many failures in this sample. Thus, despite the fact that many of these students were not "college bound," the academic areas still dominated their failure areas. The problem areas of dropouts and graduates were similar. The general finding was that pupils with poor performance records had the greatest difficulty with English, and that they seemed to fail other academic subjects somewhat uniformly.

While the distribution of failures for graduates and dropouts across subjects was similar, dropouts tended to perform more poorly than graduates in each subject. This was particularly true later in their academic career, starting clearly after the sixth grade.

What were the patterns of failure among graduates and dropouts by subject and year? Figure 1 presents a year-by-year analysis of the performance patterns for graduates and dropouts, for English, mathematics, science, and social studies. For Grades 1 through 3, the graduates received a higher percentage of low marks than the dropouts in both English and mathematics; in science and social studies the pattern was very similar. In grades 4 through 6, the percentages diverged, with the dropouts having a greater number of failures in every subject. This difference was lost at the sixth grade, and it was at this point that the graduates tended to level off somewhat, while the dropouts began to increase progressively in their number of failures. In every subject (except math, 10-12) from Grades 7 through 12, the graduates had a substantially lower percentage of poor marks over the last 5 years. It is clear that the differences, while not large, consistently favored the graduates in terms of better performance.

Two more specific analyses of these data were possible. First, were there differences among the various subgroupings of courses for the first 3 years? Second, were there differences in the F to D mark ratios of the two groups for the last 7 years? With respect to the first three grades, no significant differences were found between graduates as compared and dropouts in English (spelling reading, or oral and written usage) and in the various combinations of English, math, social studies, and science. Second, analysis of the ratio of Ds and Fs for dropouts versus graduates indicated that there was no noticeable difference in patterns. Both patterns showed a gradually increasing percentage of Fs, with the dropouts having a greater slope. However, in terms of the actual ratio of Ds and Fs, dropouts did have a larger percentage of Fs than graduates.

In what grade did dropouts leave school? Of the 270 students in the sample, 72 dropped out for various reasons. The majority of dropouts (56%) were boys. A general overview of the dropout rate in this sample indicated a fairly even

Figure 1.
Percentage failing by subject by year for dropouts and graduates.

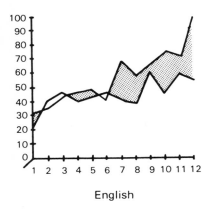
English

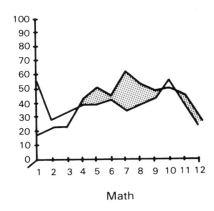
Math

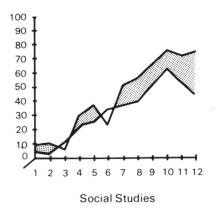
Social Studies

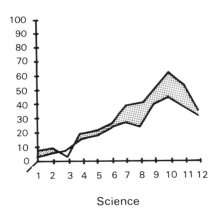
Science

attrition rate (see Table 4)—29% in the tenth grade, 36% in the eleventh grade, and 31% in the twelfth grade. (The 4% dropout figure in the ninth grade was disregarded here because few ninth grade students were old enough to drop out in accordance with State law.) However, such an overview hid some interesting distinctions between boys and girls in their dropout pattern. Boys tended to drop out at a fairly even rate from tenth grade through the twelfth grade. In contrast, the girls' dropout rate was much more varied during the four periods, with a high of 20 percent leaving during the twelfth-grade school year.

The reasons given for dropping out in some of the records may help explain this. Boys were most likely to drop out because of an alleged lack of interest in school and a desire to work. Girls, by contrast, frequently left for sudden personal reasons such as marriage, pregnancy, or illness, which occurred during the school year.

What were the interrelationships between early performance and midpoint performance? It seemed reasonable to assume that if a child performed poorly early in his academic career, his failures there might have contributed to later

Table 4
Density of Dropouts among Boys and Girls for Grades 9-12

Grade	Boys[a]	Girls[b]
9 (Summer)	2	1
10 (School year)	4	4
10 (Summer)	8	5
11 (School year)	9	6
11 (Summer)	9	2
12 (School year)	8	14

[a] $N = 40$.
[b] $N = 32$.

difficulties in subjects requiring similar skills. For example, poor reading ability which was first recorded in elementary school English might have later contributed to failure in social studies where reading was also required. It seemed useful, however, to study first the relationships between early failure and midpoint failure (rather than failures as late as the twelfth grade), for there might be important relationships which were lost in the longer range comparisons. In addition, data from the junior high school level were more uniform since all students were required to take English, mathematics, science, and social studies. The chi-square (x^2) test for two independent samples (see Siegel, 1956, p. 107) provided an appropriate technique to examine this hypothesis. A variety of early and midpoint performance patterns were compared in this manner.

It already was evident that the areas of greatest interest for early year failures included English and math. Later failures in all areas were worth exploring. The following chi-square values were obtained.

Early performance in reading showed a significant relationship to later performance in English, social studies, and science; similarly, early performance in mathematics related significantly to later performance in ninth-grade English. Analysis of the actual distributions of these individual groups was necessary to interpret the reasons for this significance. Visual inspection of the data suggested that failure to pass reading in the fourth grade corresponded significantly to future failure in both social studies and English; however, passing reading in the fourth grade was no guarantee of later passing of courses. In short, poor reading might provide a real barrier to social studies proficiency, but the presence of early reading ability did not forecast later social studies success.

Turning to the question of the relationship between fourth-grade reading and later performance in science, inspection of the distribution suggested that failure to pass English corresponded to later failure in science, but again passing English was not a guarantee of later passing science courses. Interestingly enough, failure

Table 5 Chi-Square Values for Early to Midpoint Performance	Early performance	Midpoint performance	x^2	$r\phi$
	Reading	English	17.34**	.283
	Reading	Social studies	8.14**	.175
	Reading	Science	5.34*	.142
	Reading	Mathematics	1.04	
	Oral/Written	Social studies	2.37	
	Oral/Written	Science	2.86	
	Oral/Written	Mathematics	3.38	
	Mathematics	Social studies	.95	
	Mathematics	Science	2.67	
	Mathematics	English	13.00**	.226

*$p < .05$.
**$p < .01$.

to pass reading did not seem to assure failure in science, as much as passing English related to passing science. Thus, relative to science, there was a premium on passing reading, but less concern about failing it.

Finally, early mathematics performance showed a highly significant relationship with later English performance. Very few students (27 versus 85) failed early mathematics and later passed English. However, passing mathematics, again, was no assurance of passing English. This finding was examined in light of the correspondence of early reading to later English. A chi-square value of 17.34 (significant at the .01 level) suggested that the correspondence of early mathematics to later English might be more a function of grading procedures at the fourth-grade level than of subject-grade interdependency.

What were the interrelationships between midpoint single and multiple failures at the seventh-grade level and graduation? The second stage of analysis was concerned with how midpoint failures were related to graduation versus dropout among poorly performing students (see Table 6). The seventh grade was chosen for two reasons. First, it was at this point that nongraduates seemed to demonstrate increasing difficulty with their studies. Second, it might still be possible to do something to avert future dropout if the problem area could be identified and dealt with. Chi-square values were determined in the following areas as they corresponded to dropout versus graduation.

Table 6 shows that performance in the seventh grade on a variety of measures bore a significant relationship to later graduation (or dropout). Inspection of the individual courses of English, mathematics, and social studies revealed that failure in seventh-grade mathematics corresponded most significantly to dropout at a later point in time. A significant number of dropouts failed this subject (62%), as contrasted with poorly performing students, 34% of whom failed the

Table 6
Chi-Square Values for the Performance of Dropouts versus Graduates on a Series of Single and Multiple Seventh Grade Performance Measures

Early performance of dropouts versus graduates	x^2	$r\phi$
English	4.31*	.126
Mathematics	16.06**	.244
Social studies	2.48	
Science	3.00	
English/Social studies	11.12**	.208
English/Mathematics	6.83**	.161
English/Science	6.90**	.163
English/Social studies/ Mathematics	10.70	.202

*$p < .05$.
**$p < .01$.

course but still managed to graduate. The same was true of English, but the tendency was not so pronounced. Social studies and science performance, however, failed to yield significant relationship with graduation.

Inspection of the relationship between multiple failures in the seventh grade with consequent graduation, not surprisingly, showed a highly significant pattern (all beyond the .01 level). However, inspection of the data indicated that this increased significance resulted in large part from the behavior of the graduates rather than that of dropouts; that is, a large number of graduates failed to do poorly in two or three courses simultaneously. About 50% of the dropouts did poorly in paired courses; that is, almost as many dropouts did well as did poorly and yet still dropped out. Thus, the multiple predictions must then be viewed with some skepticism. The source of significance was the disproportionate number of graduates who passed at least one of the subjects rather than a large number of dropouts who failed in two or three subjects.

Discussion

The majority of students who did poorly in their high school careers could have been identified early in elementary school. By the second grade, 50% had already experienced their first failure, 75% by the fourth grade, and 90% by the seventh grade. The critical areas of initial difficulty were English and mathematics.

Over 40% of the student records reflected a spread pattern, (initially failing in only one or two areas) indicating that many children gave early warning signs of future academic problems in many areas. Not surprisingly, the majority of spreads began with English courses.

The distribution of failures over the years was fairly consistent. Beginning with an initial increase in failures in the first 3 years, the subject failures then showed a more gradual rise over the years, reaching a high point in the ninth and tenth grades. A comparison of dropouts with low achieving graduates suggested that these graduates tended to do somewhat better than dropouts; and that the dropouts tended to fall further and further behind after the seventh grade. Over the years, English was the most frequently failed subject, with mathematics and social studies being the next most difficult areas. However, the finding that the onset of initial failures among dropouts was later than for poorly performing graduates suggested that factors other than low ability contributed to actual dropping out of school. While the pattern of performance for graduates and dropouts across the subjects and across the years was similar, dropouts generally performed at a lower level.

Concerning the grade-subject interdependencies, it was found that the reading performance in the fourth grade corresponded to performance in ninth-grade English, social studies, and science; also, fourth-grade mathematics proved to correspond to ninth-grade English. With respect to the relationship between seventh-grade subject performance and ultimate graduation, English and mathematics separately, as well as when combined with social studies and science, showed significant correspondence to subsequent graduation or dropout.

These findings confirm some rather commonly held beliefs among educators. First, students having trouble in their high school academic performance usually have had a history of performance problems. This problem frequently goes back to early elementary school. Furthermore, this study indicated that while early failures may appear in only one or two areas (e.g., English or math subjects), many students (40% of this sample) show a downstream spread of effects (e.g., to social studies and sciences). Where many early failures existed, multiple failures tended to be continued. Most initial failures tended to be continued. Most initial failures occurred in English or math, but this may be in part an artifact of the grading system since social studies and science were not objectively scored for the first 3 years. However, even when objectively scored, these subjects were only infrequently the source of initial failure.

English language development appears to be a focal point of academic failures throughout the scholastic career. For many years there has been a stress on the importance of verbal and reading-writing fluency. The findings of this study support the validity of the assumption that early general communication skills are vital to consequent success in other academic areas such as social studies and sciences.

While early performance is an excellent indicator of academic difficulty it is not as good an indicator of a student's dropout potential. The findings of this study indicate that poorly performing graduates produce a majority of early failing grades, which is consistent with the belief that dropouts do not come primarily from any "lower ability group." Many of the dropouts performed better than the poorly performing graduates in early years. Over 25% had no failures prior to the sixth grade. However, it should be remembered that dropouts are contrasted here to poorly performing graduates, not to the entire graduating class. Lack of ability alone cannot account for dropping out of school. Early assistance to poorly performing students may not, in this case, help students who will be dropping out of school, but it may increase the later performance of nondropouts who are already doing poorly. It should be remembered, however, that by the seventh grade most students who will have difficulty in high school (dropouts included) will already have difficulty, and at this point academic assistance (plus other types) can be applied.

Given these considerations, the following points seem applicable to educational planning:

1. Where students are found to be having difficulty with school during the first four grades, special remedial programs should be initiated to help improve the quality of their achievement.

2. Particular attention should be paid to the areas of English and math failures, not only to improve later performance in these areas, but to prevent a spread of performance difficulty to other areas.

3. Since in many cases dropouts fail to show performance difficulty in the first 2 or 3 years, such programs should be primarily directed toward improving quality of students' performance rather than toward increasing the quantity of graduates.

4. Where students show a reasonable performance pattern for the first few years, and then begin a sharp decline in performance, this should be taken as a warning sign of potential dropout. Assistance directed toward students at the seventh-grade level should be concerned not only with quality improvement, but also with increasing the number of graduates. Three types of assistance may be called for: academic assistance, in-school counseling assistance, and possible home visitation assistance (perhaps directed toward helping parents to provide better home atmosphere for study, and reinforcement of the child's improved performance where such occurs).

This study does not imply that preschool experiences (Bloom, 1965) and experiences during the school years which occur outside the school setting (Cervantes, 1965) are not vitally important in their influence on both the quality of graduates and the incidence of dropout. In many cases, changes in nonschool factors are decisive in changing the educational development of a given student.

There are three approaches to dealing with improvement of student performance and completion of schooling: education programs (e.g., Title 1 programs from the Elementary and Secondary Education Act), home and neighborhood improvement programs (e.g. various Office of Economic Opportunity programs), and selected combinations of the two. Application of educational programs alone implies that a problem is basically one of altering conditions in the school environment (through remedial instruction, better equipment, special counseling, etc.); that if one can alter certain characteristics of a student which affect his performance academically, he will improve and/or graduate.

Application of home and neighborhood programs alone to a problem (with associated school implications) implies that the problem is basically one of altering conditions in the home and peer environment external to the school (on both a preschool and during school basis), and that this will make it possible for the student to respond better to an already adequate education program.

Applying both educational and environmental programs in combination implies that effective change in student performance, and/or decreasing dropout rates, involves change in both the school and nonschool environment.

Determination of where either or both types of assistance is required is very difficult. It requires identification of the major contributing factors to poor student performance. In addition, the relative importance of both areas needs to be weighted and their interaction assessed. Will "curing" of one area be sufficient to override the negative impact of the other area? Conversely, is the interaction of both educational and environmental programs required to overcome performance problems and the likelihood of dropout for a given student?

Experience of Head Start Officials has shown that all of the effort to improve the preschool learning environment of the student may be futile unless there is follow-up in the schools (e.g., project "Follow Through").* Environmental improvement alone may be insufficient to improve student performance where the school fails to provide minimum necessary educational support. Conversely, administrators in the school who provide remedial instruction (and do a good job of this) may be totally frustrated when the nonschool environment negates their efforts (as in settings where there is severe malnutrition or where highly negative attitudes are held by parents and peer groups outside the school).

All this points to the conclusion that joint educational and environmental aid offer the greatest potential to improve student quality and decrease the dropout rates. Unfortunately, both limited congressional funding and lack of clear understanding on the part of educators of the interaction between school and nonschool factors make it unlikely that many disadvantaged area schools will have the luxury of coordinated aid to cope with both school and nonschool problems.

*Indeed, even the Head Start Program may be unable to overcome certain prenatal and infancy deprivation experiences.

Assume that in many instances the only aid which will be received by many local school districts will be educational assistance; that while it would be highly desirable to have environmental programs which would counteract negative home and neighborhood influences, such aid is unavailable. What should the educational program direction then be?

The findings of this study have suggested that aid early in the academic career has more impact than later aid. This is because it seems that many students who initially perform poorly only in one or two areas, may show a spread of failures in later years (thus compounding the problem). This is also consistent with the longitudinal research which suggests that the further along in a student's career, the greater is the amount of "end-career" variance already accounted for.

The findings of recent research by Coleman (1966) and Urdal (1965) both point to the importance of attitudes toward school. While home and peer groups (both preschool and during school) undoubtedly have great influence on attitudes, it seems reasonable that the experiences in the school setting contribute in their own right to attitudes regarding school. What happens when the student encounters dissonant attitudes and behaviors? Suppose that his experiences in school contradict what his family emulates and his friends say? Certainly this is better than when both school and nonschool experiences coincide in the negative!

It is hypothesized that assistance to students provided by educational programs should be both content and process oriented, and explicitly so. By process is meant the extent to which the teacher focuses on the student's feelings while the student is learning (Does a teacher reinforce improved performance? Is there peer group support for "doing better"? Are students helped to see how progress in school relates to improving their own lives? Is there genuine concern for the student's personal development and does the student sense this concern?). It is true that all of the "process" in the world will not help a student to perform better if he cannot read. However, although learning to read is essential, it is not a guarantee of improved performance. Therefore, a substantial part of the educational program emphasis should focus on *how* particular processes are to be used by teachers and counselors, that is, *programs should be oriented toward changing attitudes* (through positive experiences) as well as abilities. If attitudes toward school are vital, then programs must do more than change knowledge; they must change attitudes toward school.

References

Bloom, B. *Stability and change in human characteristics.* New York: Wiley, 1965.

Cervantes, J. S. *The dropout.* Ann Arbor: University of Michigan Press, 1965.

Coleman, J. S. *Equality of educational opportunity.* Washington, D. C.: U. S. Office of Health, Education, and Welfare, 1966.

Schreiber, D. *The school dropout.* Washington, D. C.: National Education Association, 1963-1964.

Siegel, S. *Non-parametric statistics.* New York: McGraw-Hill, 1956.

Urdal, L. B. *Dropouts. An analysis of personal variables within the school situation.* Olympia, Wash.: State Superintendent of Public Instruction, 1963.

Recommended Additional Readings

Hathaway, S. R., Reynolds, P. C., & Monachesi, E. D. Follow-up of the later careers and lives of 1000 boys who dropped out of high school. *Journal of Consulting and Clinical Psychology,* 1969, **33**, 370-379.

Hathaway, S. R., Reynolds, P. C., & Monachesi, E. D. Follow-up of 812 girls 10 years after high school dropout. *Journal of Consulting and Clinical Psychology,* 1969, **33**, 383-390.

Schreiber, D. *Profile of the school dropout.* New York: Vintage, 1967.

Language as a Symbolic Process in Communication: A Psychological Perspective

Robert M. Krauss

Orientation

The individual's acquisition of the language of the culture within which he lives is perhaps the single most dramatic and exciting event in the entire panorama of human ontogeny. This is true because language is the most unique behavior distinguishing human beings from all other animals, and because this incredible behavioral repertoire is achieved without any formal instruction or tutoring (imagine the absurdity of developing school-type lesson plans for the teaching of language to a one-year-old). Not surprisingly, therefore, the topic of language development is one of the most vigorous areas of research within contemporary developmental psychology.

The following article by Krauss deals with an aspect of language most notable in childhood but also evident in later life, perhaps even in adulthood; this aspect is the egocentricity that plagues one person's communications to another. Egocentrism, or the inability to "decenter" or take the viewpoint of someone else, frequently is found in the speech productions of children. The child is unable, or perhaps simply unwilling, to consider his listener's point of view and thereby adapt his verbal communication to the other person's informational needs. The child (and sometimes the egocentric adult) simply assumes that the listener knows exactly what he is talking about, and thus takes no effort to ascertain whether his message makes any sense to the person to whom he is speaking. As mentioned above, this phenomenon of egocentric (or "private") speech is not entirely limited to children; all too frequently adults assume that their listeners comprehend exactly whatever words and ideas they select for conversation. As a concrete example, probably every neophyte instructor soon learns, with considerable embarrassment, that his class presentations are incomprehensible to his students; he is guilty of talking to himself and of not taking into account whether his audience can make any sense at all of whatever he is trying to communicate.

The article by Krauss explores this aspect of

Reprinted from *American Scientists,* 1968, 56, 265-278, by permission of the author and the publisher.

A talk delivered to the Seminar on Immediate Symbolic Processes in Communication of the Institute for Arts and Humanistic Studies, Pennsylvania State University on April 9, 1968. Some of the research discussed was supported by Public Health Service Grant HDMH-01910 to Princeton University *speech in detail. Krauss and his co-workers have devised an ingenious experimental procedure that permits careful examination of the communication process in two-person interactions. Although their research has been involved mostly with children, the procedure they employ is equally fruitful for use with adults of all ages.*

Robert M. Krauss is a professor of social psychology at Columbia University.

24

The late A. J. Liebling, an astute observer of our culture and mores, once wrote:

> There are two kinds of writers of news in our generation. In inverse order of wordly consideration, they are:
> (1) The reporter, who writes what he sees.
> (2) The interpretive reporter, who writes what he sees and what he construes to be its meaning.
> (3) The expert, who writes what he construes to be the meaning of what he hasn't seen.

Liebling continued,

> To combat an old human prejudice in favor of eyewitness testimony ... the expert must intimate that he has access to some occult source or science not available to either reporter or reader. His is the Priest of Eleusis, the man with the big picture. ... All is manifest to him, since his conclusions are not limited by his powers of observation. Logistics ... favor him, since it is possible not to see many things at the same time.[1]

Liebling was writing about newspapers, but we can readily see the applicability of his principle to other areas. It strikes me that by inviting a psychologist rather than a poet or a novelist to talk about language, you have explicitly sought "the man with the big picture." In short, I confess to being an expert—but only in Liebling's sense of the term.

I must also confess to feeling somewhat less at home here, talking to you, than I would addressing a group of my fellow "experts." The focus of humanistic concern is with the creative or literary uses of language, and my work has involved its more mundane functions. A very minor poet once wrote, in a work depressingly familiar to us all: "Poems are made by fools like me / But only God can make a tree." Now I tend to marvel less at the tree and more at the poem—not at this particular poem, perhaps—but at the poem which exploits the resources of language to bring booming into our awareness that which we had only sensed before. About this sort of feat of mind and soul, a psychologist has little to say.

There are many approaches to the study of language, each dealing with a different aspect and each having its own intrinsic significance and validity. The linguist views language as a highly complex structure with its own system of logic and organization. The humanist is primarily concerned with the aesthetic and expressive components of language use. The formalizations so dear to the heart of the linguist are of little interest to him. The psychologist regards language—one is tempted to say the gift of language—as one of the impressive abilities of man. He is interested both in the mechanisms which make language use possible and in the role this capacity plays in the over-all functioning of the organism. His interests will at times overlap with those of the linguist, but the specific focus of his concerns will be somewhat different.

Hence—and this will be my only *Caveat auditor*—what we psychologists have to offer is a particular perspective and, like all perspectives, it is based upon a particular point of view. It chooses to ignore some things in the hope of getting a better grasp on other things. As one plausible perspective, it deserves to be presented clearly and considered seriously. I take the former as my assignment; I hope that I can convince you the latter is worth your while.

From one psychological point of view, language is of interest because of its capacity to represent experience in symbolic form. To introduce a technical term, language permits us to perform a particular kind of *encoding*, This simple fact has two consequences of enormous significance. First, because we can encode experience via language and because the symbols we use for such encodings have socially shared meanings, we can transmit our encodings to others and, by this means, make others privy to thoughts, feelings, and perceptions which they themselves have not directly experienced. In a word, we can *communicate.* And because the symbol system we use (that is to say, the language) is so rich and so flexible, we can communicate on a level of complexity unachievable by any of the other symbol systems available to us.

But another and, for the psychologist, equally important function of language is its role in cognition—in the operation of mental processes. It seems obvious that language is intimately related to man's competence as a communicator. But how is language implicated in the life of the mind? Perhaps the strongest position in this regard is the doctrine of "Linguistic Relativity," most closely associated with the anthropologist Edward Sapir and his student and later colleague, Benjamin Lee Whorf. For both Sapir and Whorf, language was not simply the mapping of a symbol system onto an external, objective reality, in the way that patterns of dots and dashes in the Morse Code are mapped onto the twenty-six letters of the alphabet. Rather, the linguistic relativist views language as the major shaper and determiner of subjective experience. As Whorf put it:

> We dissect nature along the lines laid down by our native language. The categories and types that we isolate from the world of phenomena we do not find there because they stare every observer in the face; on the contrary, the world is presented in a kaleidoscope flux of impressions which has to be organized by our minds—and this means by the linguistic system in our minds. We cut up nature, organize it into concepts, and ascribe significance to it as we do largely because we are parties to an agreement to organize it in this way—an agreement that holds throughout our speech community and is codified into the patterns of our language. The agreement is, of course, an implicit and unstated one, but its terms are absolutely obligatory; we cannot talk at all except by subscribing to the organization and classification of data which the agreement decrees.[2]

Such a position is an extreme one and potentially of profound and far-reaching significance. For what Whorf seems to be saying is that our categories of experience are largely formed by our language, and that we cannot "know" at all except as the relevant categories of knowledge are available to us in our language. Like many scientific notions of fundamental importance, the Whorfian hypothesis (as it has come to be called) is difficult to test directly. No one has devised a single experiment or set of experiments that provide a definitive basis for accepting or rejecting it. Whorf did manage to come up with many interesting and ingenious examples, drawn mainly from his extensive knowledge of American Indian languages and cultures, but the nature of such examples is to demonstrate the plausibility of the hypothesis rather than to establish its truth beyond cavil.

Most psychologists and linguists, I think, do not accept uncritically the extreme view implicit in the Whorfian hypothesis. The attempts to test it empirically have provided only weak and partial support, and, as several critics have noted, Whorf's own ability to apprehend the categories of experience available to speakers of such exotic languages as Hopi or Navajo itself belies the notion of an inflexible dominion of language over thought. But this is not to say that language and cognitive processes are *un*-related. Let me discuss the relations in terms of a simple example and then go on to describe some experiments based on this line of reasoning.

A common attribute of objects is color. It is universally present to all whose visual receptor system is not defective, and by now a good deal is known about the physiological mechanisms which make color vision possible. If one takes a simple prism and diffracts sunlight through it onto a white surface, one sees several bands of color—violet, blue, green, yellow, orange, and red. (A nicer way to do this experiment is by looking at a rainbow, but that isn't so readily available for laboratory demonstration.) Now those of us who took elementary physics—and a good many of us who didn't—know that spectral colors are a simple consequence of differences in the wave length of the light striking the white surface. The wave length of the visible spectrum ranges from about 380 to 720 millimicrons and its variations across this range are smooth and continuous. There is no sharp break which differentiates blue from green or orange from red. Yet, when we look at the spectrum our dominant perceptual response is to see six bands or color categories, not smooth variation running from violet to red. Our partitioning of the spectrum consists of the arbitrary imposition of a category system upon a continuous physical domain.

And of course this business of arbitrarily imposing categories upon continuous spectra is something we do quite often. An *A* student is one whose grade falls into the interval between ninety and one hundred per cent and a *B* student is one whose grade falls into the eighty to ninety per cent interval. A student whose number grade is ninety-one per cent probably is closer in ability to an eighty-eight per cent student than to one who gets ninety-nine per cent. And yet we consider two of them as *A* students and one as *B*. It's convenient to carve up

Figure 1.
An approximate representation of the way speakers of three languages partition the spectral colors. Taken from Henry A. Gleason, Jr., An Introduction to Descriptive Linguistics *(rev. ed.).* © *1955, 1961 by Holt, Rinehart and Winston. Reproduced by permission of Holt, Rinehart and Winston, Inc.*

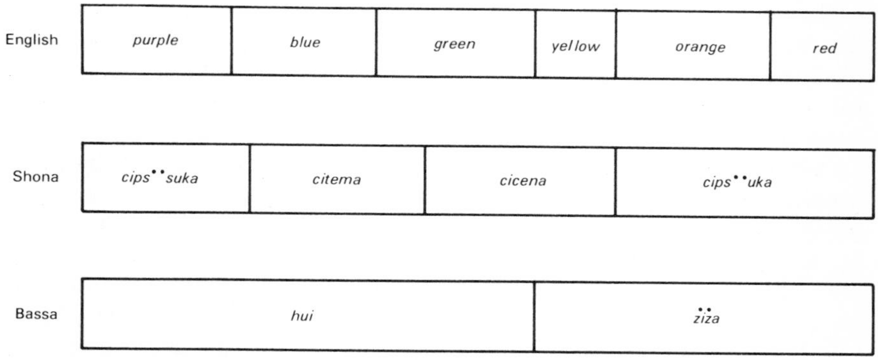

English	purple	blue	green	yellow	orange	red

Shona	cips̈uka	citema	cicena	cips̈uka

Bassa	hui	ziza

this domain in some fashion, and the imposition of letter grades has proved useful. Students have been known to challenge this assertion.

Back to colors. If it is truly the case that the categories we impose upon the visible spectrum are arbitrary, then it seems reasonable to expect that somewhere in the world there should exist a people, speaking a language unrelated to our own, who partition the spectrum differently. As the illustration in Figure 1 indicates, this indeed is the case. The top band on the figure illustrates roughly how we (i.e., speakers of English) partition the spectrum. Below that are shown the partitionings for speakers of Bassa, a language spoken in Liberia, and for speakers of Shona, a language of Rhodesia. Note two things: First, the *number* of partitionings differs among the three languages. Both languages contain fewer than the six categories we employ. Second, the locations on the spectrum at which *category breaks* occur also differ. The Shona speaker forms a color category from what we call *orange, red,* and *purple,* giving them all the same utterly unpronounceable name. But he also makes a distinction within the band we term *green.* Here we have a clear case of speakers of different languages slicing up the perceptual world differently. And, of course, it is also the case that the kinds of slices one makes are related to the names for the slices available in his language.

Now if one looks not at the spectrum but at a large assortment of discrete color specimens such as can be found on the sample charts that paint companies distribute, one finds some colors that are easy to name and some that are difficult to name. Some evoke in most viewers an unequivocal response: *red, yellow, purple.* Other samples, however, will be difficult to name, producing responses which often seem labored or contrived. In one of my own experiments such a color elicited the name *disgusting green.* But my favorite name was gleaned by another investigator, whose respondent called one sample *the color of the sweater of a not-thin girl.* As you can see, it's rather difficult to conjure up a good image of what such colors might be like. We refer to this property of colors as their *codeability,* the ease or difficulty with which they can be encoded verbally. Some colors are highly codeable, others are not.

Two psychologists, Roger Brown and Eric Lenneberg, exploited this simple fact in what seems to me one of the most elegant and interesting experiments relating language to cognitive processes.[3] I will first describe the experiment, because part of its elegance lies in its simplicity, and then discuss some implications of its outcome. Brown and Lenneberg first derived codeability scores for a carefully selected set of twenty-four colors. They did this simply by asking people to name each one and then calculating the degree of consensus among the names each color elicited. A color that elicited the same name for a large portion of the respondents is a highly codeable color, while one that evoked a high proportion of different names is low in codeability. They then had a different set of subjects (this is what participants in psychological experiments are called) perform a recognition task. A color was exposed, then taken away, and the subject had to select the color he had seen from among a large set of colors. For some subjects the recognition task was *immediate,* a very short interval of time intervened between the exposure of the "target" color and the subject's attempt to locate it in the large set. For others, the procedure followed was *delayed* recognition: The target color was exposed, then taken away, and only after some time had passed was the subject allowed to look for it.

Brown and Lenneberg found that under conditions of delayed recognition there was a strong relationship between the codeability score of a color and the accuracy with which the subjects were able to locate it. The more codeable a color was, the greater was the likelihood that it would be correctly chosen under delayed recognition. But for the immediate recognition subjects, those whose search for the target color directly followed its exposure, codeability and accuracy of recognition were unrelated. Low codeability and high codeability colors were equally well selected from the larger set.

Now what does such a result tell us? And I suppose I should point out here that the bare facts gleaned from an experiment are themselves relatively unimportant; it is the inferences these facts permit us to draw that are of primary interest to the investigator. The interpretation Brown and Lenneberg made of their findings, and one that strikes me as entirely plausible, is as follows: When one sees a color for a few moments and then has it removed from his field of vision, the visual representation of the color does not vanish immediately. In a sense one can still "see" the color, conjure up its image, for some short time following exposure. In an immediate recognition task, where the interval between exposure and recognition is minimal, the image of the color is still present. Roughly speaking, one can scan the large set of colors in the recognition array and compare each with the representation of the target color he has stored in his memory. But under delayed recognition, things are not so simple. With the passage of time, the stored visual representation of the color fades or becomes distorted and so is no longer useful for recognition. In order to be able to recognize the color one must encode it in some form other than visual. And for colors, it would appear, the form of this encoding is linguistic. It is as though the subject, working under conditions of delayed recognition, tells himself the name of the color and then sets out to find a color which corresponds to the name. But if colors vary in the degree to which they are codeable—that is to say, in the extent to which it is *possible* to derive good names for them—then they should also vary in how accurately they can be selected in the delayed recognition task. More important, this variation should be directly related to their codeability. And this, you will recall, is precisely what Brown and Lenneberg found.

I should point out here that this relationship between codeability and recognition is not limited to situations in which color is the stimulus material. An experiment involving the recognition of facial expressions produced similar results.[4] Nor is it limited to speakers of English. In an experiment conducted on the Yucatan Peninsula, involving subjects who spoke Spanish and subjects who spoke Yucatec, a local Indian dialect, a similar relationship between codeability and recognition was found.[5] Spanish speakers were most accurate at recognizing colors that were most highly codeable in Spanish and Yucatec speakers were most accurate at recognizing the colors that were most highly codeable in Yucatec. *But these were not necessarily the same colors.* Apparently the two languages map names onto colors in a somewhat different fashion, and the performance of the subjects on the recognition task reflected these differences.

These experiments, considered together, clearly establish that for at least one important type of cognitive operation—memory—language plays an important role. A good case can be made for its importance in other types of cognitive operations as well. But note, the kinds of classifications which language imposes upon my experience are not my own unique inventions. They are built into the language which I have learned and which I habitually speak. And because you and I speak roughly the same language, we also have roughly the same categories of experience. As a result, when I say *red* you have a pretty good idea of what I mean, perhaps not precisely the same red I had in mind, but close enough for most purposes. And if finer distinctions become necessary, we can dig deeper into our common color lexicon for categories like *crimson* or *rose*, which are available to a great many speakers of English. And if we were members of a specialized sub-culture for whom small differences in color names assumed special importance, such as paint manufacturers or textile designers, the likelihood is great that we would share a specialized color vocabulary containing such arcane terms as *cyan* and *teal*, which would be meaningful to us, but utterly opaque to outsiders.[6]

This raises a problem. It is not the case that an object, an event, or an experience has a single name uniquely associated with it. For reasons that a little thought will make obvious, it would be inefficient to construct languages on such a basis. Rather, any object can take a multiplicity of names, depending upon various sorts of circumstances. One and the same two-legged, feathered animal might appropriately be called *bird, ringdove, Streptopelia Risoria,* or *Frankie,* depending upon who is talking to whom and upon the context in which their conversation is set. But how does one know how to refer to a category of experience, to select from the lexical store with which our language has so generously provided us? Perhaps the most striking thing is that this so seldom seems a problem. Except in special circumstances, when we talk we are rather unselfconscious about our selection of words. To digress a moment, it may be worth our while to consider the "special circumstances" under which we *are* especially careful about our selection of words. Such occasions tend to occur, I think, when our attention is focused upon the *expressive* component of language—what our messages reveal about us as individuals. I can refer to a young lady as a *girl* of a *broad,* both terms being available in my lexicon, and in a given situation either term will convey the idea of who it is I am talking about. But my choice of terms will also tell you something about *me* and if I am concerned about the sort of impression I convey, I will be appropriately prudent in my choice of terms. But aside from situations in which the expressive component of language is an important consideration or in which great precision of communication is called for (as would be the case in, say, a diplomatic note) we select from our lexicons both appropriately and unselfconsciously. We don't call an automobile a *light blue 1964 two-door Dodge station wagon* when *car* will convey the necessary information. Nor do we call a bird *Streptopelia Risoria* when it's unlikely that the technical term will be familiar to our listener. At least we don't do these things when our goal is to communicate rather than to impress our listener with our eccentricity or our erudition.

But the subtle and complicated skill with which most adults employ their native language is not a characteristic of all language users. Individuals of defective intelligence and persons who have suffered certain sorts of damage to their central nervous systems typically lack certain sorts of competence for language use.[7] But a larger, and to the psychologist in some ways more interesting, group of individuals who lack full-fledged ability to communicate is that oppressed and overstudied minority, children.

Psychologists study children for a host of reasons, not the least of which is that they are fun to work with. More important is the fact that by carefully observing the process of normal development in the child, we can draw some inferences about the nature of the mechanism which makes possible the skilled performance of the adult. It is this sort of notion which has motivated my colleague Sam Glucksberg of Princeton University and me to undertake a series of studies begun three years ago. In our work, we have sought to create in the laboratory a microcosm of the natural context in which communication occurs. We have two people, usually but not always two children, we have something for them to communicate about, and we have provided a reason for them to communicate.[8]

This experimental microcosm is illustrated in Figure 2. The task is presented to our subjects as a game called Stack the Blocks. The subjects are provided with identical sets of six blocks, each block imprinted with a different design. Each child also has a stacking peg. One child, whom we call the Speaker, receives his six blocks in a dispenser which is so constructed that the blocks can only be removed one at a time. The other child, the Listener has her blocks spread out before her on the table. The two children are separated by a screen so that they cannot see each other. Stack the Blocks is played in the following way: the Speaker removes the first block from his dispenser and puts it on the peg. He then describes this block to his partner, the Listener, so that she can find the corresponding block among her set and stack it on her peg. When she has done this, the speaker removes the next block from his dispenser, describes it to his partner, and so on, until all six blocks have been stacked. The object of the game for our subjects is to build two identical stacks of blocks.

This may seem to be a rather trivial game, and so it is for school-age children—but for one fact I've failed to mention. You will recall that each block is imprinted with a different figure. In order for the subjects to perform their task successfully, the speaker must describe the figures well enough that his partner can select the corresponding one. This is not simple, because the figures imprinted on the blocks are the six designs shown in Figure 3. You will note that these figures are not easy to characterize. In this regard they are like the low codeability colors in the Brown and Lenneberg experiment, and indeed this resemblance is deliberate. It is not the case that these figures *can't* be described, but rather that to do so requires that one employ language in a creative fashion. In a very real sense, the child in our experiment must invent a way of talking about something he has never seen before, and to do so requires that he draw heavily upon his limited linguistic resources.

This then is our communicative microcosm. Why have we chosen to study communication in this fashion, rather than simply to observe children talking under natural circumstances? Two great disadvantages accrue from this sort of approach: First, in a natural situation children talk about all sorts of things and it is exceedingly difficult to make meaningful comparisons between the conversations of two children discussing their favorite television program and two who are examining the weighty problem of why girls cry more than boys. In our microcosm, we know what our subjects are talking about because we have pro-

Figure 2.
The "experimental microcosm" employed in studies of communication with children.

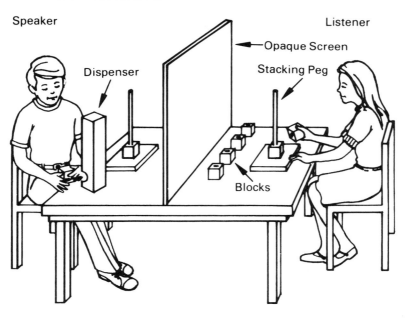

Figure 3.
The six designs used in the experiments with children.

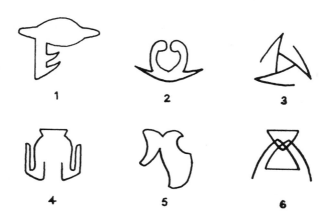

vided them with their topic of conversation. A second disadvantage of studying naturally occurring conversations is that it is often difficult, and sometimes impossible, to assess in any objective fashion the adequacy of the communication that has taken place. In our microcosm we have an objective index of how well our subjects have communicated; namely, the number of errors they make in stacking the blocks. If a pair of subjects have built identical stacks of blocks they must have communicated well, because aside from E.S.P. there is no other way they could have done so.

For adults our task is a trivial one. Even for adult subjects of meager education and lower than normal intelligence—persons who are not normally thought of as especially articulate—performance on the block-stacking task is essentially error-free. But for children—and especially for young children—this is not the case. Glucksberg and I have completed several experiments employing this procedure and I lack sufficient space to describe them all. Besides, to do so would be inappropriate; some are addressed to technical questions which are not germane to this general discussion and have yielded results which will require further research to explicate.

Rather let me describe some aspects of the performance of relatively young children in our microcosm and what we construe to be the meaning of these results—taking for the moment the role of Liebling's interpretive reporter—because I believe they have some implications of rather general relevance.

Four-and-a-half year old children—alert, clever children taken from nursery schools in Princeton, New Jersey—do very poorly at Stack the Blocks. It's not that they make some errors. Rather they do miserably, and over a long series of repetitive trials they fail to get any better. The question, of course, is why? To begin with, we can eliminate two rather mundane explanations. First, it is not the case that the game itself is too difficult for them. When the exotic novel designs are replaced by pictures of familiar things—clown, animals, and the like—the same youngsters communicate with impressive effectiveness. Nor does it seem to be the case that our young subjects are unmotivated. The results of our work with familiar figures would seem to belie this. But more compelling to me is the impression gained from watching the children at work—the serious concentration with which they apply themselves to their task and the intense disappointment evoked by the news that they had failed at it.

No, the difficulty seems to be related to the way in which our young subjects employ language and it is to this that we must turn for an explanation. Glucksberg and I spent long hours listening to tape recordings of what our subjects said during the experiment and reading transcripts of these recordings. As we listened it became increasingly clear to us that for a significant proportion of the time the Speakers—whose task it was to describe the novel figures—were employing a private language to communicate with their partners. And these messages, having been encoded in a private language, were totally opaque to the Listeners, as opaque as the screen which separated them.

What do I mean by a "private language"? An example, I think, will make this clear. At home, placed on an inconspicuous and poorly-lighted wall, hangs an abstract water color which represents my first venture into art collecting. Unfortunately, my early enthusiasm for art far exceeded my knowledge of it and my initial purchase reflects this all too clearly. When I think about this particular painting, or talk to myself about it, I refer to it as the *mistake*. For me, that is the painting's name, and for me that name is highly evocative. I know what I mean when I say it. But I would hardly expect it to communicate very much to you. Indeed, were we in my living room and were I to refer to the *mistake*, you might think that I was talking about a more recent acquisition. Others have, much to my chagrin.

Now *mistake* in this context is a private encoding—it is not socially shared— and being a competent communicator I would refrain from using it unless I had first taken pains to build up a context in which its use was meaningful. This is precisely the distinction which young children appear not to make.

We observed a striking example of this in one of our experiments. While conducting a session, sitting alongside the Listener, I heard the other child, the Speaker, refer to one of the figures as a *sheet*. The Listener was mystified by this and so was I. But over and over again, the Speaker referred to the *sheet* and each time both the Listener and I were at a loss to know what he meant. After the experiment was over, I went up to the Speaker and asked him which figure he was calling the *sheet*. Perhaps you ought to look again at the last figure, so that you can see if you are any better at decoding his message than I was. . . . The *sheet,* I was told by my four-and-a-half year old informant was the figure number three. My next question was "Why would you call that one a *sheet,* it doesn't look like a sheet to me?" The answer goes right to the heart of the matter: "Have you ever noticed when you get up in the morning the sheet is all wrinkled? Well, sometimes it looks like this."

Indeed, sometimes it does. But the connection is not apparent enough for me or for most people to see it and *that fact* makes *sheet* a bad encoding. This is a pragmatic criterion and on some esthetic grounds *sheet* may be more highly valued than some of the more banal but also more communicative terms adults use. But communication *is* a pragmatic affair. If a speaker's intention is to inform his listener, then the mark of his success is how well his listener understands. Any encoding which leads to less-than-adequate understanding must be deemed a poor encoding, no matter how interesting it may be on other grounds.

Is it the case that our subjects were really encoding privately? The *sheet* example suggests this, but taken alone it does no more than suggest. Glucksberg and I were aware of this and we devised another experiment to pursue the question further. A private encoding is one which is meaningful to the person who utters it, but which communicates little or nothing to others. Hence, one should expect that our young Speakers would be able to decode their own encodings with great accuracy although others were unable to do so. What we actually did was to ask our subjects to name each of the novel designs and then to use these names in communicating with them, with one of us playing the role of the Speaker in "Stack the Blocks" and the child himself as the Listener. The experiment confirmed our hypothesis. Our subjects' decoding accuracy was not absolutely perfect when their own encodings were used—one subject made one error. But that was all, and that is very, very good.

The time has come for summing up and a brief digression. I have tried to trace two strands in the complex net which language weaves. One strand connects language and the operation of the mind, and it emphasizes the communality of our own consciousness and the consciousness of others. The second strand of the net relates to the role of language in communication, and it stresses the awareness of the consciousness of others which is implicit in the communicative act. As communicators we are, in effect, radio transmitters finely tuned to the receiving wavelengths of our listeners. And the delicacy of the tuning of which we are capable is evidenced not simply in our encoding, but in our sensitive use of address modes, our receptiveness to minimal informational cues, and in our expressive use of language, among many other things.

But what distinguishes the Creative use of language—Creative with a capital C—from its workaday function? Certainly a poem is communication, but is it no more than that? Issac Bashevis Singer offered some sage advice on dealing with such questions. He said, "When someone asks me a question, if I know the answer, I tell it to him. If I don't know the answer, then I give him an answer to a *different* question." So let me put the question somewhat differently.

When Gerard Manley Hopkins, referring to the grandeur of God, wrote "It will flame out / Like shining from shook foil," is this essentially different from the child in our experiment calling an abstract design a *sheet?* I would say that it *is* different and it is *essentially* different. The poetic image is latent in our own awareness and it is this latent image that is triggered by the poet's metaphor. Not that we have ever thought it before necessarily, but rather that causing one to think of it has a predictable result. And this, I would argue, is quite different from what the child does. His image is formulated largely without regard to the result it will produce in his listener. Consequently, its effect on the listener is unpredictable.

Now it is true that a child will often use language in what appears to be a poetic or metaphorical fashion. But I think that these apparent instances are deceptive, for they are products of the child's private awareness formulated largely without regard for the awareness of his listener. At times they will strike a responsive chord in us and on those occasions we are likely to discern a resemblance between our offspring and the young Byron in manner, thought, and word. Parental pride is as understandable as it is inevitable, but we are, I think, on such occasions confusing coincidence with creativity. I can easily imagine Hopkins searching his consciousness, coming up with an image, then rejecting it, knowing that the image is too closely tied to his own private experience to be meaningful to anyone else. I cannot imagine a child doing this.

It would be slighting my colleagues if I failed to point out that the range of work of psychologists interested in language goes far broader and deeper than what I have been able to cover in this brief presentation. At the outset, it was my intention to provide a perspective that at least one psychologist—namely me—finds useful for thinking about language. But as was touched upon earlier, a psychological perspective is only one of several vantage points from which language may be viewed. It has its uses and its limitations as does each of the other perspectives. Psychologists have learned a great deal from the linguists and they have learned something from the humanists—not as much from the latter as they might, I suspect, but this is probably their own fault. I hope I have been able to persuade you that a psychological perspective provides a useful complement to your own in considering the role of language as a symbolic process in communication. At least I hope that you will not be moved to tell me, as Bonnie told Clyde: "Your advertising's just dandy. Folks'd never guess you don't have a thing to sell."

References

1. Liebling, A. J. *The press.* New York: Ballentine, 1961. P. 225.
2. Whorf, B. L. *Language, thought and reality.* Cambridge, Mass.: M.I.T. Press, 1956. Pp. 212-213.
3. Brown, Roger and Lenneberg, Eric H. "A Study in Language and Cognition." *Journal of Abnormal and Social Psychology,* 1954, **49**, 454-462.

4. Fridja, Nico H. and Van de Geer, John P. "Codeability and Facial Expressions: An Experiment with Facial Expressions." *Acta Psychologica,* 1961, **18**, 360-367.
5. Stefflre, Volney, Vales, Victor Castillo, and Morley, Linda "Language and Cognition in Yucatan: A Cross-Cultural Replication." *Journal of Personality and Social Psychology,* 1966, **4**, 112-115.
6. For some indication of the range of color names available in the nonspecialists' lexicon, see Alphonse Chapanis, "Color Names for Color Space," *American Scientist,* 1965, **53**, 327-346.
7. For the former group see Jordan, Thomas E. "Language and Mental Retardation: a Review of the Literature." In Schiefelbusch, Richard L., Copeland, Ross H., and Smith, James O. (Eds.), *Language and Mental Retardation.* New York: Holt, Rinehart and Winston, 1967; for the latter, Wepman, Joseph M. and Jones, L. U. "Five Aphasias." In Rioch, D. and Weinstein, E. A. (Eds.), *Disorder of Communication.* Baltimore: Williams and Wilkens, 1964.
8. Glucksberg, Sam, Krauss, Robert M. and Weisberg, Robert "Referential Communication in Nursery School Children: Method and Some Preliminary Findings." *Journal of Experimental Child Psychology,* 1966, **3**, 333-342; Glucksberg, Sam and Krauss, Robert M. "What Do People Say After They Have Learned How to Talk? Studies of the Development of Referential Communication." *Merrill-Palmer Quarterly,* 1967, **13**, 309-316; Krauss, Robert M. and Rotter, George S. "Communication Abilities of Children as a Function of Status and Age." *Merrill-Palmer Quarterly,* 1968, **14**, 161-173.

Recommended Additional Readings

Kohlberg, L., Yaeger, J., & Hjertholm, E. The development of private speech: Four studies and a review of theories. *Child Development,* 1968, **39**, 691-736.

Krauss, R. M., & Glucksberg, S. The development of communication: Competence as a function of age. *Child Development,* 1969, **40**, 255-266.

McNeill, D. The development of language. In P. H. Mussen (Ed.), *Carmichael's manual of child psychology.* (3rd ed.) Vol. 1. New York: Wiley, 1970. Pp. 1061-1161.

Vygotsky, L. *Thought and language.* Cambridge, Mass.: M.I.T. Press, 1962.

25 Individuality in Development

Riley W. Gardner

Orientation

It is clearly true that each person has certain peculiarities that make him unlike any other person. It is also true that each person has characteristics that render him similar to other people. Unfortunately, psychology knows more about and understands better the latter truth than the former. The fact of individual differences, although widely recognized, remains a troublesome reality for research psychologists, especially those of the experimental bent.

We should be quick to point out that, indeed, psychologists have expended vast efforts in the study of individual differences. The widespread development and use of intelligence and personality tests attest to this fact. However, the usual focus of scientific endeavors is to formulate general statements ("laws") which permit the description, and hopefully even prediction, of what certain elements (e.g., people) will do in given situations. Thus, typically scientists produce averages or norms as an outcome of their observations or manipulations. For example, if we wished to study the vocabulary development of three-year-olds, we would undoubtedly end up with a statistic that indicates the average number of words possessed by children of that age level. In reporting our findings, it would be too cumbersome to state the size of vocabulary of each child we observed; instead, we economize by eliminating the individual differences in vocabulary size and report only the normative figure. This procedure is scientifically useful, but we must keep in mind that "average" is always a fictional construct; its potential usefulness is determined by the amount of distortion of individual data that takes place in the averaging process.

A. Conan Doyle, the creator of Sherlock Holmes, described this dilemma very well: "While the individual man is an insoluble puzzle, in the aggregate he becomes a mathematical certainty. You can never foretell what any one man will do, but you can say with precision what an average number will be up to. Individuals vary but percentages remain constant."
The so-called Harvard Law of Animal Behavior

Reprinted from the *Bulletin of the Menninger Clinic*, 1970, 34, 71-84, by permission of the author and the publisher; Copyright © 1970 by The Menninger Foundation.

captures the essence of this problem even more succinctly: "Under the most tightly controlled of experimental conditions, the rat does as he damn pleases."

The following article by Gardner explores the issue of individual differences in personality formation that result from the reality of unique persons interacting with unique environmental situations through the course of their development.

Riley W. Gardner holds the position of senior psychologist at the Menninger Foundation in Topeka, Kansas.

25

In dealing with the problem of individuality in development, some recent work on cognitive and personality structure formation seems to expand both our clinical and experimental knowledge of individuality with respect to personality organization. The studies of the cognitive and personality organizations of adults described here include measures of defense and intellectual ability structures along with more recently developed measures of some previously unexplored aspects of individuality. These studies have provided us with a larger group of structural variables with which to consider the problem of individuality of development.

Each of the areas or aspects of behavior in which new forms of enduring individuality have been discovered has been explored in studies of both adults and children (see also, Gardner & Moriarty[1]). As I shall discuss in more detail later, the individual consistencies apparent in these areas of behavior are as evident in the performances of preadolescent children as in the performances of adolescents and adults. The developmental studies included in our work have also given us some new glimpses into the complexity of personality organization and complexity of its development.

Studies of Selectiveness of Attention

Attention could be called the key to reality contact. One of our persistent concerns in our clinical-experimental work on individuality has been that of the capacity of individuals to attend *selectively* to relevant internal or external stimuli in the face of intense stimulus competition from irrelevant, highly distracting material. This important capacity is related to the behaviors that Ernest Schachtel has described in more general terms in his discussion of "focal attention," and it is an adaptive tool of major importance. Studies here and elsewhere have shown that individuals differ widely and consistently in their capacity for selectiveness of attention in a remarkable variety of situations. We have been able to show, for example, that individuals who are unusually adept at attending to relevant materials in complex external stimulus fields are also unusually adept at extracting relevant information from confusing arrays of compelling and competing but irrelevant memories. This cognitive capacity or ability is apparent in laboratory procedures ranging from embedded-figures tests to one class of visual illusions. It is also apparent in certain aspects of response to projective tests, in the perception of persons and objects, and in a wide variety of other situations. The work of Witkin and his associates has shown that these individual consistencies are part of a broader cluster of consistencies that have implications for additional aspects of behavior.

In the course of our work, we have explored relationships between the capacity to attend selectively under conditions of competing stimuli and the defenses of isolation and repression. You may recall that Freud spoke of concentration as a key aspect of the defense of isolation. Our empirical work, however, reveals that the overlap between the cognitive control of selectiveness of attention and the defense of isolation, which involves a number of other processes and operations, is very partial, at best. The relationship is, in fact, clearer, although still quite limited, with repression than with isolation *per se*. That is, we seem to have articulated an area of cognitive consistency that is highly important to adaptation and that, at the same time, maintains a considerable independence of defense structures.

Studies of Extensiveness of Scanning

Another new area of individuality explored in these recent studies is that of the attentional scanning strategies persons employ in evaluating various stimulus situations. We have discovered in these studies that the individual is characterized by an enduring scanning strategy that adds another distinct new component to the uniqueness of his overall personality organization. We have examined the relationship of this new dimension of cognitive organization to major intellectual ability and defensive mechanisms and have found it largely independent of these other variables, although there are partial links to isolation and projection in the case of adults. Interestingly enough, these scanning strategies are also independent of selectiveness of attention. That is, this portion of our investigations has revealed that the controls of attention are multi-dimensional and of a much greater degree of potential complexity than we had anticipated. Still other aspects of individuality of attentional behavior must be explored in relationship to the dimensions already isolated before the complexity and subtlety of the controls of attentional behavior are fully understood.

Studies of Assimilative Interaction Among Percepts and Memory Traces

Another area of recent work on individual consistencies in cognitive behavior deals with what can be called assimilative interaction between memories and new percepts. Although it has not been generally recognized, individuals differ widely in the degree to which new experiences and memories interact. Individual consistencies in this aspect of cognitive behavior can be demonstrated in laboratory procedures that show adequate reliabilities over periods of years. That is, persons differ greatly in the degree to which their conscious experiences are infiltrated or colored by previous experiences.

Extreme degrees of assimilative interaction occur in certain types of illusory experiences and in one type of hallucination in which an intensely activated memory image interacts with a percept or a partial percept to produce the hallucinatory experience. The "bush-bear" phenomenon is another example of this relatively extreme kind of assimilative interaction.

Freud[2] spoke of repression as involving two major processes: (1) assimilation of new experiences to conceptually organized memories of earlier experiences that seem to the individual to be related to the new experiences; (2) counter-cathexis, the aim of which is to keep repressed material from entering consciousness. In the course of our empirical studies, we have explored the relationship between assimilative interaction, as measured in the laboratory, and repression. In the case of female subjects, both adults and children, a low but reliable degree of relationship appeared. There is a great deal of *non*-overlap, however, between assimilation and repression in these subjects, and the relationship does not appear at all for male subjects, for whom the range of repressiveness is somewhat smaller. These findings have led us to speculate that a tendency toward assimilation may be one of the prerequisites or developmental requirements for the emergence of repression as a major defense mechanism.

Studies of Conceptual Differentiation in Categorizing

Another area of our recent explorations of individuality with respect to cognitive behaviors is that of spontaneous categorizing behavior. After the earliest phases of development, every new stimulus is categorized or coded as it is experienced. The categorizing or coding process, and the anticipations involved in it, have distinct effects on the conscious experience of the stimulus itself, for example, through the process of assimilative interaction described above. Assimilation occurs, of course, only between new percepts and memories that are experienced by the individual—consciously or unconsciously—as related to the new percepts. Our interest in individual differences in spontaneous categorizing of objects, persons, and other stimuli thus represents a concern with a rather basic aspect of cognitive organization. Our method of exploring the initial hypothesis that individuals do, in fact, have consistently different categorizing styles was quite simple. In one of the original test situations, individual subjects were presented with 73 familiar objects of varied characteristics and asked to sort them in the way that seemed most natural, most logical, and most comfortable to them. In the course of our investigations in this area, we found that the range of individual differences in the degree of differentiation imposed upon such a random array of objects is remarkably great. In the first procedure of this kind that we employed, the range of performances varied from the formation of three groups to the formation of as many as thirty-three groups. We then explored the additional hypothesis that a consistent set of controls or control strategies govern the categorization of extremely varied kinds of stimulus events. For example, we developed tests in which subjects were required to categorize pictures of persons rather than objects. There were many reasons to believe that different sets of cognitive control structures would be operative in the categorization of persons, as contrasted to the categorization of objects. It is obvious, of course, that different classes of grouping concepts are involved, that very different reasons for grouping are employed, etc. The results of our studies show very clearly and reliably, however, that conceptual differentiation in categorizing is superordinate to the kinds of stimulus material with which the person is confronted. Our adult subjects who showed high degrees of conceptual differentiation in the categorization of objects showed high degrees of conceptual differentiation in the categorization of pictures of persons. Simple as it is, this seemed to us an impressive finding. Apparently there are controls of conceptual differentiation—of the categorization of experience—that are independent of the nature of the stimuli the person is confronted with and that give the stamp of individuality to a basic aspect of his behavior in a considerable realm of situations.

To explore the generality of operation of these controls still further, we then developed a number of other procedures, including what we call a behavior sorting test. In this procedure, we ask subjects to categorize a sizeable group of statements about behavior, such as "chewing gum," "going to a dance," "running for president," etc., in the way that seems most natural to them. The individual differences in conceptual differentiation that were apparent in the categorization of objects and persons were related to the individual differences in the categorization of these statements, which, indeed, was new evidence for the generality of operation of a cognitive control structure. Other studies have been conducted here and elsewhere that included such limit-testing types of stimuli as blocks of Chinese characters which were administered to groups of American students who had no familiarity with the Chinese language or symbol system. Conceptual differentiation in the categorizing of these Chinese characters was also predictable from performance in the other sorting tests.

Our explorations of categorizing styles have shown, then, that persons differ consistently in the spontaneous categorization of varied aspects of their experience. Other work has shown that, in the case of children at least, conceptual differentiation in spontaneous categorizing is related to explorativeness and, possibly, to one aspect of creativity. That is, children who form a few large groups when asked to categorize arrays of heterogeneous stimuli are more explorative and creative than children who form large numbers of small groups from the same materials.

On the face of it, the consistent formation of many small groups from such stimulus arrays seems to imply a "compulsive" approach to the ordering of experience. It is significant to note here, however, that neither the defense of isolation nor the intellectual abilities that are often prominent features of obsessive-compulsive orientations are correlated with these highly consistent and enduring categorizing styles. The same can be said of other major defense and intellectual ability structures.

Our work on categorizing behavior has, then, revealed a new dimension of cognitive structure formation that seems of rather basic importance to individual differences in the organization of experience and that should be taken account of in more comprehensive considerations of personality organization.

Studies of Tolerance for Unrealistic Experiences

We have referred to our exploration of another area of individual consistencies as studies of tolerance for unrealistic experiences. In this part of our work, we have employed such clinical procedures as the Rorschach Test, which offers the subject an implicit invitation to reveal his private thoughts, fantasies, and motives. We have also developed laboratory procedures in which the subject is confronted with stimuli that, if accepted at face value, will produce experiences that defy what he knows of conventional reality. Individuals differ widely, and consistently, in the degree to which they have such unrealistic experiences under these conditions. Response to the unreality-compelling stimulus configurations in the laboratory was predictable from the subject's general approach to the Rorschach Test, with its invitation to respond in personalized ways that are similarly fanciful or "unrealistic." Subjects whose Rorschach protocols are rated high in tolerance for unrealistic experiences tend to have the reality-defying experiences in the laboratory to a relatively great degree. Subjects who are rated as more literally conventional in their general attitude toward their Rorschach Test responses tend to have lesser degrees of the unrealistic experiences impelled by the special laboratory situations we employed.

Individual differences in this general stance toward reality-unreality seem to imply rather profound differences in the characteristic nature of waking consciousness. What is real to one person is not real to another. What can be experienced by one person cannot be experienced by another. As I shall suggest later, such basic differences in conscious experience may in themselves have important implications for empathy, communication, and other forms of mutuality in interpersonal relations.

Studies of Contrast Reactivity

Another recent area of our work has dealt with individual differences in what we call contrast reactivity, that is, the degree to which individuals appreciate and respond to the contrast between stimuli that differ greatly. Once again, highly consistent individual differences are apparent in our laboratory procedures. In a recent study, these differences in contrast reactivity also accounted for a surprising amount of variance in the evaluation of other persons. That is, contrast reactivity seems to be an aspect of behavior that reflects the presence of a consistent patterning of response within the individual in situations ranging from laboratory tests to perception and the organization of experience in everyday life. This aspect of cognitive organization is also independent, our study showed, of the other newly explored aspects of cognitive control referred to earlier. This is a significant finding, since the work on assimilative interaction deals with one aspect of response to similarity and the work on categorizing deals with another aspect of response to similarity-difference. The complete independence of these facets of response to different aspects of sameness-difference provides a meaningful glimpse into the complexity of cognitive structure formations that are key aspects of the organism's information-processing system and, as such, are vital features of overall personality organization.

Implications for Interpersonal Relationships

Like each defense mechanism, each new dimension of cognitive control has important implications for individuality of waking conscious experience.* When we consider the interactions of categorizing with assimilative interaction and other enduring features of personality organization, and when we examine the myriad combinations of multivariate patternings of structure formation characterizing our subjects, we find ourselves dazzled by the complexity of individual personality organization and the extremeness of structural difference that can characterize two persons.

Taken together, our empirical results lead directly to the argument that individuality of personality organization and individuality of conscious experience is so wide-ranging as to raise questions concerning the degree to which true empathy, communication, and tolerance is possible between extremely different persons.

When viewing the full range and complexity of individual differences, one wonders how much extra cognitive work must be performed by two persons with extremely different personality organizations in order to communicate and interact effectively with respect to more than generally agreed upon aspects of external reality. Defenses, intellectual abilities, and the newly conceptualized

---

*See R. W. Gardner: Organismic Equilibration, Cognitive Structure Formation, and Individuality of Conscious Experience. *Journal of Studies in Consciousness,* in press.

aspects of cognitive structure formation referred to here are important features of the information processing system that is a bulwark of personality organization. As such, these structures and their combinational variations may have implications for interpersonal relations that will ultimately prove to be of clinical relevance. I am pointing here to future areas of research and to findings that are not yet in hand.

It is already known that awareness of differences in belief and value has direct and immediate effects on evaluations of the worth of other persons. Newcomb[3] and others have shown with vivid clarity that a person who differs from another person with respect to important values or beliefs, including political and religious beliefs, concepts of child rearing, or, I might add, theoretical orientation, is automatically devaluated. In terms of tolerance for and acceptance of other persons, a very good case can be made that one is truly tolerant and accepting only of those who are his "identical twins" with respect to value and belief. But what of differences in the structural organization of behavior *per se,* independent of the contents of attitudes and beliefs? Here is a largely unexplored area that could hold much that is new and valuable to those concerned with interpersonal relationships. In advance of the major studies that will need to be done in this area, I would speculate that differences in personality organization will prove to have effects on interpersonal relationships that have been largely overlooked, for example, in clinical orientations in which behavior is explained entirely or almost entirely in motivational terms. I would suggest this for reasons that seem to fit what we know of evolution. That is, I would speculate that the ultimate source of intolerance of other persons of different beliefs *or* different personality organizations may be the adaptive necessity to keep cognitive effort and threat to one's own unique cognitive organization at a minimum.

It is clear by now that each individual has a relatively unique mode of processing internal and external information that is one component of his unique conscious experience and the unique psychological world in which he lives. Major structural differences would seem to make the understanding of another's personal organization inevitably difficult. The same point could be made for tolerance of or empathy with another person's work style, thinking style, or communication style. This point could also be applied to the child who—for whatever reason—has an overall personality organization that is markedly different from those of his parents or siblings. Let us consider, for a moment, the conflict that could ensue when a well-meaning parent with a high degree of sensitivity and excitability and an unusually rapid tempo of information processing interacts with a well-meaning baby with a low degree of sensitivity and excitability and an unusually slow rate of information processing. Such characteristics of the child and the parent have often been dealt with as if no genetic factors were involved. But we know from genetic studies that temperamental variables often have relatively high degrees of hereditary determination. The different overall styles characterizing this parent and child, in part as a result of multivariate temperamental differences, should produce interpersonal problems above and beyond those produced by specific motivational variables. It is simply more work and more frustration to attempt to teach a child with whom it is difficult to empathize because of the limitations imposed by one's own personality organization. Observation suggests, too, that extreme differences in personality organization are sometimes *experienced* by the persons involved as being motivated, *e.g.,* as aggressive, resistive, or uncooperative, or at least as threatening to one's own mode of organization and the worldview it yields.

The same general hypothesis could be applied to doctor-patient relationships, teacher-pupil relationships, or even relationships between groups of persons. Recognizing that similarities and differences in personality organization may attract as well as rebuff at some point on the similarity-difference continuum, or as a function of the nature and purpose of the interpersonal interaction, it seems possible that extreme difference in personality organization may evoke more effort, more learning, and more affect in the course of interpersonal interaction than does relative similarity of personality organization.

I am suggesting that tolerance-intolerance of other persons may be attributable in part to formal aspects of personality organization. I have suggested elsewhere that each individual tends to promote his own style of personality organization to other persons, including his children and patients, in a way that seems to indicate some kind of primary narcissism. I am also suggesting, however, that one's personality organization places distinct limits on his capacity to understand and empathize with others who are very different from himself, and that what appears to be narcissistic investment in one's own way of doing things may, in actuality, be an inevitable consequence of *any* particular patterning of personality organization. To *be* organized—that is, to develop an elaborate set of cognitive structures that serve to maintain overall equilibration of the personality within adaptively necessary limits—is necessary for the survival of the organism and is provided for by its evolution. But this individual evolutionary strength implies an inevitable interpersonal weakness.

Studies of Structural Development

I have described, then, a gradually broadening framework of personality variables relevant to the problem of individuality of personality development. Although the results and hypotheses concerning patterning that I have referred to thus far are relevant to both adults and children, the question remains of how individuals come to be the unique personalities they are.

As a first step in the exploration of this general question, we administered batteries of procedures originally used in our experimental studies of adults to a sizeable group of preadolescent children. This study revealed certain things about development that are surprising in the light of current theories and current lore concerning development.

First of all, the behavior of preadolescent children is characterized by clearly organized patterning that is much more similar to that of adults than was previously assumed. The *contents* of their thoughts and beliefs are quite different from those of adults, in ways that we are aware of. The *patterning* of their behavior, that is, the formal characteristics of their personality organizations, is remarkably similar to those of groups of comparable adults. The *patterning* of defenses, for example, is all but indistinguishable from that of adults. The patterning of cognitive controls is also indistinguishable from that of adults and—although no developmental theory led us to anticipate it—could have been predicted in advance from our earlier findings concerning the dimensionality of these controls in normal adult subjects. The *patterning* of intellectual abilities is also similar to that of adults. These findings stand in direct opposition to the usual assumption that children are more globally and simply organized than adults.

The second major thing we learned in our study of preadolescent children is that cognitive *development,* like cognitive organization itself, is many times more complex than has been envisioned. We found, for example, that in one area of cognitive development there is a general progression from preadolescence to adulthood toward an increase in the adaptive efficacy of performance. In other areas of cognition, however, the adaptive efficacy of performance levels off at preadolescence and changes hardly at all to adulthood. In other areas of cognition, aspects of defense and control that are statistically independent at preadolescence become part of a correlated family by the time one reaches adulthood. In other areas of cognition, links between defense and control are the same at preadolescence as at adulthood. There were even more dramatic results, however. We found, for example, that discrete performances in particular areas of cognition that have developmental curves proceeding in opposite directions nevertheless remain correlated from preadolescence through adulthood. That is, we obtained the first unique evidence we know of to show that patterning is autonomous of function. I shall not deal here with other results which show in greater detail that different aspects of cognition develop at remarkably different paces, have remarkably different developmental curves, and have quite different implications for the ultimate patterning of adult behavior.

No possible theory oriented toward stages of development in all children could cope with the organized complexity one sees in studies of individuality of personality development. Our study of preadolescent children was performed "between two levels or stages" of Piaget's valuable and illuminating theory concerning universal stages of intellectual development. Nonetheless, extremely complex, predictable patternings of personality organization were clearly evident at preadolescence, as in adulthood. Nothing in Piaget's theory would lead one to suspect the presence of such complex, enduring, individually different patternings of personal organization. Since all the children involved were at the same general stage of Piaget's theory, no reference to his theory was required to explain the findings. The complexity of behavior and of behavioral development is such that two adequate theories can be employed to account for different aspects of human development and not overlap at all.

Much more information is needed concerning the multiple developmental progressions that lead to adult personality organizations. One of our current studies is designed to fill in some of the initial gaps in our knowledge. I refer here to our five-year study of twin families, in which large amounts of data have been collected on 100 twins and their parents. This study allows us to explore the differential importance of heredity to the formation of a large number of specific personality structures. It also includes evaluations of parent-child similarities in personality organization, parental attitudes as determinants of children's structure formations, developmental changes in structure formations, developmental changes in structure formations, and a number of other major aspects of personality development. One relatively unique feature of the study is the administration to parents and children of identical procedures for the assessment of defenses, cognitive controls, and other aspects of enduring patterning in personality organization. That is, we are engaged in a large scale exploration of the *parent as a structural model*. It seems to us that the parent as a structural model and as a structural model who promotes his own personality structure must be a key factor in the child's development. It can be argued, in fact, that the structural patternings of the two parents are one major facet of the psychological world of the child. It is a universal experience that to leave home is to confront oneself with "subculture shock" in interactions with others who do things in uncomfortably different ways, operate by uncomfortably different rules, employ uncomfortably different methods of information-processing, and come to uncomfortably different conclusions about reality by different routes and processes from those one has grown up with. Part of this shock comes from the fact that the home operates according to the behavioral rules dictated by the parents' personality organizations, in addition to their specific attitudes, beliefs, and values. The formation of a relatively unique patterning of overall personality organization in such a developmental matrix leaves the individual almost totally unaware—without very special training or very special experience—that others are processing information in different ways. To be a highly repressive person who feeds the product of narrowing categorizing into a highly assimilative process of percept-memory interactions, who is also rather poor at selectiveness of attention, who is intolerant rather than tolerant of unrealistic experiences, etc., implies relative uniqueness with respect to the waking conscious experience of the individual.

Certain aspects of the difficulties we experience with children may be attributable to differences in personality organization and their consequences for conscious experience. The kinds of communication difficulties and empathy difficulties that would seem likely in the presence of different cognitive structures and different conscious experiences would almost inevitably appear as resistance in a treatment process, by our standard definitions of resistance. Similar conclusions could be drawn about the teaching and learning styles implied by different personality organizations and the need for further understanding that will lead to more effective coordination of teaching styles to learning styles.

We are currently designing studies to explore the effects of similarity-difference of overall personality organization on mutual evaluation, mutual understanding, and capacity to work together in dyads. We have hopes of opening up in a more expanded way a relatively unexplored area that may be of some importance to interpersonal relationships.

The work that I have summarized in part here points to the necessity of an ultimate theory of personality development that is considerably more complex with respect to structural organization than any theory we have envisioned in the past. It seems to those of us who work in the clinical-experimental research laboratory that adequate raw materials for such an ultimate theory will not be available for some time. We have learned new things that were not known and that in some instances are directly counter to what all major theories of development have stated or assumed to this point. But much more must be learned before we can fully outline a truly comprehensive theory of personality. This is not a reason, however, to delay the improvement of current theories along the way. I have argued elsewhere[4] that a major task confronts us with respect to using what we learn of theory construction to update any theory and, where necessary, to increase the level of abstraction at which it is pitched. The major thrust of a recent theoretical essay is that current personality theory must be organismic in nature, that it must be pitched at a more abstract and parsimonious level than the push-pull, force-counterforce level at which early psychoanalytic explanation was pitched, and that one way of accomplishing these purposes is to explain certain aspects of behavior in terms of such control structures as those explored in our studies. I would argue that we already have enough facts to make it plain that all current theories of behavior are incomplete in some ways.

Although it is a limited hope, it is our hope that a more complete understanding of the nature of structure formation and its critical role in the evolution of the organism will add new dimensions to our understanding of the implications of individuality for interpersonal relations. When applied to children's responses to particular child-rearing strategies, teaching and learning models, or therapeutic models, even our current knowledge of the wide range of individual differences in personality organization suggests that no single method or technique can hope to suit any large number of individuals. All of us who have worked in the clinic know this. The hope for the future is that we will articulate the whys and wherefores of a broadened conception of individuality so that we can understand its implications for our attempts to assist the developmental process.

References

1. Gardner, R. W., and Moriarty, A. *Personality development at preadolescence: Explorations of structure formation.* Seattle: University of Washington, 1968.
2. Freud, S. (1915) Repression. *Standard Edition,* **14**, 146-158.

3. Newcomb, T. M. *The acquaintance process.* New York: Holt, Rinehart and Winston, 1961.
4. Gardner, R. W. Organismic equilibration and the energy-structure duality in psychoanalytic theory: An attempt at theoretical refinement. *Journal of the American Psychoanalytic Association,* 1969, **17**, 3-53, 65-67.

This paper was presented at a regional meeting of the American Association of Psychiatric Clinics for Children, Topeka, Kansas, September 19, 1969. The work at the Menninger Foundation referred to in this paper was supported by the Menninger Foundation and by the following grants from the United States Public Health Service: research grants MH-25, MH-1182, MH-2454, MH-05517, and MH-14390, and Research Career Program Award and Research Scientist Award K3-MH-21-936. The author is indebted to Dr. Joseph Kovach, Miss Randine Davis, and Mrs. Diedre Michael for critical readings of this paper. In addition to Miss Davis and Mrs. Michael, current members of the research group performing the empirical work referred to include Dr. Leander J. Lohrenz, Miss Lolafaye Coyne, Miss Sue Whiffen, Mr. Wagner Van Vlack, Mr. Russell Ewbank, Mrs. Elaine Malinovitch, and Mrs. Pat Schoenfeld.

The term, "cognitive structure," in this paper refers to relatively enduring patternings, programs, strategies, or "styles" of cognitive functioning. The original model for the delineation of such patterning was contained in Freud's descriptions of defense mechanisms as enduring cognitive structures.

Recommended Additional Readings

Emmerich, W. Personality development and concepts of structure. *Child Development,* 1968, **39**, 671-690.

Escalona, S. K. *The roots of individuality.* Chicago: Aldine, 1968.

Vale, J. R., & Vale, C. A. Individual differences and general laws in psychology: A reconciliation. *American Psychologist,* 1969, **24**, 1093-1108.

Political Socialization in the Schools

Robert D. Hess

Orientation

One of the expectations a growing individual learns early in his life is that he is to become a "citizen" of his society. Accordingly, he is told that America is a democratic society, and thus he is expected to participate actively by attending meetings, voting, keeping abreast of political developments, and obeying laws. Somehow, however, all this teaching does not work very well. There is a paradox resulting from the discrepancy between what children learn about being a good citizen and what adults actually do. More than a third of the body of potential voters never exercise their right in national presidential elections, and a far greater number never bother to show up at the polls in minor elections.

The learning of the role of citizen is a complex process, probably equally as complex as other forms of socialization, including sex roles and vocational roles. The influence of schools in this process must be recognized, especially considering the importance these institutions play in the lives of children and youth. Typically schools teach abstract, vague notions about "democracy," "freedom," "responsibility," "respect for the rights of others," and other such concepts. Indeed these are noble concepts, but they are far removed from the realities of the society in which the individual is growing up. Certain ideas are taught as truth, not as relativistic assumptions and values. Just what sense a child makes of all this, and how he relates it to his immediate environment and needs, is not at all clear. Perhaps it is for this reason that heavy-handed training in citizenship in childhood has little relation to the adult's sense of social responsibility.

In the next article Hess presents a frankly polemical statement on his views of the nature of political socialization practices in schools. He is very forthright in describing what he sees as the deficiencies of this "political indoctrination." The reader is encouraged to consider his remarks in light of his own experience in school.

Robert D. Hess is a professor in the school of education at Stanford University.

Reprinted from *Harvard Educational Review,* Summer 1968, 38, 528-536, by permission of the author and the publisher; Copyright © 1968 by President and Fellows of Harvard College.

26   In its formal meaning, socialization is the process of transmitting stable patterns of behavior and values and of grooming the young for filling established adult roles in the society. Since it assumes stability and consensus in the adult population, *political* socialization is a concept which is difficult to apply to the process of political learning during periods of rapid social change and of open conflict between major segments of a society. It is of little usefulness in the United States today. The recent protests, riots, and other manifestations of division and disagreement within the society have important implications for both the conceptualizations that are relevant to the teaching and learning of political behavior in pre-adults and for the role of the educational establishment in the socializing of political values and orientations. It is in this context that the comments of the polemical paper are to be understood.

    In addition to the task of conveying information about the political systems and its processes, the school also acts to promote and teach political values and traditions. In this sense, political socialization in the school is a form of political indoctrination, designed to perpetuate the dominant values of the present system. In the past, the schools have served an acculturating, melting-pot function, providing common allegiance and values to bring together in a single country immigrant groups from different ethnic and national backgrounds. It now seems, however, that the ethnic and cultural differences within the nation cannot be easily blended into unity. Divergences and inequities which have been ignored, particularly with respect to Negroes in the society, are dramatically apparent. It is evident to many citizens that the picture of unity, equality, and freedom that is so often presented is distorted, over-simplified, and to a degree, false. Indeed, political socialization in the schools may have created an attitude of complacency, a willingness to accept the image of unity and freedom—as well as the actions of the government—and, in so doing, it may have contributed to feelings of disillusionment and the consequent climate of protest. It is by no means assured that the schools can now deal with the issues of political socialization that these new conditions present or that adequate changes can be effected which would provide more relevant teaching of political attitudes, values, and behaviors.

Empirical evidence indicates that in the United States, young children view the nation, government and its representatives, and other elements of the political system as wise, powerful, and benevolent (Greenstein, 1960, 1965; Hess & Easton, 1960; Hess & Torney, 1967). These attitudes are accompanied by an acceptance of the values, norms, and customs of the nation to a degree that contributes to the political stability the United States has experienced (Easton & Hess, 1961). The image of the United States, then, is one of a country which is an effective force for world peace, in which the laws are fair, justice prevails, people who break laws are usually apprehended, and what goes on in government is all for the best. This is the nature of the socialization that children receive both in the home and at school. Accompanied by feelings of strong attachment and loyalty, this image of the nation is designed to encourage pride in one's country and a desire to maintain it and support its policies. In transmitting this orientation to the political world, the school represents the convictions of the community. This view of the United States and its government is both distorted and incomplete, and in the past few years these inaccurate representations have been seriously challenged, particularly by college students and other young adults who have recently been through the socializing experience of the schools. A number of factors alter the idealism of the young child; subsequent general experience, an effective communication network, continuing civil rights demonstrations, and specifically the prolongation of the Vietnam war have brought disillusionment and a demand for change. The increasing volume of protest reflects the new realism; young people no longer find either government action or social and economic reality congruent with the national ideology and rhetoric of morality, civil rights, equality of opportunity, or desire for peace. It is not these values themselves that are under attack but the failure of representatives of the society to recognize the disparity between ideals and reality and take some appropriate action. In short, the spirit of protest (which may be expressed in a range of constructive and destructive forms) is based on a feeling of having been misled and on frustration over the lack of effective action. As distinguished from violence against persons and property *per se,* the targets of these protests are the institutions of the society and their representatives. These feelings are focused on the question of whether military destruction of the Vietnamese countryside and its population is justified on any grounds (together with distaste for public statements protesting that our forces there are defending the Vietnamese against aggression) and on our treatment of visible minorities—including our present policy with respect to American Indians, our treatment of Indians during the Caucasian invasion of the country (of which Theodora Kroeber's *Ishi in Two Worlds,* a story of Ishi and the Yahi of California, is a depressing example), or our current policies toward Negroes, Mexican-Americans, Puerto Ricans, Chinese, and other ethnic groups.

In other sectors of activity, there is growing dissent and protest over the inability of the industrial society to prevent pollution of basic natural resources, and over the degree to which the political and economic system we have been taught to cherish has created slums of cities, has failed to educate huge numbers of the population, and has been willing to allocate enormous sums of money for dubious military ventures abroad or for conquest of space while permitting gross poverty to exist in both rural and urban areas.

The internal difficulties of this country have created divisions along three natural or structural fault lines in the society—between ethnic groups and the dominant society; between the affluent and the poor; and between generations. Although these conflicts are concerned to some degree with issues of redistribution of wealth and other material resources, the focus of antagonistic interaction is over the division of power.

The point of this now-familiar accounting of the profound troubles of this nation is not to view them with alarm but to inquire both about the participation of the school in their etiology and about the implications of these conditions for educational policy, especially in the areas of civics, government, history and other aspects of political socialization. *It is the argument of this paper that the schools have contributed to divisions within society by teaching a view of the nation and its political processes which is incomplete and simplistic, stressing values and ideals but ignoring social realities.* In teaching the virtues of the system we fail to train the young to take effective constructive political action, leaving them with few routes to legitimate influence within institutions. This failure has contributed to the emergence of an anti-establishment ideology which has developed largely outside the institutions of the society. Itself an object of attack and criticism, the school may be unable to change adequately to make effective response.

Several features of political socialization are relevant to the role of the school in our present difficulties and to the question of what modifications might be usefully considered for the teaching of political values and behavior in the public schools. First, in the past, central goals of citizenship training have often been building character, instilling obedience to rules and laws, and developing respect for the authority of the school and society (Hess & Torney, 1967). The emphasis has been upon acceptance and on relatively passive concurrence with the status quo. This stance is an outgrowth both of a desire to minimize disciplinary problems and of a belief in the nation and its government as for the most part powerful, wise, and of good intent. One outcome of this orientation is the feelings that there is no urgency about change and the solution of problems.

A second feature of political socialization is a tendency to emphasize that the vote is the major route of influence of citizen to government and to de-emphasize or neglect to train young citizens in techniques for constructive, effective influence upon governmental representatives after they have been elected. There is little teaching of the realities of political influence, either at the community, city, or federal level. The young child sees the vote as a sort of personal clout to which all have equal access and which makes his opinion roughly equal to that of any other individual. The routes of persuasion by group action, by alignment of institutions and voluntary organizations, are relatively unknown to him; he overestimates his own power until he attempts to have an effect upon politics or institutions in government.

A third feature of political socialization which probably interacts with development in other areas is the distaste of children for conflict and division within the society (Hess & Torney, 1967). Elementary school children like to think that political cleavages are healed when the election is over. There is a reluctance to confront open conflict and to accept the possibility that acrimony may reflect something other than transitory, productive competition carried on within the accepted structure of the political game. Conflict and divisiveness are easily dismissed, or ignored. This reluctance is shared by teachers, often, particularly since discussion of truly controversial topics has hazards for their relationship with colleagues and community.* There is thus a tendency to avoid candid consideration of unpleasant social and political facts.

The fourth feature of political socialization follows from the first three. Political values and principles are often learned as slogans rather than as concepts to be applied to social issues. Students who have learned the Bill of Rights may still oppose the right of a speaker from an unpopular ideology, such as Communism, to speak in the town square; they may be able to recite the phrases about freedom of religion but be opposed to hiring an atheist as a teacher in the school (Remmers & Radler, 1957). These responses suggest that teaching of the phrases has not sufficiently involved a comprehension of the underlying principles nor of the long-term consequences that will follow from ignoring basic rights. *Because they apparently were assured, we have felt less urgency to teach understanding of the consequences of a departure from these traditional values.* This lack of cognitive cohesiveness among attitudes is illustrated in research on children's attitudes toward political objects and principles by low levels of correlation among items which, on their face value, should be highly related (Torney, 1965). The most consistently interrelated cluster of feelings and attitudes stems not from expressions or opinion or belief but from alignment with political parties. This suggests that much of the meaning of political socialization, in the past, derived from early alignments with parties and nationalistic identification and loyalty rather than from adherence to democratic principles.

In short, much of the political socialization that takes place at elementary and high school levels is lacking in candor, is superficial with respect to basic issues, is cognitively fragmented, and produces little grasp of the implications of principles and their application to new situations.†

---

*There are a number of exceptions to this pattern, such as the material produced by the Harvard Social Studies Project (Olicer & Shaver, 1966).

†The process of socialization is related to social class and intelligence, with the result that the children with less understanding of the phrases or the underlying concepts are working

What is new in this picture, of course, is the strength of protest that comes from within the society and the focusing of attack upon institutions of the society. This in itself is a sign of vigor and of belief, or at least a hope, that some remedy can be found short of full revolution. The discrepancy between the espoused principles and values and the reality of discrimination and poverty has been evident for a long time; what is new is that there are serious, widespread attempts to take effective action. The degree of violence involved is itself a measure of the severity of the injustice and of the degree of our present intransigence and resistance to change.

What then is the political socialization that should be taking place in the public schools during what will surely be a long period of crisis? This is a problem which obviously has no easy solutions and none are offered here. Some observations and comments about the nature of the problem, however, may be appropriate.

Political socialization involves both the content to be taught and the procedures for making the teaching or training useful. One of the complexities of teaching about human interaction, particularly if such teaching is also training for future behavior, is that the basic elements of human behavior cannot be easily brought into the classroom. In many fields of instruction, the raw materials about which the course is structured are more or less available for inspection and experimentation. The natural and biological sciences, while not able always to replicate the pattern and extent of the phenomena, can nonetheless deal with materials which react in much the same way in the laboratory as in the outside world. Human behavior presents quite a different situation. Political behavior, whether at the local community or national level, has strong components of *cognition* (ideas), *emotions,* and *action.* Our programs are typically oriented to teaching children to think about political matters and events, but do not give them the experience of confrontation of persons or groups in political interaction, with the powerful mixture of emotion, action, and ideas that are part of the political process. In a sense, they are taught to be observers, not actors. This observer role is further reinforced by the failure to inform them fully about the routes to political influence and about the steps that are necessary to effect change.*

The experience of conflict, the experience in which one not only observes but

class and of low or moderate educational attainment. This means that, given the relatively lower level of education of minority and poor segments of the population, there is less understanding of the basic principles of operation of the system among those groups that tend to be excluded from the resources and rights of the society. Were this not true, there would have been in all likelihood much more protest much sooner. It is no accident that a primary source of protest has been the student groups and that the leaders of the civil rights "revolution" have been persons of considerable formal training and experience with ideas and the relevance of ideas to action.

*This does not assume that we have prepared teachers to deal effectively with these issues or that they will be permitted to do so (Zigler, 1966).

also confronts the emotions and actions of others (student protests on campus are one example; militant black power groups in interaction with whites are another; to a less-intense degree, community politics and elections provide comparable events) is one for which our present practices do not prepare the student. He is not typically in a position to understand the feelings and responses of others, or of himself, in such events. In short, the *teaching of social and political interaction omits both the components of emotion and of action*—the two elements that are most likely to effect change.

It is no longer effective, perhaps, to think of socialization in terms of transmitting the norms of the system; a more useful perspective is the teaching of principles which underlie the normative statements. In times of change, policies are challenged and it becomes necessary to re-examine the rationale and the underlying principles on which they are based. Very few children are taught the issues underlying the origin of the principles of freedom of speech or why it has to be protected; few have clear ideas about why there should be separation between church and state or why it is necessary to protect freedom of the press even when the press disagrees with one's own views. There is a logic of how a complex system works that is difficult to teach. Exposure to such rhetoric as that of the Bill of Rights does very little to bring about effective understanding of how an individual may interact with the system to support or change it. There is also a purpose to be served by making the child aware that there are non-rational components of the political process and that these features of political behavior, as well as knowledge and understanding of the system, affect the outcome of political events. For example, alignment with political groups often occurs by modeling, imitation, and other forms of identification. Therefore, in addition to teaching analytic concepts, we must examine and explain the dynamics of emotional attachment to, and support of, political groups.

Political socialization assumes an acceptance of the system and a degree of agreement that its goals are appropriate and should be pursued. The role of the school becomes quite different if there are serious questions about the goals of the society. If there is widespread division with respect to some contemporary values and behavior, what is the function of the school in terms of political socialization? One example of this dilemma posed by diversity is the difficulty the school has in teaching minority groups. In the context of a lack of consensus within the society about integration, busing, open occupancy, etc., what is the role of the school in preparing children to deal with such issues: what can we in honesty teach children of Mexican-American, Negro, Indian, Chinese, or Japanese origin, about justice, political action, and equality of opportunity? Perhaps the most significant objective now is to teach involvement and understanding, using experiences which combine action and feeling with ideas of how a complex political and bureaucratic system works. Such an approach should:

1) be candid and explicit about social and political realities and disagreements;

2) emphasize both psychological and structural (sociological) processes such

as the role of anger and hate in race relations, and how institutions have contributed to racism inadvertently and deliberately;

3) provide channels for effective change and action—a knowledge of how the system works and how it may be influenced.

Perhaps our schools, which usually can act only on the basis of a consensus within the middle-class community, are not adequate to this type of training for interaction with the system. Perhaps the community will not permit the school to do an effective job. In any event, it is faced with a dilemma. If it inculcates traditional ideas, does it not deepen and polarize the conflict? If it does not build support and loyalty for the majority point of view, does it not teach dissent and intensify the conflict by opening the entire structure to question?

The position of the school as socializing agent for the community and its need for consensual support often forces it to be conservative, to evade the most sensitive issues and conflicts and by so doing, to promote apathy and ignorance. This evasion makes more likely the possibility of attack by dissenting groups because of the lack of response to obvious problems. There is some question whether these issues can be taught in the school and whether teachers are prepared to deal with complexities of this kind. The prospect is bleak, but perhaps the first step is candor and a reappraisal of the degree to which political socialization in the schools is congruent with political and social reality.

References

Easton, D., & Hess, R. D. Youth and the political system. In S. M. Lipset & L. Lowenthal (Eds.) *Culture and the social character: The work of David Riesman reviewed.* New York: Free Press, 1962.

Greenstein, F. I. The benevolent leader: children's images of political authority. *American Political Science Review,* 1960, **54,** 934-943.

Greenstein, F. I. *Children and politics.* New Haven: Yale University Press, 1965.

Hess, R. D., & Easton, D. The child's changing image of the president. *Public Opinion Quarterly,* 1960, **24,** 632-644.

Hess, R. D., & Torney, J. *The development of political attitudes in children.* Chicago: Aldine, 1967.

Kroeber, T. *Ishi in two worlds.* Berkeley: University of California Press, 1967.

Oliver, D., & Shaver, J. *Teaching Public Issues in the High School.* Boston: Houghton Mifflin, 1966.

Remmers, H. H., & Radler, D. H. *The American teenager,* Indianapolis: Bobbs-Merrill, 1967.

Torney, J. V. Structural dimensions of children's political attitude-concept systems: A study of developmental and measurement aspects. Unpublished doctoral dissertation, University of Chicago, 1965.

Zigler, H. *The political world of the high school teacher.* Center for the Advanced Study of Educational Administration, University of Oregon, 1966.

Recommended Additional Readings

Adler, N., & Harrington, C. *The learning of political behavior.* Glenview, Ill.: Scott, Foresman, 1970.

Campbell, A. Politics through the life cycle. *Gerontologist,* 1971, **11,** 112-117.

Hess, R. D., & Torney, J. V. *The development of political attitudes in children.* Garden City, N. Y.: Anchor, 1967.

27 Psychoanalysis and Education

Bruno Bettelheim

Orientation

The influence of the theorizations of Sigmund Freud upon American psychology, in particular, and psychology, in general, has been enormous. Yet paradoxically, few American psychologists today refer to or speak fondly of Freudian psychodynamic theory. This strange anomaly may be accounted for by a number of reasons, but it seems safe to say that many Freudian concepts have nevertheless been incorporated, in many ways, into general psychological thinking and research.

In the following article by Bruno Bettelheim, an unabashed though not uncritical Freudian psychotherapist, emphasis is placed upon the importance of the historical view for the understanding of the self, of the schooling process, and of the society within which one lives. Bettelheim calls for the recognition —by both teachers and students—of the past, both bad and good. This author makes a special plea for the view that learning occurs in children because of fear—a fear of anxiety, a fear of punishment, a fear of loss. Bettelheim's view contrasts sharply with the perspective taken by many contemporary "humanistic" psychologists and educators, such as Rogers and Neill, who insist that the most important learning throughout life takes place because of spontaneous enjoyment and excitement over the process of discovery and learning.

Bettelheim makes the point that significant

Reprinted from *The School Review,* 1969, 77, 73-86, by permission of the author and the University of Chicago Press. Copyright 1969 by The University of Chicago.

learning occurs chiefly as a result of hard work, although he does not deny the importance of rapture and joy in this process. He has been criticized by many for his emphasis upon the self-interest (egocentric) aspect of moral and attitudinal learning; his critics typically hold a view of human nature that assumes a basic cooperative nature of women and men. It should be pointed out, however, that Bettelheim insists this initial fear and its resultant learning may lead to courage and moral reform, at least among certain members of society. The reader may wish to consider whether this view is a necessary and all-encompassing belief, or rather one which merely justifies the existing social order.

Bettelheim's arguments relate directly to the issue of student activism and alienation (a topic to be explored in greater depth in article 28 by Keniston); it is clear that he does not hold student militants in high regard. The reader is encouraged to consider this point of view in comparison with positions held by other psychologists and educators—and, of course, by the student militants themselves.

Bettelheim further calls for the synthesis of psychoanalytic theory and educational practice. He insists that each camp has too long ignored the valuable contributions of the other. He may very well be right.

Bruno Bettelheim is the Rowley Distinguished Service Professor of Education at the University of Chicago and the director of the Sonia Shankman Orthogenetic School.

27

Education is man's oldest and best means of shaping future generations and of perpetuating his particular society. Psychoanalysis is our newest body of theory for understanding and modifying human behavior. How strange, then, that we are still without any psychoanalytic theory of learning.

Psychoanalysis has a great deal to offer education and much also to learn from it. Unhappily the relation between them has been most neurotic up to now, like a marriage where both partners are aware of their mutual need but do not really understand one another and therefore cannot pull together as one.

another and therefore cannot pull together as one. Disappointed, they come to ignore each other and go their own separate ways, though professing great mutual respect. The offspring of such a badly managed union might be likened to a bastard child. I refer to our present-day effort at integrating psychoanalysis and education. Like its parents, this child is torn by the contradictions between them—too sickly to thrive, too schizophrenic to realize the inner split that ails it.

Now I could talk at length about the loss to psychoanalysis because it failed to enrich itself with the insights of education. But I think the reader here is more interested in the other side of the coin—in what psychoanalysis has to offer that education has not yet made its own.

Let me begin with a suggestion of Freud's that education never took seriously —one that psychoanalysis, too, has done little to support. What Freud suggested, since the psychoanalysis of all children did not seem feasible, was that psychoanalytic insight be brought to all teachers. For the teacher, this would mean regaining her own repressed childhood memories, her reaching a better understanding of what shaped her in her own infancy and childhood, just as all children are shaped. Only then could she educate children in a way that was neither a repressive molding of the young nor an acting out on them of old fears and resentments.

The reason Freud's advice fell on deaf ears is that both the psychoanalyst and those educators interested in psychoanalysis have thus far been carried away by analysis as a method of treatment. What both have neglected is to take a serious look at the goals of psychoanalysis and the image of man they are based on. Because, while education and psychoanalysis may be similar in their aims— namely, to enable man to reach his own highest potentials, and in doing so create a truly human society—their methods and intermediate goals are very different and are often opposed to each other.

Perhaps the confusion arises because psychoanalysis is many different things, two of which count especially here. First, it is a method of treatment—and here the intermediate goal is to uncover the unconscious. But it is also a body of theories on human development, or more particularly on misdevelopment. Psychoanalysis is at its best in explaining what went wrong (and why) in human development. Education is very short on this but is at its best in developing the intellect when nothing goes wrong.

Even in the first psychoanalytic treatise on education, written in 1928, Bernfeld realized how different were the two distinct methods.[1] The educational process is meant to induce the child to accept and perpetuate the very best in existing society, whereas psychoanalysis (as therapy) is meant to enable the individual to fight free of much that society has imposed. That is, while education tries to perpetuate the existing order of the outer world, psychoanalysis tries to revolutionize the existing organization of the inner world. Hence their methods are opposite, though their ultimate goals may be similar.

Yet it is psychoanalysis as a method of therapy, a method designed for treating neurotic persons, that some educators and many psychoanalysts have sought to apply more or less intact to the education of normal children. If, instead, they had looked closer at psychoanalysis as a body of theory, one with a particular view of human development from infancy to maturity, both educators and analysts might have reached for their educative goals in ways more germane to education.

Freud himself stated clearly that the purpose of psychoanalysis is to see to it that "where there was id there should be ego." This basic goal, correctly applied to education, would have humanized and enriched it. Instead, and if anything of psychoanalysis was applied to education, it was the model of how to uncover and free the unconscious, the id. That is, educators influenced by psychoanalysis tried to expand the relative sphere of the id instead of that of the ego. But while freeing the id may be a needed step in treatment, it is often contrary to the methods of education.

The resultant influence of psychoanalysis on education has satisfied no one, though the disenchantment of the educators is rarely openly stated or clearly recognized. Most of those who for one reason or another went on trying to apply psychoanalysis to education behaved as if to say, "Since our efforts at freeing the unconscious don't seem to do much for ego, let's forget about the ego and just promote the id." It is an attitude that reveals itself most typically by finding laudable almost any disorganized thing the child does, because it reveals something. Or by viewing what he does as "creative," even when it is basically just an expression of the id—such as a smearing with paint, or a scribbling, or an outpouring in words of some formless inner pressures.

Now there is nothing wrong with the child's being able to mess and smear with paint, or to voice aloud his chaotic feelings. At the right time and in the right place, it may be very good for him to enjoy the chance to let go, be uncontrolled. It becomes damaging if the educator, who should know better, and should guide the child toward ego achievement, does the opposite. And he does that when he fools himself and the child into believing that if something has meaning as id expression it is therefore ego and superego correct (contains a meaningful message to others), though neither one is true. It is very much the task of education to see that the sphere of the ego should grow and be strengthened. But to do that, the teacher must know clearly what is ego correct and what is not. We stunt the child's growth if we view id expressions as creative, instead of being satisfied to recognize their possible value—that they may offer the child temporary relief and the teacher deeper insight into what is going on in the child. Thus, a dream may be very revealing of what is going on in the unconscious. But to dream it is hardly an act of creation, nor will it advance intellectual growth.

Even when later scribbling or drawing becomes more expressive, it is still solipsistic. It becomes meaningful to the child and possibly valuable to others only if—through a slow process of education, through observing and appreciating the efforts of others, through criticism, self-criticism, and the use of appropriate standards—his mere self-expression is transformed into a meaningful message to others, that is, becomes an ego achievement. The difference between an education that is informed by psychoanalysis and one that we might call a reactionary, authoritarian education is that the first one considers seriously the inner meanings of an outward-directed behavior. The second counts only the achievement, and never mind its inner meaning or its cost to the total personality. Not that the nonsense of educational obscurantists really concerns us. Only our own does. And what I call our own nonsense is the belief, for example, that education (or psychoanalysis) can proceed without taking due account of where the pupil stands in his ego development. This same issue explains why some of the best educators who become interested in psychoanalysis stop teaching children and become psychoanalysts. Certainly that was true for the very first psychoanalytic educators—Bernfeld, Aichorn, Anna Freud. Though inspired and inspiring teachers, they gave up teaching children once they became engrossed in psychoanalysis. To *practice* analysis and education just does not seem compatible.

Interestingly enough, the same seems to be the fate of recent educational reformers. To cite only one example, there is Kohl[2] who, as soon as he succeeded with some inspired teaching of his thirty-six Negro children, gave up classroom teaching. It was not (as he wishes to believe) because the existing educational system makes good teaching impossible but because (as I believe) he failed to understand the real problem. While he succeeded in interesting his pupils in some learning, the issue was not to get them to learn what he could make interesting to them—which he did—but to use education as a means of reorganizing their personalities so that learning would become a way of life for them—which he did not. In short, neither he nor we have yet solved the problem of how education can fortify the child's inner world to serve learning. But education cannot do this unless it applies the full psychoanalytic model of man and not just id or superego psychology.

How then does the child move from id expression to ego achievement? To begin with, let me stress that all efforts to educate have a great deal to do with the conscience, or superego. Children do not settle down to learning, do not sublimate because of ego tendencies alone, as we seem to want to believe. Witness the fact that our underprivileged children know perfectly well that it would be ego correct to learn in school if they want to get better jobs. But this they often cannot do. The reason is that in the arduous task of learning their ego lacks support from a strong superego.

What is overlooked is that much of learning is not just a pleasurable experience but hard work. And there is no easy transition from pleasure to hard work. If one has learned to enjoy both, then one can combine them. If not, one can do only the first and not also the second. Or to put it psychoanalytically: we do not reach ego achievements on the basis of id alone either. Id motivation gains us only what in some fashion pertains to the id.

While we can learn, on the basis of our emotions, what for one reason or another we want to learn, we can learn only that. Such learning can and does take place. That is why educators who reach their students this way, as Kohl did, are amazed at how fast and how much their children learn. It is also why they quit in disappointment when everything breaks down, which happens as soon as learning can no longer proceed on the basis of pleasure alone. All other learning (which means most of it) can occur only when we have learned to apply ourselves even when it gives no immediate pleasure satisfaction—that is, when we have learned to function on the basis of the reality principle. Because learning that gives no immediate pleasure means having learned to accept displeasure at the moment, and for some time to come, in the hope of gaining greater satisfaction at a much later time. And with modern education, this later time becomes very late, perhaps some fifteen years later. Indeed, the more the reality principle is taxed, the more likely it is to give way. If it does, then the pleasure principle becomes dominant again—unless the superego is very powerful, which it no longer is for most of our children. That is why the longer the period of schooling, the greater the rate of the dropouts, even for our nice middle-class children, even for college students, and the more likely they are to seek the easy way out that drugs seem to offer.

Here, let us not forget that fifteen years not so long ago was half the span of a man's life. To be able to postpone reaping the harvest for such a span of time calls for a powerful domination over the pleasure principle by the sense of reality. And the longer the span of time spent on education, the more dominant the reality principle must be for any learning to take place. This, in terms of educational practice, is what is meant when teachers speak of the need for discipline, attention, and concentration.

Fortunately for education as it now exists, most middle-class children still enter school with the reality principle dominant, with the ability to postpone pleasure over long time spans well established. Because of it we can still believe that our system works and that all children are fed into it to their profit. And indeed it still works, but percentagewise for fewer and fewer, and their number is constantly declining. Partly this is because the time spent on education has so increased. But more importantly this is because we no longer live with scarcity but in theoretical affluence. When scarcity reigns (at least in modern Western society), the reality principle seems the only way of life assuring survival. But the image of the affluent society plays havoc with the puritanical virtues.

At a time when one's entire life was swept up in the idea of working now for rewards in the hereafter, as it was in colonial days, then postponement of pleasure was in the very nature of things. But even then such an overweening ascendance of the reality principle had to be supported by the immense pleasure satisfaction people expected to gain, in that era, for such tremendous postponement: only the promise of heaven made it possible to so prolong the waiting for satisfaction; as only the fear of damnation could account for such a powerful superego.

Obviously a valid application of psychoanalysis to education would require us to assess the degree to which a child coming to school today has made the reality principle his own. And if he has not done so enough, then all educational efforts must be geared toward helping him accept it as more valid than immediate pleasure. So again: How do middle-class children make the transition from pleasure to hard work? The answer is that at first and for a long time to come, they do it with support from the superego—on the basis of fear. And the reason they need this support has to do with the slow development of reason, of true thought.

Because, while the ego is that institution of the mind which enables us to be rational, it is a feeble institution. And the younger the child, the feebler the ego. As Freud said, the voice of reason is very soft. It is easily drowned out by the noisy clamor of the emotions. Nevertheless, because it is weak, because the child knows himself unable to cope with full reality yet, unable to really do for himself, his fear of desertion by those who do care for him leads to the forming of a superego. This tells him he must reckon with and obey those powerful figures on the outside who will, in return, protect him. Thus the superego is first established to gain us safety from the external world. (And the same is true later on for the ego.) Typically it comprises the demands of our parents that we control our behavior in return for their keeping us safe.

Now at first this superego is unreasonably domineering and says "you must do as you are told" and not "maybe yes, it all depends on the circumstances." That is why the small child who is taught to think (or whose life experience teaches him) that taking things without permission is all right on some occasions but not others will have a superego that is full of holes, one that will not later on support him toward academic achievement. To the immature mind the "maybe yes" and "it all depends on when and how" just means "I can do as I please."

But our conscience, if well-established, will, in a slow process of learning and maturation, stop being so domineering and say more and more "I must do what will be best for me in the long run."

That is, the initial conscience we develop at a very early age is largely irrational. On the basis of fear (not of reasoned judgment) it tells the child what he must do and not do. Only later does the nature ego apply reason to these do's and don'ts and subject these earliest laws slowly, step by step, to a critical judgment. It takes mature judgment to be able to "do the right thing" though one is no longer motivated by fear.

So while conscience develops on the basis of fear, learning depends on the prior formation of a conscience which, in the process of learning, is more and more modified by reason. It is true that too much fear interferes with learning. But for a long time learning does not go well unless also prompted by some fear. And this is true until such time as self-interest is enlightened enough to power learning all by itself. But rarely does this happen before late adolescence. But by that time, personality formation is essentially completed.

Now fear, according to Nietzsche, "is the mother of morality" and "morality is the rationalization of self-interest."[3,4] Certainly the psychoanalyst agrees with Nietzsche that all morality including what leads us to accept education is, if not mothered, so fathered by fear, and that (in the last analysis) the content of morality is self-interest. After all, it is self-interest that makes some persons wish for eternal salvation, just as self-interest makes others want to succeed in the rat race. The difference is that the first kind of self-interest leads to entirely different behavior from the second.

But as for morality being based on fear, we now want to remove fear from the life of the child. And as for the content of morality being self-interest, many feel that this should not be so. In short, we now want the child to obey a morality whose basic motivations we do our best to remove.

It was Darwin, as well as a Nietzsche or Freud—these great reviewers of morality who were raised on stringent and absolute demands based on fear—who could later afford to question them ever more critically, without ever losing too much of their superego to go to pieces as persons, or to withdraw from the world in disgust. It was precisely the powerful superego instilled in them as children from which they later drew the strength to try to reshape the world by their more mature concepts.

As a matter of fact, one primary motive for learning anything at all is precisely the wish to modify an irrationally demanding superego to make it more reasonable. If there is no striving for this, because there is no strong superego anxiety to reduce, a most important motive for learning is absent.

If we do not fear God, why learn about religion? If we do not fear the forces of nature, why learn about science or society? The detachment that permits hard study out of sheer curiosity, out of a desire to know more, is a stance not often arrived at till full maturity.

We no longer can or want to base *academic* learning on fear. We know how crippling a price of inhibition and rigidity it exacts. But my contention is that for education to proceed, children must have learned to fear something before they come to school. If it is not the once-crippling fear of damnation and the woodshed, then in our more enlightened days it is at least the fear of losing parental love (or later the teacher's) and eventually the fear of losing self-respect. That is why the modern way of trying to raise children—without fear and without respect for the external superego surrogates—neither equips the child to control his desires of the moment nor prepares him as well as it once did for acting on the basis of long-range goals.

Let me illustrate by means of our classroom reading which, instead of developing the ego and superego, caters only for some time to the id. In our children's stories there are no delays for good things to come, no severe hardships, no insoluble problems, no questions of good and evil. Everybody lives in Pleasant time, on Easy street, in Friendly town. Everyone loves everyone else, inside the family and out, and the future is always rosy and full of promise. Then abruptly, in the stories we give our youngsters in adolescence, there is only Present time, everybody lives on Deprivation street, in Ghastly town, or a yellow submarine. Everyone hates everyone else, inside the family and out, and the future holds no promise, is always dismal and black. In both cases we cater to the id: in our children's stories, to the pleasure id only; in the literature for adolescents, to the anxious, hostile, and persecutory fantasies of the id. Neither of the two literatures does anything to strengthen the ego, though it is the ego we expect the child to set to work—as in mathematics, for example.

The fictional world we create for our children is removed from all the realities of life. In the primers there is no need for the hardships involved in making a living. No delay, no work morality is needed to make a go of life. But having created for our children a world picture of easy life with no problems, this world is suddenly, in adolescence, made out to be the exact opposite—a world where all morality is sham, where everything is purposeless and vicious, where even the rigors of making a living are rejected as stupid. No wonder that with such an upbringing some of the young generation feel that those over thirty have nothing of importance to say to them. Having begun life in a dream world of unreality that demanded of them no sense of purpose, dedication, morality; having then been confronted in adolescence with a world picture where all morality is sham; it is a mystery to them how anyone can take this world seriously, including his own place in it. Having failed to acquire a serious attitude to life in childhood, they feel un-serious or phony inside themselves and are convinced that so is everyone else. Above all, they feel utterly confused about things and desperate about their own inner emptiness. If, then, adult spokesmen proclaim this very lack of morality (from which, deep down, they know they suffer) to be the sign of a new and deeper morality, they are indeed lost. Because if what they suffer from—their lack of direction, inner strength, hope for the future, in short their lack of ego strength—if this is the new morality, then indeed there is no hope for them, and they might as well drop out or seek oblivion in drugs.

Even worse, if the emptiness they suffer from is also blamed on society and its evils, if their failures are ascribed not to inner weakness but to family or societal wrongs, then things are doubly hopeless. As long as they could hope to save themselves through personal striving, there was hope—and soon also the effort. But if all is society's doing, then there is no hope and no point in making efforts.

With such confusion about what kind of personality structure, what kind of commitment the good life requires, we create, from a very early age, the so-called credibility gap that we so suffer from between the generations. The gap reflects how differently young people were taught to see the world from the way mature adults see it and how different are the things they were taught to expect of their worlds. Eventually, it is true, the majority of the young still learn to shed their adolescent views of such a never-never land. But often only after a hard struggle that bruises them for life. Others never make it. By dropping out, they reject a world that failed to live up to their childish expectations.

Let me enlarge here on the problem of acquiring a superego and the difficulties it poses to modern children. First of all, and once infancy is past, the child's parents are not enough. He needs, in addition to his parents, superego representatives to help him build up his inner controls.

How difficult this has become for the child may be illustrated by that superego figure who aroused so much controversy last August—the police. For children, the policeman is the first and most visible symbol of social control. While restraining the individual at particular moments, the policeman is supposed to benefit him and all citizens in the long run.

Now nothing can so hinder us from forming and accepting a rational view of the police than to have built up in childhood an image of them that is wholly out of keeping with the functions they perform. Certainly the police are not maintained by society to find lost children or to help elderly people across the street. The function of the police is not to be everybody's friend. Their function is to see that law and order is maintained and the transgressor punished. If, then, the middle-class child is taught in infancy that the policeman is everybody's friend and helper, while the reality that confronts him in the growing-up process turns out to be quite different, then the disenchantment is devastating. It leads to charges against the police which, while partly justified by reality, owe their special emotional venom to a reality that just does not conform to a childish dream one wants to hold on to. The strange result is that, because the childish image is clung to, the reality is viewed as persecution.

If, on the other hand, as is still true in England and was true in many European countries before World War I, the police are viewed as the stern but just organs of law and order who protect where protection is warranted and punish where punishment is warranted, then police reality does not contradict any childish image. It is just accepted for what it is. Since the police were always expected to be enforcers of law and order, they are first accepted and then internalized as such. But being accepted as such, they are implicitly obeyed, and they see no need to apply force in order to carry out their functions. Since they do not use force, they do not aggravate the resentment we all feel when we are forced to bend our desires to the common interest. Most of all, though, we never respect those who fail to live up to our inner expectations, however unreasonable they are. Large segments of the population—who could not exist for any length of time without the police—find them unacceptable because of unrealistic images formed about them in childhood.

The police, for their part, do not feel themselves respected for the service they perform for society. Those in the population who cling to an image of the police as everybody's best friend view them as callous brutes, as the enemy, the persecutors. Since the police cannot feel respected, they turn to force in order to gain some respect and obedience, much to the utter detriment of the entire population, including policemen themselves. Because only a police that is respected will develop self-respect. And because of it, they will behave in self-respecting ways that gain them further respect from all but the criminal elements of society. (If, as Mead[5] has taught us, we see ourself as others see us, then, given the way some segments of society view the police, it is amazing that they can restain themselves so well from acting the way they are accused of acting when dealing with these segments of society. Of course, it could be argued that the police should have egos so strong that they resist viewing themselves as others view them. They would then resist the push to see segments of the population as their enemies because such persons see the police as their enemies. But I doubt that even in the best of societies the average policeman will have so strong an ego that the image held of him by those he deals with will not affect his actions, or his self-image.)

Since this is not the place to discuss how to reform our police, but how a superego is formed, it follows that the public outcry about the police is devastating to the small child's sense of security in this world. It not only weakens his superego but, because of it, his sense of reality, his ego. Unfortunately, it is only a strong ego that would later enable him to work for the needed reform of the police.

I do not think it needs further elaboration to show how next-to-impossible we have made it for our children to develop an adequate superego since we have deprived them, one by one, of all these private and public figures who used to serve as figures to be internalized as superegos. This goes all the way from the President who is subject to vilification, to the police who are accused of venality and brutality, to the teacher who is criticized for not doing a good job, to the father who is ridiculed as incompetent on TV and in the comics.

If we want our children to feel that life in our society is worthwhile, we must see that it comes across to them, when they are young, that things are essentially all right, though sometimes difficult and in need of improvement. And it must come across to them through symbols like the President or the police, symbols they can easily grasp in visual form. (Here again we must not give the impression that society is just here to give lollipops to little children.) This may encourage them to want to grow up so they can try to effect those improvements that are needed. But if things are pictured as almost all bad, the task seems impossible, and they may as well give up trying—or just blow it all up. How, for example, can small children grow determined enough to develop mastery, to become competent enough to straighten out those things that need fixing, if from the start we criticize all symbols of authority in society?

Given all I have said about forming a superego, it should be clear by now why the psychoanalytic model is that of a slowly developing ego. It should also be clear how different that is from the model that reactionary educators seem to embrace nowadays—be it the model based on conditioned responses reinforced by punishment (which bypasses any ego development), or the other which holds that the best way to educate is to forget about ego and id altogether and put the child under the total domination of a rigidly construed superego. What is wrong with such a method of education is not that it relies on the superego for its ends but that its ends are a personality dominated by the superego. Because the goal of education ought to be a well-balanced personality where both id and superego are subordinated to reality, to the ego. About these reactionary efforts I shall say nothing more since the altering of behavior through punishment or conditioning is not what this audience believes in. But that they have any following at all can be understood in part as a reaction to the hedonism that attracts and disappoints so many who believed that catering to the id, that subjecting the child to no controls, would automatically insure the good life.

Nothing automatically assures ego growth, neither punishment nor reward. The only thing that assures it is having the right experiences to stimulate and foster growth at the right time, in the right sequence, and in the right amount. And in the classroom, this can only happen if the teacher has a true understanding of the human personality, both intellectual and emotional, and of how and why personality is formed.

What I have been trying to suggest here is that while education in the past, and often today too, seems to concentrate on superego forces alone, lately some educators have erred in the other direction by stressing id satisfaction too much. Actually, sound education should utilize both forces—those of the id and the superego—in developing the ego.

So far I have dealt with this mainly in terms of two theoretical constructs of psychoanalysis, namely, the continua of id-ego-superego, and of the pleasure versus the reality principle. But there is at least one more of the psychoanalytic continua to consider, the one that is usually called the genetic premise of psychoanalysis. This premise states that each of the various ages and stages of development centers on a different developmental task, one that dominates our life until mastered. Or to use Erikson's concepts, these are different psychosocial crises that have to be resolved. In many respects, I feel that education has not yet made this body of genetic theories its own. It is true that education, more than any other social institution, knows there are sequences in development, that there is no point in expecting children to appreciate Shakespeare before they have gained sufficient mastery of reading comprehension. But our education, at home and in school, seems to be very deficient in applying this principle to anything but academic learning. We fail to apply it to the development of the emotions or (as I have tried to show) to personality formation, where superego learning must precede ego learning. Our schools expect both to be fully developed, without any need to promote their development. Our children are expected to have ego strength, to accept the need to work hard at learning, been strengthened enough to support it.

I spoke of our primers in which everyone loves everyone else, including, of course, the new baby. I spoke of how, without any transition, we then read of how everyone hates his mother, his father, and everyone else in the family. Maybe if we were to recognize sibling rivalry and teach it as one outcome of our deep attachment to our parents, we would teach good and bad hand in hand. At least it would be truer to life than teaching first that all is rosy, and then that all is bleak.

Unfortunately, we make the same error in our teaching about society. Like the new baby, the policeman is at first everyone's best friend, and in the next moment is corrupt, if not utterly brutal. Why should a child spend a great deal of effort on developing a personality that will help him come to terms with reality if that reality includes organized brutality by the very organ of society that is supposed to protect children? If that is how the world is, why waste one's energy? Why not drop out instead, leave the world and withdraw into oneself, or later into drug-induced dreams?

I have dwelt on the superego because of a common misunderstanding of Freud: the vague feeling among educators that children should never be made to feel frightened at breaking a moral command. Of *course* Freud stressed how damaging an overstrong conscience can be. That was where the shoe pinched in his time. But because of it he did not speak of the other extreme with its even greater dangers to man: that he will surely fail in life, and succumb to despair, if his superego is too poorly developed to back up the ego in its work of restraining the id.

In closing, I would like to stress that so far education has not asked the right questions of psychoanalysis. But neither have psychoanalysts provided education with the information it must absolutely have, if it is to make use of all the new findings of psychoanalysis.

I spoke of sibling rivalry, as an example of an overpowering emotion, and how our primers pretend it does not exist. I could go on with a catalogue of difficult emotions that haunt our children. But while the psychoanalyst tells the teacher they exist, he makes no suggestions as to how education could help the child deal with them. Consider violence, for example, since I talked of police actions in our city. What have our children been taught about it? Either to discharge it (as in competing for grades or in sports) or to repress it (as in our readers where violence just does not exist). But neither discharge nor repression teaches the child anything about his own tendencies to violence—what incites them or how to master them.

We are all familiar with the jealousy boys feel about the powers and prerogatives of girls, and the girls about the boys. We know too how anxious both are about how they will stack up as females or males. But where does education recognize these existential dilemmas? Where is it helping to guide children to master them? True, we try to teach what menstruation is, physically. But where are the efforts to explain the psychological anxiety it creates, both in boys and in girls? We teach why drugs are dangerous, but where do we teach for what psychological reasons people turn to drugs?

Essentially, progress in education has never been anything but a turning the lights on over the once hidden, of making it readily understandable to everyone who wanted to learn it. I think, above everything, the troubles of our young people show their need for being taught how to find their way in the hidden world they feel their own soul to be. This, as Freud said, cannot be done by analyzing all youngsters but by giving their teachers the wherewithal to do what the best of them long to do more than anything else: to help their children find their way out of the darkness in which they hide their true selves, to help them toward a rational understanding of themselves and a world that they vaguely know exists but in which they remain strangers without the right travel instructions.

Now, this lifelong journey of discovery cannot be charted in three easy lessons. It is not a lesson we can program for teaching machines. It is not a pleasure trip either, but becomes as demanding of soul, even of body, as the search for the white whale. But no one will ever begin such a journey who is taught that the only trouble with the world is the way his parents or society run it, or that a better production or distribution of wealth will take care of the existential agonies of man. As I have tried to make clear, a better police force will not do away with the need for a better superego, though a better superego may very well lighten the work of our police.

I believe that what education needs most today is the conviction that, while each of today's different generations was fathered by society, present-day man must change himself in his innermost core. Otherwise he will father as imperfect a society as the one that fathered him. To prevent this is the task of education and psychoanalysis combined; neither one of them can do it alone. Together, they might well bring into being the better man who will call into being a new and better society.

References

1. Bernfeld, S. *Sisyphos.* Leipzig and Vienna: Internationaler Psychoanalystischer Verlag, 1925.
2. Kohl, H. R. *36 Children.* Chicago: American Library Association, 1967.
3. Nietzsche, F. *Beyond good and evil.* New York: Modern Library, 1917.
4. Nietzsche, F. *The genealogy of morals.* New York: Macmillan, 1897.
5. Mead, G. H. *Mind, self, and society.* Chicago: University of Chicago Press, 1934.

Recommended Additional Readings

Eastman, G. The critical need for a new education—Part I: Philosophical substructure. *Educational Theory,* 1971, 21, 178-186.

Endleman, R. Oedipal elements in student rebellion. *Psychoanalytic Review,* 1970, 57, 442-471.

Feuer, L. S. *The conflict of generations: The character and significance of student movement.* New York: Basic Books, 1969.

Rogers, C. R. *Freedom to learn.* Columbus, Ohio: Merrill, 1969.

Student Activism, Moral Development, and Morality

Kenneth Keniston

Orientation

As long as there have been philosophers there has been concern about morals and ethics. Just what constitutes a moral is a matter for debate and deliberation among those who deal with lofty philosophical matters. However, in the not-so-lofty world of ordinary people it is universally believed that morals exist and must be ascribed to. What may be moral for one person may be immoral or perhaps just unimportant to someone else; in any case, everyone somehow acquires a set of moral and ethical beliefs in the course of his development.

Psychologists have only recently become concerned with how the growing individual acquires moral beliefs and principles. Freud's postulation of the superego (i.e., the conscience) was perhaps the first systematic effort to outline the processes by which a child attains an understanding of good and bad, right and wrong. Freud's superego, however, was a harsh, punitive entity and was not well received by many other psychologists. During the 1920s Jean Piaget proposed a theory of moral development as a result, primarily, of observing his own children. The first "morality" to appear within the child, said Piaget, was a "heteronomous" morality—standards of good and bad are set by others, the adult world. Later, at about the age of eight, a new morality evolves—"autonomous" morality. This is a higher level of understanding; moral principles are individually held and they are characterized by reciprocity and the Golden Rule.

Through most of the twentieth century, American psychologists have assumed that moral principles are learned. By observing and imitating the behavior of adults, the child gradually copies and then internalizes adult standards and values. However, another approach to moral development is gaining support rapidly; this is the theory espoused and developed by Lawrence Kohlberg. Kohlberg calls his approach the "cognitive developmental approach"; for him, moral development is not simply a matter of learning, in a passive manner, adult standards of right and wrong.

Reprinted from *American Journal of Orthopsychiatry*, 1970, 40, 577-592. Copyright © 1970, the American Orthopsychiatric Association, Inc. Reproduced by permission of the author and the publisher.

This paper is a slightly revised version of one presented at the 1969 annual meeting of the American Orthopsychiatric Association, New York, N. Y.

The issue is much more complex. The child is an evolving, changing organism, and as he evolves he interacts with the environment in different ways. Because of these changing interactions, several distinct stages or levels of morality ought to be distinguishable. Indeed, Kohlberg and his followers have produced impressive evidence that this is the case.

Regarding the moral development of the adolescent and young adult, certain theorists would maintain that his basic morality is well established by this time; others would say that his morality can be changed by instructional or training programs; still others (e.g., Kohlberg) would say that the adolescent's morality is capable of evolving to even higher levels of sophistication, but that formal training in moral principles is futile. This last position needs emphasis. It is the belief (or perhaps hope) of many educators, clergy, and parents that all that needs to be done about what they perceive to be the shoddy morality and character of many of today's youth is simply to provide them with stern discipline and rigorous moral education (sometimes also called "character education"). But, as with sex education and drug education, this instructional approach simply does not work; at least, the outcome of didactic training at the adolescent level is not that intended by the supervising adults. Individuals do not learn values and morals by mere passive instruction; these principles are acquired, in the main, by an active interaction process, with the individual acting upon his environment, and the environment acting upon the individual. Indeed, as many researchers have clearly demonstrated, morality is typically situation-specific. A person who would never cheat on an examination in a university classroom might have no qualms about cheating on his yearly income tax forms.

The business of schooling and formal education has historically been tied to moral training. A long-standing reason for the perpetuation of compulsory education upon all children in this society is that proper training must be given to young people in order that they will be imbued with sound habits and principles; eventually the products of this training, it is thought, will be sound adult citizens. The article to follow by Keniston explores a number of the issues involving the schooling process, modern society, and the current forms of student activism and alienation from adult society.

Kenneth Keniston is professor of psychology in the department of psychiatry, Yale University School of Medicine.

28

To discuss student activism today without immediately becoming involved in moral issues seems almost impossible. From the rhetoric of politicians to empirical research, judgments of moral praise or condemnation enter into (and frequently dominate) reactions to student protest. Political tracts, novels, research studies, biographies, and autobiographies that deal with student activism today generally emphasize the conflict between youth's personal morality (or immorality) and the immoral (or moral) practices of the surrounding world. Even the most thoughtful and scholarly analyses of contemporary student dissenters usually place them near one of two poles: "amoral-and-neurotic rebels"[2,11,41] or "fine-young-idealists-who-may-save-us-all."[35,40] Whether we like it or not, the phenomenon of youthful protest seems to stimulate intense moral concerns in the beholder.

In the comments that follow, I will discuss data and interpretations concerning the moral development of politically active young men and women at a particular stage of life that coincides roughly with college and graduate school age. I will call this stage of life "youth." By speaking of "youth" instead of late adolescence, I mean to suggest that the experience of those whom we awkwardly term "late-adolescents-and-young-adults" is in many respects different from the experience of younger adolescents but, at the same time, that it differs profoundly from that of adults. In other writings [18-21] I have argued that one of the characteristics of post-industrial societies is that they are beginning to sanction a previously unrecognized stage of development that intervenes between the end of adolescence proper and the beginning of adulthood. Not everyone passes through this stage: traditionally, most young men and women have had little real adolescence at all; and those few who have experienced adolescence as a developmental stage have usually entered adulthood immediately thereafter. But today, for a rapidly growing minority of young Americans—mostly college students, graduate students, members of the New Left, hippies, or in some cases military recruits—a previously unlabeled stage of development is opening up. This stage is defined sociologically by postadolescent disengagement from the adult society, developmentally by continuing opportunities for psychological growth, and psychologically by a concern with the relationship of self and society. It is this stage of life which I term the stage of youth.

In considering youthful activism and moral development, it is necessary to underline that moral development is not only an essential sector of development in its own right, but also a battleground upon which conflicts whose origins lie elsewhere are fought out. It is in fact arbitrary to isolate moral development from identity development, from ego development, from psychosexual development, from the development of intimacy, from new relationships with parents and peers, and from intellectual development. Nonetheless, moral development during youth has been more carefully studied than any other sector. In the work of Erikson,[6-8] Lawrence Kohlberg,[22-26] William Perry,[36,37] and Smith, Block, and Haan,[3,14,16,42,43] have accumulating evidence about the relationship between moral development and the often disruptive, idealistic, moralistic and anti-conventional behaviors of modern youth.

Moral Development and Socio-Political Activism

The early psychoanalytic account of superego development, though it still provides an essential underpinning for any study of the psychology of morality, clearly omits or neglects many of the dynamic and structural complexities of moral development in later life. In so far as the classical psychoanalytic account stresses only the formation of the superego through the introjection of the same-sex parent at the conclusion of the Oedipus complex,[10,13] it leaves out many subsequent changes in the superego and in morally determined behavior. Recently, psychoanalysts and others have shown greater interest in these changes. For example, it is now commonly recognized that during normal adolescence there can occur a "rebellion against the superego," by which the individual rejects not only his parents but that part of his own superego which is based upon unreflective internalization of their standards.[4,5] Other students of adolescence have emphasized the increasing integration of the superego and ego which can occur during this stage and the greater elaboration of self-accepted moral principles, which form part of the ego ideal.[4,32]

Psychoanalysis has largely dealt with the genetic and dynamic aspects of superego development. Jean Piaget's account of moral development,[38] in contrast, emphasizes changes in the logic or structure of moral reasoning throughout childhood. And more recently, Lawrence Kohlberg, in a series of brilliant studies, has modified and extended Piaget's work by developing a comprehensive account of developmental changes in the structure of moral reasoning.[22, 26] Kohlberg finds that moral reasoning develops through three general stages. The earliest is the *pre-conventional* stage, which involves relatively egocentric concepts of right and wrong as that which one can do without getting caught, or that which leads to the greatest personal gratification. The pre-conventional stage is followed, usually during later childhood, by a stage of *conventional* morality, during which good and evil are first identified with the concept of a "good boy" or "good girl," and then with the standards of the community, i.e. with law and order. The individual in the conventional stage may not *act* according to his perceptions of what is right and wrong; but he does not question the fact that morality is objective, immutable, and derives from external agencies like parental edicts, community standards, or divine laws.

Kohlberg also identifies a third and final stage of moral development that is *post-conventional*—what Erikson has called the "ethical" stage.[6, 17] This stage involves reasoning more abstract than that found in earlier stages, and it may lead the individual into conflict with conventional standards. The first of two subphases within the post-conventional stage basically involves the concept of right and wrong as resulting from a *social contract*—as the result of an agreement entered into by the members of the society for their common good—and therefore subject to emendation, alteration, or revocation. Conventional moral thinking views moral imperatives as absolute or given by the nature of the universe: social contract reasoning sees rules as "merely" convenient and therefore amendable.

Kohlberg identifies the highest post-conventional phase as that in which the individual becomes devoted to *personal principles* that may transcend not only conventional morality but even the social contract. In this stage, certain general principles are not seen as personally binding though not necessarily "objectively" true. Such principles are apt to be stated in a very high level of generality: e.g. the concept of justice, the Golden Rule, the sanctity of life, the categorical imperative, the promotion of human development. The individual at this stage may find himself in conflict with existing concepts of law and order, or even with the notion of an amendable social contract. He may, for example, consider even democratically-arrived-at laws unacceptable because they lead to consequences or enjoin behaviors that violate his own personal principles.

With the development of moral reasoning (as with all other sectors of development), precise ages cannot be attached to the attainment of specific stages. But Kohlberg's research indicates that those who attain post-conventional levels generally do so during later adolescence and in the years of youth. Figure 1 extrapolates from Kohlberg's research to give a rough indication of the timing of moral development, categorized according to his three general stages. The subjects on which this figure is based are middle-class American urban males: thus, at the age of 16 they are probably college-bound; at 20, they are likely to be in college; and at 24, many are in graduate schools. It is clear from Kohlberg's data

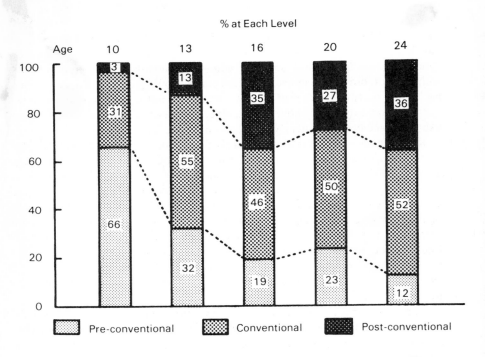

Figure 1.
Level of Moral Reasoning and Age
Ss: middle class urban American males

that the highest (post-conventional) phases are *never reached* by most men and women in American society, who remain at the conventional stage. Even at age 24, only 10% of this middle-class urban male population have reached the personal principles phase, while another 26% are at the social contract phase.

Finally, Figure 1 indicates that between the ages of 16 and 20 the number of individuals in the pre-conventional stage increases. Kohlberg accounts for this increase by the phenomenon of "moral regression" as a routine developmental occurrence in many college students. Longitudinal studies conducted by Kramer[28] have documented the occurrence of such regression in a number of late adolescent and youthful subjects. In what Kohlberg has termed the "Raskolnikoff Syndrom," the individual moving toward post-conventional morality regresses to the earlier, pre-conventional (egocentric) stage in an apparent effort to free himself from irrational and ego-dystonic guilt. Interestingly, Kramer finds that such young Raskolnikoffs eventually return to the developmental track at approximately the point where they dropped off.

The structure of moral reasoning is of course not all of moral development; conceivably an individual may reason one way, yet act in another. But several studies have demonstrated that the way a person reasons morally is closely related to his actual behavior under conditions of moral stress. Figure 2 presents some central findings about moral reasoning and behavior in situations of moral conflict. The Milgram experiment[33] is presented as an experiment in negative reinforcement. The subject is asked to administer high levels of electric shock to another experimental subject (actually a stooge). The stooge protests violently at the shock and eventually warns the subject that his heart condition makes the experiment dangerous. The great majority of college students and the noncollege population, when encouraged by the experimenter to continue to administer shock, do so despite the victim's protests. But Kohlberg[25] finds that 75% of the

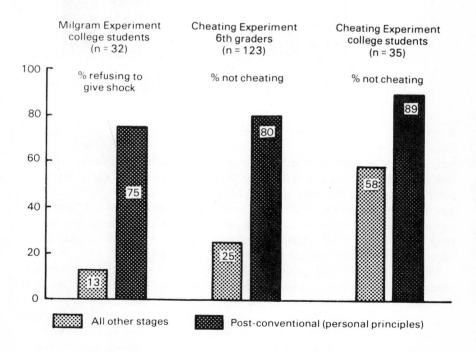

Figure 2
Level of Moral Reasoning and Behavior in Moral Stress Conditions

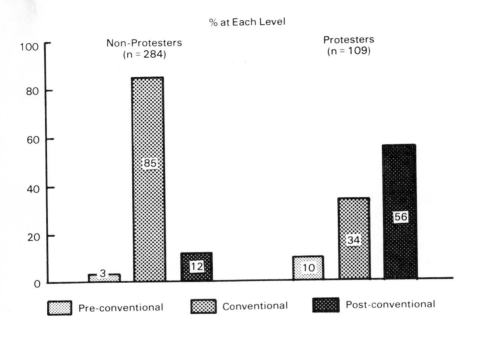

Figure 3.
Level of Moral Reasoning and Socio-Political Protest
Ss: students at U.C. (Berkeley) and S.F. State

subjects at the stage of personal principles—the highest stage—refuse to continue shocking the victim, as compared to only 13% of subjects at all earlier stages. In another experiment, studying cheating behavior in sixth graders,[29] only 25% of the conventional sixth graders did not cheat, while 80% of the post-conventionals did not cheat. In a study of college students,[9] the corresponding figures were 58% and 89%. There is strong evidence, then, that the level of moral reasoning is associated with the actual morality of behavior.

Figure 3 portrays the relationship between level of moral development and participation in student protest activities. This figure is based upon research done by Brewster Smith, Jeanne Block, and Norma Haan at the University of California at Berkeley.[14-16, 42, 43] The subjects are male and female college student freedom of speech, the war in Vietnam, or alleged racism in the uni- into two groups: (1) the protesters, who have engaged in sit-ins, peace marches, picketing, and various forms of disruption or direct action over such issues as student freedom of speech, the war in Vietnam, or alleged racism in the university or in society; (2) all nonprotesting students, including political inactivities, apolitical fraternity and sorority members, and students who engage in social service activities but do not take part in protests.

The findings of the Berkeley research are complex, but as summarized in Figure 3, they indicate a marked difference in the level of moral development of protesters and nonprotesters in this college population. A clear majority (56%) of all protesters are at post-conventional levels of morality, whereas only 12% of nonprotesters have reached this level. The nonprotesters are overwhelmingly (85%) in the conventional stage—that is, they define morality as adherence to law and order, or as involving some concept of being a "good boy" or "good girl." Only 36% of protesters are at the conventional stage. Interestingly, the proportion of protesters at the pre-conventional stage is also disproportionately large—10% of protesters as against 3% of nonprotesters. Kohlberg's writings suggest that such individuals may be in a state of moral regression (Raskolnikoffs), perhaps epitomized by certain variants of the hippie subculture.

The complexity of these data, however, are emphasized when we analyze them in a different way. Unlike Figure 3, Figure 4 distinguishes between the

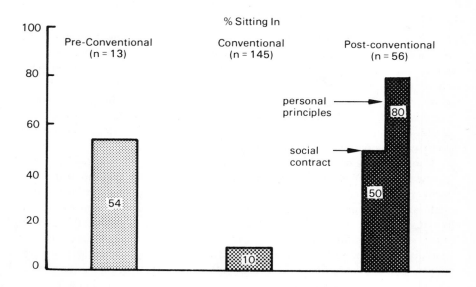

Figure 4.
Level of Moral Reasoning and
Participation in FSM Sit-in (Berkeley 1964)

behavior of those at different levels of moral development, not between the moral development of those who behave in different ways. The behavior here studied was being arrested as a result of the Free Speech Movement sit-in in Sproul Hall at Berkeley in 1964. This analysis indicates that the proportion of pre-conventionals involved was about the same as the proportion of post-conventionals, although in absolute numbers there were many fewer pre-conventionals in Sproul Hall. These findings make clear that level of moral development and socio-political activism are not correlated in a linear manner. They suggest that any protest will, depending on the issues involved, enlist supporters from several different levels of moral development.

This conclusion is supported by an unpublished study of Kohlberg's on the participants in the Harvard College sit-in in the spring of 1969.[27] Kohlberg predicted that at Harvard, unlike Berkeley, the students at the post-conventional level would *not* be overrepresented amongst those who sat in. He based this prediction on an analysis of the issues in the Harvard sit-in, which did not seem to him to involve a comparable appeal to abstract principles. His findings confirmed this prediction. These studies, then, do not indicate that high levels of moral development lead *automatically* to participation in all protests, sit-ins, confrontations, and disruptions. Rather they indicate that those who have reached higher levels of moral development are more likely to act in the service of their principles—protesting when their principles are at issue; refusing, also for reasons of principle, to take part in other protests and forms of activism.

Social Catalysts for Moral Development

I have so far presented research findings on the relationship between moral reasoning and socio-political activism. On the basis of such findings, we would predict that an increase in the proportion of the student population at post-conventional levels would also increase the likelihood of principled student socio-political activism. I will not argue, more speculatively, that modern social and historical conditions are providing new catalysts and facilitations for high levels of moral development, and that these new developmental attainments constitute *one* partial explanation of socio-political activism.

Kohlberg does not address himself specifically to the psychological or social catalysts of moral development. But his data makes clear that moral development is by no means guaranteed by aging, maturation, or socialization. Physical maturation may make possible the development of post-conventional morality, but it obviously does not ensure it. And the pressures of socialization may in many instances militate *against* the development of a principled morality that can place the individual in conflict with his socializing environment—for example, with college administrators, with political parties, with the police, or with the present American Selective Service System. If neither maturation nor socialization guarantee moral development, how can we explain it?

Haan, Smith, and Block have provided us with a first account of some of the psychological antecedents of various levels of moral reasoning in late adolescence and youth.[14-16] They report, for example, that students at the highest moral stage of personal principles "had a history of preparedness within politically liberal families who frankly experienced and examined conflict, and with parents who exercised their own rights as people, rather than the power and control that society automatically ascribes to them."[14] Their data—too complex to be summarized here—clearly indicate that family milieu during the preadolescent years plays an important role in facilitating or obstructing later moral development.

Here, however, I will not discuss the impact of these early experiences, but will consider the effects of more general social, historical, and political factors on adolescent and postadolescent changes in moral reasoning. That is, I will not consider why some individuals arrive in adolescence or youth already predisposed to develop to the post-conventional or "ethical" stages in moral development, but will discuss in a speculative way why post-conventional (ethical)

Figure 5.
Level of Moral Reasoning and Social Class

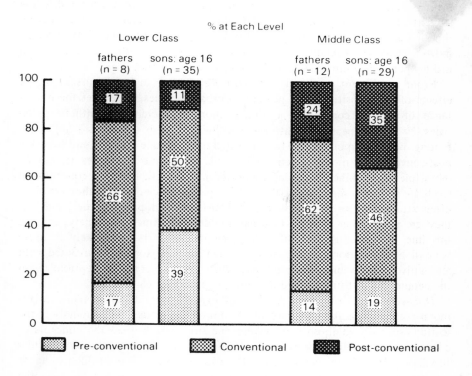

moral reasoning may characterize a growing proportion of today's college generation in America and in the other advanced nations.

Disengagement from Adult Society: A prolonged period of disengagement from the institutions of adult society seems to facilitate moral development. Conversely, immediate entry into the labor force and early marriage with responsibilities for maintaining a family tend to constrain or obstruct moral development. Kohlberg's data, reported in Figure 5, showing higher modal levels of moral development in middle-class (college-bound or college) students than in lower-class (noncollege) youth, are consistent with this hypothesis. For, insofar as an individual during or immediately after puberty or adolescence takes a job, marries, and has children, the opportunity for confronting and challenging conventional morality seems to lessen. The risks of unconventionality become greater; the price for departure from conventional morality, and especially for that moral regression which Kohlberg finds a frequent if usually temporary part of moral development, becomes too high for most individuals to pay.

Confrontation with Alternate Moral Viewpoints: William Perry, in his pioneering studies of ethical and intellectual development during the college years,[36, 37] suggests that one prime catalyst for intellectual and moral development is confrontation with relativistic points of view in professors and fellow students. Such confrontations stimulate the student to abandon simple dualistic thinking about right and wrong, good and bad, truth and falsehood. He tends to move first toward a relativistic concept of morality and truth, and later, toward making personal commitments *within* a relativistic universe.

From a different perspective, Robert Redfield,[39] in his discussion of the effects of the transition from peasant to urban societies, underlines the importance of culture contact in producing more high-level and synthetic ideologies. [17, 34] The peasant, confronted in the city with others who hold conflicting moral viewpoints, may be compelled to reexamine his own, and to seek a post-conventional moral system that stands above and reconciles traditional moral pieties. Kohlberg's findings, reported in Figure 6, that post-conventional levels are almost never attained by age 16 in peasant societies, that they are more often attained by urban middle-class students in developing societies, and that they are most often attained in urban middle-class American society, support this line of reasoning. Put differently, an individual is more likely to move beyond a conventional moral system when he is personally confronted with alternative moral values, and especially when these are concretely epitomized in the people, the institutions, and the cultures among which he lives.

Discovery of Corruption: A third catalyst for moral development, as for moral regression, is the discovery of corruption, hypocrisy, and duplicity in the world, especially in those from whom one originally learned the concepts of conventional morality. For example, disillusionment with parents—in particular the discovery of moral turpitude (or, in Erikson's[7] terms, lack of fidelity) in the parents' lives—may play a critical role in pushing the individual to reject the morality he learned from them. Obviously not all young men and women react

Figure 6.
Level of Moral Reasoning and Type of Society

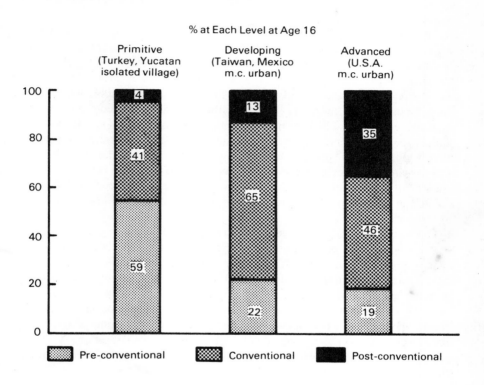

identically to such discoveries. Some may accept them without regression: they will push the individual to higher stages of moral development. This advance seems especially likely if the discovery of corruption in the world is accompanied by growing awareness of one's own potential for corruption. Other youths, however, will react with at least a temporary regression to moral cynicism, in which they behaviorally flout and intellectually reject what they consider to be "hypocritical" conventional values.

Historical Pressures Towards Moral Development

Three social factors that may stimulate moral development in youth have been considered: continuing disengagement from adult institutions, confrontation with alternative moral viewpoints, and the discovery of corruption in the world.

There is reason to believe that all three of these conditions obtain to an unusual degree today. In the extension of higher education, the cross-cultural implosion, and the pervasive reductionism of our age we have created important new catalysts, for better and for worse, for higher levels of ethicality, as for more marked moral regressions.

The Extension of Higher Education: Our own era has witnessed an historically unprecedented influx of students to colleges and universities. To cite but one statistic, during the time of the Russian student movement in the middle of the nineteenth century, there were never more than 8,000 university students in all of Imperial Russia.[12, 45] Today in America, there are 7,000,000 (almost 1000 times as many); while in Western Germany, France, and England there are by rough count 1,250,000. In most advanced nations of the world, the proportion of young people who attend colleges and universities is increasingly logarithmically; furthermore, this increase has largely occurred in the last two decades. The growing affluence of the highly industrialized nations permits them to keep millions of the young out of the labor force; the increasing need for high-level training in technological societies requires them to offer a university education to these millions.

Higher education does not, of course, inevitably entail moral growth. But one consequence of the prolongation and extension of higher education is that a massive group of young men and women have been disengaged for an increasingly protracted period from the institutions of the adult society, in particular from occupation and marriage. Freed of responsibilities of work, marriage, and parenthood, at least some find themselves in university atmospheres that deliberately challenge and undermine their preexisting beliefs and conventional assumptions. Thus, the extension of youth via prolonged education on a mass scale probably tends to stimulate the development of post-conventional moral thinking.

Culture Contact and the Cross-Cultural Implosion: The individual who attends a liberal arts college or university is very likely to confront in his daily experience both peers and professors who preach and practice a different morality from the one he was brought up to take for granted. Many universities deliberately confront students with contrasting cultures that give allegiance to alien moral concepts, and deliberately provoke students to question the unexamined assumptions of their own childhood and adolescences. They thus push the individual away from what Perry calls "dualistic" thinking,[36] away from an unthinking acceptance of conventional moral "truths," and toward a more individuated moral position, at once more personal and more abstract.

But it should also be recalled that outside the university, as within it, we live in an age of extraordinary culture contact and conflict. The electronic revolution, coupled with the revolution in transportation, enables us to confront alien values within our living room or to immerse ourselves physically in alien cultures after a flight of a few hours. The days when one could live in parochial isolation, surrounded only by conventional morality, are fast disappearing. Conflicts of ideologies, of world views, of value systems, of philosophical beliefs, of esthetic orientations, and of political styles confront every thoughtful man and woman, wherever he lives. If such confrontations stimulate moral development, then we live in an era in which technology and world history themselves provide new facilitation for moral growth. We are all today a little like Redfield's peasants who move to the city, living in a world where conflicting cultures and moral viewpoints rub against us at every turn. The urbanizing and homogenizing process has become worldwide: we live in an era of cross-cultural implosion. My argument here is that this cross-cultural implosion helps stimulate moral development.

Cynicism and Reductionism: Universities often undertake to expose the student to the gap between preaching and practice in society, in admired individuals, and even in the student himself. Whatever their many conformist pressures, universities in America and abroad also have another side: they have often been focal points of criticism of the surrounding society—institutional consciences that may collectively remind the surrounding society of its failure to live up to its ideals. To attend a university may systematically expose the student to the actual corruption that exists in the world, in representatives of the status quo, and even in himself.

But this exposure to corruption is today by no means confined to the university itself: ours is, in general, an age of skepticism with regard to traditional moral pieties and platitudes. Hypocrisy and corruption are constantly exposed at a cultural as well as an individual level. The debunking of traditional models and values is a favorite contemporary pastime; duplicity, dishonesty, compromise, and deceit are widely reported. Many of our most powerful intellectual systems are highly developed in their capacity to debunk, reduce, and explain away the ideologies, values, and convictions of others. The sociology of knowledge, psychoanalysis, Marxism, philosophical analysis, cultural relativism, and a variety of other idea systems can all be readily used (or misused) for this purpose. Thus, even if the individual does not discover corruption in his own parents or immediate world, he is still hard put to avoid confrontation with the corruption that exists in the wider society. In an age of debunking, conventional morality tends to suffer: individuals are pushed to higher levels of moral development or to moral regression. The data suggest that student protestors are disproportionately drawn from just these two groups: primarily the morally advanced, but secondarily, the morally regressed.

Ethicality and Zealotry

My argument so far has been that new educational, technological, and historical factors today facilitate the development of post-conventional morality in larger and larger numbers of young men and women. Furthermore, post-conventional morality understandably characterizes many of those who are involved in principled protests against the conventional moral order and its institutions. We may therefore interpret worldwide student protest as *partly* a result of the fact that societies like our own are stimulating more individuals than ever before to higher levels of moral development. One aspect of the student movement must be seen as a result of a psychological advance, and not as a result of psychopathology or psychological retardation.

But how should we judge the development of a morality based on a commitment to ethical principles that are maintained even when they conflict with conventional moral wisdom? Is it really an advance? In evaluating the meaning of the highest levels of moral development, we are immediately confronted with a paradox. On the one hand, Kohlberg identifies such ethical reasoning with admirable men like Socrates, Gandhi, Lincoln, and Martin Luther King—men for whom devotion to the highest personal principles was paramount over all other considerations, and who as a result were moral leaders of their time. Yet on the other hand, especially during the past two cold war decades, we have been taught to view abstract personal principles with considerable mistrust—as a part of ideology not in the Eriksonian sense but in the highly pejorative sense.[1] Such principles, it has been argued, are intimately—perhaps inevitably—related to the development of moral self-righteousness, zealotry, dogmatism, fanaticism, and insensitivity. In pursuit of his own personal principles, a man will ride roughshod over others who do not share these principles, will disregard human feelings or even destroy human life. During the period when the "end of ideology" was being announced on all sides, when instrumental and consensus politics was being extolled, we learned to identify abstract personal principles with dogmatic and destructive moral zealotry. How are we to combine these two perspectives? Do we see in Brewster Smith's findings confirmation of the view that student activists are dangerous moral zealots? Or do we adhere to Kohlberg's implication that such individuals are more likely moral heroes than despots?

The answer lies, I think, in recalling my earlier observation that the separation of the moral sector of human development from other sectors is analytic and arbitrary. Anna Freud has taught us to think in terms of an ideal "balance" between what she terms "developmental lines" (sectors of development). Yet she has also shown that such balance is never found in practice, and that in any specific individual we always find retardations or accelerations of development within different sectors. Following Anna Freud's thinking suggests that whether the highest stages of moral reasoning lead to destructive zealotry or real ethicality depends upon the extent to which moral development is *matched by development in other sectors.* The critical related sectors of development, I submit, are those which involve compassion, love, or empathic identification with others.

Most moral zealots, bigots, and dogmatists are probably best described, in Kohlberg's terms, as conventionalists, while others are perhaps permanent regressees to the Raskolnikoff Syndrome. But there are at least a few whom we know from personal experience or from history who seem truly post-conventional in moral reasoning but whose genuine adherence to the highest moral values is *not* matched by compassion, sympathy, capacity for love, and empathy.[31] In such individuals, the danger of hurting men to advance Mankind, of injuring people in order to fulfill one's own moral principles, is all too real. We see this danger realized in the pre-Nazi German Youth Movement,[30] where post-conventional morality often went hand in hand with virulent anti-semitism. Pascal put it well when he noted that "Evil is never done so thoroughly or so well as when it is done with a good conscience."

Thus, neatly to identify high levels of moral reasoning with any one kind of action, much less with human virtue, mental health, maturity, and so on is a serious mistake. What we might term "moral precocity" in youth—high moral development not attended by comparable development in other sectors of life—is often dangerous. The danger lies not in high levels of moral development in themselves, but in the retardation of other sectors of development. What is dangerous is *any* level of moral development, be it post-conventional, conventional, or pre-conventional, is the absence of a developed capacity for compassion, empathy, and love for one's fellow man.

No one phrase will adequately characterize the other developmental accomplishments that are essential to humanize the highest levels of moral reasoning. But the history of revolutions that have failed through the very ardor of their search for moral purity suggests that the combination of abstract personal principles with a humorless and *loveless asceticism* is especially likely to be dangerous. There are of course many kinds of asceticism, some of them mature, self-accepted, and benign. But there are other asceticisms that are based upon inhibition of the capacity to love, upon failure in the development of interpersonal mutuality, and upon absence of empathy. Often, these qualities are combined with ascetic self-denial based more upon unconscious fear and inhibition than upon self-accepted personal values. Lewis Feuer's recent critique of student movements[11] identifies all student protesting activity[45] with the excesses of those student movements where high principles have been combined with asceticism, e.g. the prewar German Youth Movement. But if we examine the current American student movement, we find less ground for concern: however highly principled many of today's dissenting students may be, they are scarcely an ascetic lot.

In the end, then, we reach the paradoxical conclusion that morality is necessary but not sufficient; even the highest levels of moral reasoning do not alone guarantee truly virtuous behavior. Kohlberg's research, of course, shows that men who reason at an advanced level tend to act morally as well. And the Berkeley research suggests optimism about the high level of moral development of many or most student activists. Yet what is true for most is not true for all; and historically many crimes have been committed in the name of the highest principles, sincerely held. In the end, the findings of developmental psychology in the context of youthful political activism may merely return us to ancient truisms—mercy without justice is sentimental and effusive, while justice without mercy is cold and inhuman.

References

1. Bell, D. The end of ideology. New York: Free Press, 1959.
2. Bettelheim, B. Obsolete youth. *Encounter*, 1969, **23** (3): 29-42.
3. Block, J., Haan, N., & Smith, M. B. Activism and apathy in contemporary adolescence. In J. F. Adams (Ed.), *Contributions to the understanding of adolescence.* Boston: Allyn and Bacon, 1969.
4. Blos, P. *On adolescence.* New York: Free Press, 1962.
5. Douvan, E. & Adelson, J. *The adolescent experience.* New York: Wiley, 1966.
6. Erikson, E. The Golden Rule in the light of new insight. In E. Erikson (Ed.), *Insight and responsibility.* New York: Norton, 1964.
7. Erikson, E. Youth: Fidelity and diversity. In E. Erikson (Ed.), *The challenge of youth.* New York: Anchor, 1965.
8. Erikson, E. *Identity, youth and crisis.* New York: Norton, 1968.
9. Feldman, Schwartz, Brown, & Heingartner. Moral level and cheating in college students. Cited in ref. 25.
10. Fenichel, O. *The psychoanalytic theory of neurosis.* New York: Norton, 1945.
11. Feuer, L. *The conflict of generations.* New York: Basic Books, 1969.
12. Footman, D. *Red prelude: A life of A. I. Zhelyabov.* London: Cresset Press, 1944.
13. Freud, S. *New introductory lectures on psychoanalysis.* New York: Norton, 1933.
14. Haan, N., Smith, M. B., & Block, J. Moral reasoning of young adults: political-social behavior, family background, and personality correlates. *Journal of Personality and Social Psychology* 1968, **10**, 183-201.
15. Haan, N., & Block, J. Further studies in the relationship between activism and morality. I. The protest of pure and mixed moral stages. Berkeley, Calif.: Institute of Human Development, 1969.
16. Haan, N., & Block, J. Further studies in the relationship between activism and morality. II. Analysis of case deviant with respect to the morality-activism relationship. Berkeley, Calif.: Institute of Human Development, 1969.
17. Keniston, K. Morals and ethics. *American Scholar*, 1965, **34**, 625-632.
18. Keniston, K. *Young radicals: Notes on committed youth.* New York: Harcourt, 1968.
19. Keniston, K. Notes on young radicals. *Change*, Nov.-Dec. 1969, 25-33.
20. Keniston, K. You have to grow up in Scarsdale.... *New York Times Magazine*, April 27, 1969.
21. Keniston, K. Youth as a stage of life. In *Psychopathology of adolescence*, Zubin & Freeman (Eds.) New York: Grune & Stratton, 1970.
22. Kohlberg, L. Education for justice: A modern statement of the platonic

view. Burton Lecture on Moral Education, Harvard University, Cambridge, Mass., 1968.
23. Kohlberg, L. The child as a moral philosopher. *Psychology today*, Sept. 1968, 25-30.
24. Kohlberg, L. The concept of moral maturity. Paper presented at NICHD Conference on the Development of Values, Washington, May 15, 1968.
25. Kohlberg, L. Continuities and discontinuities in child and adult moral development. Harvard University, Cambridge, Mass., 1969. (Mimeographed.)
26. Kohlberg, L. *Stages in the development of moral thought and action*. New York: Holt, Rinehart and Winston, 1970.
27. Kohlberg, L. Personal communication, 1970.
28. Kramer, R. Changes in moral judgment response pattern during late adolescence and young adulthood. Unpublished Ph.D. thesis, University of Chicago., 1968.
29. Krebs, R. Some relations between moral judgment, attention and resistance to temptation. Unpublished Ph.D. thesis, University of Chicago, 1967.
30. Lacqueur, W. *Young Germany*. New York: Basic Books, 1962.
31. Lampert, E. *Sons against fathers: Studies in Russian radicalism and revolution*. London: Oxford University Press, 1965.
32. Lidz, T. *The person*. New York: Basic Books, 1968.
33. Milgram, S. Some conditions of obedience and disobedience to authority. *Human Relations*, 1965, **18**, 57-76.
34. Mumford, L. *The city in history*. New York: Harcourt, 1962.
35. Newfield, J. *A prophetic minority*. New York: Signet Books, 1966.
36. Perry, W. Forms of intellectual and ethical development in the college years. Harvard University Bureau of Study Counsel, Cambridge, Mass., 1968. (Mimeographed.)
37. Perry, W. Patterns of development in thought and values of students in a liberal arts college. Harvard University Bureau of Study Counsel, Cambridge, Mass., 1968. (Mimeographed.)
38. Piaget, J. *The moral judgment of the child*. New York: Free Press, 1948.
39. Redfield, R. *The primitive world and its transformations*. Ithaca, N.Y.: Cornell University Press, 1953.
40. Roszak, T. *The making of a counter-culture*. New York: Doubleday, 1969.
41. Rubenstein, B., & Levitt, M. The student revolt: Totem and taboo revisited. Paper delivered at American Orthopsychiatric Association Meeting, New York, March 31, 1969.
42. Smith, M. B. Morality and student protest. In *Social psychology and human values*. Chicago: Aldine, 1969.
43. Smith, M. B. The crisis on the campus. See ref. 42.
44. Walter, R. *Student politics in Argentina*. New York: Basic Books, 1968.
45. Yarmolinsky, A. *Road to revolution*. New York: Crowell-Collier-Macmillan, 1962.

Recommended Additional Readings

Flacks, R. The liberated generation: An exploration of the roots of student protest. *Journal of Social Issues*, 1967, **23**, 52-75.

Haan, N., Smith, M. B., & Block, J. Moral reasoning of young adults: Political-social behavior, family background, and personality correlates. *Journal of Personality and Social Psychology*, 1968, **10**, 183-201.

Heath, D. H. Student alienation and school. *School review*, 1970, **78**, 515-528.

Horn, J. L., & Knott, P. D. Activist youth of the 1960s: Summary and prognosis. *Science*, 1971, **171**, 977-985.

Kohlberg, L. Moral education in the schools: A developmental view. *School Review*, 1966, **74**, 1-29.

29 On Psychosexual Development

William Simon
John H. Gagnon

Orientation

Sex is a topic of considerable interest to just about everyone. It is also one of the most problematic issues in our culture. In early adolescence, the recognition that one is a sexual creature and capable of reproducing the species causes great alteration in self concept. How easily are these necessary alterations accomplished? What is the significance of sex for the individual?

Reprinted from the *Handbook of Socialization Theory and Research*, David A. Goslin, editor, by permission of the authors and the publisher; © 1969 by Rand McNally & Company, Chicago.

A revised version of a paper presented at the National Institute of Child Health and Human Development Conference on Social Aspects of Socialization, December 8, 1967. This research was supported by United States Public Health Service Grant HD02257.

There can be little doubt that changes are occurring on the sexual scene in this society. As the following article emphasizes, it appears that the most striking change is that sex is becoming increasingly more public. Dramatic changes are evident in motion pictures, and significant changes can also be found in all aspects of the mass media. Sex is now an acceptable topic for conversation in public gatherings—at bridge parties, at public lectures, and on television. In the words of the sociologists Simon and Gagnon, the American environment has been "eroticized." Thus it would seem that for the developing individual, it is very easy to obtain information regarding sexual behaviors and styles. In fact, sexual "scripts" are readily available for the young person in movies, paperback books, and Playboy magazine.

Yet for the individual within the context of his own personal life, sex remains very much a private matter. To whom does the early adolescent turn for answers to questions he or she may have regarding personal anxieties about sexual functioning and behaviors? It is true that young people converse about sex more casually than in earlier historical periods, but it seems that this public, open discussion of sex remains at an abstract, impersonal level. It is easy to talk about the sexual behavior of "them" or about sex "in general," but what about the individual's sexual anxieties or uncertainties? A very real generation gap, at least in one sense, exists between youth and their parents; both seem unable or unwilling to engage in meaningful discussion about sexual matters with each other. The consistent reports that adolescents obtain their sex education from agencies other than their parents (typically from peers—perhaps slightly older peers—and books and other written materials) attest to this communication gap. However, although young people do engage in conversations regarding matters sexual with their friends, it appears that these discussions seldom pertain to personal concerns and behaviors. Sex is discussed in the abstract—what "other people" think and do.

With the onset of adolescence comes another potential source of anxiety. For the first time the individual is perceived by the world as a sexual being. This can be a troublesome point: Am I attractive? Am I sexually desirable? Why are my friends growing so much faster than I? Why do I have such a small penis? Will my breasts ever fill out? Many more insecurities arise. Peer relationships reinforce the cultural ideals for physical desirability (it should be emphasized that these ideals, however important, are completely arbitrary), and the individual soon develops an attitude about how well he or she meets these ideals.

It must be emphasized that parents also now recognize their offspring as a sexually mature individual; the baby fat and naiveté of childhood are gone. The adolescent may have difficulty in accommodating the fact that he is now a sexual being within the family circle. Perhaps he or she would like to ignore this fact as far as parents are concerned, but *the reality of menstruation or nocturnal emissions (those soiled bed sheets!) are concrete reminders of this dramatic change in status.*

It is essential that the student of human development acquire a clear understanding of the importance of sexual changes during early life. One's sexuality is an important part of his identity; one has little choice about this, for society forces the individual to adopt a sexual orientation. In addition, a number of crucial societal problems are related directly to this issue of childhood and adolescent sexuality: premarital pregnancies, age of marriage, the widespread incidence of venereal disease, sex education in the schools.

The next article, by Simon and Gagnon, presents a view of human sexual development that differs in several important ways from the more widely understood Freudian psychodynamic theory. In addition to exploring issues related to psychosexual development in childhood and adolescence, these authors suggest some provocative ideas about adults' orientation to and handling of sex and sexuality in a rapidly changing culture.

William Simon is affiliated with the Institute for Juvenile Research in Chicago, and John H. Gagnon is on the faculty of the department of sociology at the State University of New York at Stony Brook.

29

Erik Erikson (1963) has observed that, prior to Freud, "sexology" (a wretched and discredited term) tended to see sexuality as suddenly appearing with the onset of adolescence. From Erikson's point of view, Freud's discovery of infantile and childhood expressions of sexuality was a crucial part of his contribution. Libido—the generation of psychosexual energies—was now viewed as a fundamental element of the human experience from its very inception, at the latest beginning with birth and possibly prior to birth. Libido was conceived as something essential to the organism, representing a kind of biological constant with which forms of social life at all levels of sociocultural organization and development, as well as personality structure at each point in the life cycle, had to cope.

In Freud's view, the human infant and child behaved in ways that were intrinsically sexual and that remained in effective and influential continuity with later forms of psychosexual development (Freud, 1953; Chodoff, 1966). Implicit in this view was the assumption that the relation between available sexual energies and emergent motives and attachments would be complex—but direct. In some aspects of psychoanalytic thinking, both adolescent and adult sexuality were viewed as being in some measure re-enactments of sexual commitments developed, learned, or acquired during infancy and childhood.

This view presents both an epistemological and sociolinguistic problem. Freud's language was the language of adult sexual experience imposed upon the "apparent" behavior and "assumed" responses, feelings, and cognitions of infants and children. Acts and feelings are defined as sexual, not because of the actor's sense of the experience, but because of the meanings attached to those acts by adult observers or interpreters whose only available language is that of adult sexual experience (Schachtel, 1959). However, the assumptions implicit in the adult terminology with which such behaviors are perceived and considered must be approached with caution. The dilemma arises from the problem of distinguishing between the sources of labeling specific actions, gestures, or bodily movements as sexual. For the infant playing with his penis, the activity is not sexual in the sense of adult masturbation, but merely diffusely pleasurable in the same manner as are many other activities. The external observer imputes to the child the complex set of states that are generally associated with physically homologous adult activities. It is only through the processes of maturing and learning these adult labels that the child comes to masturbate in the fullest sense of the word. It is in the process of converting external labels into internal capacities for naming that the activities become more precisely defined and linked to a structure of sociocultural expectations and needs that define what is sexual. To suggest that infant or childhood genital play is prototypical of or determines adult patterns is to credit the biological organism with more "natural" wisdom than we normally do in areas where the biological and the sociocultural intersect. Undeniably, sexuality is rooted in biological processes, capacities, and possibly even needs. But admitting this in no way provides for a greater degree of biological determinism than is true of other areas of corresponding intersection. Indeed, the reverse may be true: the sexual area may be precisely that realm wherein the superordinate position of the sociocultural over the biological level is most complete.

It is important to note here that the very diffuse quality of most of preadolescent experience poses a number of major problems, one of which is the extreme difficulty in getting accurate data. Part of the problem centers on a kind of faulty recall which is not rooted merely in inaccurate memories. This is a source of error pointed out in the existentialist insight that rather than the past determining the character of the present, it is possible that the present significantly reshapes the past as we reconstruct our autobiographies in an effort to bring them into greater congruence with our present identities, roles, situations, and available vocabularies. The other part of the problem results from attempting to gather data from children who when interviewed are ill-equipped to report upon their own internal states or from adults who are asked to report on periods of life when complex vocabularies did not exist for them. The problem is attempting to determine what is being felt or thought when confronted with organisms whose restricted language skills may preclude certain feelings or thoughts.

Essential to the perspective of this chapter is the assumption that with the beginnings of adolescence—and with the increasing acknowledgement by the surrounding social world of an individual's sexual capacity—many novel factors come into play, and that an over-emphasis upon a search for continuity with infant and childhood experiences may be dangerously misleading. In particular, it may be a costly mistake to be overimpressed with preadolescent behaviors that appear to be manifestly sexual. In general, it is possible that much of the power of sexuality may be a function of the fact it has frequently been defined as powerful or dangerous. But this overenriched conception of sexual behavior (to the degree that it occurs) must largely follow upon considerable training in an adult language which includes an overdetermined conception of sexuality. Thus it does not necessarily follow that the untrained infant or child will respond as powerfully or as complexly to his own behaviors that appear to be sexual to an adult observer as will that observer.

Of all the experiences of the preadolescent, it would appear that the most influential in the determination of later psychosexual development is that large and complex cluster of elements that come under the heading of sex-role learning. It is perhaps better to label this process "gender-role" learning, underscoring thereby its indirect link to sexuality in many, if not most, of its aspects. In other words, we may find that the development of an emerging sense of self is significant in the determination of later sexual commitment or capacities, but that this sense of self *does not* itself derive, in most cases, from preadolescent sexual or even near-sexual experiences.

For the purposes of this chapter, we reject the unproven assumption of the "power" of the psychosexual drive as a fixed biological attribute. We feel there is little evidence to suggest that such a "drive" need find expression in specific sexual acts or categories of sexual acts. More importantly, we reject the even more dubious assumption that sexual capacities or experiences tend to translate immediately into a kind of universal knowing or wisdom: the assumption that sexuality possesses a magical ability that allows biological drives to seek direct expression in psychosocial and social areas in ways that we do not expect in other biologically rooted behaviors. This assumption can be seen in the psychoanalytic literature, for example, in which the child who views the "primal scene" is seen on some primitive level as intuiting its sexual character. Also, the term latency, in its usage by psychoanalytic theorists, suggests a period of integration by the child of prior intrinsically sexual experiences and reactions which reduces adolescence, on this level, to a mere management or organization on a manifest level of the commitments and styles already prefigured, if not preformed, in infancy and childhood experiences.

In contradistinction to this tradition, this chapter adopts the view that the point at which the individual begins to respond in intrinsically sexual ways, particularly in terms of socially available or defined outlets and objects, is a time during which something occurs which is discontinuous with previous "sexual experience" (whatever that might mean). Further, at this point in the developmental process, both seemingly sexual and seemingly nonsexual elements "contest" for influence in complex ways which in no respect assure a priority for those elements that are apparently sexual in character.

The prevailing image of the sexual component in human experience is that of a fairly intense, high-pressure drive (except, perhaps, during later stages of the life cycle) that constrains the individual to seek sexual gratification either directly or indirectly. This is clearly present in the Freudian tradition. A similar position is observable in more sociological writings. This is apparent, for example, in the thinking of Kingsley Davis for whom sex is also a high-intensity, societal constant that must be properly channeled lest it find expression in behaviors which threaten the maintenance of collective life (Davis, 1961; Durkheim, 1951).

Our sense of the available data suggests a somewhat different picture of human sexuality, one of generally lower levels of intensity or, at least, greater variability in intensity. More clearly evident among females than males (see below), there are numerous social situations or roles where reduction and even elimination of sexual activity is managed by greatly disparate populations with little evidence of direct corollary or compensatory intensification in other spheres of life (Gagnon & Simon, 1968a). It is possible that, given the historical nature of human societies, we are victim to the needs of earlier social orders. For earlier societies it may not have been a matter of having to constrain severely the powerful sexual impulse in order to maintain social stability (particularly in family life), but rather a matter of having to invent an importance for sexuality not only to insure high levels of reproductive activity but also to provide a socially available reward that might be placed at the service of many social ends. A part of the legacy of Freud is that we have all become relatively adept at seeking out the sexual ingredient in many forms of nonsexual behavior and symbolism. What we are suggesting is in essence the now three-decade old insight of Kenneth Burke (1935) to the effect that it is entirely plausible to examine sexual behavior for its capacity to express and serve nonsexual motives.

For us, then, sexual behavior is socially scripted behavior and not the masked or rationalized expression of some primordial drive. The individual learns to be sexual as he or she learns sexual scripts, scripts that invest actors and situations with erotic content. One can easily conceive of numerous social situations in which all or almost all the ingredients of sexuality are present, but which remain nonsexual in that not even sexual arousal occurs (Simon & Gagnon, 1967a).

Thus combining such elements as desire, privacy, and a physically attractive alter of the appropriate sex, the probability of something sexual occurring will, under normal circumstances, remain exceedingly small until either one or both actors organize these elements into an appropriate script. The very concern with foreplay in sexual behavior suggests something of this order. From one point of view, foreplay might be defined as progressive physical excitement or what the authors (Gagnon & Simon, 1968b) have elsewhere referred to as the "rubbing of two sticks in order to get a fire going" model. From another point of view, this activity may be defined as a method of eroticizing the body and the activity, as a method for transforming mute, inarticulate gestures and motions into a sociosexual drama.

Lastly, in these introductory comments we might continue to belabor the issue of sociocultural dominance by making some preparatory distinctions, distinctions that we would have to make in any event. Psychosexual development, while a universal component in the human experience, certainly does not occur with universal modalities. Even ignoring the striking forms of cross-cultural variability (Ford & Beach, 1951), we can observe striking differences within our own population, differences that, given the relatively low level of thinking and the sparsity of research in the area, appear to require not a description of psychosexual development, but descriptions of different developmental processes characterizing different segments of the population. The most evident of these are the large number of important differences between observable male and female patterns of sexual behavior (Maccoby, 1966). This particular difference may in some respects be partly attributable to the role played by the biological substratum (Hamburg & Lunde, 1966; Young, Goy & Phoenix, 1964). We have to account not only for the gross physiological differences and the different roles in the reproductive process that follow from these physiological differences, but must also consider differences in hormonal functions at particular ages. Yet, while our knowledge of many of the salient physiological and physiochemical processes involved is far from complete, there is still little immediate justification for asserting a direct causal link between these processes and specific differential patterns of sexual development observed in our society. The recent work of Masters and Johnson (1966), for example, clearly points to far greater orgasmic capacities on the part of females than males; however, their concept of orgasm as a physiological process would hardly be a basis for accurately predicting rates of sexual behavior. Similarly, within each sex, important distinctions must be made for various socioeconomic status groups whose patterns of sexual development will vary considerably, more impressively for males than for females (Kinsey, 1948). And with reference to socioeconomic status differences, the link to the biological level appears even more tenuous, unless one is willing to invoke the relatively unfashionable conceptual equipment of Social Darwinism. These differences, then, not only suggest the importance of the sociocultural elements and social structure, but also stand as a warning against too uncritical an acceptance of unqualified generalizations about psychosexual development.

Childhood

Obviously Erikson is correct in agreeing with Freud: we do not become fully sexual all at once. There is significant continuity with the past; nevertheless, continuity is not causality. Even in infancy experiences can occur that will strongly influence later sexual development. However, such experiences will, in all likelihood, be influential not because of their essentially sexual character, but because of their general influential character; that is, they probably influence many more things than just sexual development. It is also possible to talk about the kinds of experience that give rise to fixation, but such experience is probably as unusual as it is traumatic and can hardly be the basis for general theories. One has difficulty conceiving of situations in infancy—or even early childhood—that can be linked to psychosexual development on a level more specific than that of potentiation.

In infancy and in the interrelationships that are part of infant care, we can locate many of the pre-verbal experiences that are preparation for the development of verbal capacities which, in turn, will bind the child to the social world. In this period we can locate some of the experiences—perhaps only sensations—that will help bring about a sense of the body and its capacities for pleasure and discomfort, and we can also locate the experiences that will influence the child's ability to relate to other bodies. The key term remains potentiation: it is possible that through these primitive experiences, ranges are being established, but these ranges are sufficiently broad and overlapping that little can follow by way of specification except through the dubious route of tortured reconstruction. Moreover, if these are profoundly significant experiences to the child—and they may well be—they stand not as expressions of biological necessity nor the inherent wisdom of the body, but as expressions of the earliest forms of social learning (Gewirtz, 1961).

Unlike the period of infancy, where at best we are confronted with a universe of unnamed gestures, it is possible to conceive of activity in childhood that appears to be explicitly sexual to the observer. About half of all adults report having engaged in some form of "sex play" as children (Kinsey, 1948, 1953). And conceding a certain role to faulty recall or repression, the proportion having behaved this way may be as much as half again as large. The crucial and almost impossible question remains: What does it mean to the child? It is likely that some part of the adult definition of the activity is realized by the participants, but one suspects—as in much of childhood role-playing—that their sense of the adult meaning is fragmentary and ill-formed. It is clear that some learning about the adult world's judgments of the sexual occurs as is indicated by the fact that a high proportion of adults recall that they were concerned, while engaging in childhood sex play, over the possibility of being "found out." However, it may not be the content of sex play activity—as if they were responding to their nascent sexuality—but the mystery that enchants the child (Freud, 1964). Stated differently, the child may be assimilating external bits of information about sex for which, at the time, there are no real internal correlates.

For some small number of persons sexual activity does occur during preadolescence. Probably the largest group are those who become involved as objects of adult-initiated behavior. But for most of them, little seemingly follows. Lacking appropriate sexual scripts the experience remains unassimilated, except perhaps for those who derive a "meaning" of the experience from a subsequent clinical situation. For some it is equally clear that a severe reaction does follow from falling "victim" to the sexuality of some adult figure, but here it is debatable whether this reaction follows from the sexual act itself or from the social act, the tone, and the intensity of the reactions by others (Gagnon, 1965). In short, relatively few preadolescents become truly sexual, that is, become sexually active. For preadolescent females more often than males, this sexual activity is not immediately related to internal states—capacities for sexual arousal or sexual gratification—but to an instrumental use of sexuality to achieve nonsexual goals and gratifications. This "seductive" preadolescent female, for all her statistical rarity, may represent a significant adumbration of a more general pattern of psychosexual development: a process wherein a commitment to sociosexuality precedes a commitment to sexuality. Among sexually active preadolescent males, behavior appears to be more intrinsically sexual and is associated with a high frequency of subsequent deviant adaptations, most often homosexual (Gebhard et al., 1965).

Of considerable importance is the internalization of values—or the images that powerfully stand for values—that may not be directly or exclusively referential to sexual matters. These values or value images will constitute aspects in the construction of sexual scripts giving rise to senses of the evil, the extraordinary, and the erotic. Despite our present capacity as a society to generate high levels of public discourse about sexual matters, it is probably not unreasonable to assert that learning about sex in our society is learning about guilt; conversely, learning how to manage sexuality constitutes learning how to manage guilt. An important source of guilt is the imputation by adults of sexual capacities or qualities in children that the children may not have, but that result in—however imperfectly—children learning to act as if they had such capacities or qualities. For example, at what age do girls learn to sit with their knees together, and when do they learn that the upper part of the torso must be hidden, and what of a sexual nature do they learn from all this? Childhood learning of major themes which establish sex or gender role identities is of critical importance. Much of what appears under the heading of sex-role learning involves elements that are remote to sexual experience or that become involved with sexuality only after the latter has become salient (Sears, 1965). Here the meanings and postures of masculinity and femininity are rehearsed and assimilated in many nonsexual ways. Here, also, the qualities of aggression, deference, and dominance needs, which Maslow (1939, 1942)—however imperfect his data—persuasively argues are strongly implicated in the organization of sexual styles, are initially rehearsed, experimented with, and assimilated.

Kagan and Moss (1962) report similar findings: aggressive behavior is a relatively stable aspect of male development and dependency is a similar characteristic in female development. Significant appearances of aggressive behavior by females tended to occur most often among females from well-educated families, families that tended to be more tolerant of deviation from sex-role standards. They also find, of particular interest, that ". . . interest in masculine activities for age 6 to 10 was a better predictor of adult sexuality than was heterosexual activity between 6 and 14." Curiously, they also report that "it was impossible to predict the character of adult sexuality in women from their preadolescent and early adolescent behavior." This the authors attribute to a random factor, that of an imbalance of sex ratios in the local high school and, more significantly, to the fact that "erotic activity is more anxiety-arousing for females than for males" and that "the traditional ego ideal for women dictates inhibition of sexual impulses."

This concept of the importance of early sex-role learning for male children may be viewed in two ways. From one perspective elements of masculine-role learning may be seen as immediately responsive to—if not expressive of—an internal sexual capacity. From another perspective, we might consider elements of masculine identification merely as a more appropriate context within which the mediation of the sexual impulse—which becomes more salient with puberty—and the socially available sexual scripts can occur. Our bias, of course, is towards the latter.

The failure of sex-role learning to be effectively predictive of adult sexual activities, noted by Kagan and Moss, also may lead to alternative interpretations. Again, from one perspective, where sexuality is viewed as a biological constant for both women and men, one can point to the components of female-role learning that facilitate the successful repression of sexual impulses. The other perspective or interpretation suggests differences in the process not of handling sexuality, but of learning how to be sexual, differences between men and women that will have consequences for both what is done sexually as well as when it is done. Our thinking, once again, tends towards the latter. This position is supported by some recent work of the present authors (Simon & Gagnon, 1967b, 1967c) on female homosexuals, where it is observed that patterns of sexual career management (e.g., the timing of entry into actual sexual behavior, entry into forms of sociosexual behavior, onset and frequency of masturbation, number of partners, reports of feelings of sexual deprivation, etc.) were for lesbians almost identical with those of heterosexual women. Considering what was assumed to be the greater salience of sexuality for the lesbian—her commitment to sexuality being the basis for her entry into a highly alienative role—this seemed to be a surprising outcome. What was concluded was that the crucial operating factor was something that both heterosexual and homosexual women share: the components of sex-role learning that occur before sexuality itself becomes significant.

Social class differences also appear to be significant, although both in the work of Kinsey (1948, 1953) and that of Kagan and Moss they appear as more important factors for males than for females. Some part of this is due to aspects of sex-role learning which vary by social class. Differences in the legitimacy of expressing aggression or perhaps merely differences in modes of expressing aggression come immediately to mind (Sears, 1965; Biller & Borstelmann, 1967). Another difference is the degree to which sex-role models display a capacity for *heterosociality*. The frequently noted pattern of the sexual segregation of social life among working-class and lower-class populations may make the structuring of later sexual activity, particularly during adolescence, actually less complicated (Rainwater, 1966; Simon & Gagnon, 1966). Another aspect of social class differences is the tolerance for deviation from traditional attitudes regarding appropriate sex-role performances. Clearly, tolerance for such deviance is positively associated with social class position, and it may well stand in a highly complex and interactive relationship to capacities for heterosocial activities.

We have touched here upon only a few of the potentially large number of factors that should be related to important social class differences and to the processes of psychosexual development. In general, even during this relatively early period of life, complex elements of the ego begin to take form, including the crude outlines of what might be called a repertoire of gratifications. It seems rather naive to conceive of sexuality as a constant pressure upon this process, a pressure that has a particular necessity all its own. For us, this crucial period of childhood has significance not because of what happens of a sexual nature, but because of the nonsexual developments that will provide the names and the judgments that will condition subsequent encounters with sexuality.

Adolescence

Adolescence is a period with ill-defined beginning and end points. There is variation in its general definition, still greater variability in the application of these definitions, and the situation is additionally complicated by the variability of developmental rates of specific individuals. However ill-defined this period may be, the beginning of adolescence marks the time when society, as such, first acknowledges the sexual capacity of the individual. His or her training in the postures and the rhetoric of the sexual experience will now begin to accelerate. Most importantly, the adolescent will start to view others in his or her immediate environment—in particular, peers, but also some adults—as sexual actors and will find confirmation of this view in the definitions of others toward these actors.

As has already been indicated, for a number of individuals sexual activity begins prior to adolescence, or some portion of children begin engaging in adolescent sexual behavior before they are defined as adolescents. Thus, Kinsey (1948, 1953) reports that by age 12 about a tenth of his female sample and a fifth of his male sample had already experienced orgasm through masturbation. (There is some evidence that early entry into sexual activity is associated with alienative adjustments in later life, but this may not be a function of sexual experience per se so much as the consequence of having fallen out of the more modal socialization patterns and, as a result, having to run greater risks of not receiving appropriate forms of social support.) But this is still an atypical preface to adolescence. For the vast majority, aside from relatively casual childhood sex play and the behaviors that post-Freudians view as masked sexuality, movement into sexual experience which the actor defines and accepts as such begins with the passage into adolescence. Even for persons with prior sexual experience, the newly-acquired definition of their social status as adolescents qualitatively alters the meaning of both current and prior sexual activity; they must now integrate such meanings in more complex ways, ways that are related to both larger spheres of social life and greater senses of self. For example, it is not uncommon during the transitional period between childhood and adolescence for both males and females to report arousal and orgasm while engaging in many kinds of physical activity that are not manifestly sexual—climbing trees, sliding down bannisters, or other forms of activity where there is genital contact—without it being defined as sexual by the adolescent (Kinsey, 1948). Indeed, in many such cases there may not even follow subsequent self-explorations in order to achieve some repetition of what was, in all likelihood, a pleasurable experience.

The onset of adolescent sexual development, which really represents the beginning of what will be an adult commitment to sexuality, will be somewhat disjunctive with past experience in most cases. As we have suggested, not only are future experiences to occur in much more complex situations, but also for the first time the more explicit social implications of sexual activity will further complicate matters (Reiss, 1960). The need to manage sexuality, following from a growing sense of having a sexual status, will derive not only from the intrinsic attractions of the sexual experiences for some, but from the increasingly important role sexuality will play in the conduct of both heterosocial and homosocial relationships.

The onset of adolescence demands a separate consideration on our part of the developmental process for males and females. The one thing both genders will share at this point is the reinforcement of their competence in their new sexual status by the occurrence of a dramatic biological event: for males the discovery of the ability to ejaculate, for females the onset of menstruation. This difference can be made evident in the simplest way by pointing to the separate modal routes into sexual experience. For males the organizing event that initiates a sexual commitment is the biological event of puberty; within two years of puberty all but a relatively few males have their commitment to sexuality reinforced by the experience of orgasm, almost universally brought about by masturbation (Kinsey, 1948). The corresponding organizing event for females is not a biological but a social event: the arrival at an age that suggests a certain proximity to marriage. In contrast to male masturbation rates, we find that only two-thirds of a female population will report ever having masturbated (and then, characteristically, with much lower frequency), and cumulatively we have to go into the latter part of the third decade of life (the twenties) before a proportion is reached which is comparable to that reached by males at age 16 (Kinsey, 1948, 1953). Indeed, it is significant that about half of the females who masturbate will do so only after having their sexuality reinforced by the experience of orgasm in some sociosexual situation. This contrast between males and females, which we will elaborate below, once again points to a distinction between the developmental process for males and that for females: the movement from sexuality to sociosexuality characteristic of males is reversed for females.

When we turn to a more detailed view of the adolescent male experience, despite the fact that we have up to now worked very hard to establish the dominance of the sociocultural or social-psychological over the biological sphere, we must now briefly reverse course. There is some evidence that the early impulse to sexual expression—again initially through masturbation—is linked to high hormonal inputs during the period of puberty, producing an organism that, to state it in its simplest terms, is easily "turned on" (Kinsey, 1948). It is not uncommon for young male adolescents to report frequent erections during this period, often without the provocation of erotic preoccupations. This obviously focuses considerable attention on the genitalia. Yet, however powerful the biological origins of the propensity for the young adolescent male to masturbate, the meaning and the organization of the activity tend to occur in terms of social and psychological factors.

On one level masturbation is a guilt and anxiety provoking activity for most adolescent males. And despite changes in the rhetoric of public and semipublic discourse on the topic—a shift from a rhetoric of mental and physical destructiveness to one of vaguely inappropriate or nonsocial behavior—this is not likely to change in the foreseeable future. It may be this very element, however, that permits the sexual experience to generate an intensity of affect that may often be attributed to the powers of sexual capacity itself. This guilt and anxiety does not follow simply from the general societal disapproval of this activity. Rather it appears to derive from several sources, one of which is the difficulty encountered in presenting a sexual self to members of the immediate family, particularly parents, where the sentimental and the erotic still remain ill-defined. Another source of guilt and anxiety stems from the fact that arousal and excitement in sexual activity increasingly derives from vicarious organization of sociosexual activity—fantasies involving doing sexual "things" to others or having others do sexual "things" to oneself—or from the assimilation and rehearsal of available sexual scripts, many of which involve engaging in proscribed activities or relationships. An additional source of guilt or anxiety centers on the general sanction against masturbatory behavior, an activity which few males are willing to admit to after the period of early adolescence.

Despite guilt and anxiety, which may consequently become part of the more general commitment to sexuality for many men, masturbation remains an extremely positive and gratifying experience, as it constitutes for most males through the middle of adolescence the major source of sexual outlet and is engaged in fairly frequently during this period. This form of introduction to sexuality tends to give rise to a capacity for detached sexual activity, activity where the only sustaining motive is sexual; this may actually be the hallmark of male sexuality in our society.

Of the three sources of guilt and anxiety alluded to above, the first, the problem of managing both sexuality and an attachment to family members, is probably common across social class lines. The other two sources should display marked social class differences. The second, the problem of managing a fairly elaborate and exotic fantasy life, should most typically be a problem for higher social class levels. On lower class levels there is a less frequent use of fantasy during masturbation, and this in turn may be linked to a generalized lack of training and competence in the manipulation of symbolic materials (Kinsey, 1948). Higher social class male adolescents are presumably trained for such competence. Masturbation accompanied by fantasy should in turn reinforce the manipulation of symbolic materials. If we de-emphasize the physical component in the masturbatory act, it can be conceptualized as an activity in which the actor is, in effect, training himself to invest symbols with affect and to derive gratification from the manipulation of symbols. Successfully doing this results in what is classically the best reward or reinforcer in learning theory terms: it is the immediate, specific, and pleasurable experience of orgasm. It may well be that this behavior—which is something females tend not to engage in—plays a role in the processes by which middle-class males catch up with females in measures of achievement and creativity and then, by the end of adolescence, move out in front. This is, of course, merely a wild hypothesis—the Dr. Krankeit (sic) hypothesis (Kenton, 1958). This primary reliance upon masturbation with fantasy should also have a number of consequences beyond the capacity for a relatively detached sexual performance. One such consequence is a tendency to eroticize large parts of the world, as well as an ability to respond to a wide array of visual and auditory *stimuli*. Moreover, to the extent that Wilhelm Reich (1942) is correct in his assertion that the scripting for masturbatory fantasies has a non- and possibly anti-coital character, we might also expect both a capacity and a need for fairly elaborate forms of sexual activity. Further, in so far as the masturbatory fantasy focuses upon relationships and activities essentially preparatory to the coital act, the masturbatory experience should also reinforce an already developing capacity for sustaining heterosocial activity.

The third source of guilt and anxiety, essentially the unmanliness of masturbation, should be more typically a concern of lower social class male adolescents. Among these lower-class male adolescents the general pattern of sexual segregation of social life and the relatively narrower range of rewarding social experiences available to them should combine to constrain the adolescent to move into heterosexual relationships sooner than does his middle-class counterpart. The first condition, the sexual segregation of social life, should make it easier for him to gravitate toward a world of casual, if not exploitative, sexual relations: it is easier for him than for the middle-class adolescent to learn that he does not have to love everything he desires. The second condition—the more limited available social rewards and particularly rewards deriving from activities that will be validated by his peers—should lead to an exaggerated concern for masculinity-enhancing behavior leading, in turn, to displays of physical prowess, successful staging of aggressive behaviors, and visible sexual success. The three— physical prowess, aggressive behavior, and sexual success—are, of course, not unrelated and frequently are mutually reinforcing.

Available data suggest, in line with the above, that one of the differences between lower-class and middle-class male adolescents is the phasing of entry into heterosexual forms of sociosexuality or the phasing of movement from masturbation to coital behavior (Kinsey, 1948). In a sense, the lower-class male is the first to reach "sexual maturity" as defined by those working in an essentially Freudian tradition. That is, these lower-class males are generally the first to become exclusively heterosexual and exclusively genital in orientation. One consequence, however, is that while their sexual activity is almost exclusively heterosexual, it also tends to be more homosocial in character. The audience to which the lower-class male's sexual activity is directed will tend not to be his female partner, but rather it will be more referential to his male peers (Simon & Gagnon, 1966). Middle-class adolescent males, who will initiate coital activity at a significantly later time, will not be exempt from a need and tendency for homosocial components in their sexual lives, but the complexity of their fantasies as well as their social training in an environment less sexually segregated—a world in which distinctions between masculine and feminine roles are not as clearly drawn—will facilitate an easier withdrawal from homosocial commitments. This difference between the social classes will obviously have important consequences for the management of stable adult relations, a point which we will consider below.

One characteristic that will tend to be common to the male experience during adolescence is that while this stage provides extensive opportunity for developing a sexual commitment in one form or other, there is little training in the management of affectively-charged relations with females. The imagery and rhetoric of romantic involvement is abundantly present in the society, and it is not unlikely that a great deal of this is absorbed by the male adolescent, but it is not likely to be significantly tied to a sexual commitment. To the degree that it is related to sexuality, the connection may be quite inhibiting, as is suggested by the still operative "bad girl—good girl" distinction. This is important to keep in mind as we turn now to a consideration of the female side of this story.

The pattern of female development during adolescence is so resistant to social class variation that one is tempted to seek explanation in something as relatively immutable as the biological level (Kinsey, 1953). However, we will resist this temptation. The female, to be sure, is not provided with the same biologically-rooted incentive to begin an active sexual commitment; but, by the same token, little evidence exists that there is either a social or a biological inhibitor to the development of such a commitment. In simplest terms, the physical equipment for the generation of sexual pleasure is clearly present by puberty and before (Kinsey, 1953), but it tends not to be used by many adolescent females on any social class level. Masturbation rates are fairly low (and those who do masturbate do so fairly infrequently), arousal from explicitly sexual materials and situations is an infrequent occurrence, and there are exceedingly few reports of a self-conscious feeling of sexual deprivation during this period. Explanations for this low level of female sexual activity or interest are numerous on both functional and historical grounds. The basic element in all of these is the idea that females in our society are not encouraged to be sexual, and, indeed, it is possible that they are strongly discouraged from being sexual. As Rheingold (1964) describes this: where men have only to fear sexual failure, women have to fear failure and success simultaneously. Or, as several people have observed, while the category "bad boy" has many descriptive sub-categories, the category "bad girl" tends almost exclusively to describe sexual delinquencies. Clearly, it is both difficult and dangerous for a female to become too committed or too sexually active during adolescence.

Whether this extended period of relative sexual inactivity represents the outcome of successful repression of an elementary sexual urge or merely represents a failure to have an opportunity to learn how to be sexual, is an important question for consideration. The alternative answers have different implications for how we view the later development of a sexual commitment during late adolescence or postadolescence. The "repression answer" suggests that we approach later activity in terms of processes of de-inhibition by which the female learns to find, in varying degrees, modes of more direct expression of internally experienced feelings. It also requires a quest for the sexually determined aspects of nonsexual behavior. The "learning answer" suggests that women create or invent a capacity for sexual behavior, learning how to be aroused and learning how to be responsive. The latter approach also implies greater flexibility in overall adjustments: unlike the repression view, it makes sexuality something other than a constant that is likely to "break loose" at any point in strange or costly ways. In addition, the learning approach lessens the power of the sexual component as a variable; all at once, there is no necessarily healthy or pathological component to a particular style of sexual activity. Lastly, the appeal for use of this approach is somewhat subjective: it tends to seem less like a projection of male sexuality.

The absence of intrinsically sexual activity by adolescent females does not mean that sexual learning and training fail to occur. Curiously, women who are, as a group, far less sexually active than men, receive far more training in self-consciously conceiving of themselves as being sexual on the object level. This is particularly true for recent age cohorts. On the level of the cosmetic self, females begin relatively early in adolescence to define attractiveness in at least partially sexual terms. One suspects that the same instrumental approach that marked our preadolescent "seductress" now characterizes larger proportions of adolescent females. Parsons' (1954) language of the wife "using" sex to bind the husband to the familial unit, for all its harsh sound, may be quite accurate. In more general terms, the development of a sexual role appears to involve a need to assign to that role services other than pleasure.

To complete this picture of a lack of symmetry between the sexes, the female appears to be trained in precisely that area for which males during adolescence are least trained and for which they are least expected to display a capacity: intense, affect-laden relationships and the rhetoric of romantic love. When sexual arousal is reported by females during this period, it is more often reported as a response to representations of romantic love than as a response to erotic representations.

The movement into later adolescence and the concomitant increase in opportunities for sociosexual activity can be described as a situation in which males—committed to sexuality and relatively untrained in the rhetoric of romantic love—interact with females who are committed to romantic love and relatively untrained in sexuality. Dating and courtship may well be considered processes in which persons train members of the opposite sex in the meaning and content of their respective commitments. And while data in this area are deficient in many regards, the data that are available suggest the exchange system does not often work smoothly. Thus, as is partly suggested by Ehrmann's (1959) work and partly by our own present studies of college students, it is not uncommon to find, ironically, that the male suitor frequently becomes emotionally involved with his partner and correspondingly less interested in engaging in sexual activity with her, and that the female, whose appreciation of the genuineness of her suitor's affection allows her to feel that sexual activity is now both legitimate and desirable, becomes more interested in engaging in sexual activity with him. Data from the authors' recent study of college students, now under analysis, demonstrate that this difference in commitment is exemplified in several ways. One such item, dealing with the number of times individual respondents had intercourse with their first partner, shows the mode for males around one to three times, while the mode for females is at ten times or more. Clearly, for females, initial intercourse becomes possible only in relatively stable relationships or relationships involving rather strong bonds.

On a theoretical level, we find the male experience conforming to the general Freudian expectation with males moving from a predominantly sexual commitment to an ability to form cathectic attachments in loving relationships. In effect, this movement is reversed for females with cathectic attachments being, in many cases, a necessary precondition for coital activity. It is not surprising, perhaps, that Freud had great difficulty approaching female sexuality. This "error" in conceptualizing female sexuality—of seeing it either as being similar to male sexuality or as a mirrored image—may derive from the fact that so much of the theory construction in this area was done by males. In Freud's case, we also have to consider the very conception of sexuality that was essential to most of Victorian Europe—it was an elemental beast that had to be curbed.

In addition, it is particularly important to consider social class differences among females, if only as a way of assessing the possibility of a biological factor producing the above-discussed outcome. As we have already indicated, there are very few such differences, far fewer than may be observed among males. One, however, is particularly relevant to this question—the age of first intercourse. This varies inversely with social class and is strongly associated with similar class differences in age of first marriage (Kinsey, 1953). There is no evident basis for associating such social class differences with biological differences. A second difference, perhaps linked only indirectly to social class, is educational achievement. For this variable a single cutting point appears to separate two distinct populations: women who have attended graduate or professional schools and all other women. The former tend to be the most "successful" sexually—at least if one is willing to accept as a measure of success the relatively crude indicator of the proportion of sexual acts that culminate in orgasm. One possible interpretation of this finding derives from Maslow's (1939, 1942) work: women who survive the academic process and go on for additional training are more likely to be more aggressive and/or have strong dominance needs, and both characteristics are associated with heightened sexual commitments. Another somewhat more general interpretation would argue that in a society which still strongly encourages women to form primary allegiances to roles as wives and mothers, the decision to go on to graduate school represents something of a deviant adaptation. This adaptation represents, in turn, a failure of, or alienation from, modal female socialization processes. And, in effect, it is faulted socialization which produces both the academic or professional commitment and the sexual commitment.

For both males and females, progressively greater involvement with sociosexuality may be one of the factors that marks the end of adolescence. This is a transition about which little is really known, particularly with reference to noncollege populations. Work currently under way by the authors supported by the National Institute of Child Health and Human Development, is attempting to deal with this problem. Our present feeling about the place of sexuality in the management of this transition is that it plays a significant role. First, on a somewhat superficial level, progressive involvement in sociosexuality is important in family formation or in the entry into roles and role obligations that are more explicitly adult in character. But, perhaps on a more fundamental level, it is possible that sociosexual activity is the one aspect of identity experimentation that we associate particularly with later adolescence, a period in which the psychosocial moratorium that Erikson describes as protecting the adolescent during this period of crises and experimentation fails to operate (Erikson, 1963: Reiss, 1960). This may be partly due to the fact that the society has some difficulty protecting the adolescent from the consequences of that part of his behavior it is not prepared to admit he is engaging in. More importantly, it may be due to the fact that we have, at all age levels, great problems in talking about sexual feelings and experiences in personal terms which, in turn, make it extremely difficult to get social support for our experiments with our sexual selves (the term "pluralistic ignorance" is perhaps nowhere more applicable than in the sexual area). It may be that these experiments with sexual capacities and identities rank among the first unprotected tests of competence and the quest for a basis for self-acceptance. We suspect that success or failure in the management of sexual identity may have consequences in many more areas of personality development than merely the sexual sphere.

Adulthood

All but very few persons in our society ultimately marry. The management of sexual commitments within a marital relationship characterizes the larger part of postadolescent experience in our society. Once again, it is important to underscore the real poverty of data on this topic. Sexual adjustment presumably plays an important role in overall marital adjustment. This judgment largely derives from studies of broken marriages or marriages that are in trouble, and we really have very little sense of the degree to which sexual problems in troubled or dissolved marriages exceed those found in marriages which remain intact. It is possible that we have assumed an important role for sexuality and the management of sexuality in the maintenance of marital bonds because we have assumed sex itself to be an important part of most people's lives. This may not be true. Particularly after the formation of the marital unit, it is quite possible that sex declines in salience. It may stand as less important than alternative modes of gratification, or the weight of alternative gratifications may minimize the effects of sexual dissatisfaction. It is also possible that individuals learn to derive sexual gratification from non- or only partially sexual activities. This is not to suggest support for the concept of sublimation, but rather to point out that in the processes that follow marriage, newly learned alternative patterns of gratification may substitute for the sexual.

The main determinant of adult rates of sexual activity in our society is the level of male commitment. While interest in intercourse is highest for males during the early years of marriage, a corresponding peak in coital interest occurs much later in marriage for females (who require longer periods of time to either become de-inhibited or to learn to be sexual—depending upon your point of view). Nonetheless, coital rates in marriage decline steadily through marriage (Kinsey, 1953). This decline, it should be noted, can only be partly attributed to declines in biological capacity on the part of the male. The decrease may derive from many things. In many cases the problem is one of relating sexually to a person whose roles have become complicated by the addition of maternal functions (Freud, 1949). For lower-class males, there is a problem of not receiving homosocial support for marital intercourse, to which we might also add the disadvantage of being less trained in the use of auxiliary materials to heighten sexual interest (Simon & Gagnon, 1966). For middle-class males, the decline is less steep, owing perhaps to their ability to find sexual stimulation from auxiliary sources—literature, movies, etc. Also operative is a greater capacity for invoking and responding to fantasy. It should be noted that for about 30 per cent of college-educated males, masturbation continues as a regular source of sexual outlet in marriage and during periods when a wife is available (Kinsey, 1948). To this we might add an additional but unknown proportion who do not physically engage in masturbation but for whom the source of sexual excitement is not just coital activity alone but also the fantasy elements which accompany coital activity. But even for the middle-class male, sexual activity declines in degrees that cannot exclusively be accounted for by changes in the organism. Perhaps it is simply that the conditions under which we learn to be sexual in our society make it extremely difficult to maintain high levels of sexual performance with a single partner over long periods of time. This may remain relatively unimportant in the maintenance of the family unit or even the individual's sense of his own well-being because of the relative unimportance of sexual dissatisfaction or the relatively greater significance of other areas of life.

About half of all ever-married males and a quarter of all ever-married females will engage in extramarital sexual activity at one time or another. For females there is some suggestion of a secular trend toward increases in extramarital activity from the beginning of this century to the early fifties. This is linked to a corresponding generational rise in rates of orgasm during this same period (Kinsey, 1948, 1953). It is possible that the very nature of female sexuality may be undergoing change. Our current data will hopefully shed some light on this. For males there are strong social class differences, with lower-class males accounting for most of the extramarital activity, particularly during the early years of marriage. This may be a direct reflection of their earlier mode of assimilation of the sexual commitment. As we previously observed, it is difficult for lower-class males to receive homosocial validation from marital sexual activity (unless, of course, it culminates in conception); this is not the case for extramarital activity for which there is abundant homosocial validation.

In general, it is our feeling that sexuality and sexual activity are by and large derivative functions even during adulthood. There are only a few periods in the life cycle at which there are high rates of sexual activity and/or sexual activity that is complicated by passion and high intensity of affective investment. These are usually adolescence in the male, the early and romantic years of marriage for both men and women, and the highly charged extramarital experiences that can be called affairs. Most of the time sex is really a relatively docile beast, and it is only the rare individual who through the processes of self-invention or alienation from the normal course of socialization is prepared to risk occupation, present comfort, wife and children, or the future for the chancy joys of sexual pleasure.

From this point of view it might be more proper to suggest that, contrary to the Freudian point of view that sex manifests itself in some form in other types of conduct or that other conduct is symbolic of sexual conflict, reports of sexual conflict may in fact stand for difficulties in the more conventional zones of life. Thus, the married couple who come for counseling because of a sexual problem may be merely reporting the conventional rhetoric of the society about what they think the sources of a marital difficulty ought to be. Ovesey (1950) reports that homosexual dreams (overtly homosexual, not merely symbolic) of heterosexual men really relate to occupational problems and that the submissiveness required occupationally appears in the convenient symbolism of the "purported" femininity of homosexual relations. Indeed, many forms of both heterosexual and homosexual acting-out seem to be related to stress reactions to other life situations rather than having specifically sexual motivations. Thus, studies of sex offenders often are overly concerned with the sexual life of the offender when this may merely be the symptom of disorders of other kinds.

Conclusion

It is only a fairly recent development in the history of man that he could begin to conceive of the possibility of social change, that he could begin to understand that his time and place did not represent the embodiment of some eternal principle or necessity, but was only a point in an on-going, dynamic process. For many it is still more difficult to conceive of the possibility of the nature of man himself changing, and particularly changing in significant ways (Van den Berg, 1964). Much of this conservative view of many still permeates contemporary behavioral science. Thus, for many social theories, a view of man as a static bundle of universal needs supplies the necessary stability not available elsewhere in the flux of social life. A conception of man as having relatively constant sexual needs is a necessary part of this point of view. As a contrast to this conservative view, we have attempted to offer a description of sexual development as a variable sociocultural invention, an invention that in itself explains little and requires much explanation.

References

Biller, H. B., & Borstelmann, L. J. Masculine development: An integrative review. *Merrill-Palmer Quarterly*, 1967, **13**, 253-294.
Burke, K. *Permanence and change.* New York: New Republic, Inc., 1935.
Chodoff, P. Critique of Freud's theory of infantile sexuality. *American Journal of Psychiatry.* 1966, **123**, 507-518.
Davis, K. Sexual behavior. In R. K. Merton & R. Nesbitt, *Contemporary social problems.* (2nd Ed.) New York: Harcourt, 1961.
Durkheim, F. *Suicide.* New York: Free Press, 1951.
Ehrmann, W. *Premarital dating behavior.* New York: Holt, Rinehart and Winston, 1959.
Erikson, E. H. *Childhood and society.* (2nd Ed.) New York: Norton, 1963.

Ford, C. F., & Beach, F. A. *Patterns of sexual behavior.* New York: Harper & Row, 1951.
Freud, S. The most prevalent form of degradation in erotic life. *Collected papers,* Vol. 4. London: Hogarth, 1949.
Freud, S. Three essays on sexuality. *Complete psychological works.* (Std. Ed.) Vol. VII. London: Hogarth, 1953. pp. 135-245.
Freud, S. Analysis terminable and interminable. *Complete psychological works.* (Std. Ed.) Vol. XXIII. London: Hogarth, 1964. Pp. 216-253.
Gagnon, J. H. Female child victims of sex offenses. *Social Problems,* 1965, **13,** 176-192.
Gagnon, J. H. & Simon, W. The social meaning of prison homosexuality. *Federal Probation,* 1968. (a)
Gagnon, J. H. & Simon, W. Sex education and human development. In P. J. Fink (Ed.), *Human sexual function and dysfunction.* Philadelphia: F. A. Davis, 1968. (b)
Gebhard, P. H., et al. *Sex offenders.* New York: Harper & Row, 1965.
Gewirtz, J. L. A learning analysis of the effects of normal stimulation, privation, and deprivation in the acquisition of social motivation and attachment. In B. M. Foss (Ed.), *Determinants of infant behavior.* New York: Wiley, 1961. Pp. 213-290.
Hamburg, D. A., & Lunde, D. T. Sex hormones in the development of sex differences in human behavior. In E. Maccoby (Ed.), *The development of sex differences.* Stanford, Calif.: Stanford University Press, 1966, Pp. 1-24.
Kagan, J., & Moss, H. A. *Birth to maturity.* New York: Wiley, 1962.
Kenton, M. (pseud. of Southern, T. & Hoffenburg, M.) *Candy.* Paris: The Olympia Press, 1958.
Kinsey, A. C., et al. *Sexual behavior in the human male.* Philadelphia: Saunders, 1948.
Kinsey, A. C., et al. *Sexual behavior in the human female.* Philadelphia: Saunders, 1953.
Maccoby, E. (Ed.) *The development of sex differences.* Stanford, Calif.: Stanford University Press, 1966.
Maslow, A. H. Dominance, personality, and social behavior in women. *Journal of Social Psychology,* 1939, **10,** 3-39.
Maslow, A. H. Self esteem (dominance feeling) and sexuality in women. *Journal of Social Psychology,* 1942, **16,** 259-294.
Masters, W. H., & Johnson, V. E. *Human sexual response.* Boston: Little, Brown, 1966.
Ovesey, L. The homosexual conflict: An adaptational analysis. *Psychiatry,* 1950, **17,** 243-250.
Parsons, T. The kinship system of the contemporary United States. In *Essays in sociological theory.* (Rev. Ed.) New York: Free Press, 1954.
Rainwater, L. The crucible of identity: The Negro lower class family. *Daedalus,* 1966, **95,** 172-216.

Reich, W. The function of the orgasm. In *The discovery of the Orgone*. Vol. I. New York: Orgone Institute Press, 1942.

Reiss, A. J., Jr. Sex offenses: The marginal status of the adolescent. *Law and Contemporary Problems*, 1960, **25**, No. 2.

Rheingold, J. C. *The fear of being a woman*. New York: Grune & Stratton, 1964.

Schachtel, E. *Metamorphosis*. New York: Basic Books, 1959.

Sears, R. R. Development of gender role. In F. A. Beach (Ed.), *Sex and behavior*. New York: Wiley, 1965. Pp. 133-163.

Simon, W., & Gagnon, J. H. Heterosexuality and homosociality: A dilemma of the lower class family. 1966. (Mimeographed)

Simon, W., & Gagnon, J. H. Pornography, raging menace or paper tiger? *Trans-Action*, 1967, **4**, 8, 41-48. (a)

Simon, W., & Gagnon, J. H. The lesbians: A preliminary overview. In J. H. Gagnon & W. Simon (Eds.), *Sexual deviance*. New York: Harper & Row, 1967. (b)

Simon, W., & Gagnon, J. H. Femininity in the lesbian community. *Social Problems*, 1967, **15**, 212-221. (c)

Van den Berg, J. H. *The changing nature of man*. New York: Dell, 1964.

Young, W. R., Goy, R., & Phoenix, C. Hormones and sexual behavior. *Science*, 1964, **143**, 212-218.

Recommended Additional Readings

Brown, D. G., & Lynn, D. B. Human sexual development: An outline of components and concepts. *Journal of Marriage and the Family*, 1966, **28**, 155-162.

Chodoff, P. A. A critique of Freud's theory of infantile sexuality. *American Journal of Psychiatry*, 1966, **123**, 507-518.

Kreitler, H., & Kreitler, S. Children's concepts of sexuality and birth. *Child Development*, 1966, **37**, 363-378.

The Generation Gap: Beliefs about Sexuality and Self-Reported Sexuality

Paul Cameron

Orientation

We live in an age-stratified society. In addition to our categorization of people according to variables such as sex, ethnicity, income level, and social class, people can easily be distinguished according to the age segments (or cohorts or generations) to which they belong. In fact, differences in values between cohorts are probably at least as dramatic as differences between social-class and ethnic groups. All of these differences are produced as the outcome of similar reasons: variation in upbringing and variation in present life conditions. Indeed, one could invoke the term "cohort-centric" (analogous to egocentric and ethnocentric) to refer to the apparent inability of one cohort to understand and empathize with the values held by another cohort. In sum, the fact of the succession of generations, in addition to its basic significance for the social structure of our society, has given rise to age-specific values, roles, and bahavioral expectations. All of this has brought about the creation of highly cohesive bonds within each age stratum but has restrained relations between age strata.

In the short article to follow, Cameron presents some rather surprising findings regarding the perceived generational differences in sexual knowledge, skill, and desirability. The student is invited to compare these findings with the theorizations of Simon and Gagnon in the preceding article; further, the survey questions and procedure developed by Cameron provide the student with an excellent starting point for research activities of his own implementation.

Paul Cameron is an associate professor of psychology in the department of psychology at the University of Louisville.

A briefer version of this article originally appeared in *Developmental Psychology*, 1970, 3, 272; reprinted by the permission of the author and the American Psychological Association.

Which generation believes itself the sexiest and which is believed the sexiest by the other generations? Although numerous investigators have measured various indices of sexuality for varying samples of adults to date, samples of the generations are lacking. The present study was performed to partially rectify this situation.

Method

Subjects:

In order to determine which ages should be included in the generations "young adult," "middle-aged," and "old," a pilot study was conducted which empirically established that, linguistically speaking, persons are young adults aged 18 to 25 inclusive, middle-aged from 40 to 55, and old from 65 to 79 (Cameron, 1969). Desiring to represent these age-groups as adequately as possible, and realizing that there might be sex differences in the results, six strict area samples of the city of Detroit (excluding the inner city), for Caucasians only, were performed. Three hundred and seventeen Ss were approached at home at least twice to assure a low rejection-rate, and were paid \$2 for the interview. Through a system of almost unlimited call-backs we ended up with 19 rejections (i.e., a rejection rate of $< 6\%$). Interviewers were 12 trained, paid college students. Slightly over 50% of the data were recalled and verified both to clarify ambiguities, or correct mistakes, and to assure that no "dry labbing" had occurred.

Interview: S was approached by asking him to fill out a questionnaire concerning his beliefs about adults. He was assured anonymity, asked to fill out the interview in another part of the residence, and then, when finished, to seal the interview in an envelope provided for the purpose. The interviewer: (1) induced the S to cooperate, (2) delivered the materials, and (3) answered any questions that the S might have (there were relatively few). Whenever asked questions about the questionnaire, the interviewer read the question aloud for the respondent, and in a few cases provided minor definitional clarification. When leaving, the interviewer induced the S to record his phone number on the envelope, and when outside, recorded the address. Whenever a S had neglected to answer any question(s), a person other than the interviewer reinterviewed the S by phone or in person and requested completion. The data were thus made over 99% complete. Each interviewer interviewed an equal number of males and females evenly distributed over the three age-groups to control for any idiosyncratic interviewer effects. Half of the subjects in each sample were administered the "opinions" part of the questionnaire first, and the other half the "self-report" part first.

Questionnaire: Ossorio's (1966) analysis of our linguistic system suggests that a person's behavior is the resultant of six independent variables. For a person to perform a behavior he must: (1) *know about* the behavior, (its consequences and effects), (2) *want* to do it, (3) *know how* to do it, (4) have the *capacity* to do it, (5) must make the attempt, and (6) must have the opportunity to do it. Further, the incidence of any given behavior will vary as a function of all these variables (for instance, a person with limited social access to a behavior may engage in it more than once with great access due to a greater number of attempts). If Ossorio's common-language-related analysis is correct, then we would expect a questionnaire based upon it to approximate maximal intelligibility to Ss, as well as maximal intelligibility of results whether to professional or laymen. Accordingly, the following 14 items were constructed in as close conformity to Ossorio's analysis as the author could achieve. All responsibility for adequate representation is, of course, the author's. The "opinions" part of the questionnaire was introduced: "How do you believe the three age-groups (young adults, middle-aged, and old) compare in:

1) Knowledge about sexual techniques? Know the most (*fill in*)? Know the least (*fill in*)?
2) Sexual desire? Want sex the most (*fill in*)? Want sex the least (*fill in*)?
3) Skill in performing sexual activities (aside from any physical differences)? The most skilled (*fill in*)? The least skilled (*fill in*)?
4) Frequency of attempts to have sexual relations? Try the most (*fill in*)? Try the least (*fill in*)?
5) Physical capacity for sexual activity (aside from any differences in skill)? The most able (*fill in*)? The least able (*fill in*)?
6) Frequency of physical sexual activity? Do the most (*fill in*)? Do the least (*fill in*)?
7) Access to sexual partners? Have the most opportunities (*fill in*)? Have the fewest opportunities (*fill in*)?

The "self-appraisal" part of the questionnaire was a mirror of the above, consisting of a Lickert scale (above average, average, below average). It asked: "How would you compare yourself with *All Other Adults of Your Sex* in:

1) Knowledge about sexual techniques?
2) Sexual desire?
3) Skill in performing sexual activities (aside from any physical differences)?
4) Frequency of attempts to have sexual relations?
5) Physical capacity-ability for sexual activity (aside from any differences in skill)?
6) Frequency of physical sexual activity?
7) Access to sexual partners?

S was also asked his age, sex, his income to the nearest $500 for the past year, and the highest grade he completed in formal schooling. SES was computed by multiplying the number associated with income (e.g., 1 = under $5,000; 2 = $5,000-$7,999; 3 = $8,000-$11,999; 4 = $12,000+) by 2 and adding it to the number associated with education (1 = completed less than 8th grade; 2 = 8th-11th; 3 = H.S. diploma; 4 = some college; 5 = B.A.; 6 = M.A.; 7 = Ph.D.). All analyses broke SES into two levels, those scoring 9 or higher and those scoring 8 or less.

The mean ages of the samples were 21.0 for the young adults, 48.2 for the middle-aged, and 70.2 for the old. SES levels were 8.89, 9.39, and 6.02 respectively (educational levels were 3.52, 3.25, and 2.49 with the young and middle-aged, thus falling between the "H.S. diploma" and "some college" categories, while the old fell between the "8th-11th" and "H.S. diploma" categories; for income, the groups averaged 2.68, 3.10 and 1.79, with the young thus averaging under $8,000/yrs, the middle aged over $8,000 and the old under $5,000).

Results

Knowledge about sex: All generations agreed that the middle aged know the most and the old the least about sex. A three-way fixed effects analysis of variance (3 generations X set X 2 levels of SES) revealed no generational, sexual, or SES main effects—this belief appears societywide.

Sexual desire: The generations, sexes, and SES levels agreed—the young have the greatest and the old the lowest sexual desire.

Sexual skill: The middle aged were universally considered the most sexually skillful and the old the least skillful.

Sexual attempts: Again, essential unanimity that young adults make the most and the old adults the fewest sexual advances.

Physical sexual capacity: The young were generally regarded as possessing the most sexual capacity and the old possessing the least.

Social opportunity: The generations agreed that the young have the most and the old the least access to sex partners.

Sexual frequency: There was general agreement that the young achieve the highest and the old the lowest frequency of sexual activity.

Self-Reported Sexuality

Since S report of "above average" was scored a 3, "average" a 2, and "below average" a 1, mean scores could range from 1 (low) to 3 (high). This range can be divided into three equal segments, from 1.0 to 1.67, 1.67 to 2.33, and 2.33 to 3.0, with the lowest sub-range considered below average and the highest as above average. Similarly, the "average" range can be divided into 3 equal segments, with points of division at 1.88 and 2.11 for low and high average.

Knowledge about sex: Young adults considered themselves more knowledgeable about sex than the old considered themselves, and tended toward greater confidence in their knowledge than the middle-aged. While the young considered themselves high average and the middle-aged average in knowledge, the old considered themselves low average. Though there are no SES differences, at every age level men felt themselves more knowledgeable than women, even though each sex was asked to rate itself relative to their own sex.

Sexual desire: The young considered themselves high average in sexual desire, the middle-aged exactly average, and the old below average. Males claimed generally greater desire than females.

Sexual skill: The young and middle-aged considered themselves quite average in sexual skill, while the old felt themselves below average. Females considered themselves less skillful, while greater feelings of skill were associated with SES.

Physical capacity: The young and middle-aged considered themselves average and the old below average in felt physical capacity. SES was associated with self-appraised sexual capacity.

Sexual attempts: The young adults and the middle-aged felt their number of attempts at sex were average, while the old felt themselves below average in this respect.

Social opportunity: The young felt they had more social opportunity than the middle-aged (average vs. low average), who felt they had more than the old.

Sexual frequency: The young and middle-aged judged themselves average in sexual frequency, while the old considered themselves quite below average.

Discussion

Our population's opinions about the sexuality of the generations appear stereotypic, with no variation by age, SES, or sex. The middle-aged are believed to be cognitively superior sex-wise, knowing more about sex and being more sexually skilled. The young are believed more virile, desiring, attempting, more physically capable, and achieving more sex. Further, the young are provided the greatest access to sex partners. The old are seen as well-nigh sexless—desiring little, knowing little, attempting little, and getting less.

"Common sense" psychology appears rather robust in this instance. These stereotypes seem to have considerable basis in fact. Though the self-reported rank-order of means always ran from young adult down to old, if we consider the stereotypes predictions, our sample "hit" on 38 of 42 tries.

Since the old have lived longer, and presumably have therefore sexed more, their below-average appraisal of their sexual knowledge and skill is puzzling. Have the old, who after all *produced* via sexing, forgotten what they once knew, or do they feel they were culturally deprived in their fertile years? Has their lack of sexual desire and/or opportunity caused what they presumably once knew to "leak" from their minds, or are we dealing with a cultural-gap in sexual knowledge?

When newsmen or evangelists want to get older people annoyed with the young, tales of youthful sexuality seem "sure fire." Why? Possibly the "sex-gap" between the young and old is not mainly developmental as has been generally supposed (e.g., Kuhlen, 1964; Christenson & Gagnon, 1965), but cultural. Perhaps the old feel "gypped" relative to what "they could have done at your age." If the old didn't claim so much sexual ignorance, we might more readily suspect they had adjusted their desires and attempts to the "way things are." But their claimed ignorance makes the "cultural-gap" interpretation attractive.

Women of all ages claimed less sexual desire, knowledge, skill, and physical capacity than men. This generalized lower female "sexiness" has been found in just about all previous studies (Anderson, 1959). Though much has been said of the recent liberation of women's sexuality, nary a hint of such turned up in the results.

Higher SES Ss rated their sexual skill and physical capacity higher. Their feelings of greater physical capacity seem consonant with the generally better health of the financially better off members of our social system. That they should feel more sexually skillful would seem appropriate in light of their supposedly more varied sexual experience (Kinsey, Pomeroy, & Martin, 1948).

It is common knowledge that people forget *skills* that fall into disuse, but many theories of learning posit eternal retention of abstract knowledge. Though it would still probably cost him his academic career, it would be interesting for a researcher to find out to what degree, if any, the old are really less sexually knowledgeable.

References

Anderson, J. E. The use of time and energy. In J. E. Birren (Ed.), *Handbook of aging and the individual*. Chicago: University of Chicago Press, 1959. Pp. 769-796.

Cameron, P. Age parameters of young adult, middle-aged, old, and aged. *Journal of Gerontology*, 1969, **24**, 199-200.

Christenson, V., & Gagnon, J. H. Sexual behavior in a group of older women. *Journal of Gerontology*, 1965, **20**, 351-356.

Kinsey, A. C., Pomeroy, W. B., & Martin, C. E. *Sexual behavior in the human male*. Philadelphia: Saunders, 1948.

Kuhlen, R. G. Developmental changes in motivation during the adult years. In J. E. Birren (Ed.), *Relations of development and aging*. Springfield, Ill.: Charles C. Thomas, 1964. Pp. 209-246.

Ossorio, P. G. *Persons*. Los Angeles: Linguistic Research Institute, 1966.

Recommended Additional Readings

Berezin, M. A. Sex and old age—a review of the literature. *Journal of Geriatric Psychiatry*, 1969, **2**, 131-149.

Cameron, P. Age parameters of young adult, middle aged, old, and aged. *Journal of Gerontology*, 1969, **24**, 201-202.

Rubin, I. The "sexless older years"—a socially harmful stereotype. *Annals of the American Academy of Political and Social Science,* 1968, **376**, 86-95.

31 The Foreshortened Life Perspective

Robert Kastenbaum

Orientation

There is little question that most persons are fascinated with the concept of time. This fascination or preoccupation is manifested in many forms, including systematic entries into diaries and "baby books" and photo albums, the study of history, reminiscences about past life, planning for the future, and so on. Sometime within an individual's life there occurs the incredible discovery that one's existence will extend into the distant future, but this understanding is tempered by a recognition of the possibility that one's life could be snuffed out very quickly by some unanticipated, sudden accident of daily living.

There are certain, unavoidable events within one's lifetime that insistently remind him of his personal movement through a limited period of historical time. For young children, perhaps the entrance into school is suggestive of a new stage of one's development. Other major life events are graduation from high school (or, for some, making the decision to drop out of school), getting that first job, or leaving home and traveling to college. For adults, the bidding of farewell to the last of the offspring as he leaves for his first job or for college will mark a new stage in life. It appears that many middle-aged adults, for the first time, recognize that these events spell a new perspective on personal time: A switch is made from the orientation of time-since-birth to the orientation of time-left-to-live. Significant life events include the onset of the reproductive menopause and the retirement from one's life vocation. Other dramatic events that may induce a change in time perspective are physical ailments, such as heart attacks, strokes, and cancerous growths.

Reprinted from *Geriatrics*, 1969, 24, 126-133, by permission of the author and the publisher.

The author gratefully acknowledges the support of the University of Michigan—Wayne State University Institute of Gerontology.

One other major reality of living is the ultimate recognition of the end of personal existence—death. Usually this understanding comes as a result of the death of a relative or a close friend. The death of a loved one cannot help—except in the case of extreme, blind denial—but remind one of the limitation of personal time and the immediacy of death.

The next article, the last in this book of readings, deals explicitly with the concept of time perspective and the nature of its change across the life span. As this article makes clear, from the earliest unlimited view of futurity to the final imminence of death, human beings must ultimately and inevitably place their lives within the context of the passage of time.

Robert Kastenbaum is a professor in the department of psychology at Wayne State University and is the editor of the journal, Aging and Human Development.

31

The elderly person is marked in more ways than one. His face, hands, and all those body parts which are significant in social communication have become unmistakably engraved with age. Before he speaks, he has already identified himself to others as a person who occupies an extreme position in the spectrum of life. Should his words and actions also betray those features we associate with advanced age, then we are further encouraged to mark him down as one who is strikingly different from ourselves.

He is also marked by numbers, of course. In the case of the elderly person, the statistics relentlessly intersect and pursue. Begin any place. Begin by tracing the declining function of this organ system or that one. Begin by measuring changes in the musculoskeletal system or the speed of central nervous system activity. Begin by locating the elderly person within the actuarial charts. Wherever we begin, it is clear that the numbers have a common bias; they are all against him.

These markings, however, do not tell the whole story. As a matter of fact, they provide only the background and props. It is true enough that any of us might contrive the story of any elderly person's life, based upon these externals. We could manufacture suppositions about what he is experiencing within these biological and statistical markings—how he regards the past, how he views the future, and all the rest. The pity of it is—we do this sort of thing much of the time, without realizing that what we are hearing is not his story, but merely the sound of our own voices. It is so easy to suppose that he feels the way we think we would feel if we were in his situation, or simply that he must feel the way we think it proper for an elderly person to feel. Any time we begin with such a misstep we are likely to accumulate even further distortions. We are likely to generalize without correctives. We are likely to develop set ways of dealing with the way we think he is.

How does the elderly person actually view his own life? What is his perspective and how did he happen to develop it? What functions does it perform for him? In what ways might his perspective affect the course of his own life and the lives of others? Of his total life experiences—past, present, and potentially future—what has he included? What has he excluded? Has he settled upon this perspective as a fixed, permanent vantage point, or are other orientations to come? Most basically, has he managed to create a symbolic structure that comes to terms with his total existence at the very time that this existence itself has become so vulnerable?

To gain perspective on the life perspective of aged people, it might be helpful to back up all the way to infancy. Does the infant have a life perspective? Quite on the contrary—he is almost totally engrossed in life. He experiences the moment. He does not reexperience the past or preexperience the future, at least not in that sense which depends upon the development of symbolic structures. One of the most profound differences between infant and adult is the raw experiencing of the moment, an experiencing that lacks the protection afforded by perspective. Although this point may be an obvious one, it should be emphasized because it is important in a different light in the phenomenological world of the aged.

Very quickly, the infant comes to appreciate the difference between a presence and an absence. At first, this awareness does not distinguish between temporal and spatial dimensions. Either something (for example, smiling-mother-presenting-lunch) is both here-and-now or it is absent—totally absent. By contrast, the adult differentiates "absence" into several alternatives that have differential meanings to him: something exists now, but not here, in this space; or, something will be in this space at a later time, or something has been in this space, but at a previous time, or, again, something is neither here nor there, now, then, or ever.

Such distinctions, and many others that are crucial to the development of a life perspective, come later in life, but the first gropings toward a perspective begin in infancy. There is a clear directional movement. The infant becomes increasingly liberated from its biology, on the one side, and its immediate environment, on the other. The directional thrust is a general process that must be distinguished from what might be called the solidified achievements of human development. This is perhaps the most fundamental basis for challenging the notion that the aged person and the child have a great deal in common. Although certain similarities do exist, the fact remains that it is only the child who is being carried forward on the tide of psychobiological development. All of his behavior and experience is marked by the directional thrust.

The surge of general development continues with great vigor throughout the childhood years and is manifested in physical, social, and psychological changes. Although all of these developments contribute to the emergence of life perspectives, two of the most salient psychological discoveries that children make are:

1) The discovery of futurity. This discovery has several components. First, there is the discovery of future time in the sense that "when this moment is over, there will be some more time coming." Secondly, there is the discovery of future time as qualitatively different from any other kind of time—it is fresh, unused time that can bring forth new experiences and events. This discovery implies a dawning appreciation of possibility and uncertainty. Third, there is the discovery of world or objective future time. Implied in this discovery is the realization that one does not really possess magical control over the universe and cannot really "take time out." Additionally, this is one of the insights that prepared the child to appreciate that, for all his precious and self-evident individuality, he is simply one of many fellow creatures, all of whom dance (or drag) to the music of time.

2) The discovery of mortality. Some rudimentary appreciation of nonexistence may be achieved in early childhood, perhaps even before the conquest of language, but many additional years of development are required before the child can frame the concept of personal mortality. The proposition "I will die" is intimately related to the sense of futurity. It will be the continuing task of the adolescent and the adult to define the nature of this relationship for himself, to integrate the concepts of more time—fresh, new time—possibility, hope, trust, and uncertainty with the concepts of certain death.

The available evidence suggests that adolescence is usually the time during which the developing person begins seriously to create his life perspective. He has had many of the elements previously but now, for the first time, he also has the intellectual equipment to forge these elements into a perspective—and the psychosocial readiness to venture forth as his own self. Children have their notions—but it is adolescents who have ideologies. It is the transformation in thought that underlies the adolescent's changes in social behavior. He now can think about thought, compare ideal with reality, shatter the world as it is presented to him with his new tools of intellectual analysis, and at least try to put the pieces together again in a new and more satisfying manner.

Adolescence, then, is the time of life in which the act of trying to develop perspectives is dominant.

Other characteristics of adolescence are: First, the adolescent has a strong sense of moving into the future. This is not at all the same thing as planning for or visualizing the future; rather, it is a restless experiencing of the developmental current running within oneself.

Second, the adolescent typically projects his thought and feeling intensively into a fairly narrow sector of the future. It is the proximate future that counts, that decisive and eventful time which is just around the corner. Old age is so remote and unappealing a prospect that it hardly can be located at all on his projective charts.

Third, the adolescent often neglects the past, especially his personal past. Neglect may be too passive a word to describe this phenomenon. I have the impression that many adolescents are waging an active battle against the past, trying to put psychological distance between who-I-used-to-be and who-I'm-going-to-be.

Finally, there is the adolescent's way of coming to terms with finality. The prospect of death, like the prospect of aging, often is regarded as a notion that is so remote as to have no relevance to one's own life. Death is avoided, glossed over, kidded about, neutralized, and controlled by a cool, spectator type of orientation. This is on the level of what might be called self-conscious, socially communicated thought. However, more probing and indirect methods of assessment suggest that many adolescents are extremely concerned about death—both in the sense of attempting to fathom its nature and meaning and in the sense of confronting the actual prospect of their own demise. We are no longer surprised when we come across an adolescent whose behavior is influenced by the expectation that he may die at an early age. Indeed, a foreshortened life perspective is by no means the special prerogative of the aged person.

What happens to life perspective during the adult years? We know less about mature perspectives than any other sort, but I think the life perspective of a mature adult has the following characteristics:

It is, first of all, a genuine perspective. This means that the individual has been able to subdivide his life-space into multiple points which stretch away in both directions from the present moment. He is able to locate himself at any one of these points and utilize the other points to achieve the effect of perspective. He might, for example, evaluate the immediate situation in terms of its possible future consequences. A more complex perspective consists of evaluating the immediate situation in terms of both past and future circumstances. More complex still is the perspective in which the individual flexibly shifts the emphasis among past, present, and future standpoints, with all three orientations always involved but varying in relationship to each other. At one moment his pivotal concern may be with past events; thus he calls upon his immediate observations and future projections to provide a context of meaning around the past. At another moment he locates himself in the future and scans his past and present for what clues they can yield that would help him to comprehend his future self.

Upon closer inspection, his perspective will prove to be a structure that includes a variety of subperspectives. These might be visualized as operating in an umbrella type of arrangement. Opened slightly, the perspective system permits the individual to gain coverage of his proximate past and future. This could be called the yesterday-and-tomorrow framework. Opened more broadly, he now has perspective on a larger period of time, but this is still only a small range within his total life-span, where he has been and where he is going, relative to where he is now.

A mature use of the life perspective involves good judgment in deciding when it is appropriate to use a particular subperspective. It involves the ability to scan time in two distinctly different ways—the axiological and the probabilistic. In projecting future, for example, the individual identifies his hopes, fears, and values. This is the axiological orientation. But he also is capable of reading the future in a more objective style, trying to establish the most likely pattern of expectancies. The ability to sweep through time in both axiological and probabilistic styles seems to be one of the hallmarks of a mature life perspective that is maturely employed. Furthermore, there will be an optimal balance between perspectives-already-established and fresh perspective-seeking activities. A flexible life perspective makes it possible to identify and integrate the novel or unexpected event without scuttling the more enduring perspectivistic structure.

Just as important as the life perspective itself, however, is the ability to let go, to know when it is in one's best interests to become totally engrossed in a situation. All perspective and no engrossment makes for a barren, abstracted sort of life.

A mature life perspective is the type that permits a person to make constructive use of his past experiences without becoming enslaved to them and to confront his future, including the prospect of death, without capitulating in that direction either. Many people fail to develop a functional and versatile life perspective, however. In some cases we see a distorted or dysfunctional perspective; in other cases we are struck by the absence of perspective. These different psychological orientations cannot be expected to lead to the same situation when the individuals involved reach advanced age.

In exploring what has been learned and what has yet to be learned about life perspectives in the aged, we should examine the disengagement theory. This is not just a courtesy call to respect the contributions of Elaine Cumming and William E. Henry—it happens that the disengagement theory is one of the few conceptual orientations to make something of life perspectives in later adulthood. Everybody knows by now that the hypothetical process of disengagement involves a gradual and mutual withdrawal of the aging individual and his society. It is said to be an inevitable and normal developmental process. It is said to occur universally, or at least to occur universally under favorable conditions. Obviously, this is an important proposition. Is it also a true proposition? That is another question and one which would take us beyond the scope of this discussion.

But there is a relevant question here. How does the disengagement process itself get started? Cumming and Henry have suggested that disengagement begins with an event that takes place within ourselves, or more specifically, within our life perspectives. As we approach the later years of our lives we come to realize that our future is limited. There is not enough time left to do everything we have hoped and planned. Eventually we also realize that time is not only limited, but it is running out. Death comes into view as a salient prospect. Do Cumming and Henry mean that without this altered life perspective there would be no disengagement? They say: "It seems probable that disengagement would be resisted forever if there were no problem of the allocation of time and thus no anticipation of death. Questions of choice among alternative uses of time lead to curtailment of some activities. Questions of the inevitability of death lead to introspective reflections on the meaning of life."

Although this formulation emphasizes the importance of the individual's inner framework for organizing his experience and, in particular, the role of death anticipations, the formulation appears to be at variance with the facts. Although our knowledge of life perspectives is far from adequate, I believe that enough has been learned to indicate that the disengagement hypothesis has only limited application.

The disengagement hypothesis assumes that everybody has just about the same kind of perspective as they approach the later years of life. This generalization is not tenable. It is already clear that there are significant individual variations, even within particular subgroups in our own society. Some people, for example, never develop the complete umbrella of perspectives described earlier. They move through their life-span within a narrow shell of time, almost day-by-day. This kind of person does not wake up one morning and gasp, "My God, I have only a finite number of years ahead; I had best reallocate my time." The sound of distant drums never had much influence over him, and it may not get to him now, either. Many people in their seventh, eighth, and ninth decades maintain a well-entrenched narrow perspective.

By contrast, there are other people who have been brandishing a wide-open perspective umbrella ever since their youth. The use of time and the prospect of death are factors which have influenced their lives every step of the way. Such people confront different challenges than do those who may be first awakening to intimations of mortality, or those whose limited perspectives have been little influenced by the passing years.

Many people do not experience the altered outlook on time and death that Cumming and Henry proposed as the psychological trigger for disengagement, but even among those who do confront this prospect within their life perspectives, there are important variations. The disengagement theorists have stated that "The anticipation of death frees us from the obligation to participate in the ongoing stream of life. If there is only a little time left, there is no point in planning for a future and no point in putting off today's gratification."

On the contrary, many people intensify their participation in life in order to obtain the greatest possible yield from the time remaining to them. This orientation can persist well beyond the sixth and seventh decades. In studying the psychology of dying and death within a population of very aged patients in a geriatric hospital, we have encountered many who came to terms with approaching death by investing themselves solidly in the network of interpersonal life.

Furthermore, there is reason to believe that the aged person who does clamber out of "the ongoing stream of life" may be doing so for a different reason. Our research interviews suggest that in many cases the individual is not gracefully disengaging to enjoy today's gratification because the future is too short to support long-range plans. Rather, he is more likely to feel that he is no longer capable of making good use even of the limited time that is available to him. It is a sense of inner depletion, impotence, and frustration coupled with the appraisal that his environment offers very little that is inspiring or rewarding.

Perhaps Cumming and Henry have projected into the minds of elderly people the sort of outlook on time and death that they themselves believe to be reasonable and appropriate. This is one of the pitfalls of those who deal with the aged, but most aged people are not theoreticians and simply do not develop the kind of perspective that comes naturally to a theoretician's mind.

Also, we have learned from a number of aged people that they are likely to experience a double-bind regarding time—there is an awareness that future time is scarce but also a heavy sense of oppression at the hands of the clock, too much time that they cannot put to satisfying use. Even a heartfelt lament about the uselessness of future time is not identical with a will-to-die.

Finally, for at least some aged people, the qualitative nature of the future has changed radically. It is no longer the time in which exciting, fresh, novel events are to be expected. The future, in a sense, may be regarded as "used up" before it occurs. The past wends its way forward into the future.

Other points that have emerged from research and clinical experience include:

1) A foreshortened perspective at any age is likely to increase the probability of premature death. The specific pathway of lethality may be through suicide or accident, but particular attention should be given to what might be called psychosomatic or subintentional suicides, in which the individual's physical vulnerabilities are self-exploited to hasten his death.

2) The balance between perspective and engrossment becomes increasingly difficult to maintain with advanced age. An environment that truly shelters the aged person, that truly protects him during his periods of special vulnerability, would make it possible for him to enjoy the spirit-replenishing experience of engrossment more frequently. We become more vulnerable when we are engrossed. We could help our elders if we developed ways of enabling them to drop the burden of their perspectives from time to time without excessive physical or social danger.

3) The perspective of the aged person may become more diffuse or even collapse. Changes in the direction of simplification may be appropriate and beneficial to some people. But there is the danger that the entire perspective may become dysfunctional and contribute to an unnecessarily steep decline in social integration and behavioral competency. There are things we can do that are likely to have a bolstering effect on the aged person's perspectivistic system. For example, we could enter his past as an active force, a sort of participant-observer. Too often, the aged person's preoccupation with his past chases us away—he is snubbing us by focusing upon a scene in which we had no role. We

can develop a sort of semirole in his past and, through this, help him to link his past with the present that all of us share and the future that most of us expect. We are also likely to gain something ourselves through this interpenetration of life perspectives.

4) Both our formal and informal socialization processes emphasize personal growth and expansion during the early years of life. "The System" ill prepares us for living within limits, living with losses, and living with the prospect of death. When the achievement-oriented socialization system gets to work on a person who is growing up in a deprived or ruptured environment, he is alerted to the incongruity between the ideal and the reality. His reaction may take the form of a refusal to accept the socially-sponsored perspective, in the first place, or a rapid aging of the perspective if he does try it out. The person who is growing up in an environment that makes the goals of "The System" appear attractive and feasible is, of course, more likely to develop a life perspective that is centered around individual achievement in the usual sense of the term. Both kinds of people would be better served if our socialization processes—including the classroom—offered a broader, more versatile, and more humane model from which the individual could fashion his own life perspective.

Recommended Additional Readings

Cottle, T. J. Adolescent perceptions of time: The effect of age, sex, and social class. *Journal of Personality*, 1969, **37**, 636-650.

Kubler-Ross, E. *On death and dying.* New York: Crowell-Collier-Macmillan, 1969.

Pearson, L. (Ed.) *Death and dying: Current issues in the treatment of the dying person.* Cleveland: Case Western Reserve University, 1969.

"Well, Doris, the children are grown and gone, and now it's just you versus me."

Drawing by H. Martin; ©1969 The New Yorker Magazine, Inc.